Computing

EXPLORING
SYSTEMS R
APPROACHI

Exploring Information System:
students in the Information S
arranged in sections that refl
research approaches. The boc
and the conclusions which c
communication technologies
society worldwide.

The articles selected have b
within an approach (e.g., sing
industries). Each section is p
context of other, similar resea

■ research method employed
■ focus and perspective of th
■ technology being employe(
■ findings and overall contrib

Each introduction also hig
studying each of the articles i
tions suitable for doctoral res

Robert D. Galliers is Provos
Economics, UK, where he wa
Department of Information S

M. Lynne Markus is the Jo
Bentley College, USA, and w
pore, and the 2006 Fulbrigh
University, Canada.

Sue Newell is the Cammarata Professor of Management at Bentley College, USA, and Visiting Professor of Information Management at Warwick Business School, UK.

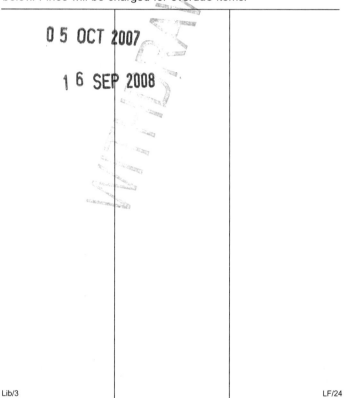

Exploring Information Systems Research Approaches

Readings and reflections

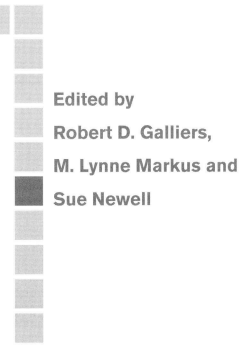

Edited by
Robert D. Galliers,
M. Lynne Markus and
Sue Newell

LONDON AND NEW YORK

First published 2007
by Routledge
2 Park Square, Milton Park, Abingdon, Oxon OX14 4RN

Simultaneously published in the USA and Canada
by Routledge
270 Madison Avenue, New York, NY 10016

Routledge is an imprint of the Taylor & Francis Group, an informa business

© 2007 Selection and editorial matter, Robert D. Galliers, M. Lynne Markus
and Sue Newell; individual readings, the contributors (as specified in the
acknowledgements page).

Typeset in Amasis and Akzidenz Grotesk by
Bookcraft Ltd, Stroud, Glos
Printed and bound in Great Britain by
TJ International Ltd, Padstow, Cornwall

All rights reserved. No part of this book may be reprinted or reproduced or utilised in
any form or by any electronic, mechanical, or other means, now known or hereafter
invented, including photocopying and recording, or in any information storage or
retrieval system, without permission in writing from the publishers.

British Library Cataloguing in Publication Data
A catalogue record for this book is available from the British Library

Library of Congress Cataloging in Publication Data
Exploring information systems research approaches : readings and reflections/edited
by Robert D. Galliers, M. Lynne Markus and Sue Newell.
  p. cm.
  Includes bibliographical references and index.
  1. Information technology–Management. 2. Telecommunication. 3. Management
  –Communication systems. 4. Management information systems. I. Galliers, Robert,
  1947– II. Markus, M. Lynne. III. Newell, Susan.
HD30.2.E986 2006
658.4'038011–dc22   2006021735

ISBN10: 0-415-77196-X (hbk)
ISBN10: 0-415-77197-8 (pbk)

ISBN13: 978-0-415-77196-2 (hbk)
ISBN13: 978-0-415-77197-9 (pbk)

Contents

Preface ix
Acknowledgements xv

PART 1 THE IMPACT OF GROUP SUPPORT SYSTEMS ON DECISION-MAKING AND GROUP CREATIVITY   1

Introduction   3

Group Polarization and Computer-mediated Communication:
Effects of Communication Cues, Social Presence, and Anonymity   9
C.-L. Sia, B. C. Y. Tan and K.-K. Wei

Group Support Systems in Hong Kong: An Action Research Project   33
R. Davison and D. Vogel

Understanding Computer-mediated Discussions:
Positivist and Interpretive Analyses of Group Support System Use   47
E. M. Trauth and L. M. Jessup

Participation in Groupware-mediated Communities of Practice:
A Socio-political Analysis of Knowledge Working   82
N. Hayes and G. Walsham

PART 2 THE IMPACT OF IT ON ORGANIZATIONS   101

Introduction   103

Paradox lost? Firm-level Evidence on the Returns to
Information Systems Spending   109
E. Brynjolfsson and L. Hitt

Information Technology and the Nature of Managerial Work:
From the Productivity Paradox to the Icarus Paradox?   128
A. Pinsonneault and S. Rivard

IT Value: The Great Divide between Qualitative and Quantitative
and Individual and Organizational Measures — 152
Y. E. Chan

The Impact of Information Technology Investment on Enterprise
Performance: A Case Study — 183
M. K. Cline and C. S. Guynes

PART 3 INTER-ORGANIZATIONAL SYSTEMS AND PROCESS IMPROVEMENTS — 191

Introduction — 193

Competing though EDI at Brun Passot: Achievements in France and
Ambitions for the Single European Market — 199
T. Jelassi and O. Figon

Business Value of Information Technology: A Study of Electronic
Data Interchange — 215
T. Mukhopadhyay, S. Kekre and S. Kalathur

The Performance Impacts of Quick Response and Strategic
Alignment in Specialty Retailing — 235
J. W. Palmer and M. L. Markus

Coordination and Virtualization: The Role of Electronic
Networks and Personal Relationships — 257
R. Kraut, C. Steinfield, A. P. Chan, B. Butler and A. Hoag

PART 4 THE IMPACT IT ON MARKETS — 281

Introduction — 283

Do Electronic Marketplaces Lower the Price of Goods? — 288
H. G. Lee

Next-generation Trading in Futures Markets: A Comparison of
Open Outcry and Order Matching Systems — 298
B. W. Weber

Trust, Technology and Transaction Costs: Can Theories Transcend
Culture in a Globalized World? — 311
K. Kumar, H. G. van Dissel and P. Bielli

Reengineering the Dutch Flower Auctions: A Framework for
Analyzing Exchange Organizations — 341
A. Kambil and E. van Heck

PART 5 GLOBAL AND SOCIETAL ISSUES   363

Introduction   365

Cross-cultural Software Production and Use: A Structurational Analysis   372
G. Walsham

Key Issues in Information Systems Management: An International Perspective   392
R. T. Watson, G. G. Kelly, R. D. Galliers and J. C. Brancheau

The Global Digital Divide: A Sociological Assessment of Trends and Causes   412
G. S. Drori and Y. S. Jang

Information Technology and Transitions in the Public Service:
A Comparison of Scandinavia and the United States   430
K. V. Andersen and K. L. Kraemer

Index   447

Preface

Exploring Information Systems Research Approaches is a book intended for supervisors and research students in the Information Systems and related fields. It provides a collection of thought-provoking articles, arranged in sections that reflect the broadening nature of the field, and that provide examples of a range of research approaches. The book is designed for research seminars to focus on alternative research approaches – their strengths, limitations, and impacts on the nature of, and conclusions drawn from, research utilizing different approaches. The book is thus an *exploration* of research approaches suitable for studying the impact of information and communication technologies on groups, on organizations, between organizations, on markets, and on society worldwide.

Exploring Information Systems Research Approaches has its genesis in a number of previous contributions to the field. Primary amongst them is the first IFIP Working Group 8.2 Conference on the topic held in Manchester, England, some twenty years ago. This was certainly among the first occasions when the international research community in Information Systems came together to discuss the growing range of approaches to, and underlying philosophy of, what was then a developing locus of interest in a relatively new field. Further IFIP conferences on the topic – in Copenhagen, Denmark, in 1990 (Nissen *et al.*, 1991) and in Philadelphia, USA, in 1997 (Lee *et al.*, 1997) – have built on the early contributions made by the Manchester conference. In addition, the twentieth anniversary conference was held, once again in Manchester, in 2004 (Kaplan *et al.*, 2004).

In addition, *Exploring Information Systems Research Approaches* builds on a book that one of the current co-editors had published back in 1992. That book – *Information Systems Research: Issues, Methods and Practical Guidelines* – aimed to contribute by bringing together papers on key aspects of actually doing research in the field. Its goal was to assist those setting out on their research agenda, and their research supervisors, by providing a grounding for their work. Rather than jumping headlong into the fray, the idea was to provide food for thought about a range of topics that might require the attention of budding researchers, and the various approaches that might be adopted in undertaking that research.

Both prior to and following the publication of *Information Systems Research: Issues, Methods and Practical Guidelines*, a number of other extremely valuable contributions have been made. Notable amongst these are: formerly, the various contributions in the Harvard Business School series on *The Information Systems Research Challenge*; and more recently, Eileen Trauth's *Qualitative Research in IS: Issues and Trends*, and Michael Myers and David Avison's *Qualitative Research in Information Systems*. While each of these volumes has made a significant contribution to our understanding of the various challenges and approaches to Information Systems research work, they deal with a *particular* set of approaches only. Additionally, we have seen in relatively early contributions, such as Warren McFarlan's *The Information Systems Research Challenge*, Dick Boland and Rudy Hirschheim's *Critical Issues in Information Systems Research*, and more recently in Wendy Currie and Bob Galliers' *Rethinking Management Information Systems*, books that aimed to provide grist to the mill as regards the *topics* worthy of our attention. Additionally, we have seen some excellent contributions to the field in recent years from a social and philosophical perspective, notably, John Mingers and Leslie Willcocks' *Social Theory and Philosophy for Information Systems*, and Chrisanthi Avgerou and colleagues' *The Social Study of Information and Communication Technologies*.

Exploring Information Systems Research Approaches takes an alternative route. In a single volume, we aim to provide an indication of the range of approaches that might be adopted in researching different aspects of our field. As will be seen, and despite calls for greater focus of our research on the information technology artifact itself,[1] the territory is broader and more varied than has been the case heretofore. As such, a broad range of approaches is available to the Information Systems researcher, each with its own underlying philosophy and assumptions about the world, and each presenting quite different challenges. In an earlier contribution, one of the co-editors of this book proposed a taxonomy of research approaches applicable to the field.[2] This taxonomy is reproduced as Table 1 below. As can be seen, it would appear that some approaches are more appropriate than others given the topic under study. In addition, we have seen a much greater incidence of interpretive approaches, championed by such journals as *Information and Organization*,[3] in particular since the IFIP Working Group 8.2 conference in 1984 (Mumford, *et al.*, 1985). *Exploring Information Systems Research Approaches* provides further evidence to back up the need for a careful choice of approach by highlighting common approaches adopted when research efforts are focused at various levels, from individual users through to societal and global issues.

Object of study	Interpretive approaches				Positivist approaches			
	Laboratory Experiment	Field Experiment	Survey	Case Study	Forecasts/ Futures research	Simulation	Reviews	Action Research
Society	x	x	xxx	xx	xxx	xx	xxx	xx
Organization/ Groups	xx	xxx	xxx	xxx	xxx	xxx	xxx	xxx
Individuals	xxx	xxx	xx	xx	xx	xxx	xxx	xx
Technology	xxx	xxx	xx	x	xxx	xxx	xx	xxx
Methodology	x	xx	xxx	xxx	x	xxx	xxx	xxx

Table 1 Information Systems research approaches: an outline taxonomy.

Key: xxx most applicable ↔ x least applicable
Source: amended from Galliers and Land, 1987: 1901; Galliers, 1991: 339; 1992: 159

Exploring Information Systems Research Approaches uses a taxonomy such as that in Figure 1 as the basis for its structure. Following currently popular research topics, we consider five levels of analysis: groups, organizations, inter-organizational relationships, markets, and society.[4] In each case we focus on research that explores the *impact* of information technology (IT). In other words, we have identified papers that consider the impact of IT on groups, on organizations, on inter-organizational relations, on markets and society, in each case ensuring that we include papers that utilize a wide range of research methodologies. We commence in Part 1 with articles that focus attention on the impact of IT on group interaction. Part 2 widens our horizon to a consideration of the use and impact of IT in and on organizations and their performance. We move on to inter-organizational systems in Part 3. Part 4 widens our focus still further, with a consideration of IT's impacts on markets. Global and societal issues are considered in the final section – Part 5. The broadening focus of attention is illustrated in Figure 1.

Each of the articles in each section of the book has been chosen to represent an approach to research, or an alternative design within an approach (e.g. single case versus multiple cases; survey within

PREFACE

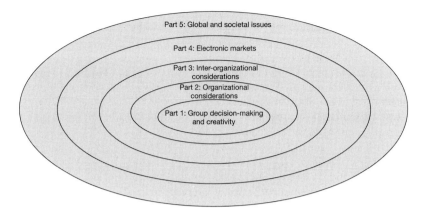

Figure 1 The book in outline: the broadening focus of Information Systems research considerations.

industry versus survey across industries) in the particular context of that aspect of Information Systems research under consideration – as illustrated in Figure 1. Thus, the research approaches used in the four articles chosen for Part 1, dealing with the impact of group support systems on decision-making and group creativity, include action research, case studies, and laboratory experimentation – as one would expect from Table 1. Each section is preceded by an introduction that attempts to place the chosen articles in context of other, similar research, and each introduction provides a summary of the articles in terms of the:

- research method employed;
- focus and perspective of the research;
- technology being employed;
- findings and overall contribution of the work, as illustrated in Table 2.

Author(s), and year of publication	Article 1	Article 2	Article 3	Article 4
Research method				
Focus and perspective				
Technology				
Findings and contribution				

Table 2 Summary framework for each of the articles.

Each introduction also highlights various issues and factors that we wish the reader to consider when studying each of the articles in the section. In this way, we hope to facilitate the exploration of Information Systems research approaches available to us. Ideas for further reading are also provided, as are discussion questions suitable for research seminars.

We hope that in putting together a text such as this, research students – and indeed researchers – in the field of Information Systems, no matter their particular topic of interest, will find the content helpful when they consider the myriad research approaches available to them ... *in context*. In other words, rather than simply choosing approaches with which they may be most familiar, they may reflect on the efficacy of different approaches in light of their particular focus and subject matter (cf. Figure 1). In addition, in reflecting on relevant previous research in their topic area, consideration may be given not only to the findings and conclusions but also to the approaches adopted in coming to these conclusions. Thus, a more critical stance may be taken in reviewing the extant literature – from both an epistemological as well as an ontological perspective. Read together with such texts or articles that deal with the 'how to' of undertaking research (e.g. Philips and Pugh, 1987; Grover, 2001; Robey, 2001; Sekaran, 2003; Oates, 2006), the philosophical underpinnings of IS research (e.g. Mingers and Willcocks, 2004), or particular schools of thought (e.g. Myers and Avison, 2002; King and Lyytinen, 2006), *Exploring Information Systems Research Approaches* is designed to provide means by which thoughtful consideration may be given to the choice of approach, and the likely impacts of this choice on their findings.

Bob Galliers, Lynne Markus and Sue Newell
Bentley College, April 2006

NOTES

1 See, for example, Benbasat and Zmud (2003); see also Galliers (2003) and *Communications of the Association for Information Systems* 12 (30–41) for counter arguments.
2 Galliers (1991); see also Galliers and Land (1987).
3 Formerly published under the title *Accounting, Management and Information Technologies*.
4 In addition, issues associated with globalization are incorporated in this section.

BIBLIOGRAPHY

Avgerou, C., Ciborra, C. and Land, F. (2004) *The Social Study of Information and Communication Technology: Innovation, Actors, and Contexts*, Oxford: Oxford University Press.

Benbasat, I. (ed.) (1989) *The Information Systems Research Challenge: Experimental Research Methods* vol. 2, Boston, MA: Harvard Business School.

Benbasat, I. and Zmud, R. (2003) "The identity crisis within the IS discipline: defining and communicating the discipline's core properties", *MIS Quarterly*, 27(2), June: 183–94.

Boland, R. J. and Hirschheim, R. A. (eds) (1987) *Critical Issues in Information Systems Research*, Chichester: Wiley.

Cash Jr., J. I. and Lawrence, P. R. (eds) (1989) *The Information Systems Research Challenge: Experimental Research Methods* vol. 1, Boston, MA: Harvard Business School.

Communications of the Association for Information Systems (2003), 12(30–41), <http://cais.isworld.org>.

Currie, W. L. and Galliers, R. D. (eds) (1999) *Rethinking Management Information Systems: An Interdisciplinary Perspective*, Oxford: Oxford University Press.

Galliers, R. D. (1991) "Choosing appropriate information systems approaches: a revised taxonomy", in H.-E. Nissen, H. K. Klein and R. Hirschheim (eds), *Information Systems Research: Contemporary Approaches and Emergent Traditions*, Amsterdam: North-Holland, 327–45.

Galliers, R. D. (ed.) (1992). *Information Systems Research: Issues, Methods and Practical Guidelines*, Oxford: Blackwell Scientific Publications.

Galliers, R. D. (2003) "Change as crisis or growth? Toward a trans-disciplinary view of Information Systems as a field of study – a response to Benbasat and Zmud's call for returning to the IT artifact", *Journal of the Association for Information Systems*, 4(6), November: 337–51.

Galliers, R. D. and Land, F. F. (1987) "Choosing an appropriate information systems research methodology", *Communications of the ACM*, 30(11), November: 900–2.

Grover, V. (2001) "10 mistakes doctoral students make in managing their program", *Decision Line*, March: 11–13.

Kaplan, B., Truex III, D. P., Wood-Harper, A. T. and DeGross, J. I. (eds) (2004) *Information Systems Research: Relevant Theory and Informed Practice*, Boston, MA: Kluwer Academic Publishers.

King, J. L. and Lyytinen, K. (2006) *Information Systems: The State of the Field*, Chichester: Wiley.

Lee, A. S., Liebenau, J. and DeGross, J. I. (eds) (1997) *Information Systems and Qualitative Research*, London: Chapman & Hall.

McFarlan, F. W. (ed.) (1984) *The Information Systems Research Challenge: Proceedings of the Harvard Business School Research Colloquium*, Boston, MA: Harvard Business School Press.

Mingers, J. and Willcocks, L. (eds) (2004) *Social Theory and Philosophy for Information Systems*, Chichester: Wiley.

Mumford, E., Hirschheim, R. A., Fitzgerald, G. and Wood-Harper, A. T. (eds) (1985) *Research Methods in Information Systems*, Amsterdam: North-Holland.

Myers, M. D. and Avison, D. (eds) (2002) *Qualitative Research in Information Systems: A Reader*, London: Sage Publications.

Nissen, H.-E., Klein, H. K. and Hirschheim, R. A. (eds) (1991) *Information Systems Research: Contemporary Approaches and Emergent Traditions*, Amsterdam: North-Holland.

Oates, B. J. (2006) *Researching Information Systems and Computing,* Thousand Oaks, CA: Sage.

Phillips, E. M. and Pugh, D. S. (1987) *How To Get A PhD: Managing The Peaks And Troughs of Research*, Milton Keynes: Open University Press.

Robey, D. (2001) "Answers to doctoral students' frequently asked questions (FAQs)", *Decision Line*, March: 10–12.

Sekaran, U. (2003) *Research Methods for Business: A Skill Building Approach*, 4th edn, New York: John Wiley & Sons, Inc.

Trauth, E. M. (ed.) (2001) *Qualitative Research in IS: Issues and Trends,* Hershey, PA: Idea Group Publishing.

FURTHER READING

Avgerou, C., Siemer, J. and Bjorn-Andersen, N. (1999) "The academic field of Information Systems in Europe", *European Journal of Information Systems*, 8(2): 136–53.

Bhattacharjee, S., Tung, Y. A. and Pathak, B. (2004) *Communications of the Association for Information Systems*, 13(37), June: 629–53.

Cornford, T. and Smithson, S. (1996) *Project Research in Information Systems: A Student's Guide*, Basingstoke: Macmillan Press Ltd.

Cotterman, W. W. and Senn, J. A. (eds) (1992) *Challenges and Strategies for Research in Information Systems Development*, Chichester: Wiley.

Cummings, L. and Frost, P. (eds) (1985) *Publishing in the Organizational Sciences*, Homewood, IL: Richard D. Irwin.

Easterby-Smith, M., Thorpe, R. and Lowe, A. (1991) *Management Research: An Introduction*, London: Sage Publications.

Freeman, L. A. S., Jarvenpaa, S. L. and Wheeler, B. C. (2000) "The supply and demand of information systems doctorates: past, present, and future", *MIS Quarterly*, 24(3): 255–380.

Frost, P. and Stablein, R. (eds) (1992) *Doing Exemplary Research*, Newbury Park, CA: Sage Publications, Inc.

Galliers, R. D. and Meadows, M. (2003) "A discipline divided: globalization and parochialism in Information Systems research", *Communications of the Association for Information Systems*, 11: 108–17.

Galliers, R. D. and Whitley, A. A. (2002) "An anatomy of European Information Systems research", ECIS 1993 – ECIS 2002: some initial findings, in S. Wrycza, (ed.) *Proceedings: 10th European Conference on Information Systems*, Gdansk, Poland: 3–18.

Girden, E. R. (2001) *Evaluating Research Articles From Start to Finish*, 2nd edn, Thousand Oaks, CA: Sage Publications, Inc.

Gummesson, E. (1988) *Qualitative Methods in Management Research*, Lund, Sweden: Studentlitteratur/Chartwell-Bratt.

Howard, K. and Sharp, J. A. (1983) *The Management of a Student Research Project*, Aldershot: Gower.

Lawler III, E. E., Mohrman Jr., A. M., Mohrman, S. A., Ledford Jr., G. E., Cummings, T. G., and Associates (1985) *Doing Research That Is Useful For Theory and Practice*, San Francisco, CA: Jossey-Bass Publishers.

Locke, L. F., Spirduso, W. W. and Silverman, S. J. (1999) *Proposals That Work: A Guide for Planning Dissertations and Grant Proposals*, 4th edn, Thousand Oaks, CA: Sage Publications, Inc.

Mason, J. (1996) *Qualitative Researching*, London: Sage Publications Ltd.

Myers, M. D. (1997) "Qualitative research in information systems", *MIS Quarterly*, 21(2): 241–2.

Myers, M. D. (1997) "Qualitative research in information systems", *MISQ Discovery*, <http://www.misq.org/misqd961/isworld/>.

Myers, M. D. (living) "Qualitative research in information systems: living version", *MISQ Discovery*, <http://www.auckland.ac.nz/msis/isworld>.

Orlikowski, W. J. and Baroudi, J. J. (1991) "Studying information technology in organizations: research approaches and assumptions", *Information Systems Research*, 2(1): 1–28.

Peffers, K. and Ya, T. (2003) "Identifying and evaluating the universe of outlets for Information Systems research: ranking the journals", *Journal of Information Technology Theory and Application*, 5(1): 63–84.

Whitley, E. A., Sieber, S., Cáliz, C., Darking, M., Frigerio, C., Jacucci, E., Nöteberg, A. and Rill, M. (2004) "What's it like to do an Information Systems PhD in Europe? Diversity in the practice of IS research" <http://cais.isworld.org/articles/default.asp?vol=13&art=21>, *Communications of the Association for Information Systems*, 13(21), March: 317–35.

Wolcott, H. F. (2001) *Writing Up Qualitative Research,* 2nd edn, Thousand Oaks, CA: Sage Publications, Inc.

Yin, R. K. (1982) *The Case Study Strategy: An Annotated Bibliography*, Washington, DC: Case Study Institute.

Yin, R. K. (1989) *Case Study Research: Design and Methods*, Newbury Park, CA: Sage Publications, Inc.

Yin, R. K. (1993) *Applications of Case Study Research*, Newbury Park, CA: Sage Publications, Inc.

Acknowledgements

We gratefully acknowledge all the many contributors to this book, not least the authors of the papers and the publishers who willingly gave permission for us to use their material in this way.

Special mention should also be made of Jacqueline Curthoys, Emma Joyes and Elisabet Sinkie of Taylor and Francis for their help in bringing this project to publication, and to Diane Whelan of Bentley College for her cheerful, expert and unstinting assistance for this and all the other myriad tasks on hand at any one time.

Bob Galliers, Lynne Markus and Sue Newell

The authors and the publisher would like to thank the following publishers:

INFORMS for permission to reprint Sia, C.-L., Tan, B. C. Y. and Wei, K.-K., "Group Polarization and Computer-mediated Communication". From *Information Systems Research* 13(1): 70–90. Copyright 2001 by The Institute for Operations Research and the Management Services (INFORMS).

Blackwell Publishing Ltd for permission to reprint Davison, R. and Vogel, D., "Group Support Systems in Hong Kong". From *Information Systems Journal*, 10(1): 3–20. Copyright 2000 by Blackwell Science Ltd.

Management Information Systems Research Centre (MISRC), for permission to reprint Trauth, E. M. and Jessup, L. M., "Understanding Computer-mediated Discussions". From *MIS Quarterly* 24(1): 43–79 (March). Copyright 2000 by Management Information Systems Research Centre (MISRC), University of Minnesota.

Elsevier for permission to reprint Hayes, N. and Walsham, G., "Participation in Groupware-mediated Communities of Practice". From *Information and Organization*, 11: 263–88. Copyright 2001 by Elsevier Science Ltd.

INFORMS for permission to reprint Brynjolfsson, E. and Hitt, L., "Paradox Lost?" From *Management Science*, 42(4): 541–58. Copyright 1996 by The Institute for Operations Research and the Management Services (INFORMS).

Management Information Systems Research Centre (MISRC) for permission to reprint Pinsonneault, A. and Rivard, S., "Information Technology and the Nature of Managerial Work". From *MIS Quarterly* 26(3): 287–311 (September). Copyright 1998 by Management Information Systems Research Centre (MISRC), University of Minnesota.

M. E. Sharpe Publishers for permission to reprint Chan, Y. E., "IT Value". From *Journal of Management Information Systems*, 16(4): 225–61. Copyright 2000 by M. E. Sharpe Inc.

CRC Press for permission to reprint Cline, M. K. and Guynes, C. S., "The Impact of Information Technology Investment on Enterprise Performance". From *Information Systems Management*, 18(4). Copyright 2001 by CRC Press.

Management Information Systems Research Centre (MISRC) for permission to reprint Jelassi, T. and Figon, O., "Competing through EDI at Brun Passot". From *MIS Quarterly* 18(2): 337–52 (December). Copyright 1994 by Management Information Systems Research Centre (MISRC), University of Minnesota.

Management Information Systems Research Centre (MISRC) for permission to reprint Mukhopadhyay, T., Kekre, S. and Kalathur, S., "Business Value of Information Technology". From *MIS Quarterly* 19(2): 137–56 (June). Copyright 1995 by Management Information Systems Research Centre (MISRC), University of Minnesota.

INFORMS for permission to reprint Palmer, J. W. and Markus, M. L., "The Performance Impacts of Quick Response and Strategic Alignment in Specialty Retailing". From *Information Systems Research* 11(3): 241–59. Copyright 2000 by The Institute for Operations Research and the Management Services (INFORMS).

INFORMS for permission to reprint Kraut, R., Steinfield, C., Chan, A. P., Butler, B. and Hoag, A., "Coordination and Virtualization". From *Organization Science*, 10(6): 722–40. Copyright 1999 by The Institute for Operations Research and the Management Services (INFORMS).

ACM Publications for permission to reprint Lee, H. G., "Do Electronic Marketplaces Lower the Price of Goods?" From *Communications of the ACM*, 41(1): 73–80. Copyright 1998 by Association for Computing Machinery (ACM).

M. E. Sharpe Publishers for permission to reprint Weber, B. W., "Next-generation Trading in Futures Markets". From *Journal of Management Information Systems*, 16(2): 29–45. Copyright 1999 by M.E. Sharpe Inc.

Management Information Systems Research Centre (MISRC), for permission to reprint Kumar, K., van Dissel, H. G. and Bielli, P., "Trust, Technology and Transaction Costs", an extended version of "The Merchant of Prato – revisited". From *MIS Quarterly*, pp. 199–226 (June). Copyright 1998 by Management Information Systems Research Centre (MISRC), University of Minnesota.

INFORMS for permission to reprint Kambil, A. and van Heck, E., "Reengineering the Dutch Flower Auctions". From *Information Systems Research* 9(1): 1–19. Copyright 1998 by The Institute for Operations Research and the Management Services (INFORMS).

Management Information Systems Research Centre (MISRC) for permission to reprint Walsham, G., "Cross-cultural Software Production and Use". From *MIS Quarterly* 26(4): 359–80 (March). Copyright 2000 by Management Information Systems Research Centre (MISRC), University of Minnesota.

M.E. Sharpe Publishers for permission to reprint Watson, R. T., Kelly, G. G., Galliers, R. D. and Brancheau, J. C., "Key Issues in Information Systems Management". From *Journal of Management Information Systems*, 13(4): 91–116. Copyright 1997 by M. E. Sharpe Inc.

Sage Publications, Inc. for permission to reprint Drori, G. S. and Jang, Y. S., "The Global Digital Divide". From *Social Science Computer Review*, 21(2): 144–61. Copyright 2004 by Sage Publications.

Palgrave Macmillan for permission to reprint Andersen, K. V. and Kraemer, K. L., "Information Technology and Transitions in the Public Service". From *European Journal of Information Systems*, 4(5): 51–63. Copyright 1995 by Operational Research Society Ltd.

PART ONE

The impact of group support systems on decision-making and group creativity

INTRODUCTION TO PART ONE

This first section of the book deals with the innermost ring of Figure 1.1: research that explores the ways in which IT, designed to support group working, actually impacts group decision-making, problem identification and problem solving. The unit of analysis in this work is therefore the team – how is the group impacted by the use of IT? In the past, team work was largely confined to face-to-face work since the only alternatives for communication were either face-to-face or the asynchronous broadcasting of documents (Huber, 1990). Given that team work is defined in terms of interactivity (Hackman, 2002), team work had to involve face-to-face interaction. With the development of different types of computer-assisted communication (CMC) the situation is very different today. Technologies such as email, computer conferencing, video-conferencing and other forms of specialist computer-assisted decision-aiding technologies (often referred to as groupware or group support systems (GSS) are now available making it possible for groups to interact virtually, both synchronously and asynchronously. Much research in the IS field has been done to consider how these new forms of technology impact group processes and decision-making in particular. Questions include, for example, the impact of groupware on levels of participation in decision-making or on the amount of time absorbed by team meetings, or on the quality of decisions. The rationale behind using technology to mediate team work is that potentially it increases anonymity, reduces visual cues and so can influence group processes.

The papers that have been selected in this section have been chosen because they help to illustrate the diversity of methodological approaches that have been adopted as well as theoretical perspectives. Two of the papers are actually themselves focused on exploring methodological issues associated with research exploring the impact of IT on group processes. The other two papers use an experimental

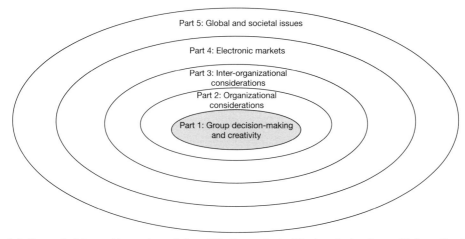

Figure 1.1 Group decision making and creativity within the context of the broadening focus of Information Systems research considerations.

method and a case study and adopt a positivist and an interpretive methodological approach respectively. Reading these different papers, then, provides a contrast in terms of what can be achieved using different methods and different approaches. Table 1.1 contrasts the four papers in terms of the framework that we are using in this book. Next, the papers will briefly be described and discussed, highlighting similarities and differences that can be the focus of reflection and group debate.

The first paper by Sia et al., (2002) uses an experimental set-up to explore how computer-mediated communication influences group polarization through being able to allow anonymous participation or through removing visual cues. By conducting two different experiments, the second building on the results of the first, they are able to conclude that both anonymity and the removal of visual cues increase the likelihood of group polarization through lowering social presence that in turn is influenced by the psychological processes of social comparison and persuasive arguments. This study is thus very much based on a traditional scientific model of doing research: hypotheses are developed based on an extensive review of previous literature; experiments are designed which allow these hypotheses to be tested, i.e. refuted; results are presented to support or refute these hypotheses; the results are discussed in terms of how they support, contradict or build-upon existing theory. Given the need to develop specific hypotheses, there is a clear focus to the study – the impact of CMC on group polarization, which is a group-level phenomenon that previous research has shown can have a profound impact on decision-making. Competing theories exist as to why group polarization occurs and the use of the experimental method is able to tease out which of these different theories is more or less helpful in explaining the phenomenon. The experiment is designed to control all factors that might influence the phenomenon (the dependent variable, which here is group polarization), other than the independent variables of interest – here verbal and visual cues and anonymity. By careful design in the first experiment they were able to show that verbal cues are not responsible for group polarization while visual cues are; in the second experiment they were able to show that anonymity can substitute for the absence of visual cues in reducing social presence, which in turn affects the extent of group polarization.

We can contrast this very carefully designed experimental study with the Davison and Vogel (2000) paper. In many ways they were interested in a similar phenomenon – how CMC (albeit in this paper they refer to this as group support systems or SMS) influences group decision-processes – but here they use a very different methodology, action research. Given the action research methodology adopted, they are not able to test specific hypotheses about what types of group decision-processes CMC (or SMS) will influence. Instead they assume a flexible approach, adapting the research plan as the results from the earlier stages of the intervention unfold. For example, the CIO assumed a very dominant approach and this restricted the participation of other team members. The action researcher then assumed the process facilitation role in a bid to diffuse the impact of this domineering style. The authors conclude that the GSS technology, along with the researcher's action research interventions, produced results that would not have been produced had a GSS not been used. In their own words, they conclude: "Therefore, it can be asserted that the GSS did exert a significant and positive influence on the meeting process." However, they also reflect upon the fact that in the situation they were studying, the GSS could not be used in the way the literature suggests. For example, during the team meetings the GSS was only used about 50 percent of the time because it was found to be acceptable by the team members only for tasks that involved idea generation. For the development of consensus, face-to-face interaction was preferred.

There are a number of questions that arise from contrasting these two papers that can help readers to think about when and why different research methodologies are more or less appropriate:

- What is the impact of having more or less control over the research process on the development of theory?
- What is the impact of having more or less control over the research process on the development of practical recommendations?
- Do you find the conclusions of the two papers equally convincing? Explain why?

	Sia, Tan and Wei (2001)[1]	Davison and Vogel (2000)[2]	Trauth and Jessup (2000)[3]	Hayes and Walsham (2001)[4]
Phenomenon of interest	Influence of computer-mediated communication on group polarization.	How action research can facilitate the effective adoption of GSS in organizational settings.	Differences between positivist and interpretive analyses of GSS use.	How groupware influences knowledge working.
Perspective or theory	How communication cues, social presence and anonymity influence group polarization.	Methodology – action research.	Comparison of methodologies using the same data-set.	Uses a communities of practice perspective to explore the impact of groupware.
Research method	Two lab experiments.	Action research.	Case study but analyzed using two different perspectives.	Case study in a pharmaceutical company.
Technology	Groupware – computer-mediated communication.	Groupware – group support system.	Groupware – group support system.	Groupware.
Findings and contributions	Through experimental control, the study is able to analyze the causes of group polarization and so contribute to differentiating between competing explanations of the group polarization phenomena.	The action research methodology demonstrated that GSS can have a positive influence on group processes. However, the research also demonstrated that in practice GSS cannot always be used in the way that is predicted by theory.	Interpreting the results of a single study through two different lenses demonstrates how it is possible to "tell two different stories from the same facts." The study contributes to our understanding about the significance of taking into consideration the interpretive lens that is used.	Participation in the use of groupware, designed to encourage knowledge sharing, depends on power relations present in different contexts; more specifically use depended on whether the context was perceived as 'safe' or 'political'. The study contributes to our understanding of how the impact of groupware depends on the context of use.

Table 1.1 Comparison of chapters in Part 1.

Notes
1. Sia, C.-L., Tan, B. C. Y. and Wei, K.-K. (2001) "Group polarization and computer-mediated communication: Effects of communication cues, social presence, and anonymity". *Information Systems Research,* 13(1), 70–90.
2. Davison, R. and Vogel, D. (2000) "Group Support Systems in Hong Kong: An action research project". *Information Systems Journal,* 10(1), 3–20.
3. Trauth, E. M. and Jessup, L. M. (2000) "Understanding computer–mediated decisions: Positivist and interpretivist analyses of group support systems". *MIS Quarterly,* 24(1), 43–79.
4. Hayes, N. and Walsham, G. (2001) "Participation in groupware-mediated communities of practice: A socio-political analysis of knowledge working". *Information and Organization,* 11(4), 263–88.

- Do you agree with the contention of Davison and Vogel that "flexibility permitted me to intervene in a manner impossible with other, more structured methodologies?" If yes, what is the implication of this for research that can contribute to practical problem-solving in organizations?
- How can the findings reported in the Davison and Vogel paper, help us to understand the limitations of the findings reported in Sia *et al.*, both from a theoretical and a practical perspective?

So a comparison of the first two papers raises a whole host of questions about what different methodological approaches can and cannot help us to understand, both theoretically and practically. We can build on this by considering the third paper in this section by Trauth and Jessup (2000). This is a very original paper in that it used one method but two different analytical perspectives to explore the data that was derived from this method. Thus, the same data-set (transcripts from four discussions about gender equity at a US university that were conducted using a GSS) was explored from two different ontological perspectives: a positivist approach and an interpretive approach. First the data are analyzed using a positivist approach and then the same data are analyzed using an interpretive approach. So, on the one hand, the phenomenon of interest might be described as the impact of GSS on the generation and evaluation of ideas in order to come to a consensus around specific solutions, in this instance how to improve gender equity at this particular university. On the other hand, the real phenomenon of interest is whether an interpretive analysis of GSS uses results in a different understanding to that provided by a positivist analysis. As the authors state, the paper demonstrates how it is possible to "tell two different stories from the same facts". While above we have contrasted the Sia *et al.*, (2000) paper that uses a positivist approach with the Davison and Vogel (2000) paper that uses an interpretive approach, in the Trauth and Jessup paper we see that it is not simply the method adopted that leads to different understandings of the same phenomenon but also how the findings from any given method are actually perceived.

In the Trauth and Jessup (2000) paper the results of the two different analyses lead to rather different conclusions about the impact of the GSS sessions on problem-solving at this particular university. The positivist conclusion is that the GSS sessions were helpful in relation to the stated goals of effective information exchange, generation of solutions and consensus around solutions. From the interpretive perspective, however, the conclusion is that the GSS-mediated discussions were, at best, only partially successful. The sessions did facilitate the exchange of information and lead to a number of suggestions about solutions, but there was no real consensus around the issues, even for the relatively selective group who actually participated in the sessions. Given the highly politicized university context, described as being one "fraught with tension," the authors conclude: "The insertion of the GSS sessions into this setting was, at best, like dropping a pebble into a stream to build a dam; at worst these sessions were mere public relations" (p. 64).

Considering this paper, both in its own right and in comparison with the two proceedings papers, a number of further questions can be asked:

- Which analysis do you find most convincing in the Trauth and Jessup paper in relation to theory development?
- Which analysis do you find most convincing in the Trauth and Jessup paper in relation to practical advice on using GSS?
- Do you agree that the results from the two different analyses in the Trauth and Jessup paper are really complementary and supportive? Or do you think that they are different and contradictory?
- Do you agree with Trauth and Jessup that it would be useful to conduct a follow-up positivist study at a State University or a positivist replication of this study at another university using the coding categories that resulted from this interpretive analysis?
- Both the Trauth and Jessup (from the interpretive perspective) and the Davison and Vogel papers identify the influence of the context on group processes. Thus, we can ask, how useful is it to focus on the group as the unit of analysis, in isolation from the organizational context (or perhaps even the individual make-up of the team) as does the Sia *et al.* paper?

- Power and politics clearly mediated the impact of the GSS in both the Trauth and Jessup and the Davison and Vogel papers. Can the mediating influence of power and politics be explored in an experimental setting?

The final paper that we introduce in this section is by Hayes and Walsham (2001). This paper uses a case study methodology to explore how the introduction of a groupware system affects knowledge work. As with the Trauth and Jessup paper, the case study methodology chosen leads to the provision of a great deal of rich contextual information. The case is interpreted using a communities of practice (Brown and Duguid, 1991; Lave and Wenger, 1991) perspective, which views learning as being situated and developed through practice. The case study is set in a pharmaceutical company and is longitudinal, with interviews being conducted at two points in time, separated by four months. This longitudinal element was seen to be important to track changes in attitudes and actions over time. The authors stress that the informal interactions during their time spent in the company was as important as the interviews, in helping them to acess the "rich under life" that helped them to develop their analysis.

The main focus of the case study is to track how the introduction of a groupware system in the case company facilitated participation within different communities. More specifically, the groupware system was introduced to encourage improved group working and knowledge sharing across the various functional groups within the company. In understanding the impact of groupware the authors find it helpful to distinguish between safe and political enclaves because they found that participation varied depending on the power relations present in different contexts of groupware use. They argue that within political enclaves, groupware created a homogenization of perspectives, as users used the virtual space to reflect the views of dominant groups. This happened because the groupware system was used for political purposes – for managers to engage in surveillance activities and for employees to engage in career-enhancing impression management tactics. This undermined the extent to which the groupware facilitated genuine participation across the different communities. In safe enclaves more genuine participation occurred. However, safe enclaves were found to be rarer than political enclaves of use. They use the findings of their study to develop the communities of practice perspective, adding a political dimension which emphasizes how learning is shaped by the socio-political context of practice, in this case practice within the groupware space.

Considering the Hayes and Walsham paper (2001) in its own right, as well as in comparison with the other papers raises more questions about both method and interpretive frameworks:

- How helpful did you find the adaptive communities of practice perspective as an interpretive lens through which to understand the case in the Hayes and Walsham paper?
- Would it have been possible to develop the critical insights related to legitimate peripheral participation using quantitative research methods? Would it have been possible to use quantitative methods to test the suggestion arising from this paper that learning is shaped by the socio-political context? Or that it is possible to distinguish between political and safe groupware enclaves? If yes, how?
- More generally, this paper highlights the importance of the socio-political dimension in understanding behaviour in organizations. What is the best way to research this dimension? And are there certain methods that will miss this dimension?
- Hayes and Walsham discuss how they moved from a structured, through semi-structured to unstructured interview over the course of their study. Why do you think they did this and how, if at all, in your opinion does this affect the validity of their conclusions?
- Hayes and Walsham point to the importance of the informal interactions with participants in helping them better to understand the groupware impact in which they were interested. Do you feel these informal interactions add to, or distract from, the quality of this research?
- Tape recorders were not used in this study. What do you see as the advantages and disadvantages of this strategy?
- How important is the contextual information provided in the Hayes and Walsham paper, as well as in the Trauth and Jessup paper, in understanding the results from a positivist as well as an interpretive analysis? In many positivist papers, including papers based on case study methods, this level of

contextual detail is missing. Do you think that this is a problem in relation to understanding the results and conclusions from a given study?
- Do you agree with those who argue that positivist and interpretive approaches are mutually exclusive or do you agree with those who argue that these two approaches can be mutually supportive?
- How do the conclusions from this study relate to the impact of groupware, a form of CMC, compared and contrasted with the findings from the other three papers in this section? Why do you think these differences exist?
- More specifically, having read these four papers, do you feel that the impact of CMC on group processes is positive, negative or neutral?
- Can you explain why the results of previous research on the impact of CMC on group processes is sometimes contradictory? What does this lead you to conclude about the following key features of research design – the selection of the unit of analysis, the selection of a particular method, the interpretive frame for the analysis, the selection of a theoretical framework?

REFERENCES

Hackman, R. (2002) *Leading Teams: Setting the Stage for Great Performances*. Boston, MA: Harvard Business School Press.

Huber, G. (1990) "A theory of the effects of advanced information technologies on organizational design, intelligence and decision-making". *Academy of Management Review*, 15(1): 47–71.

FURTHER READING

Benbunan-Fich, R., Hiltz, S.R. and Murray, T. (2003) "A comparative content analysis of face-to-face versus asynchronous group decision-making". *Decision Support Systems*, 34(4): 457–70.

Cappel, J. and Windsor, J. (2000) "Ethical decision-making: a comparison of computer-supported and face-to-face groups". *Journal of Business Ethics*, 28(2): 95–108.

Group Polarization and Computer-mediated Communication: Effects of Communication Cues, Social Presence, and Anonymity

Choon-Ling Sia, Bernard C. Y. Tan and Kwok-Kee Wei

Department of Information Systems, City University of Hong Kong, 83 Tat Chee Avenue, Kowloon, Hong Kong, iscl@cityu.edu.hk
Department of Information Systems, National University of Singapore, 3 Science Drive 2, Singapore 117543, Republic of Singapore, btan@comp.nus.edu.sg
Department of Information Systems, National University of Singapore, 3 Science Drive 2, Singapore 117543, Republic of Singapore, weikk@comp.nus.edu.sg

Abstract

Group polarization is the tendency of people to become more extreme in their thinking following group discussion. It may be beneficial to some, but detrimental to other, organizational decisions. This study examines how computer-mediated communication (CMC) may be associated with group polarization. Two laboratory experiments were carried out. The first experiment, conducted in an identified setting, demonstrated that removal of verbal cues might not have reduced social presence sufficiently to impact group polarization, but removal of visual cues might have reduced social presence sufficiently to raise group polarization. Besides confirming the results of the first experiment, the second experiment showed that the provision of anonymity might also have reduced social presence sufficiently to raise group polarization. Analyses of process data from both experiments indicated that the reduction in social presence might have increased group polarization by causing people to generate more novel arguments and engage in more one-upmanship behavior. Collectively, process and outcome data from both experiments reveal how group polarization might be affected by level of social presence. Specifically, group discussion carried out in an unsupported setting or an identified face-to-face CMC setting tends to result in weaker group polarization. Conversely, group discussion conducted in an anonymous face-to-face CMC setting or a dispersed CMC setting (with or without anonymity) tends to lead to stronger group polarization. Implications of these results for further research and practice are provided.

Keywords: group polarization; computer-mediated communication; social presence; communication cues; anonymity; persuasive argumentation; social comparison

1. Introduction

From hundreds of empirical studies, scholars have observed that people are susceptible to phenomena that have profound impact on organizational and societal decision making (Rajecki 1990). One such pervasive phenomenon is *group polarization,* the tendency of people to become more extreme in their thinking following group discussion (Isenberg 1986). For example, people that are moderately risk seeking (or risk averse) can emerge from group discussion exhibiting very risky (or cautious) thinking. In other words, the phenomenon of group polarization is about extremity of decisions (which is the focus of this study) rather than quality of decisions or commitment to decisions.

Unfortunate incidents that have been blamed on group polarization include escalation of the Vietnam War by the Johnson administration (McCauley 1989), disruption of racial harmony following the assassination of Martin Luther King (Riley and Pettigrew 1976), and risk-taking ethos of NASA that led to the Challenger disaster (Janis 1989). Group polarization is also thought to be the reason behind gang delinquency (Cartwright 1975) and escalation of investment in failing business ventures (Brockner 1992, Whyte 1993). Notwithstanding these negative examples, there are situations that can benefit from group polarization. Examples are participation in social support systems (Festinger et al. 1956), prosocial discussion about donations (Muehleman et al. 1976), and mutual counseling in self-help groups (Toch 1965). Organizations that seek to encourage innovation through risk taking may also promote group polarization during discussion (Argyris 1992). In short, it is desirable to sometimes raise and, other times, lower group polarization during discussion. A possible way that group polarization may be altered is with the use of computer-mediated communication (CMC) (El-Shinnawy and Vinze 1998).

CMC may allow people to engage in group discussion with reduced social presence (Rice 1993) compared to face-to-face communication. Williams (1977) observes that a reduction in social presence arising from the removal of communication cues raised group polarization. Through empirical studies, Siegel et al. (1986) found that dispersed groups communicating electronically with reduced social presence had greater group polarization than proximate groups communicating face-to-face. This result could be due to the increased uninhibited remarks generated (e.g., Kiesler et al. 1985, Siegel et al. 1986) when people communicated in dispersed CMC settings with reduced social presence. It could also be due to the more unique and high-quality ideas produced (Valacich et al. 1994a, 1994b) when people worked together via CMC in dispersed settings with reduced social presence. Although these findings suggest that a reduction in communication cues may change social presence, thereby altering group polarization, it is not clear what communication cues contributed to group polarization.

CMC also allows people to engage in anonymous discussion. Anonymity, a key feature offered by CMC, can potentially raise group polarization by restricting the exchange of social cues (Connolly et al. 1990) and reducing social presence among group members (Short et al. 1976). This promotes uninhibited behavior (Jessup et al. 1990) and elicits interesting arguments (Connolly et al. 1990) during group discussion. However, it is not clear whether and how anonymity may moderate the impact of reduced social presence arising from the removal of communication cues on group polarization. Without knowing the effects of each feature offered by CMC (Dennis and Kinney 1998), it is not possible to fully understand how group polarization may be affected.

While prior studies have provided clues about how CMC may affect group polarization, the precise (and combined) impact of each feature offered by CMC on group polarization and the underlying processes through which such impact may come about are not well understood. This study advances our knowledge in this area by unraveling the precise and combined effects of communication cues and anonymity (key features of CMC) on group polarization, and generating insights on the underlying processes that bring about such effects. The first experiment compares an unsupported setting, a face-to-face CMC setting, and a dispersed CMC setting to assess the effects of reduced communication cues on group polarization. Building on results of the first experiment, the second experiment compares the face-to-face CMC setting and the dispersed CMC setting under identified and anonymous conditions to investigate the moderating effects of anonymity on group polarization. In each experiment, meeting logs for each treatment were analyzed to yield a better understanding of the underlying processes through which group polarization was impacted. Collectively, these two experiments yield insights on how the selection of CMC features to support organizational decision making may affect the extent of group polarization.

2. Group Polarization

Group polarization, the phenomenon that is generally considered a decision bias, has been reported to affect human thinking under numerous contexts. Examples include political decisions (Janis 1989), jury decisions (Myers and Kaplan 1976), investment decisions (Whyte 1993), donation decisions (Muehleman et al. 1976), gambling behavior (Blascovich et al. 1975a), aggressive behavior (Yinon et al. 1975), racial attitudes (Riley and Pettigrew 1976), community conflict (Coleman 1957), gang delinquency (Cartwright 1975), religious fellowship (Festinger et al. 1956), and mutual counseling (Toch 1965). Given its pervasive and profound effects on human thinking, a better understanding of group polarization may enable people to exploit it when it is helpful and avoid it when it is harmful. To be effective, such understanding must be anchored on theories of group polarization. Among the attempts to explain group polarization, two theories that have been widely accepted by scholars and practitioners (Rajecki 1990) are social comparison theory (SCT) (Sanders and Baron 1977) and persuasive arguments theory (PAT) (Kaplan 1977).

SCT is a normative explanation for group polarization that has received empirical support (Isenberg 1986). It argues that group polarization occurs because people are motivated to present themselves in a socially desirable light during discussion (Brown 1965). People do this by continually comparing their opinions with others and adjusting their opinions in the direction valued by others. Two mechanisms that facilitate group polarization in this manner are one-upmanship and pluralistic balance (Isenberg 1986). *One-upmanship* is the tendency of people to try to outdo each other in the socially valued direction. When people are exposed to mutual positions during group discussion, they will try to outdo each other by changing their positions in the direction valued by the group. This causes group polarization (Fromkin 1970). *Pluralistic balance* is the desire of people to achieve a compromise between their preferred positions and the positions thought to be favored by others. During discussion, people commonly present their positions as compromises between what they prefer and what they think others prefer. When people discover that they have inaccurately assessed the collective position (which is closer to their preferred position than they originally thought), they will change their positions in the direction of the collective position. This leads to group polarization (Pruitt 1971). Proponents of SCT have demonstrated its value by showing that mere exchange of mutual positions among people was sufficient to produce group polarization (e.g., Baron and Roper 1976, Blascovich et al. 1975b).

PAT is an informational explanation for group polarization that has been supported by empirical studies (Isenberg 1986). It posits that people arrive at their positions based on known arguments for and against issues underlying those positions. People will change their positions as they are exposed to arguments from others (Vinokur and Burnstein 1974). The persuasiveness of each argument (the extent to which it can cause people to change positions) depends on its validity and novelty (Burnstein 1982, Vinokur and Burnstein 1978). *Validity* is the degree to which people feel that the argument is correct and accurate. *Novelty* is the degree to which people feel that the argument is new and interesting. During discussion, when people are presented with valid or novel arguments from others supporting the collective position, they will change their positions in the direction of the collective position. This causes group polarization (Burnstein 1982). Proponents of PAT have demonstrated that group polarization was increased when people exchanged arguments considered to be valid (Burnstein 1982) or novel (Vinokur and Burnstein 1978).

Two decades of empirical research suggest that the processes of social comparison and persuasive argumentation occur simultaneously during group discussion to produce group polarization (Butler and Crino 1992, Isenberg 1986, Rajecki 1990). Typically, persuasive arguments generated during discussion cause people to move towards the collective position. This triggers social comparison that causes people to move beyond the original collective position. Given the complementary nature of these theories, SCT and PAT constitute a useful theoretical platform, based on which the nature of group polarization can be studied and better understood.

Group polarization is commonly measured using choice shift and preference change (Zuber et al. 1992). *Choice shift* is the difference between the average pre-meeting position of everyone involved and the final collective position. Group polarization (measured at a group level) occurs when the collective position is more extreme than the average premeeting position of everyone. *Preference change* is the average difference between the premeeting and the postmeeting positions of each person involved. Group polarization (measured at an individual level) occurs when the discussion has caused people to change their premeeting positions in the direction of the collective position. While choice shift and preference change were found to correlate significantly in empirical studies (e.g., Dubrovsky et al. 1991, McGuire et al. 1987), the magnitude of choice shift was typically stronger than that of preference change (Whyte 1993).

3. Social Presence and Communication Cues

Social presence is the degree to which people establish warm and personal connections with each other in a communication setting (Short et al. 1976). It is related to the degree of salience attached to others involved in an interaction. Settings that are high in social presence encourage people to treat each other as social beings with feelings rather than objects that can be ignored. Social presence is not an entirely objective quality of a communication setting (Short et al. 1976). Social group membership may affect the level of social presence (Bhappu et al. 1997). As group members interact together over time, the level of social presence can be raised through a process of social construction (Carlson and Zmud 1999, Walther 1995).[1]

Changes in the level of social presence can affect group communication. For example, a reduction in social presence may result in the pursuit of self rather than group interests (Walton and McKersie 1965) and difficulties in arriving at mutually agreeable decisions (Lewicki and Litterer 1985). Short et al. (1976) summarized the results from a series of experiments investigating the effects of social presence on group communication. They concluded that the extent of opinion change reported was inversely proportional to the level of social presence in a communication setting. The lower the level of social presence, the greater the extent of opinion change. This conclusion could partly be attributed to the ability of people in communication settings with low social presence to change their opinions without a loss of face. It could also partly be due to the greater salience of arguments, and the tendency of people to be influenced by these arguments, in communication settings with low social presence. Similar results were obtained by Siegel et al. (1986), who found that communication settings with low social presence helped to focus human attention on arguments presented.

As a fundamental aspect of discussion, the various types of communication cues that people exchange can alter the level of social presence, thereby changing human behavior (Johansen et al. 1991, Short et al. 1976, Williams 1977). The three main types of communication cues are verbal, visual, and textual cues (McGrath 1984). *Verbal cues* refer to information conveyed vocally, including tone and loudness of voice, and rate of speech (Cook and Lallijee 1972, Daft et al. 1987, McGrath 1984). *Visual cues* include visual orientation and facial expressions, such as smiles, frowns, nods, and other types of body language. *Textual cues* are information embodied in written and printed texts and graphics. These include information transmitted electronically via CMC (Poole and Jackson 1993). Different types of communication cues have been found to yield different levels of social presence in various communication settings (Sproull and Kiesler 1986). For example, unsupported face-to-face communication settings had been found to result in higher social presence than dispersed computer-mediated communication settings (e.g., Rice 1993, Chidambaram 1996, Straub and Karahanna 1998). Communication cues that typically yield higher social presence are those that convey immediacy (Latane 1981) of others. Immediacy is the psychological distance between people who are communicating (Wiener and Mehrabian 1968, Short et al. 1976). Face-to-face communication settings usually have higher social presence because verbal and visual cues are better at conveying immediacy of others.

Dispersed communication settings usually have lower social presence because textual cues are usually not good for conveying immediacy of others (Poole and Jackson 1993, Straub and Karahanna 1998).

Empirical studies have provided clues about how people may behave in communication settings with different levels of social presence. In dispersed CMC settings with lower social presence, people were reported to generate more uninhibited remarks than in unsupported face-to-face settings with higher social presence (Kiesler et al. 1985, Siegel et al. 1986). People working together in dispersed CMC settings have also been found to produce more unique and high-quality ideas than people working together in unsupported face-to-face settings (Valacich et al. 1994a, 1994b). While such findings may be attributed to differences in the level of social presence in various communication settings, it is not clear which types of communication cues (verbal or visual) have a greater impact on social presence and human behavior.

CMC can aid group discussion in a face-to-face setting or a dispersed setting (e.g., Tan et al. 1998). When used in a face-to-face setting, CMC allows people to replace (totally or partially) verbal cues with electronic textual cues. This reduction of verbal cues can lower social presence among people and affect the processes of persuasive argumentation and social comparison. Lower social presence may reduce evaluation and communication apprehension (Nunamaker et al. 1991), thereby causing people to be more willing to contribute novel and valid arguments. The greater salience of these valid and novel arguments in a communication setting with lower social presence (Griffith et al. 1998) can encourage people to change their positions and move towards the collective position, leading to group polarization. Social comparison occurs as people change their positions. The reduced social presence resulting from the reduction of verbal cues allows people to change their opinions without a loss of face (Short et al. 1976). In normal everyday situations, Salancik (1977) noted that people were less committed to their positions if they had not made a verbal declaration of their positions. In their quest for pluralistic balance, lower social presence may cause people to be more ready to forgo their positions in favor of the collective position. Lower social presence may also bring about depersonalization, which encourages uninhibited behavior (Diener et al. 1976). Less inhibition can promote one-upmanship among people in their pursuit of self-interests. In short, the reduction of verbal cues should reduce social presence, leading to stronger group polarization.

Hypothesis H1a. Choice shift will be significantly stronger in a face-to-face CMC setting than in an unsupported setting.
Hypothesis H1b. Preference change will be significantly stronger in a face-to-face CMC setting than in an unsupported setting.

When CMC is used for group discussion in a dispersed setting, all verbal and visual cues are replaced by electronic textual cues. The effects of removing visual cues in addition to verbal cues may be manifested through a further reduction in social presence (Rice 1984). As discussed above, this may cause people to generate even more novel and valid arguments through even lower communication and evaluation apprehension and make the arguments generated even more salient (Griffith et al. 1998). This also allows people to change their positions even more rapidly without losing face. Salancik (1977) noted that people were even less committed to their positions when others could not see them. Thus, people may be even more ready to forgo their positions in favor of the collective position when trying to achieve a pluralistic balance. The very low social presence may further promote depersonalization, which encourages impulsive behavior (Diener et al. 1976). Siegel et al. (1986) observed "flaming" behavior in a dispersed CMC setting. Lower levels of inhibition may further encourage one-upmanship behavior among people. In short, removal of visual cues on top of verbal cues should further reduce social presence, thus further raising group polarization.

Hypothesis H2a. Choice shift will be significantly stronger in a dispersed CMC setting than in a face-to-face CMC setting.
Hypothesis H2b. Preference change will be significantly stronger in a dispersed CMC setting than in a face-to-face CMC setting.

4. Experiment 1: Impact of Communication Cues

This experiment had a three-cell simple design involving an unsupported setting, a face-to-face CMC setting, and a dispersed CMC setting (see Figure 1).[2] In the *unsupported setting,* people could use verbal, visual, and textual cues during discussion. In the *face-to-face CMC setting,* people were not allowed to communicate with verbal cues. They were limited to visual and textual cues during discussion.[3] The experimental administrator, who was a graduate student trained to conduct experiments in an unobtrusive and noninterfering way, ensured that no verbal cues were exchanged in all experimental sessions. In the *dispersed CMC setting,* people could neither see nor hear each other because they were separated by wall-to-ceiling partitions. Their discussion was conducted solely via electronic textual cues. The experimental administrator did not observe any instance of visual or verbal cues being exchanged in all experimental sessions. Differences between the unsupported and face-to-face CMC settings were mainly due to verbal cues (and the paralinguistic visual cues such as hand and head movements and facial expressions that accompanied the act of speaking). This formed the basis for testing H1a and H1b. Differences between the face-to-face CMC and dispersed CMC settings could mainly be attributed to visual cues. This formed the basis for testing H2a and H2b. All experimental sessions for the three communication settings were conducted in the same room (which was configured differently for each communication setting).[4] The main purpose of Experiment 1 was to determine which types of communication cues (verbal or visual) had a greater impact on group polarization.

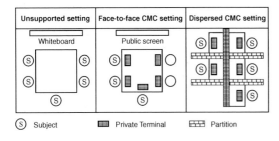

Figure 1 The experimental settings.

A pilot study involving nine groups of five subjects each (three groups under each setting) was carried out to make sure the experimental treatment, procedure, and task did not cause major problems. Minor changes were made to the experimental procedure following the pilot study. In the actual study, each of the three settings had 13 groups of five subjects each. There was no anonymity in any of the three settings.

4.1. Subjects

The subjects were 195 senior information-systems undergraduates from a large public university. Their average age was 21.4 years. About 40% were females and 60% were males. As part of their coursework, all subjects had experience with group discussion. They were randomly assigned to groups, which were randomly assigned to treatments. This helped to alleviate the possibility of individual characteristics affecting the results. None of the groups formed for this study had a prior history of working together. Subjects received course credit for their participation. Those who chose not to participate could do an alternative assignment for the same amount of course credit.

4.2. Experimental task

Experiment 1 employed a choice-dilemma task (Wallach et al. 1962), used in numerous prior studies on group polarization (Isenberg 1986, Pavitt 1994). It required people to play the role of captain of a college football team. They had to decide on a probability of success, beyond which they would be willing to choose a risky strategy against a traditional rival (see Appendix 1).

4.3. Experimental procedure

Each session began with subjects reading the task description for 10 minutes. The process of group discussion, which comprised a series of rounds, was explained to them. The communication network employed in this research was the wheel topology (Shaw 1978). The communication strategies adopted were simultaneous opinion generation and sequential opinion presentation. During each round of decision making the subjects simultaneously generated their positions with underlying arguments and then sequentially presented their views to others. For the unsupported setting, subjects wrote their positions and arguments on paper at the same time and then took turns presenting their views by reading aloud while the experimental administrator recorded their names and views on a whiteboard.

A Macintosh version of the SAMM groupware was used for electronic communication for both the CMC settings. Specifically, the idea-gathering tool of this groupware was used. This groupware tool could be considered very simple relative to the experience of the subjects in using similar technologies. Using such a simple groupware tool minimized the training required for subjects and reduced the possibility that training overheads might confound with experimental treatments. For both the CMC settings, subjects entered their opinions and arguments at the same time through individual private terminals. These views were sent to the server terminal through a local-area network. When all the subjects had contributed their views, the experimental administrator would activate the server terminal to display the names and views of each subject. In the case of the face-to-face CMC setting, such information would be displayed on a big public screen. In the case of the dispersed CMC setting, such information would be displayed on a separate window at the individual private terminals.

Based on the results of each round, subjects proceeded to the next round and did likewise. The session ended when all subjects had arrived at the same position (a consensus) or when a maximum of four rounds had been completed.[5] When subjects could not reach a consensus after four rounds, their average position was taken as the collective decision. Although this process of group discussion appeared to be restrictive, it ensured that all subjects had the opportunity to present their views and to consider the views of others. It also helped to enforce experimental controls (Tan et al. 1998) by regulating the amount of group discussion, a potential confounding factor affecting group polarization (Kaplan 1977, McGuire et al. 1987). Before each session ended, subjects restated their positions individually and responded to questions for control and manipulation checks.

4.4. Measuring group polarization

Group polarization is widely measured using choice shift and preference change (Zuber et al. 1992). As in numerous prior studies, choice shift and preference change were measured as follows:

Let:
F_i = Individual position of subject i in the first round (before discussion);
L_i = Individual position of subject i in the last round;
A_i = Individual position of subject i after discussion;
n = Number of subjects per session (which was five in this study);

Collective decision $(G) = \sum_{i=1}^{n} L_i / n$;

Choice shift $= \left| \sum_{i=1}^{n} (G - F_i) \times 10 / n \right|$;

Preference change $= \left| \sum_{i=1}^{n} (A_i - F_i) \times 10 / n \right|$.

4.5. Control and Manipulation Checks

Besides stating their age and gender, subjects completed questions assessing their group experience, task experience, and task knowledge. F-tests found no significant differences for subjects in the three treatments in terms of age, group experience, task experience, and task knowledge. A Kruskal-Wallis test found no significant differences in gender ratio of subjects across the three treatments. Control over subject characteristics through randomization seemed to be successful.

Two questions, both anchored on seven-point scales ranging from strongly disagree (1) to strongly agree (7), assessed the manipulation on verbal cues: "I could communicate using verbal cues" and "I could express my views using verbal cues". Subjects in the unsupported setting agreed on both questions significantly more than subjects in the face-to-face CMC and dispersed CMC settings ($F = 7.73, p < 0.01$ and $F = 8.10, p < 0.01$, respectively). The manipulation on verbal cues seemed to be successful. No manipulation checks were carried out for visual cues because in the dispersed CMC setting, it was physically difficult (if not impossible) for people in the ad-hoc groups of this study to raise their social presence by seeing mutual expressions.[6]

Communication setting	Choice shift	Preference change
Unsupported	0.46 (0.39)	0.32 (0.36)
Face-to-Face CMC	0.91 (0.50)	0.53 (0.30)
Dispersed CMC	1.62 (0.47)	1.34 (0.48)

Table 1 Mean (standard deviation) of group polarization for experiment 1.

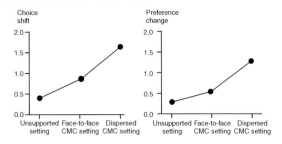

Figure 2 Graphical presentation of group polarization for Experiment 1.

4.6. Hypotheses tests

Descriptive statistics for the three communication settings are presented in Table 1 and shown graphically in Figure 2. Statistical tests were carried out based on a 5% level of significance. Where necessary, the data was transformed (Weisberg 1985) to meet homogeneity and normality requirements of parametric tests. Given that choice shift and preference change were significantly correlated ($R^2 = 0.77$, $p < 0.01$), a MANOVA test involving the independent variable (communication setting) and both dependent variables (choice shift and preference change) was carried out. It detected a significant effect for communication setting ($F = 3.53$, $p < 0.02$). This finding allowed F-tests to be applied separately to choice shift and preference change.

F-tests detected significant effects for communication setting on choice shift ($F = 6.13$, $p < 0.01$, power = 0.86, effect size = 0.25) and preference change ($F = 8.43$, $p < 0.01$, power = 0.95, effect size = 0.32). Fisher's LSD revealed that the dispersed CMC setting had significantly greater choice shift and preference change than the face-to-face CMC and the unsupported settings (see Table 1 and Figure 2). H1a and H1b were not supported, but H2a and H2b were supported. These results showed that replacement of verbal cues by electronic textual cues in the face-to-face CMC setting did not produce a significant impact on group polarization. However, the removal of visual cues in the dispersed CMC setting resulted in significantly stronger group polarization.

4.7. Meeting log analyses

To obtain additional insights on the mechanisms that contributed to group polarization, all meeting logs for the three communication settings were analyzed based on the psychological processes posited by PAT and SCT. Based on PAT, the amount of persuasive argumentation could be assessed by counting the number of novel and valid arguments generated (see Appendix 3), and these counts might be related to group polarization. Based on SCT, the amount of social comparison could be assessed by counting the number of occasions that subjects engaged in one-upmanship and pluralistic balance behavior (see Appendix 3), and these counts might be related to group polarization. Table 2 summarizes the amounts of persuasive argumentation and social comparison for the three communication settings in Experiment 1.

F-tests were conducted for each process measure. These tests revealed significant effects for communication setting on novel arguments ($F = 32.32$, $p < 0.01$, power = 1.00, effect size = 0.64)

Communication setting	Novel arguments	Valid arguments	One-upmanship behavior	Pluralistic balance behavior
Unsupported	1.15 (0.69)	16.08 (1.32)	0.31 (0.48)	0.31 (0.48)
Face-to-Face CMC	1.15 (0.90)	14.54 (2.70)	0.46 (0.52)	0.69 (0.85)
Dispersed CMC	4.23 (1.59)	16.46 (5.08)	1.46 (0.88)	1.08 (1.26)

Table 2 Mean (standard deviation) of process measures for experiment 1.

Process measure	Choice shift	Preference change
Novel arguments	$r = 0.62, p < 0.01**$	$r = 0.73, p < 0.01**$
Valid arguments	$r = 0.31, p = 0.06$	$r = 0.28, p = 0.09$
One-upmanship behaviorg	$r = 0.67, p < 0.01**$	$r = 0.70, p < 0.01**$
Pluralistic balance behavior	$r = 0.34, p < 0.04*$	$r = 0.33, p < 0.05*$

Table 3 Pearson's correlation scores for experiment 1.
Notes
**$p < 0.01$
*$p < 0.05$,

and one-upmanship behavior ($F = 12.06, p < 0.01$, power = 0.99, effect size = 0.40). Communication setting had no significant effects on valid arguments ($F = 1.16, p = 0.33$, power = 0.24, effect size = 0.06) and pluralistic balance behavior ($F = 2.27$, $p = 0.12$, power = 0.43, effect size = 0.11). Fisher's LSD revealed that the dispersed CMC setting had significantly more novel arguments and one-upmanship behavior than the face-to-face CMC and the unsupported settings (see Table 2). In addition, these two process measures (novel arguments and one-upmanship behavior) were significantly correlated with the two measures of group polarization (see Table 3). Based on these results, both PAT (particularly generation of novel arguments) and SCT (particularly one-upmanship behavior) seemed to be useful for explaining group polarization. The removal of visual cues appeared to stimulate group polarization by encouraging people to contribute more novel arguments and engage in more one-upmanship behavior. This might be a result of lower social presence in the dispersed CMC setting.

5. Social Presence and Anonymity

When people employ CMC for group discussion, the two main types of anonymity that they can invoke are process anonymity and content anonymity (Valacich et al. 1992). With *process anonymity*, people do not know who is contributing and who is not. With *content anonymity*, people cannot associate comments to contributors. This study focuses on content anonymity[7] because it potentially affects group polarization (El-Shinnawy and Vinze 1998). When people communicate via CMC, anonymity provides people with a low threat setting that may alter human behavior (Nunamaker et al. 1991). Anonymity reduces social cues (Connolly et al. 1990) and the immediacy of others (Latane 1981) in a communication setting, thereby lowering social presence (Short et al. 1976).

Anonymity can affect the processes of persuasive argumentation and social comparison. By lowering social presence, anonymity may reduce evaluation and communication apprehension (Nunamaker et al. 1991). This may encourage people to be more open in sharing their views, and to contribute novel and valid arguments. Some empirical results support these observations. Siegel et al. (1986) found that anonymous groups contributed about four times more remarks than identified groups. Anonymous groups have also been found to produce more unique ideas (Jessup et al. 1990; Valacich et al. 1992) and generate more novel solutions (Connolly et al. 1990) than identified groups. The novel and valid arguments generated in such a communication setting with low social presence may also be more salient (Griffith et al. 1998) as people tend to focus on arguments rather than presenters (Nunamaker et al. 1991). This can cause people to move towards the collective position more readily, leading to group polarization.

Social comparison occurs as people change their positions. Salancik (1977) reported that people were less committed to their positions in a communication setting with low social presence, particularly if others could not associate them with their positions. This is the case with anonymity. Hence, as people seek a pluralistic balance, anonymity may cause them to be more ready to give up their positions in favor of the collective position. By reducing social presence, anonymity lowers the behavioral constraints of people, causing them to be more uninhibited during discussion (Jessup et al. 1990). This can promote one-upmanship behavior among people. In short, anonymity should lead to stronger group polarization.

The provision of anonymity and the removal of communication cues, both of which can reduce social presence in a communication setting, affect the processes of persuasive argumentation and social comparison in very much the same way. Therefore, the impact of providing anonymity on group polarization should mirror that of removing communication cues. But because both factors are likely to trigger group polarization by lowering social presence, the combined effects are unlikely to be additive. Instead, having one factor on top of the other may not bring about increased group polarization.

Hypothesis H3a. Choice shift will be significantly stronger in an identified dispersed CMC setting than in an identified face-to-face CMC setting.

Hypothesis H3b. Preference change will be significantly stronger in an identified dispersed CMC setting than in an identified face-to-face CMC setting.

Hypothesis H4a. Choice shift will not be significantly different between an anonymous dispersed CMC setting and an anonymous face-to-face CMC setting.

Hypothesis H4b. Preference change will not be significantly different between an anonymous dispersed CMC setting and an anonymous face-to-face CMC setting.

Hypothesis H5a. Choice shift will be significantly stronger in an anonymous face-to-face CMC setting than in an identified face-to-face CMC setting.

Hypothesis H5b. Preference change will be significantly stronger in an anonymous face-to-face CMC setting than in an identified face-to-face CMC setting.

Hypothesis H6a. Choice shift will not be significantly different between an anonymous dispersed CMC setting and an identified dispersed CMC setting.

Hypothesis H6b. Preference change will not be significantly different between an anonymous dispersed CMC setting and an identified dispersed CMC setting.

6. Experiment 2: Impact of Anonymity

In this experiment, the unsupported setting was dropped for two reasons. First, results of Experiment 1 showed that the removal of verbal cues did not have a significant impact on group polarization, possibly because social presence had not been sufficiently reduced. Second, a full factorial design would entail investigation of anonymity in an unsupported setting, which would be highly artificial (George et al. 1990)[8] By using a different task, Experiment 2 also helped to assess the generalizability of the significant result obtained in Experiment 1 (Dickson 1989).

This experiment is designed to investigate the impact of two features, available in CMC meetings, on group polarization. It employed a 2 × 2 full factorial design that manipulated communication setting (face-to-face CMC versus dispersed CMC) and anonymity (identified versus anonymous). The face-to-face CMC and the dispersed CMC settings were similar to Experiment 1 (see Figure 1). In the *identified setting,* people had their identities tagged to their positions. In the *anonymous setting,* people were disassociated from their positions by not having their identities linked to their positions. Each of the four settings had 10 groups of five subjects each.

6.1. Subjects

The subjects were 200 senior information-systems undergraduates from the same university as in Experiment 1. On average, they were 21.5 years old. About 35% were females and 65% were males. As in Experiment 1, they were randomly assigned to groups, which were randomly assigned to treatments, to alleviate possible confounding effects due to individual characteristics. Like Experiment 1, none of the groups formed had a prior history of working together. Subjects received course credit for their participation. Those who chose not to participate could obtain the same amount of course credit by doing an alternative assignment. The pools of subjects for the two experiments came from different classes with minimal overlap in enrollment. Nevertheless, care was taken to ensure that no subjects participated in both experiments.

6.2. Experimental task

Experiment 2 employed a business-risk task (Paese et al. 1993) used in recent studies on group polarization. It required people to play the role of a senior executive of a computer manufacturing company. They had to select a scheme from among a number of alternative schemes to increase the market share of their company. The probabilities of success for alternative schemes were paired with payoffs in an inverse fashion so that the expected value (product of probability and payoff) of all alternative schemes remained constant. The probabilities of success for alternative schemes ranged from a very risky 0.1 (with a payoff of 40.0%) to a very cautious 0.9 (with a payoff of 4.4%) at intervals of 0.1 (see Appendix 2). Although this is a hypothetical business-risk task, it can elicit human responses as for an actual business-risk task of a comparable nature (Wiseman and Levin 1996).

6.3. Experimental procedure

Each session commenced with subjects reading the task description for 10 minutes. As in Experiment 1, the process of group discussion, which comprised a series of rounds, was explained to the subjects. Under the identified condition, the mode with which subjects generated and presented their views in the face-to-face CMC and the dispersed CMC settings was similar to Experiment 1. Compared to the identified condition, the only difference in the anonymous condition was that the experimental administrator randomly displayed the views of subjects without their names so that no one could associate people with their positions.

The session ended when subjects had attained consensus or when a maximum of four rounds had been completed (in this case, their average position was taken as the collective decision). As for Experiment 1, this restrictive process of group discussion helped to enforce experimental controls by regulating the amount of group discussion, a potential confounding factor affecting group polarization. Before each session ended, subjects restated their positions individually and responded to questions for control and manipulation checks.

6.4. Measuring group polarization

The method for computing choice shift and preference change was similar to that used in Experiment 1 (see § 4.4).

6.5. Control and manipulation Checks

ANOVA tests detected no significant differences for subjects in the various treatments in terms of group experience, task experience, and task knowledge. However, an ANOVA test found that age of subjects was significantly different across communication setting ($F = 8.59, p < 0.01$) but not across anonymity. Mann-Whitney tests found no significant differences in gender ratio of subjects across communication setting and anonymity. Other than age, control over subject characteristics through randomization seemed to be successful. Because average age of subjects differed across some treatments, significant results obtained in subsequent data analyses were confirmed with additional analyses using average age of subjects as covariate (Neter et al. 1990).

Two questions, both anchored on seven-point scales ranging from strongly disagree (1) to strongly agree, served as manipulation checks on anonymity: "I could trace each position to the person involved" and "I was aware of the position of each person." Subjects in the identified setting agreed on both questions to a significantly greater extent than subjects in the anonymous setting ($t = 72.04, p < 0.01$ and $t = 13.50, p < 0.01$, respectively). The manipulation on anonymity seemed to be successful. Like Experiment 1, no manipulation checks were carried out for visual cues because in the dispersed CMC setting, it was physically difficult (if not impossible) for people in ad hoc groups of this study to raise their social presence by seeing mutual expressions.

6.6. Hypotheses Tests

Descriptive statistics for the four communication settings are summarized in Table 4 and Figure 3. Statistical tests were conducted based on a 5% level of significance. Where necessary, the data was transformed (Weisberg 1985) to meet homogeneity and normality requirements of parametric tests. Since choice shift and preference change were significantly correlated ($R^2 = 0.88, p < 0.01$), a MANOVA

Communication setting	Anonymity	Choice shift	Preference change
Face-to-Face CMC	Identified	0.38 (0.28)	0.26 (0.32)
	Anonymous	1.36 (0.59)	1.06 (0.46)
Dispersed CMC	Identified	1.52 (0.55)	1.22 (0.55)
	Anonymous	0.98 (0.49)	0.60 (0.44)

Table 4 Mean (standard deviation) of group polarization for experiment 2.

test involving both independent variables (communication setting and anonymity) and both dependent variables (choice shift and preference change) was carried out. It detected an interaction between both independent variables ($F = 4.38$, $p < 0.03$). This result was confirmed by a MANCOVA test ($F = 4.44$, $p < 0.02$) using average age of subjects as covariate. This finding allowed ANOVA tests to be applied separately to choice shift and preference change. Results of these tests (together with results of confirmation ANCOVA tests using average age of subjects as covariate) are reported in Table 5.

ANOVA tests detected significant interactions involving both independent variables on choice shift ($F = 5.16$, $p < 0.03$) and preference change ($F = 8.84$, $p < 0.01$). These results were confirmed with ANCOVA tests using average age of subjects as covariate (see Table 5). For each dependent variable, two sets of simple effects analyses (Keppel 1991) were conducted to examine the interaction. First, the data was first split along anonymity. Under the identified condition, the dispersed CMC setting yielded significantly greater choice shift ($t = 8.14$, $p < 0.02$, power = 0.77, effect size = 0.31) and preference change ($t = 7.82$, $p < 0.02$, power = 0.75, effect size = 0.30) than the face-to-face CMC setting. H3a and H3b were supported. Under the anonymous condition, the dispersed CMC and the face-to-face CMC settings did not differ significantly in terms of choice shift ($t = 0.39$, $p = 0.54$, power = 0.09, effect size = 0.02) and preference change ($t = 1.98$, $p = 0.18$, power = 0.27, effect size = 0.10). The low power scores (Cohen 1988) of these tests indicated that the null hypotheses could be accepted with reasonable confidence. H4a and H4b were supported.

Second, the data were split along communication setting. Under the face-to-face CMC setting, the anonymous condition produced significantly

Source of Variation	Choice shift		Preference change	
	ANOVA	ANCOVA	ANOVA	ANCOVA
Communication setting	$F = 1.70$	$F = 2.76$	$F = 0.97$	$F = 0.58$
	$p = 0.20$	$p = 0.11$	$P = 0.33$	$p = 0.45$
Anonymity	$F = 0.50$	$F = 0.54$	$F = 0.50$	$F = 0.48$
	$p = 0.48$	$p = 0.47$	$P = 0.48$	$p = 0.49$
Communication setting X Anonymity	$F = 5.16$	$F = 4.89$	$F = 8.84$	$F = 8.69$
	$p < 0.03$*	$p < 0.04$*	$P < 0.01$**	$p < 0.01$**
Average age of subjects (Covariate)		$F = 1.28$		$F = 0.09$
		$p = 0.26$		$p = 0.77$

Table 5 ANOVA and ANCOVA tests on group polarization for experiment 2.

Notes

**$p < 0.01$;
*$p < 0.05$.

Treatment	Choice shift	Preference change
Identified	Dispersed CMC > FTF CMC (H3a) $t = 8.14, p < 0.02^*$, power = 0.77, effect size = 0.31	Dispersed CMC > FTF CMC (H3b) $t = 7.82, p < 0.02^*$, Power = 0.75, effect size = 0.30
Anonymous	Dispersed CMC = FTF CMC (H4a) $t = 0.39, p = 0.54$, power = 0.09, effect size = 0.02	Dispersed CMC = FTF CMC (H4b) $t = 1.98, p = 0.18$, power = 0.27, effect size = 0.10
Face-to-Face (FTF) CMC	Anonymous > Identified (H5a) $t = 5.11, p < 0.04^*$, power = 0.57, effect size = 0.22	Anonymous > Identified (H5b) $t = 8.85, p < 0.01^{**}$, power = 0.80, effect size = 0.33
Dispersed CMC	Anonymous = Identified (H6a) $t = 1.08, p = 0.31$, power = 0.17, effect size = 0.06	Anonymous = Identified (H6b) $t = 2.45, p = 0.13$, power = 0.28, effect size = 0.10

Table 6 Simple effects analyses on group polarization for experiment 2.

Notes

$^{**}p<0.05$;
$^*p<0.01$.

greater choice shift ($t = 5.11, p < 0.04$, power = 0.57, effect size = 0.22) and preference change ($t = 8.85, p < 0.01$, power = 0.80, effect size = 0.33) than the identified condition. H5a and H5b were supported. Under the dispersed CMC setting, the anonymous and identified conditions did not differ significantly in terms of choice shift ($t = 1.08, p = 0.31$, power = 0.17, effect size = 0.06) and preference change ($t = 2.45, p = 0.13$, power = 0.28, effect size = 0.10). Again, the low power scores of these tests indicated that the null hypotheses could be accepted with reasonable confidence. H6a and H6b were supported. Results of these simple effects analyses are depicted in Table 6. Collectively, these results illustrated that the identified face-to-face CMC setting had significantly lower choice shift and preference change than the other three communication settings (see Table 4 and Figure 3).

6.7. General Conclusions

Three conclusions could be drawn from the results of Experiment 2. First, under the identified condition, the absence of visual cues in the dispersed CMC setting resulted in significantly stronger group polarization, possibly by lowering social presence. Given that Experiment 2 employed a different task, this finding added generalizability to the significant result of Experiment 1. Second, in the face-to-face CMC setting, the provision of anonymity led to significantly stronger group polarization. Again, this might be due to the lower social presence that resulted when people did not reveal their identities. Third, the provision of anonymity did not further raise group polarization in the dispersed CMC setting. In other words, the impact of providing anonymity and the impact of removing communication cues on group polarization were not additive but were substitutive (Howell et al. 1986) in nature. As hypothesized, either factor resulted in stronger group polarization but having both factors together would not further raise group polarization. This might be because both factors helped to lower social presence, which could in turn lead to stronger group polarization.

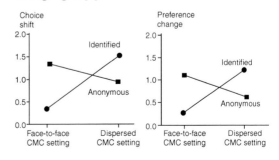

Figure 3 Graphical presentation of group polarization for Experiment 2.

Comm. setting	Anomymity	Novel arguments	Valid arguments	One-upmanship behavior	Pluralistic balance behavior
Face-to-Face	Identified	1.20 (1.32)	15.00 (2.31)	0.10 (0.32)	2.90 (1.20)
CMC	Anonymous	4.70 (1.49)	16.80 (2.15)	1.00 (0.82)	2.80 (1.32)
Dispersed	Identified	6.20 (1.40)	16.70 (2.71)	1.00 (0.67)	2.60 (1.35)
CMC	Anonymous	5.30 (1.95)	15.40 (4.33)	1.00 (0.82)	3.00 (1.33)

Table 7 Mean (standard deviation) of process measures for experiment 2.

6.8. Meeting Log Analyses

As in Experiment 1, additional insights on the mechanisms that led to group polarization were obtained by analyzing all meeting logs for the four experimental treatments based on the psychological processes posited by PAT and SCT. The methods of counting the novel and valid arguments (as posited by PAT), and one-upmanship and pluralistic balance behavior (as posited by SCT) were identical to those in Experiment 1 (see Appendix 3). For the anonymous condition, the groupware tracked the views contributed by each subject in each round. However, the subjects were not aware that the groupware was tracking their views. Table 7 records the amounts of persuasive argumentation and social comparison for the four experimental treatments in Experiment 2.

ANOVA tests were carried out for each process measure. These tests detected significant interactions involving both independent variables on novel arguments ($F = 19.94$, $p < 0.01$) and one-upmanship behavior ($F = 4.31$, $p < 0.05$). The interpretation of significant interactions should take precedence over the interpretation of significant main effects (Keppel 1991). No significant results were found for valid arguments and pluralistic balance behavior (see Table 8). To investigate the interactions, two sets of simple effects analyses were conducted for novel arguments and one-upmanship behavior. First, the data was first split along anonymity. Under the identified condition, the dispersed CMC setting had significantly more novel arguments ($t = 8.23$, $p < 0.01$, power = 1.00, effect size = 0.79) and one-upmanship behavior ($t = 3.86$, $p < 0.01$, power = 0.95, effect size = 0.45) than the face-to-face CMC setting. Under the anonymous condition, the dispersed CMC and the face-to-face CMC settings did not differ significantly in terms of novel arguments ($t = 0.77$, $p = 0.45$, power = 0.11, effect size = 0.03) and one-upmanship behavior ($t = 0.00$, $p = 1.00$, power = 0.05, effect size = 0.00).

Source of variation	Novel arguments	Valid arguments	One-upmanship behavior	Pluralistic balance behavior
Communication setting	$F = 32.29$	$F = 0.03$	$F = 4.31$	$F = 0.02$
	$P < 0.01**$	$P = 0.88$	$P < 0.05*$	$P = 0.90$
Anonymity	$F = 6.96$	$F = 0.07$	$F = 4.31$	$F = 0.13$
	$P < 0.02*$	$P = 0.79$	$P < 0.05*$	$P = 0.72$
Communication setting X Anonymity	$F = 19.94$	$F = 2.67$	$F = 4.31$	$F = 0.37$
	$P < 0.01**$	$P = 0.11$	$P < 0.05*$	$P = 0.55$

Table 8 ANOVA Test on Process Measures for Experiment 2.

Notes

**p<0.01;
*p<0.05.

Treatment	Novel arguments	One-upmanship behavior
Identified	Dispersed CMC > FTF CMC $t = 8.23, p < 0.01**$, power = 1.00, effect size = 0.79	Dispersed CMC > FTF CMC $t = 3.86, p < 0.01**$, power = 0.95, effect size = 0.45
Anonymous	Dispersed CMC = FTF CMC $t = 0.77, p = 0.45$, power = 0.11, effect size = 0.03	Dispersed CMC = FTF CMC $t = 0.00, p = 1.00$, Power = 0.05, effect size = 0.00
Face-to-Face (FTF) CMC	Anonymous > Identified $t = 5.56, p < 0.01**$, power = 1.00, effect size = 0.63	Anonymous > Identified $t = 3.25, p < 0.01**$, power = 0.87, effect size = 0.37
Dispersed CMC	Anonymous = Idendified $t = 1.19, p = 0.25$, power = 0.20, effect size = 0.07	Anonymous = Identified $t = 0.00, p = 1.00$, power = 0.05, effect size = 0.00

Table 9 Simple effects analyses on process measures for experiment 2.

Notes

**$p<0.01$;
*$p<0.05$.

Second, the data were split along communication setting. Under the face-to-face CMC setting, the anonymous condition produced significantly more novel arguments ($t = 5.56, p < 0.01$, power = 1.00, effect size = 0.63) and one-upmanship behavior ($t = 3.25, p < 0.01$, power = 0.87, effect size = 0.37) than the identified condition. Under the dispersed CMC setting, the anonymous and identified conditions did not differ significantly in terms of novel arguments ($t = 1.19, p = 0.25$, power = 0.20, effect size = 0.07) and one-upmanship behavior ($t = 0.00, p = 1.00$, power = 0.05, effect size = 0.00). Results of these simple effects analyses are reported in Table 9. Together, these results suggested that the identified face-to-face CMC setting produced significantly less novel arguments and one-upmanship behavior than the other three communication settings (see Table 7). Just like Experiment 1, there were also significant correlation between these two process measures (novel arguments and one-upmanship behavior) and both measures of group polarization (see Table 10). Again, these results suggested that both PAT (particularly generation of novel arguments) and SCT (particularly one-upmanship behavior) appeared to be useful for explaining group polarization. The removal of visual cues or the provision of anonymity appeared to stimulate group polarization mainly by encouraging people to contribute more novel arguments and engage in more one-upmanship behavior.

Process measure	Choice shift	Preference change
Novel arguments	$r = 0.40, p < 0.01**$	$r = 0.42, p < 0.01**$
Valid arguments	$r = 0.07, p = 0.68$	$r = 0.07, p = 0.66$
One-upmanship behavior	$r = 0.40, p < 0.01**$	$r = 0.37, p < 0.02*$
Pluralistic balance behavior	$r = 0.12, p = 0.47$	$r = 0.02, p = 0.92$

Table 10 Pearson's correlation ccores for Experiment 2.

Notes

**$p<0.01$;
*$p<0.05$.

7. Discussion and Implications

7.1. Discussion of Results

Zuber et al. (1992) found that choice shift and preference change were correlated in an unsupported setting. McGuire et al. (1987) reported the same correlation in a dispersed CMC setting. In Experiments 1 and 2 of this study, similar correlations were obtained. These results show that the correlation between choice shift and preference change can be extended to a face-to-face CMC setting. Given the robustness of this correlation across communication settings, future studies may want to adopt either variable (instead of both variables) to measure group polarization.

Siegel et al. (1986) found that dispersed CMC settings (identified or anonymous) yielded greater group polarization than an unsupported setting. Results of this study provide some insights on, and add to the findings of Siegel et al. (1986). Specifically, the results of Experiment 1 reveal that the stronger group polarization in a dispersed CMC setting results from the removal of visual cues rather than verbal cues. This may be a result of lower social presence in a dispersed CMC setting. The results of Experiment 2 show that the provision of anonymity, just like the removal of visual cues, leads to stronger group polarization. Both the provision of anonymity and the removal of visual cues may have produced this impact by lowering social presence. Given that the impact of providing anonymity and the impact of removing visual cues are substitutive in nature, this impact can only be seen in a face-to-face CMC setting and not in a dispersed CMC setting.

Dubrovsky et al. (1991) and Weisband (1992) compared choice shift in an unsupported setting and a dispersed CMC setting, where communication took place via synchronous electronic mail. Because the identities of people were tagged to their electronic mails, there was no anonymity involved. Both studies found no significant differences in choice shift between the two communication settings. These results contradict those of Experiments 1 and 2 in this study, where the dispersed CMC setting yielded significantly stronger choice shift than the unsupported setting. A plausible factor contributing to these differences in results is interactivity (Carlson and Zmud 1999). In this study, people in the dispersed CMC setting followed a controlled procedure when exchanging their views and had little time to build up interactivity. However, in Dubrovsky et al. (1991) and Weisband (1992), people in the dispersed CMC setting exchanged their views interactively over a longer period of time. This interactivity might have allowed people to build up their social presence over time (Carlson and Zmud 1999, Walther 1995). The resulting increase in social presence may have alleviated group polarization. This contention can be tested in future research.

The meeting log analyses, anchored on psychological processes of PAT and SCT, yielded insights about how CMC may impact group polarization. Based on Experiment 1, the removal of visual cues (rather than verbal cues) appears to lower social presence so that people contribute more novel arguments and engage in more one-upmanship behavior. Based on Experiment 2, the removal of visual cues or the provision of anonymity appears to invoke the same mechanism of lowering social presence to allow people to contribute more novel arguments and engage in more one-upmanship behavior. When working with others in a group, people usually desire to be valuable and decisive (McGrath 1984). Contributing novel arguments may generate more value compared to contributing valid arguments (e.g., reinforce the arguments of others). One-upmanship behavior is perceived to be more related to decisiveness than pluralistic balance behavior because the former behavior may move the group to a final decision faster. But when social presence is high, people may be deterred from demonstrating value or decisiveness due to communication and evaluation apprehension. Therefore, it may be easiest for people to contribute novel arguments or engage in one-upmanship behavior when social presence is low.

In both experiments, the result of having more novel arguments and more one-upmanship behavior is stronger group polarization. PAT and SCT posit four psychological processes (novel arguments, valid arguments, one-upmanship behavior, and pluralistic balance behavior) that may increase group polarization. Results of this study suggest that CMC (specifically, the restriction of communication cues or the provision of anonymity) may alter group polarization through two of these processes (novel arguments and one-upmanship behavior).

To enforce strict experimental controls, this study was staged using small ad hoc groups that had neither a history nor a future, tasks in which the subjects had no stake, and an artificial communication setting that was short term in nature. In other words, both experiments aimed to achieve high internal validity at the expense of external validity. Consequently, the recommendations based on the results of this study are extremely limited in terms of generalizability. Caution must be exercised when attempting to apply these results to organizational decision makers, tasks, and communication settings. Ways to extend these findings include replicating this study across a variety of tasks such as the choice-dilemma tasks (Wallach et al. 1962), subjects such as organizational executives or MBA students, and settings in actual organizations.

7.2. Implications for Research

Several related avenues of further research can be undertaken. First, although this study involved both the communication process and outcome, future studies can investigate the communication process at a greater level of detail (e.g., Poole et al. 1991, Todd and Benbasat 1987, Zigurs et al. 1988). The analyses of the communication process in this study can serve as a starting point for this line of research. In-depth analyses of the nature of materials exchanged (McGuire et al. 1987), together with post-discussion interviews with subjects to find out how these materials have affected their thinking, would yield more precisely knowledge about the relationship between the communication process and communication outcome. Combined positivist and interpretivist analyses of this nature (e.g., Trauth and Jessup 2000) can yield more precise insights on how CMC may impact group polarization.

Second, some studies have reported that the differences in impact produced by different communication settings may diminish over time as people learn to overcome restrictions of certain communication settings (e.g., Walther 1995). Therefore, this study can be extended longitudinally by examining the communication process and outcome of people over multiple sessions (e.g., Chidambaram 1996, Chidambaram and Bostrom 1993). If people in a dispersed CMC setting can raise their social presence through interactivity over time (Carlson and Zmud 1999), their tendency towards stronger group polarization may diminish accordingly. Also, if people have sufficient time to learn the idiosyncrasies of each other, they may be able to associate most messages with the people sending them. In such situations, the provision of anonymity may not lower social presence sufficiently for people to experience stronger group polarization in a face-to-face CMC setting.

Third, group size can be examined to see whether and how it may moderate the impact of communication cues and anonymity on group polarization. Social impact theory (Latane 1981) states that social influence experienced by an individual depends, in part, on the number of people involved. This implies that as more people are involved, the PAT and SCT psychological processes may produce stronger impact on each individual, thereby yielding stronger group polarization. Dennis et al. (1990) and Valacich et al. (1992) observed that the effects of CMC are more pronounced in larger than in smaller groups. This suggests that as more people are involved, the effects of communication cues and anonymity observed in this study may be magnified. Future studies can include more people in each experimental session to verify such observations.

Fourth, this study may be replicated in other cultural settings. The extent to which people exhibit group polarization tendencies after discussion depends on the initial inclinations of the people involved. Risk-seeking (or risk-averse) people tend to emerge from their discussion with like-minded people to become even more risk seeking (or risk averse), due to group polarization (Isenberg 1986, Rajecki 1990). Therefore, group polarization tendencies and the concomitant impact of CMC may be more pronounced in populations that are risk seeking or risk averse rather than in populations that are risk neutral. Such risk behavior of populations are affected by uncertainty avoidance (Keil et al. 2000), a cultural dimension that differentiates populations according to their tolerance for risks (Hofstede 1991). Future studies can verify whether the effects of communication cues and anonymity observed in this study are stronger in cultures that are very high or very low on uncertainty avoidance, or weaker in cultures that are more neutral on uncertainty avoidance.

7.3. Implications for Practice

Results of this study provide some understanding about how CMC may be associated with group polarization. These results may be useful to decision-makers in various situations, some of which may benefit from less group polarization and others of which may benefit from more group polarization. Examples of situations that can benefit from less group polarization are discussion involving racial attitudes (Riley and Pettigrew 1976), community conflicts (Coleman 1957), and investments in failing business ventures (Brockner 1992). This study reveals two settings in which group polarization tends to be low: first, when people engage in discussion in an unsupported communication setting (without CMC) and second, when the identities of contributors are tagged to comments during a face-to-face discussion via CMC (perhaps for ease of record keeping). In situations where people have to discuss in a dispersed CMC setting, they can perhaps interact actively over time so as to raise their social presence, as in Dubrovsky et al. (1991) and Weisband (1992).

Examples of situations that may benefit from more group polarization are discussions on social work (Festinger et al. 1956), donations (Muehleman et al. 1976), and mutual counseling efforts (Toch 1965). Group polarization may also be used to break deadlocks when making political decisions (Janis 1989), jury decisions (Myers and Kaplan 1976), or investment decisions (Whyte 1993). Organizations that wish to promote innovation and entrepreneurship via risk taking may also benefit from group polarization (Argyris 1992). This study uncovers two settings in which group polarization tends to be high. The first is when discussions are conducted in a dispersed CMC setting with people connected together via a computer network. The second is when discussions are held in either face-to-face CMC or dispersed CMC settings with anonymity so that people cannot associate specific comments to contributors. These two settings may especially benefit organizations that are encouraging innovation and entrepreneurship because, as this study illustrates, both settings may cause people to generate a greater number of novel arguments.

Choice shift and preference change have been found to be correlated in a wide range of communication settings, by this and other studies. Since choice shift is a group-level measure and preference change is an individual-level measure (see §2), this correlation implies that the post-discussion positions of people tend to be close to the collective position. In other words, people are likely to have internalized the collective position in situations involving group polarization. Hence, if CMC can impact group polarization, it can also change the conviction of people for the collective position. Since organizational decisions are typically implemented by people who make the decisions and that success of implementing the decisions may depend on the conviction of these people, care should be exercised when using CMC to manage group polarization.[9]

8. Conclusion

Through two research experiments, this study makes a modest contribution to the literature by providing some insights into the processes that trigger group polarization. The results show that the removal of visual cues or the provision of anonymity through CMC may help to lower social presence. This causes people to contribute more novel arguments and engage in more one-upmanship behavior, which leads to stronger group polarization (as posited by PAT and SCT). In other words, strong group polarization may be associated with CMC. These results have useful implications for research and practice.

The increasing complexity of the business environment has made it necessary for many strategic organizational decisions to be made collectively rather than individually (Davis 1992, Hart et al. 1985). As a result, these critical organizational decisions may be prone to the effects of group polarization, one of several very pervasive social phenomena. The spread of computer networks allows organizational decisions to be made through CMC. Besides being a convenient means of discussion, CMC may impact group polarization. Given the profound impact of group polarization on critical organizational decisions, it is useful to understand how group polarization may be affected in meetings supported by CMC. Research along the direction taken by this study can help to advance our understanding on this important issue.

NOTES

1. The impact of social construction on group polarization, which tends to be more pronounced with prolonged interactivity, is outside the scope of this study. Therefore, the groups do not have a past history of working together and have not been given an extended period of time to perform their task. Instead, they have to complete their task within a reasonably short time period (determined using a pilot study).
2. An unsupported dispersed setting (e.g., telephone conference calls) was omitted because of the need to maintain comparability among different treatments. As an example, in all three experimental settings, a screen that held the decisions of everyone involved served as group memory. But at the time of this study, there was no reliable technology to project such a screen in an unsupported dispersed setting.
3. Although it seemed unrealistic, this setting had theoretical and practical significance (Tan et al. 1999). In theory, it allowed the maximum impact of removing verbal cues to be measured. In practice, people could choose not to use verbal cues at all if it was beneficial for them to do so.
4. Many prior CMC studies that compared face-to-face with dispersed conditions had also adopted this approach. These studies had not reported any impact of physical differences (e.g., seating arrangements or partitions) on the results (Fjermestad and Hiltz 1999). Thus, it is unlikely that physical arrangements would affect the results.
5. Results of the pilot study showed that group polarization tendencies were very clearly manifested after four rounds. People who did not exhibit group polarization tendencies after four rounds were very unlikely to do so thereafter.
6. Manipulation checks would be useful mainly in situations where physical controls could not be carried out (Dickson 1989).
7. Henceforth, anonymity would refer to content anonymity unless otherwise stated.
8. Griffith and Northcraft (1994) operationalized anonymity in an unsupported setting by getting people to exchange textual cues (e.g., pieces of paper) that did not carry any form of identification. However, the experimental procedure of this study did not permit such a treatment to be included without compromising comparability among different treatments. Nevertheless, future studies should include such a setting when the experimental procedure permits. If such a setting is indeed found to be beneficial, technologies may be implemented to support such a setting.
9. Attempts to manage group polarization may raise ethical concerns about manipulation of organizational decisions. Nevertheless, it is useful for managers to understand how CMC may impact group polarization with different communication settings.

ACKNOWLEDGMENTS

The authors thank the Associate Editor and three anonymous reviewers for their many useful suggestions and comments on earlier versions of this paper.

REFERENCES

Agyris, C. 1992. *On Organizational Learning*. Blackwell Publishers, Cambridge, MA.

Baron, R. S. and G. Roper. 1976. Reaffirmation of social comparison views of choice shifts: Averaging and extremity effects in an autokinetic situation. *J. Personality Soc. Psych.* 33(5) 521–30.

Bhappu, A. D., T. L. Griffith and G. B. Northcraft. 1997. Media effects and communication bias in diverse groups. *Organ. Behavior and Human Decision Processes* 70(3) 199–205.

Blascovich, J., G. P. Ginsburg and R. C. Howe. 1975a. Blackjack and the risky shift, II: Monetary stakes. *J. Experiment. Soc. Psych.* 11(3) 224–32.

——, ——, T. L. Veach. 1975b. A pluralistic explanation of choice shifts on the risk dimension. *J. Personality Soc. Psych.* 31(3) 422–9.

Brockner, J. 1992. The escalation of commitment to a failing course of action: Toward theoretical progress. *Acad. Management Rev.* 17(1) 39–61.

Brown, R. 1965. *Social Psychology*. The Free Press, New York.

Burnstein, E. 1982. Persuasion as argument processing. H. Brandstatter, J. H. Davis, and G. Stocker-Kreichgauer, eds. *Contemporary Problems in Group Decision-Making*. Academic Press, New York, 103–24.

Butler, J. K. and M. D. Crino. 1992. Effects of initial tendency and real risk on choice shift. *Organ. Behavior and Human Decision Processes* 53(1) 14–34.

Carlson, J. R. and R. W. Zmud. 1999. Channel expansion theory and the experiential nature of media richness perceptions. *Acad. Management J.* 42(2) 153–70.

Cartwright, D. S. 1975. The nature of gangs. D. S. Cartwright, B. Tomson and H. Schwartz, eds. *Gang Delinquency*. Brooks/Cole, Monterey, CA 1–22.

Chidambaram, L. 1996. Relational development in computer-supported groups. *MIS Quart.* 20(2) 143–65.

——, R. P. Bostrom. 1993. Evolution of group performance over time: A repeated measures study of GDSS effects. *J. Organ. Comput.* 3(4) 443–70.

Cohen, J. 1988. *Statistical Power Analysis for the Behavioral Sciences.* Lawrence Erlbaum Associates, Hillsdale, NJ.

Coleman, J. S. 1957. *Community Conflict.* Free Press, New York.

Connolly, T., L. M. Jessup and J. S. Valacich. 1990. Effects of anonymity and evaluative tone on idea generation in computer-mediated groups. *Management Sci.* 36(6) 689–703.

Cook, M., M. Lallijee. 1972. Verbal substitutes for visual signals in interaction. *Semiotica 6* 212–21.

Daft, R. L., R. H. Lengel and L. K. Trevino. 1987. Message equivocality, media selection, and manager performance: Implications for information systems. *MIS Quart.* 11(3) 355–66.

Davis, J. H. 1993. Some compelling intuitions about group consensus decisions, theoretical and empirical research, and interpersonal aggregation phenomena: Selected examples, 1950–1990. *Organ. Behavior and Human Decision Processes 52*(1) 3–38.

Dennis, A. R. and S. T. Kinney. 1998. Testing media richness theory in the new media: The effects of cues, feedback, and task equivocality. *Inform. Systems Res.* 9(3) 256–74.

——, J. F. Nunamaker and D. R. Vogel. 1990a. A comparison of laboratory and field research in the study of electronic meeting systems. *J. Management Inform. Systems* 7(3) 107–35.

——, J. S. Valacich and J. F. Nunamaker. 1990b. An experimental investigation of the effects of group size in an electronic meeting environment. *IEEE Trans. Systems, Man, and Cybernetics* 25(5) 1049–57.

Dickson, G. 1989. A programmatic approach to information systems research: An experimentalist's view. I. Benbasat, ed. *The Information Systems Research Challenge: Experimental Research Methods.* Boston, MA 147–70.

Diener, E., S. Fraser, A. L. Beaman and R. T. Kelem. 1976. Effects of deindividuating variables on stealing by Halloween trick-or-treaters. *J. Personality Soc. Psych.* 33(2) 178–83.

Dubrovsky, V. J., S. B. Kiesler and B. N. Sethna. 1991. The equalization phenomenon: Status effects in computer-mediated and face-to-face decision-making groups. *Human-Computer Interaction.* 6(2) 119–46.

El-Shinnawy, M. and A. S. Vinze. 1998. Polarization and persuasive argumentation: A study of decision making in group settings. *MIS Quart.* 22(2) 165–98.

Festinger, L., H. W. Riecken and S. Schachter. 1956. *When Prophecy Fails.* University of Minnesota Press, Minneapolis, MN.

Fjermestad, J., S. R. Hiltz. 1999. An assessment of group support systems experimental research: Methodology and results. *J. Management Inform. Systems* 15(3) 7–150.

Fromkin, H. 1970. Effects of experimentally aroused feelings of undistinctiveness upon valuation of scarce and novel experiences. *J. Personality Social Psych.* 16(3) 521–9.

George, J. F., G. K. Easton, J. F. Nunamaker and G. B. Northcraft. 1990. A study of collaborative group work with and without computer-based support. *Inform. Systems Res.* 1(4) 394–415.

Griffith, T. L. and G. B. Northcraft. 1994. Distinguishing between the forest and the trees: Media, features, and methodology in electronic communication research. *Organ. Sci.* 5(2) 272–85.

——, M. A. Fuller and G. B. Northcraft. 1998. Facilitator influence in group support systems: Intended and unintended effects. *Inform. Systems Res.* 9(1) 20–36.

Hart, S. L., M. Boroush, G. Enk and W. Hornick. 1985. Managing complexity through consensus mapping: Technology for the structuring of group decisions. *Acad. Management Rev.* 10(3) 587–600.

Hofstede, G. 1991. *Cultures and Organizations: Software of the Mind.* McGraw Hill, London. U.K.

Howell, J., P. Dorfman and S. Kerr. 1986. Moderator variables in leadership research. *Acad. Management Rev.* 11(1) 88–102.

Isenberg, D. J. 1986. Group polarization: A critical review and meta-analysis. *J. Personality Social Psych.* 50(6) 1141–51.

Janis, I. L. 1989. *Crucial Decisions: Leadership in Policymaking and Crisis Management.* Free Press, New York.

Jessup, L. M., T. Connolly and J. Galegher. 1990. The effects of anonymity on GDSS group process with an idea-generating task. *MIS Quart.* 14(3) 313–21.

Johansen, R., D. Sibbet, S. Benson, A. Martin, R. Mittman and P. Saffo. 1991. *Leading Business Teams: How Teams Can Use Technology and Process to Enhance Group Performance.* Addison-Wesley, Reading, MA.

Kaplan, M. F. 1977. Discussion polarization effects in a modified jury decision paradigm: Informational influences. *Sociometry* 40(3) 262–71.

Keil, M., B. C. Y. Tan, K. K. Wei, T. Saarinen, V. Tuunainen and A. Wassenaar. 2000. A cross-cultural study on escalation of commitment behavior in software projects. *MIS Quart.* 24(2) 299–325.

Keppel, G. 1991. *Design and Analysis: A Researcher's Handbook.* Prentice-Hall, Upper Saddle River, NJ.

Kiesler, S., D. Zubrow, A. M. Moses and V. Geller. 1985. Affect in computer-mediated communication: An Experiment in synchronous terminal-to-terminal discussion. *Human-Comput. Interaction* 1(1) 77–104.

Latane, B. 1981. The psychology of social impact. *Amer. Psych.* 36(4) 343–56.

Lewicki, R. J. and R. Litterer. 1985. *Negotiation.* Irwin, Homewood, IL. McCauley, C. 1989. The nature of social influence in groupthink: Compliance and internalization. *J. Personality Social Psych.* 57(2) 250–60.

McGrath, J. E. 1984. *Groups: Interaction and Performance.* Prentice-Hall, Englewood Cliffs, NJ.

McGuire, T. W., S. B. Kiesler and, J. Siegel. 1987. Group and computer-mediated discussion effects in risk decision making. *J. Personality and Soc. Psych.* 52(5) 917–30.

Muehleman, J. T., C. Bruker and C. M. Ingram. 1976. The generosity shift. *J. Personality Soc. Psych.* 34(3) 344–51.

Myers, D. G. and M. F. Kaplan. 1976. Group-induced polarization in simulated juries. *Personality Soc. Psych. Bull.* 2(1) 63–6.

Neter, J., W. Wasserman and M. H. Kutner. 1990. *Applied Linear Statistical Models: Regression, Analysis of Variance, and Experimental Designs.* Irwin, Homewood, IL.

Nunamaker, J. F, A. R. Dennis, J. F. George, J. S. Valacich and D. R. Vogel. 1991. Electronic meeting systems to support group work. *Comm. ACM* 34(7) 40–61.

Paese, P. W., M. Bieser and M. E. Tubbs. 1993. Framing effects and choice shift in group decision making. *Organ. Behavior and Human Decision Processes* 56(1) 149–65.

Pavitt, C. 1994. Another view of group polarizing: The 'reasons for' one-sided oral argumentation. *Comm. Res.* 21(5) 625–42.

Poole. M. S. and M. H. Jackson. 1993. Communication theory and group support systems. L. M. Jessup and J. S. Valacich, eds. *Group Support Systems: New Perspectives.* Macmillan, New York 281–93.

——, M. Holmes and G. DeSanctis. 1991. Conflict management in a computer-supported meeting environment. *Management Sci.* 37(8) 926–53.

Pruitt, D. G. 1971. Choice shifts in group discussion: An introductory review. *J. Personality Soc. Psych.* 20(3) 339–60.

Rajecki, D. W. 1990. Group discussion and the polarization of attitudes. *Attitudes, Themes and Advances.* Sinauer Associates, Sunderland, MA 183–214.

Rice, R. E. 1984. *The New Media: Communication, Research, and Technology.* Sage, Newbury Park, CA.

——. 1993. Media appropriateness: Using social presence theory to compare traditional and new organizational media. *Human Comm. Res.* 19(4) 451–84.

Riley, R. T. and T. F. Pettigrew. 1976. Dramatic events and attitude change. *J. Personality Soc. Psych.* 34(5) 1004–15.

Salancik, G. R. 1977. Commitment and the control of organizational behavior and belief. B. M. Staw and G. R. Salancik, eds. *New Directions in Organizational Behavior.* St. Clair, Chicago, IL 1–54.

Sanders, G. S. and R. S. Baron. 1977. Is social comparison irrelevant for producing choice shift? *J. Experiment. Soc. Psych.* 13(4) 303–14.

Shaw, M. E. 1978. Communication networks fourteen years later. L. Berkowitz, ed. *Group Processes.* Academic Press, New York, 351–62.

Short, J., E. Williams and B. Christie. 1976. *The Social Psychology of Telecommunication.* John Wiley, New York.

Siegel, J., V. Dubrovsky, S. B. Kiesler and T. W. McGuire. 1986. Group processes in computer-mediated communication. *Organ. Behavior and Human Decision Processes* 37(2) 157–87.

Sproull, L. and S. Kiesler. 1986. Reducing social context cues: Electronic mail in organizational communication. *Management Sci.* 32(11) 1492–12.

Straub, D. and E. Karahanna. 1998. Knowledge worker communications and recipient availability: Toward a task closure explanation of media choice. *Organ. Sci.* 9(2) 160–75.

Tan, B. C. Y., K. K. Wei and R. T. Watson. 1999. The equalizing impact of a group support system on status differentials. *ACM Trans. Inform. Systems* 17(1) 77–100.

———, ———, ———, D. L. Clapper and E. R. McLean. 1998. Computer-mediated communication and majority influence: Assessing the impact in an individualistic and a collectivistic culture. *Management Sci. 44*(9) 1263–78.

Toch, R. 1965. *The Social Psychology of Social Movements.* Bobbs-Merrill, Indianapolis, IN.

Todd, P., I. Benbasat. 1987. Process tracing methods in decision support systems research: Exploring the black box. *MIS Quart. 11*(4) 493–512.

Trauth, E. M. and L. M. Jessup. 2000. Understanding computer-mediated discussions: Positive and interpretive analyses of group support system use. *MIS Quart. 24*(1) 43–79.

Valacich, J. S., A. R. Dennis and T. Connolly. 1994a. Idea generation in computer-based groups: A new ending to an old story. *Organ. Behavior and Human Decision Processes 57*(3) 448–67.

———, ———, J. F. Nunamaker. 1992. Group size and anonymity effects on computer-mediated idea generation. *Small Group Res. 23*(1) 49–73.

———, J. F. George, J. F. Nunamaker and D. R. Vogel. 1994b. Physical proximity effects on computer-mediated group idea generation. *Small Group Res. 25*(1) 83–104.

Vinokur, A., E. Burnstein. 1974. Effects of partially-shared persuasive arguments on group-induced shifts: A group problem-solving approach. *J.Personality Soc. Psych. 29*(2) 305–15.

———, ———. 1978. Novel argumentation and attitude change: The case of polarization following group discussion. *Euro. J. Soc. Psych. 8*(3) 335–48.

Wallach, M. A., N. Kogan and D. J. Bem. 1962. Group influence on individual risk-taking. *J. Abnormal Soc. Psych. 65*(2) 75–86.

Walther, J. B. 1995. Relational aspects of computer-mediated communication: Experimental observations over time. *Organ. Sci. 6*(2) 186–203.

Walton, R. E. and R. B. McKersie. 1965. *A Behavioral Theory of Labor Negotiation.* McGraw-Hill, New York.

Weisband, S. P. 1992. Group discussion and first advocacy effects in computer-mediated and face-to-face decision making groups. *Organ. Behavior and Human Decision Processes 53*(3) 352–80.

Weisberg, S. 1985. *Applied Linear Regression.* John Wiley, New York.

Whyte, G. 1993. Escalating commitment in individual and group decision making: A prospect theory approach. *Organ. Behavior and Human Decision Processes 54*(3) 430–55.

Wiener, M. and A. Mehrabian. 1968. *Language Within Language: Immediacy, a Channel in Verbal Communication.* Appleton Century Crofts, New York.

Williams, E. 1977. Experimental comparisons of face-to-face and mediated communication: A review. *Psych. Bull. 84*(5) 963–76.

Wiseman, D. B. and I. P. Levin. 1996. Comparing risky decision making under conditions of real and hypothetical consequences. *Organ. Behavior and Human Decision Processes 66*(3) 241–50.

Yinon, Y., Y. Jaffe and S. Feshbach. 1975. Risky aggression in individuals and groups. *J. Personality and Soc. Psych. 31*(5) 808–15.

Zigurs, I., M. S. Poole and G. DeSanctis. 1988. A study of influence in computer-mediated group decision making. *MIS Quart. 12*(4) 625–44.

Zuber, J. A., H. W. Crott and J. Werner. 1992. Choice shift and group polarization: An analysis of the status of arguments and social decision schemes. *J. Personality Soc. Psych. 62*(1) 50–61.

APPENDIX 1.
EXPERIMENTAL TASK FOR STUDY 1

Read the following scenario carefully. Decide on the required *probability of success* (ranging from 0.01 to 1.00) for you to recommend taking the *risky option*. For example, a probability of 1.00 would mean that the risky option would have to be a sure success for you to recommend it. Conversely, a probability of 0.01 would mean that the risky option is so attractive that it is worth going for it in the face of almost certain failure.

Task Scenario. You are the captain of the university football team, which is in the final seconds of a game with its traditional rival. You may choose a cautious play that would almost certainly result in a draw. You may also choose a risky play that would almost certainly lead to victory if successful or defeat if unsuccessful.

What must the probability of success be before you are willing to choose the risky play?

APPENDIX 2. EXPERIMENTAL TASK FOR STUDY 2

Task Scenario. You are the vice-president of a computer manufacturing company. It has maintained a sizeable market share by providing reliable computers at affordable prices. Recently, your research and development team developed some proprietary technology that makes it possible to produce and market a new computer model. This new model is faster and more affordable than the model currently marketed by your company. The President of your company hints to you that your career prospects will depend on how well you exploit this opportunity. A number of short-term alternatives to exploit a gain in market share are open to you, as follows:

- Alternative A: 10% probability of increasing your market share by 40.0%.
- Alternative B: 20% probability of increasing your market share by 20.0%.
- Alternative C: 30% probability of increasing your market share by 13.3%.
- Alternative D: 40% probability of increasing your market share by 10.0%.
- Alternative E: 50% probability of increasing your market share by 8.0%.
- Alternative F: 60% probability of increasing your market share by 6.7%.
- Alternative G: 70% probability of increasing your market share by 5.7%.
- Alternative H: 80% probability of increasing your market share by 5.0%.
- Alternative I: 90% probability of increasing your market share by 4.4%.
- *Which alternative would you recommend?*

APPENDIX 3. ANALYSES OF MEETING LOGS

All the meeting logs were analyzed by two graduate students who had been trained to identify instances of novel arguments, valid arguments, one-upmanship behavior, and pluralistic balance behavior. Both graduate students had total agreement for all instances of one-upmanship behavior and pluralistic balance behavior. Their few instances of disagreement for novel arguments and valid arguments were resolved through discussion.

A *novel argument* is one that contained facts in support of the collective position and yielded fresh insights (i.e., it was not related to arguments presented earlier). Examples of novel arguments (extracted from different meeting logs) are:

1 "I think we should not be overly cautious here. New technology becomes old technology very quickly. So I suggest we go for high-risk options. But not to the extent of options A, B, or C."
2 "We are not in a situation where we have to choose a safer option or else we will LOSE EVERYTHING. We are now talking about taking advantage of the fact that we have something new. 4.4% isn't really that great. Sales gimmicks may even make that figure. Try something that will make us a MONOPOLY in the short term."
3 "This situation is a matter of how sure we are to gain something. Since this product is proprietary, we need to gain a pool of customers. Once we have 'locked' people into using it, the product will be in high demand and support for it WILL DEFINITELY come."

A *valid argument* is one that contained facts in support of the collective position and reinforced other arguments (i.e., it was related to arguments presented earlier). Examples of valid arguments (extracted from different meeting logs) are:

1 "I note from the description that these alternatives are 'short term.' So I also choose Alternative C, as the main aim here is to increase market share. The 0.3 probability is acceptable to me."
2 "Someone said that the market is large! It makes no sense to be cautious and I prefer Alternative A :-) 10% of anything big is still substantial."
3 "Like I said, Option G is quite good but Option F is slightly better for me. But, well, the stuff some of you guys said about new technology is true. Yup, we should ensure our investment is worth it."

A person was considered to have engaged in *one-upmanship behavior* if he or she moved in the direction of the collective position beyond the average collective position in the previous round, regardless of the arguments given. For example, if the collective group position for the previous round was 0.7 and a person changed his or her position from 0.6 to 0.9 in the current round, then the person would be considered to have exhibited an instance of one-upmanship behavior.

A person was considered to have engaged in *pluralistic balance behavior* if he or she moved in the direction of the collective position but not beyond the average collective position in the previous round, and provided no facts in support of the collective position (e.g., simply stated personal preference). For example, if the collective group position for the previous round was 0.8 and a person changed his or her position from 0.6 to 0.7 in the current round without providing arguments in support of the collective position, then the person is considered to have exhibited an instance of pluralistic balance behavior.

The counting scheme used in this study tended to undercount instances of pluralistic balance behavior. For example, if a subject changed his/her decision due to both novel or valid arguments as well as pluralistic balance behavior, the novel or valid arguments would be counted, but not the pluralistic balance behavior. It was not possible to identify all instances of pluralistic balance behavior based on the information captured by the groupware. Nevertheless, prior research had found that in such circumstances, novel and valid arguments tended to have a stronger influence on human decisions than pluralistic balance behavior (Isenberg 1986).

Group Support Systems in Hong Kong: An Action Research Project

Robert Davison and Doug Vogel*

*Department of Information Systems, City University of Hong Kong, Tat Chee Avenue, Kowloon, Hong Kong, email: isrobert@is.cityu.edu.hk and *isdoug@is.cityu.edu.hk*

Abstract

The last dozen years have seen a considerable investment of resources into the research and development of group support systems (GSS) technology. This paper describes how GSS was used to support a process improvement project in a Hong Kong accounting firm. Although the project encountered many difficulties, the application of action research facilitated the adaptation of the GSS to the shifting circumstances, and the project was successfully completed. A variety of lessons concerning the use of GSS are presented, while increased use of action research in complex organizational contexts is recommended.

Keywords: action research, group support systems, motivation, participation, process interventions

Background

The last dozen years have seen a considerable investment of resources into the research and development of group support systems (GSS) technology in organizations. GSS have been applied to a wide variety of circumstances, including requirements engineering (Liou and Chen, 1993), strategic management (Tyran *et al.*, 1992), software inspections (Mashayekhi *et al.*, 1993), capacity building (Jones *et al.*, 1998) and international diplomacy (Lyytinen *et al.*, 1993). This literature paints a somewhat mixed picture of GSS, with the general conclusion that GSS, when appropriately applied, can benefit adopters.

This paper describes how a GSS (GroupSystems for Windows v.1.1d, distributed by Ventana Corporation, USA) was used in support of a process improvement project in an accounting firm in Hong Kong. The task, its setting and its actors are described, followed by a justification for the use of action research as a guiding methodology. Two sample cycles of activity are presented, while key problems encountered and the measures taken are described. The lessons learned, and their implications for researcher and methodology alike, are discussed.

Task and Task Setting

Background and project objectives

This paper describes how a GSS was used by a small project team in Zeta (Zeta is a pseudonym – at the firm's request), a medium-sized, international accounting firm, to improve the firm's client billing process. Zeta maintains offices around the world, employing around 200 people in Hong Kong. Shortly before this study was initiated, a Chief Information Officer (CIO) was appointed with the remit of bringing Zeta into the IT age, while also revamping business processes within the firm in order to improve its competitiveness.

The objectives of the project, as defined by the CIO, were (a) to re-engineer the billing process (The CIO's use of terminology was sometimes potentially misleading. He referred to business process re-engineering on a number of occasions, but did not communicate this effectively to the other team members, i.e. they did not feel empowered to introduce radical changes in the billing process. Despite the CIO's misuse of some terms, his discourse has generally been preserved, as it provides important indicators of his style of interaction and control) and (b) to learn about the appropriateness of GSS technology in support of such a re-engineering process. The first objective involved examining the existing billing process and re-engineering its requirements. The new billing process, and a plan for its implementation, would be submitted to Zeta's Strategy Review Group (SRG). Table 1 charts the CIO's proposed meeting schedule.

Week no.	Proposed activity
1	Identify problems with the existing billing process
2	Identify goals for the new billing process
3–4	Understand the existing billing process
5–6	Devise an ideal billing process
7–8	Develop a practical new billing process, together with an implementation plan, for submission to the SRG
9	Review what had been learned about the appropriateness of the various process review tools and techniques applied
10	Develop an action plan for the new billing process on the basis of feedback received from the SRG

Table 1 Proposed schedule of billing process meetings.

Team membership

The project team was headed by the CIO and had six other members (see Table 2 for details). All team members, including the CIO, were in their late 20s or early 30s, and used a variety of information systems in the course of their regular work. An executive sponsor was identified for the project, but contact with him was conducted entirely through the CIO, as he never attended meetings.

Title	Background and education	Experience
CIO	He had recently arrived in Hong Kong from the UK with degrees in information systems and statistics. Current MBA student	Recently appointed at Zeta, with little work experience in Hong Kong
Manager – Tax	He was born in Hong Kong, but worked and studied in Australia for many years	Recently appointed in Zeta
Manager – Insolvency	He was born and educated in the UK, is a chartered accountant, but has low computer literacy	Recently appointed in Zeta
Manager – Audit	He was born in Malaysia, educated in the UK and is computer-literate	Several years' working experience in Zeta
Manager – Business Services	He was born in Hong Kong, is educated to A-level (grade 13/form 7) standard and is computer-literate	Many years' working experience in Zeta
Manager – Secretarial Services	She was born in Hong Kong, has a bachelor's degree from a local university, and is computer-literate	Many years' working experience in Zeta
Administrative Officer (EDP)	She was born in Hong Kong, is educated to O-level (grade 11/form 5) standard and is computer-literate	About 4 years' working experience in Zeta

Table 2 Team membership.

Week	1	2	3	4	5	6	7	8	9	10	11
Date	26/1	2/2	16/2	23/2	2/3	16/3	30/3	6/4	13/4	26/5	17/6

Table 3 Timeline of meetings (26 January – 17 June 1997).

The action researcher and Zeta

(Throughout this paper, the action researcher is referred to in the first person singular. To clarify, this is the paper's first author.) Educated in the UK, I had lived in Hong Kong for 6 years, was working as a lecturer at the City University of Hong Kong and had considerable experience in the facilitation of meetings using GSS. I introduced the principles of GSS at a seminar organized jointly by the Hong Kong Management Association and the Macquarie Graduate School of Management, which the CIO attended. My motivation to work with Zeta stemmed largely from my appreciation of the potential value of applying GSS to complex organizational circumstances – an application that is often called for in the research literature, but encountered relatively infrequently in practice.

The CIO described the proposed project, and I agreed that it had potential for GSS support. No formal protocol governing my involvement was specified, although senior partners at Zeta were informed of my presence. The CIO initially nominated me as a technical facilitator, although later this was extended to include process facilitation. Ventana Corporation authorized me to install the GSS software at Zeta for the duration of the project at no charge, provided that I supervised the use of the software and subsequently shared research findings with Ventana.

I was the sole facilitator at Zeta, but I was neither bound by any formal contract nor remunerated for my work, which was restricted to this one project; furthermore, I was never requested to deliver a solution that would support any particular viewpoint. The key ethical dilemma I faced related to the dominance of the CIO. This I addressed by attempting to bring the other team members into discussions, and successfully persuading the CIO to relinquish some of his responsibilities. At all times, I attempted to ensure that the techniques I used in team interactions did not conflict with local cultural sensitivities (style of communication for instance). My formal involvement with Zeta lasted from January to July 1997, during which time I visited Zeta's offices primarily to facilitate meetings (see Table 3), although on two occasions, I visited solely to fix software-related problems.

Project reports and operationalization

The CIO reported project progress to the executive sponsor on a regular basis. When the project was completed, I wrote up a report describing the discussions and the agreements reached, distributing this to all team members. The CIO was given access to an early version of the report from which this paper has been drawn. He responded critically and assertively to many instances of his described behaviour, vigorously defending himself, rejecting some arguments and demanding that the text be toned down. In this paper, the CIO's legitimate concerns are addressed, but the accuracy of the descriptions is maintained.

The CIO realized that participation in the process review was essential to the subsequent implementation of the new billing process in Zeta. He also recognized that the participants were too busy to commit much of their time to face-to-face interactions, thus requiring a process that permitted them to work remotely. The CIO believed that a GSS could support this process, facilitating productive group discussions and enabling all team members to benefit from their shared interaction. While the GSS was continuously available for 'remote' activities, it was not always used during face-to-face meetings. On some occasions, a whiteboard was used to draw team members' attention to system details. At other times, verbal discussion was found to be more effective than text input. Overall, the GSS was used for $\approx 50\%$ of meeting time, although the over-riding concern was to ensure that usage was appropriate to the context.

Methodology, GSS Review and Project Structure

Improving the client billing process was a complex task involving actors from different departments with differing vested interests. Consequently, the gathering and analysis of substantial rich information and the asking of probing 'why', 'how' and 'how to' questions was mandated. At the same time, it was desirable that the team members should play as active a role in the project as possible, as their insights, and those of their colleagues, should be relevant to the re-engineering process. Furthermore, it was believed that this participation would increase the likelihood that the new client billing process would be implemented effectively in all departments.

Given the task context, both the CIO and I agreed that action research would be the most suitable methodology to use. Not only would action research mandate my active intervention and empower the team members, but its cyclic structure would provide a framework for ensuring that the problem would be reconsidered continuously throughout the project. Furthermore, given action research's emphasis on reflection, the opportunity to glean lessons – both about the problem and its solution, and about how GSS can be applied in organizations – would be provided, thus linking theory and practice.

Action Research

Elden and Chisholm (1993) explain that action research is a change-oriented research methodology that seeks to introduce changes with positive social values, the key focus being on a problem and its solution. Baskerville and Wood-Harper (1996, p. 239) point out that 'the ideal domain of the action research method' is one where: "the researcher is actively involved, with expected benefit for both researcher and organisation; the knowledge obtained can be immediately applied ... ; the research is a cyclical process linking theory and practice." Furthermore, problems for which previous research has provided a validated theory are particularly appropriate for the application of action research, as the researcher can intervene in the problem situation, before applying and subsequently evaluating the value and usefulness of the theory. Such practice enables the researcher both to validate or improve upon existing theories and to introduce practical improvements in the problem situation investigated (Checkland, 1981; Heller, 1993). However, Eden and Huxham (1996) note that such an intervention may result in changes within the organization and so threaten the status quo. They also emphasize (Eden and Huxham, 1996, p. 84) that action research must have implications "beyond those required for action ... in the domain of the project. It must be possible to envisage talking about the theories developed in relation to other situations."

The practice of action research is cyclical. A researcher starts with planning what action to take, continues to intervention with the action, observes the effects of that intervention and, finally, reflects upon the observations in order to attempt to learn how better to plan and execute the next cycle. The reflections also inform the assessment of theory that must take place subsequently.

Group Support Systems

GSS (Nunamaker et al., 1991) are networked, computer-based systems designed to facilitate structured, interactive discussion in a group of people communicating face to face or remotely, synchronously or asynchronously. Group members type their contributions into the system, which immediately makes each contribution available to all other participants. Thus, nobody forgets what they want to say while waiting for a turn to speak. It is also possible for a group to enter ideas anonymously, if that is thought appropriate, e.g. if members feel unwilling to submit ideas that are considered abnormal, unusual or unpopular.

Project Structure and Protocols

In conjunction with the CIO, I initiated planning for the project. This helped me to understand the CIO's objectives and the project's limitations as imposed by organizational structures. The CIO demonstrated that he had a clear idea of what he wanted to achieve, his forceful style of interaction dominating discussions. It was agreed that each meeting would be planned in advance and its progress discussed afterwards – often these discussions took place through email. It was also planned that the GSS would be used for idea generation/categorization and for forging consensus. The literature (e.g. Nunamaker et al., 1997) suggests that GSS should work well for these two activities. Team members were informed in advance of forthcoming activities to give them preparation time; they were also reminded to meet task deadlines.

The CIO suggested that the project team should first unravel the details of the existing billing process, before considering the requirements for and design of the new process. This would involve looking at the various actors involved (clients, managers of the various departments, secretaries, partners) and analyzing data flows and decision-making policies. Finally, the project team would review the effectiveness of the techniques used in the process review and analyze their suitability for other process reviews in the firm.

The CIO intended that the team members would meet face to face once a week. After these meetings, he would prescribe 'homework' for the team members to work on during the week, i.e. asynchronously and in a distributed fashion as their offices were located in two buildings a mile apart in Hong Kong's central business district. This 'timetable' was frequently interrupted, with up to 6 weeks separating some meetings (see Table 3 for the actual timeline of meetings).

Through the 11 cycles of meetings and other project activities, I collected data from a variety of sources. These included:

- my own subjective observations and discussions with team members before, during and after meetings about project progress and problems;
- unstructured telephone and email interviews/discussions, initiated by me, with (a) individual team members conducted regularly throughout the project (typical questions covered team member comfort with the technology, satisfaction with the project process and belief that the project was managed in an appropriate manner); (b) the CIO two or three times per week to consider completed and current activities, as well as future plans;
- company and project documentation;
- an instrument (see Appendix), developed by Davison (1997, 1998), used to collect data concerning team members' perceptions regarding the following meeting processes: ease and effectiveness of communication; discussion quality; status effects experienced; team work; efficiency.

Although the CIO agreed that the questionnaire should be completed after each meeting, this was confounded by team members who refused to do so on a number of occasions, with data collected only in the first, second, fourth, fifth and seventh cycles.

The First Two Cycles

In order to illustrate how the action research protocol was operationalized, the first two cycles of activities are described briefly.

Cycle one

Planning

The CIO and I agreed in advance that the first meeting should be primarily introductory, lasting about 1 hour, familiarizing the team members with the task and explaining my role, while also describing the GSS software and how it had been used by other organizations. An initial application of the GSS to explore relevant issues would follow, with the team members being asked to complete the meeting process questionnaire. Subsequent analysis of these data would determine how the second meeting should be organized.

Intervention

After the CIO's introductions, team members vigorously challenged the anonymity of the GSS, observing that the process review would threaten the status quo in Zeta and would therefore be controversial. Consequently, their participation would be conditional upon the confidentiality of their input. The CIO reassured them with respect to confidentiality, clarifying that he would take personal responsibility for the project.

Familiarization with the GSS was achieved by asking the team members to brainstorm issues relevant to the billing process, new features of the GSS interface being introduced progressively. This illustrated how the GSS could be used to handle various aspects of the billing process. After some 20 minutes, 20 ideas and 15 comments had been entered into the GSS (two or three pages of single-spaced A4 paper).

At this point, the CIO wrapped up the meeting, requesting members to continue to input ideas remotely during the week: a total of 30 ideas and

70 comments were generated in this first week. However, the meeting process instrument was not completed immediately, members pleading fresh meetings to attend.

Observations

I noted the team members' healthy level of initial interest in the GSS and the task, as well as their perception that the issues involved would be controversial. They did not appear to experience task- or GSS-related difficulties, contributing many sensible and thoughtful ideas. The CIO privately confided that immediate completion of the questionnaire would be better, yet professed himself unable to require compliance.

Reflections

Subsequent data analysis revealed few problems, team members disagreeing that the language used (English) prevented their participation, although they only weakly disagreed that they were reluctant to contribute ideas. Discussions were seen to be reasonably meaningful, open and appropriate. Importantly, members evinced comfort with the technology itself, sensing that it did facilitate their work. Finally, although the first meeting ran smoothly enough, it was apparent that the CIO's positional power was limited.

Cycle Two

Planning

The CIO and I held a 90-minute reflecting and planning session before the second meeting. It was noted that the ideas generated in the first cycle were wide-ranging, but lacked focus. To remedy this, the CIO proposed that the second meeting should focus on the scope and objectives of the new billing process, with the GSS being used to support the necessary idea generation.

Intervention

Two team members were unable to attend this meeting, and were substituted by colleagues unfamiliar with the software. After reacquainting themselves with the material, the CIO initiated a verbal discussion of the issues. This lasted some 30 minutes, but was little more than a conversation between the CIO and the Insolvency Manager. At my suggestion, all team members verbally discussed and categorized each of the 30 ideas as lying inside or outside the scope of the billing process. The discussion was initially dominated by the CIO, but I intervened to ensure that all team members had the opportunity to participate.

The GSS functioned in this meeting as a form of team 'memory'. All members had access to the 30 ideas and 70 comments, using them to inform their discussion. I 'moved' ideas in real time from the main list to one of two new 'categories' – 'inside scope' and 'outside scope'.

Observations

After this meeting, I discussed project issues with the CIO and the Insolvency Manager, both of whom felt that the other team members were very unwilling to get involved in discussions, not appreciating either the purpose of the project or their involvement. This attitude stemmed in part from Zeta's organizational culture, which did not encourage, let alone reward, innovation or working outside one's immediate task environment. The CIO explained that he wanted to encourage participation, with all team members having a fair go at contributing to the processes, yet he also felt the occasional need to be autocratic to ensure that things did get done.

Reflections

Clear problems emerged in this meeting that related to the interaction between the CIO and the team members. Although the team members were competent to perform the work required, their willingness to take responsibility appeared to be low. Indeed, they appeared to have little vested interest in, or motivation for, the problem they had been assigned to tackle. Instrument data corroborate this, revealing a general apathy towards the meeting processes, with a lack of team spirit and diminished discussion quality. This lack of interest was augmented by the CIO's failure either to communicate why the review process was important or to allocate a sufficiently high priority to the task himself.

Key Problems Encountered During the Project

Five key problem areas were encountered during the project.

Chargeable Time

It became apparent that a key demotivator for the team's participation was that they could not charge the time so spent to any account. As in many similar firms, Zeta uses a system of 'chargeable time', but the time team members spent on the project was not chargeable. A senior partner in the firm was confronted with this inconsistency, but he merely replied that the firm did value the time and effort committed by the team members. The lack of substantive evidence to support this assertion made it very difficult to believe that the firm was sincere in its intent.

Empowerment

The CIO clearly identified the need to empower the team members and entrust them with the responsibility for re-engineering the billing process. However, he did not explicitly communicate his rationale to them, nor could he appreciate their lack of interest in solving the task. Instead, he supplied them with weighty texts on re-engineering, which remained unread. Privately, some team members expressed the view that the new system might improve some aspects of the billing process but, if they found that it obstructed their work, they would simply ignore it and create work-arounds. Indeed, they had little interest in being empowered (none of them having volunteered to work on the project), believing that the task of re-engineering was more properly the CIO's. None of the team members commented on or contributed to the final project report. Indeed, their unwillingness to undertake any more work than absolutely necessary meant that they could not be considered as co-researchers in the project.

Misappropriation of Anonymity

The CIO misappropriated the anonymity of the GSS, projecting large numbers of his own ideas without their authorship being positively attributable. He freely admitted to me that he sometimes submitted wild or provocative ideas in order to see what he could get away with. Team members noted individually to me that they would prefer that the communications using the GSS be identified, as this might improve the value and sincerity of the discussions. The CIO, however, vetoed this proposal.

Conflict Between The CIO and The Researcher

Initially, the CIO identified my responsibility as being primarily technical, advising him on how to apply the GSS most effectively in order to achieve his objectives in running the project. This became an increasingly untenable role to play as the CIO's dominance of meetings increased. On one occasion, for example, I intervened in a heated debate between the CIO and team members to suggest that some cultural confusion might underlie the discussion and its lack of progress. The CIO took grave offence at this intervention and subsequently reprimanded me. I was careful not to align myself with either the CIO or the other participants, in order not to be accused of bias. However, my dissonant stance had clearly aggravated the CIO.

Data Quality

The instrument used was designed to collect perceptions of the participants concerning five key meeting processes, namely discussion quality, communication, efficiency, teamwork and status effects. While useful, the instrument only measured perceptions within a single meeting. After the third meeting, the CIO criticized the instrument design, arguing that it would be more useful to make comparative measurements from week to week. In consequence, it was redesigned.

Reflections and Actions

Assessment of the situation

The role of an action researcher is to work with team members in order to devise the most appropriate solution to the problem being tackled. Although problems were intertwined in this project, many related to the CIO's paradoxical dominance of activities, but lack of positional power.

The team members recognized that the CIO was frustrated by the tortuous progress of the project, but also suggested that his own preference for open and rigorous debate contrasted sharply with their reticence to express opinions publicly and preference for private discussions. They also recognized that there was a need for a neutral moderator who could draw out the views of the team members, while neutralizing the excesses of the CIO. Ironically, the GSS offered the opportunity to communicate ideas anonymously, but a culture of cautiousness hampered the situation: the manager from the Tax Department commented that the female members, in particular, would typically not contribute ideas if they were unsure of their accuracy. I observed such behaviour on a number of occasions.

The CIO was not at all reticent about making his views known. Although he recognized from the outset that participation might turn out to be a problem, his apparent lack of awareness of the importance of chargeable time for the team members was an unexpected revelation. His comments, however, are most insightful, if a little beguiling: "For me personally, chargeable hours are not of importance as I do not do client work. Although I knew that chargeability is used as a measure of performance for client service providers, I did not fully appreciate the influence that it has."

With the approval of the CIO, I took a leading role in meetings from cycle seven onwards. Having observed the negative effects of the CIO's style of leadership, as well as the fact that the GSS was used effectively for discussion of billing process problems, but ineffectively for consensus development, I consciously attempted to:

- imbue meetings with a more friendly style of communication;
- ensure that tasks were rich in opportunities for idea generation and discussion;
- reduce the normative influence of the CIO by increasing the involvement of team members in specific tasks;
- identify learning opportunities that the project presented for team members.

The original instrument for collecting data on participants' perceptions of meeting processes did not incorporate any element of comparison with previous meetings and hindered the analysis of processes in longitudinal meetings. During the third cycle, it was agreed that some of the instrument items should be realigned to reflect changes from meeting to meeting. Thus, team members were asked whether a process (e.g. discussion openness) had improved, deteriorated or remained the same when compared with the previous meeting. This instrument revision is seen as a critical element of the action research, in that tools and techniques should only be used for as long as they are useful. Realigning the focus of the instrument was a joint solution that helped to ensure useful data collection.

The comparative data collected proved very useful for future planning of meetings. It also enabled me to observe, for example, that while the level of conformance pressure fluctuated through the project, the levels of intimidation and influence dropped as I took a more active role in meeting processes. Despite these fluctuations, the extent to which team members actually felt inhibited from participating was initially low and remained unchanged throughout the project. Certainly, participation was problematic, but the data collected appear to confirm the earlier finding that notional willingness to participate was high.

Discussion and Lessons Learned

The organizational circumstances and group dynamics encountered in this research setting were taxing. The fact that the CIO lacked the positional power to enforce any of his requests contributed considerably both to the length of the project and to his own frustration with the meeting process. However, the CIO failed either to increase his own prioritization of the task or to seek direct power from the executive partner in order to complete project tasks more effectively.

Senior partners ostensibly approved the initiation of the project, yet failed to give the team members any overt recognition of the value of their work, and it is certain that the system of chargeable time mitigated against their involvement; that the firm and the CIO failed to take account of this when the project was approved was short-sighted at best. At the end of the project, the project's executive sponsor expressed some appreciation for the work that the team had accomplished, but did not indicate that changes in Zeta's policy concerning chargeable time would be forthcoming. Although the GSS literature pays little attention to the issue of participant motivation, the facilitation literature is more forthcoming with the suggestion that the motivation of a group to accomplish its task, and the need to keep it focused on its outcomes, are critical (Clawson et al., 1993; Niederman et al., 1996).

In the following discussion, the roles of the two key aspects of this research, namely the action research methodology and the GSS technology, are evaluated. This evaluation will lead to the identification of key lessons that can be applied in similar contexts.

Reflections on Action Research

Motivation to participate was always low, with features of the organizational setting and reward mechanisms playing a contributory role. The direction of the project was sometimes unclear, and the CIO might have abandoned the notion of involving the team members at all, attempting to re-engineer the processes himself; equally, the team members might have totally ignored the CIO and ceased attending meetings or contributing. The fact that these did not happen, despite the CIO's wry comment that the team members probably autodeleted his email, testifies to the robustness of a methodology that supports flexibility, continuous intervention and appropriate change.

A central tenet of action research is that a researcher should act to introduce changes with positive social values, even if these conflict with the status quo, while considering the needs of the clients. The team members were consulted as to their views on the progress of the project, the appropriateness of the goals and the suitability of the CIO as a project leader. Feedback from these consultations was usefully incorporated into the way that the project was run, with the result that changes were made and, most critically, the project continued to run. However, these actions that I took should be seen as essential modifications to what was an intentionally tenuous initial plan. When the project was initiated, there was no long-term plan about how to proceed, nor indeed should there have been one. Checkland (1981, p. 153) notes that action research "cannot be wholly planned and directed down particular paths." Lévi-Strauss has made similar comments about field research, observing that one should 'go along with the lie of the land' rather than expecting to stick to predetermined techniques and styles of enquiry (cited in Descola, 1996, p. 40).

The researcher's flexibility is an essential component of action research and, thus, not a weakness in the research design. If one instrument or technique proves inadequate, it should be revised or discarded to be replaced by a better one. I refined the data collection instrument midway through the project, redefined my own and the CIO's roles, and reallocated responsibilities accordingly. It is conceivable that motivation-related questions could have been included in the instrument but, in other meetings, a different factor could contribute negatively. While the instrument was valuable, therefore, in eliciting certain key problems and supplementing the richer information obtained in interviews, it was not designed to identify all potential problems; my own intervention was essential for this. None of these experiences causes us to dispute the value of planning as a key element of the action research cycle, but does lead us to caution that such plans should incorporate flexibility, an attribute of facilitation identified as being important by Niederman et al. (1996). This flexibility permitted me to intervene in a manner impossible with other, more structured methodologies. In consequence, the solution attained was probably better than would have been the case had no intervention been performed.

Checkland (1981) notes that a researcher should not remain an observer outside the subject of investigation, but become a participant in the action. Mid-way through the project, participants suggested that I take a moderating role in the project discussions and so neutralize some of the CIO's excesses. When I changed my role in response to this request, attempting to increase the involvement of participants and decrease the normative influence of the CIO, the private feedback from participants was positive, and the levels of perceived status influence diminished. It has been suggested (Gersick, 1991) that a radical change in group organization can disrupt dysfunctional behaviours and so improve a group's productivity.

Reflections on GSS

The CIO observed early on in the project that he could re-engineer the billing process himself, but needed the participative co-operation of the whole team to ensure that the final system would be acceptable to different departments. The history of GSS use elsewhere (e.g. Grohowski *et al.*, 1990) suggests that its application in complex organizational circumstances could indeed be successful. Thus, there was reason to think that use of GSS was warranted and would be received openly. Indeed, the literature (e.g. Nunamaker *et al.*, 1997) suggests that GSS can enhance effectiveness and efficiency, resulting in higher levels of participant satisfaction than might be attained in the absence of technological support.

If the CIO had tackled the project without either the GSS technology or my interventions, it is unlikely that such a process would have produced results acceptable to Zeta's employees. Therefore, it can be asserted that the GSS did exert a significant and positive influence on the meeting process. What became apparent early in the project, however, was that the GSS could not be used in the same way that the literature suggests. Although team members appreciated being able to use the GSS for tasks that required substantial idea generation, given that it markedly improved their productivity and enabled them to contribute ideas at their own pace, it was definitely not appreciated for tasks requiring the development of consensus or fine points of detail. These findings encouraged me to vary the application of GSS to fit the characteristics of the team more effectively. Previous research in GSS seldom mentions the importance of matching GSS tools with tasks, but evidence from meeting facilitators does indicate its importance (Niederman *et al.*, 1996).

Furthermore, anonymity was used in a manner not described in the literature, with the CIO taking advantage of it to cloak his own many contributions with a quasi-team authorship. Anonymity is often seen as a vital aspect of GSS research and practice, the usual rationale being that it promotes unbiased and task-focused discussions, while diminishing the negative effects caused by domination and intimidation. Lyytinen *et al.* (1993), however, observed that its use may not always be appropriate, citing the example of international diplomacy, where it is essential for all contributions to be identified for them to be meaningful. In this project, it is certain that anonymity contributed negatively, with team members commenting that discussions would have been more frank and sincere had contributions been identifiable.

Conclusions

This research underscores the difficulties encountered in applying GSS in organizational contexts in the presence of dysfunctional circumstances. Issues that affected motivation in this project, including chargeable time, leadership style and sense of responsibility for the project's outcomes, should all have been accounted for before the project was initiated. Of particular concern are organizational culture issues. The GSS was used to mediate some of the dysfunctions, demonstrating that key stakeholders can be involved in the improvement of organizational processes and that tools such as GSS can be useful.

Action research enabled us to investigate the key issues underlying some of the organizational problems. The combination of GSS and action research supported an exploration and analysis of ideas that enabled the team to achieve some semblance of success. The project was tentatively completed, and a pilot system was running 9 months later in one department, although considerable angst was experienced during the review process. The modest outcomes, while better than what could realistically have been attained had neither GSS nor action research been used, could nonetheless have been achieved less painfully and more efficiently in more ideal circumstances.

The results reported here have implications for future work with GSS in a number of respects. GSS can be used effectively for longitudinal meeting contexts as long as the GSS facilitator/researcher uses the technology flexibly when tackling problems. Participants in this project, despite their lack of familiarity with GSS software, expressed comfort with the technology, believing that it facilitated their work.

For Zeta to make any significant, IT-related improvements in the future, significant changes to its organizational culture would be necessary. Barriers to participation, concerning chargeable time and the reward system, must be addressed. Although I sought to introduce positive social changes during this study, these were countered by Zeta's organizational culture, which did not appear to value motivational issues.

Research and practice that draws upon the synergy of GSS and action research is not encountered frequently yet, as this paper demonstrates, has much potential. Researchers and practitioners are encouraged to explore this domain further in different cultures and organizational contexts.

ACKNOWLEDGMENTS

The authors would like to thank Maris Martinsons, Choon-Ling Sia and Sabine Hirt for helpful comments and suggestions made on earlier drafts of this paper. We particularly value the constructive reviews from the two anonymous referees and the associate editor that have enabled us to make significant improvements to the paper.

References

Baskerville, R. L. and Wood-Harper, A. T. (1996) A critical perspective on action research as a method for information systems research. *Journal of Information Technology*, 11, 235–46.

Checkland, P. (1981) *Systems Thinking, Systems Practice*, John Wiley & Sons, New York.

Clawson, V. K., Bostrom, R. P. and Anson, R. (1993) The role of the facilitator in computer-supported meetings. *Small Group Research*, 24, 547–65.

Davison, R. M. (1997) An instrument for measuring meeting success. *Information and Management*, 32, 163–76.

Davison, R. M. (1998) *An Action Research Perspective of Group Support Systems: how to improve meetings in Hong Kong*. Unpublished PhD Dissertation, City University of Hong Kong.

Descola, P. (1996) *The Spears of Twilight (Les Lances Du Crépuscule)*. The Free Press, New York.

Eden, C. and Huxham, C. (1996) Action research for management research. *British Journal of Management*, 7, 75–86.

Elden, M. and Chisholm, R. F. (1993) Emerging varieties of action research: introduction to the special issue. *Human Relations*, 46, 121–42.

Gersick, C. J. G. (1991) Revolutionary change theories: a multilevel exploration of the punctuated equilibrium paradigm. *Academy of Management Review*, 16, 10–36.

Grohowski, R., McGoff, C., Vogel, D. R., Martz, B. and Nunamaker, J. F. (1990) Implementing electronic meeting systems at IBM: lessons learned and success factors. *Management Information Systems Quarterly*, 14, 369–83.

Heller, F. (1993) Another look at action research. *Human Relations*, 46, 1235–42.

Jones, N., de Vreede, G. J. and Mgaya, R. (1998) A new driving force behind capacity building in Africa: group support systems. *Proceedings of the 31st Hawaii International Conference on System Sciences*, Kona, Hawaii, VI, 705–14.

Liou, Y. I. and Chen, M. (1993) Using group support systems and joint application development for requirements specification. *Journal of Management Information Systems*, 10, 25–41.

Lyytinen, K., Maaranen, P. and Knuuttila, J. (1993) Unusual business or business as usual: an investigation of meeting support requirements in multilateral diplomacy. *Accounting, Management and Information Technology*, 3, 97–117.

Mashayekhi, V., Drake, J. M., Tsai, W. T. and Riedl, J. (1993) Distributed, collaborative software inspection. *IEEE Software*, 10, 66–75.

Niederman, F., Beise, C. M. and Beranek, P. M. (1996) Issues and concerns about computer supported meetings: the facilitator's perspective. *Management Information Systems Quarterly*, 20, 1–21.

Nunamaker, J. F., Dennis, A. R., Valacich, J. S., Vogel, D. R. and George, J. F. (1991) Electronic meeting systems to support group work. *Communications of the ACM*, 34, 40–61.

Nunamaker, J. F., Briggs, R. O., Mittleman, D. D., Vogel, D. R. and Balthazard, P. A. (1997) Lessons from a dozen years of group support systems research: a discussion of lab and field findings. *Journal of Management Information Systems*, 13, 163–207.

Tyran, C. K., Dennis, A. R., Vogel, D. R. and Nunamaker, J. F. (1992) The application of electronic meeting technology to support strategic management. *Management Information Systems Quarterly*, 16, 313–34.

ABOUT THE AUTHORS

Robert Davison is an Assistant Professor in the Department of Information Systems at the City University of Hong Kong. His current research interests involve an examination of the impact of information systems on group decision-making, communication and learning, particularly in cross-cultural and developing country settings. His work has been published by *Information and Management*, *Communications of the ACM* and *Decision Support Systems*.

Douglas R. Vogel is Professor of Information Systems at the City University of Hong Kong. He has been involved with computers and computer systems in various capacities for over 30 years. His interests bridge the business and academic communities in addressing questions of the impact of management information systems on aspects of business process improvement, group problem-solving, education and organizational productivity. Professor Vogel is especially active in introducing group support technology into enterprises. His particular focus emphasizes integration of audio, video and data in interactive distributed group support.

APPENDIX

The instrument developed by Davison (1997, 1998) includes five constructs measuring communication in meetings (C), discussion quality in a meeting (D), status effects experienced in meetings (S), team work in a meeting (T) and efficiency of meeting processes (E). Two versions of the instrument were developed. The first, referred to as the absolute version, is designed for one-off meetings or initial meetings in a series. The second, referred to as the relative version, is designed to be used when it is necessary for respondents to compare their perceptions with those made in previous meetings. Each item within a construct is followed by a code, e.g. C1. The letter refers to the construct, i.e. C = communication in meetings. The digit refers to the nominal order of the item within the construct.

Absolute version of the instrument

With regard to your own participation in the meeting, please indicate to what extent you agree with the following statements:

The language of the meeting prevented you from participating (C1)
Strongly agree ☐ ☐ ☐ ☐ ☐ Strongly disagree

You found it hard to understand other group members when they talked (C2)
Strongly agree ☐ ☐ ☐ ☐ ☐ Strongly disagree

You experienced problems expressing yourself (C3)
Strongly agree ☐ ☐ ☐ ☐ ☐ Strongly disagree

You felt reluctant to put forward your own ideas (C4)
Strongly agree ☐ ☐ ☐ ☐ ☐ Strongly disagree

You experienced pressure, either to conform to a particular viewpoint or not to contradict others (S4)
Strongly agree ☐ ☐ ☐ ☐ ☐ Strongly disagree

With regard to all meeting members as a whole, how would you rate the discussions in the meeting in terms of the following scales?
Meaningful   ☐ ☐ ☐ ☐ ☐ Meaningless (D1)
Appropriate   ☐ ☐ ☐ ☐ ☐ Inappropriate (D2)
Open   ☐ ☐ ☐ ☐ ☐ Closed (D3)
Imaginative   ☐ ☐ ☐ ☐ ☐ Unimaginative (D4)

Please indicate to what extent you agree with the following statements:

Other members appeared willing to answer questions when asked (T1)
Strongly agree ☐ ☐ ☐ ☐ ☐ Strongly disagree

Members worked together as a team (T2)
Strongly agree ☐ ☐ ☐ ☐ ☐ Strongly disagree

Members had sufficient access to the information they needed so as to participate actively in and fully understand the meeting (T3)
Strongly agree ☐ ☐ ☐ ☐ ☐ Strongly disagree

The time spent in the meeting was efficiently used (E2)
Strongly agree ☐ ☐ ☐ ☐ ☐ Strongly disagree

Issues raised in the meeting were discussed thoroughly (E3)
Strongly agree ☐ ☐ ☐ ☐ ☐ Strongly disagree

Some group members tried to intimidate others, e.g. by talking loudly, using aggressive gestures, making threats, etc. (S1)
Strongly agree ☐ ☐ ☐ ☐ ☐ Strongly disagree

Some group members tried to use their influence, status or power so as to force issues on the other group members (S2)
Strongly agree ☐ ☐ ☐ ☐ ☐ Strongly disagree

You felt inhibited from participating in the discussion because of the behaviour of other meeting members (S3)
Strongly agree ☐ ☐ ☐ ☐ ☐ Strongly disagree

What percentage of meeting time do you think was spent on serious discussion? _____ % (E4)

To what extent would you say that this meeting was result oriented? (E1)
Strongly result oriented ☐ ☐ ☐ ☐ ☐ Weakly result oriented

Relative Version of The Instrument

Compared with previous meetings of this team, do you feel that:

Your ability to participate in the meeting (C1)
Improved ☐; Stayed about the same ☐; Deteriorated ☐

Your understanding of the typed comments from other group members (C2)
Improved ☐; Stayed about the same ☐; Deteriorated ☐

Your ability to express yourself (C3)
Improved ☐; Stayed about the same ☐; Deteriorated ☐

Your willingness to put forward ideas (C4)
Increased ☐; Stayed about the same ☐; Decreased ☐

The pressure you experienced, either to conform to a particular viewpoint or not to contradict others (S4)
Increased ☐; Stayed about the same ☐; Decreased ☐

Compared with previous meetings of this team, do you feel the discussions improved, stayed the same or deteriorated on the following scales:
Meaningful (D1) Improved ☐; Stayed about the same ☐; Deteriorated ☐
Appropriate (D2) Improved ☐; Stayed about the same ☐; Deteriorated ☐
Openness (D3) Improved ☐; Stayed about the same ☐; Deteriorated ☐
Imaginative (D4) Improved ☐; Stayed about the same ☐; Deteriorated ☐

Compared with previous meetings of this team, do you think that:

The willingness of other members to answer questions when asked (T1)
Increased ☐; Stayed about the same ☐; Decreased ☐
The extent to which members worked together as a team (T2)
Increased ☐; Stayed about the same ☐; Decreased ☐

The extent to which members had sufficient access to the information they needed so as to participate actively in and fully understand the meeting (T3)
Increased ☐; Stayed about the same ☐; Decreased ☐
The time in the meeting was used (E2)
More efficiently ☐; As efficiently ☐; Less efficiently ☐
Ideas were discussed (E3)
More thoroughly ☐; As thoroughly ☐; Less thoroughly ☐

The extent to which some group members tried to intimidate others, e.g. by talking loudly, using aggressive gestures, making threats, etc. (S1)
Increased ☐; Stayed about the same ☐; Decreased ☐

The extent to which some group members tried to use their influence, status or power so as to force issues on the other group members (S2)
Increased ☐; Stayed about the same ☐; Decreased ☐

The extent to which you felt inhibited from participating in the discussion because of the behaviour of other meeting members (S3)
Increased ☐; Stayed about the same ☐; Decreased ☐

What percentage of this meeting's time do you think was spent on serious discussion _____% (E4)

To what extent would you say that this meeting was result oriented? (E1)
Strongly result oriented ☐ ☐ ☐ ☐ ☐ Weakly result oriented

Understanding Computer-mediated Discussions: Positivist and Interpretive Analyses of Group Support System Use[1]

Eileen M. Trauth and Leonard M. Jessup

Management Information Systems, College of Business Administration, Northeastern University, 214 Hayden Hall, Boston MA, 02115, U.S.A., trauth@neu.edu
Department of Accounting and Information Systems, Kelley School of Business, Indiana University, Tenth and Fee Lane, Room 560, Bloomington, Indiana 47405-1701, U.S.A., ljessup@indiana.edu

Abstract

This research considers whether interpretive techniques can be used to enhance our understanding of computer-mediated discussions. The case study considered in this research is the use of a group support system (GSS) to support employee discussions about gender equity in a university. Transcripts of the four discussions were analyzed using two analysis techniques: a positivist approach, which was focused on the GSS sessions themselves, and an interpretive approach which broadened the scope to include contextual considerations as well. What emerged from the positivist analysis was the conclusion of effective group behavior directed toward consensus around alternative solution scenarios. What emerged from the interpretive analysis was evidence of multiple, rich types of information at three levels: cognitive, affective, and behavioral. The interpretive analysis also uncovered the absence of shared consciousness about the issue and imbalanced participation in the sessions. Comparison of the results of both approaches showed that, while the positivist analysis provided useful information, the interpretive analysis provided a different understanding of the same evidence and new information not found in the positivist analysis of the group discussions. This research adds to the body of knowledge concerning the effects of virtual group meetings on the type of information that is shared and the value of a combination of positivist and interpretive analyses of GSS data.

Keywords: anonymity, computer-mediated communication, ethnography, gender, group decision making, group decision support system, hermeneutics, information richness, interpretive methods, IS research methodologies, positivist methods, virtual group

ISRL Categories: AA0901, AC0402, AI0112, AI0116, AI0801, AI0802, HA0301

Introduction

As the technologies for computer-mediation evolve, businesses are increasingly relying on computer-supported forms of communication, collaboration, and coordination. One such technology, the group support system (GSS), is popular for contexts in which co-workers desire to engage in joint problem solving at the same time and in the same place (Jessup and Valacich, 1993). Research on GSS suggests that, for certain situations, GSS provides some advantages over face-to-face discussion and other forms of computer-mediated communication. Among these potential advantages are anonymity (Connolly *et al.*, 1990) and process structuring (Dennis *et al.*, 1996; Wheeler and Valacich, 1996).

Early publications on GSS called for research using multiple methods and conducted within multiple contexts (e.g. DeSanctis and Gallupe 1985, 1987; Jessup 1987). Toward this end, a number of early research efforts in the GSS area were case studies of GSS implementations that were focused either directly or indirectly on organizational issues (e.g., Dennis et al. 1990; DeSanctis et al. 1991, 1992; Nunamaker et al. 1989). Others have devised and/or used robust coding schemes to measure group interaction (e.g., Zigurs et al. 1988), studied groups over time (Chidambaram 1996; Chidambaram and Bostrom 1993; Chidambaram et al. 1990), supplemented quantitative, a priori measures with post-hoc analysis of unusual groups (Gallupe et al. 1988), used post hoc, qualitative findings to uncover unintended consequences of GSS use (Watson et al. 1988), proposed useful alternatives for theorizing about and studying GSS use (DeSanctis and Poole 1994), used interpretive methods to study GSS (DeSanctis et al. 1993; Rebstock Williams and Wilson 1997), and theorized on the interplay between technology and organizational form (Fulk and DeSanctis 1995). Similarly, there have been a number of interpretive studies focused on the implementations of computer conferencing (Orlikowski et al. 1995), groupware (Orlikowski 1996), and other computer-mediated communication systems within organizational contexts.

Nevertheless, a great deal of GSS research has been conducted within a positivist[2] paradigm (e.g., Connolly et al. 1990; Dennis and Valacich 1993, 1994; Dennis et al. 1990; Gallupe et al. 1992; Jessup et al. 1990, 1996; Jessup and Tansik 1991; Valacich et al. 1992, 1994, 1995; Valacich and Schwenk 1995). This research has generally relied on a specific set of assumptions about the technology and about the research conducted. Much of this research follows the technological determinism approach, which assumes that GSS are developed to improve group and organizational productivity. The intent of GSS research, according to this approach, is to understand whether the technology can be helpful, how the technology can best be deployed, and what the marginal contributions of the various components of the technology are to productivity (DeSanctis 1993; DeSanctis and Poole 1994). This research has relied to a great extent on the structured analyses of transcripts from GSS sessions. Comment categories determined a priori are designed to capture the useful ideas generated during the sessions. It reflects the positivist paradigm in that it aims to quantify social reality, subjecting it to experimental controls and hypothesis testing (Lee 1991).

While this has been a fruitful thread of research, it is not the only possible one. Another approach is suggested by the use of an interpretive paradigm, and there have been examples of this within the GSS literature as cited above. The recent IS literature gives evidence of an interpretive movement in IS research in general (Avison and Myers 1995; Harvey 1997; Harvey and Myers 1995; Kaplan and Maxwell 1994; Lee 1991; Myers 1997; Prasad 1997; Walsham 1995), arguing that methods such as case study, grounded theory, ethnography and hermeneutics, which facilitate deeper probing into the subtleties of context, are appropriate methods for mainstream IS research. Interpretive methods have been used to study both the introduction and management of IT (Davidson 1997; Davies 1991; Davies and Nielsen 1992; Hughes et al. 1992; Orlikowski 1991; Preston 1991; Simonsen and Finn 1997; Trauth et al. 1993). Therefore, an alternative research approach for understanding GSS discussions would be to engage in an interpretive analysis of them.

A variety of research benefits can derive from adopting a different research stance. Since each research method has different assumptions and procedures, one method can complement another. This triangulation is the rationale for mixed methods research (Trauth and O'Connor 1991), particularly that which employs both quantitative and qualitative methods (Greene et al.

1989; Jick 1979; Kaplan and Duchon 1988). This viewpoint is consistent with Lee's (1991) argument that positivist and interpretive approaches need not be viewed as mutually exclusive. Rather, they can become mutually supportive.

In addition to the contribution of mixed method approaches to IS research in general, there are benefits to GSS research in particular. Because interpretive methods were developed and are typically used to analyze face-to-face or written communication and, hence, have a different set of assumptions and procedures, they may help us to understand GSS in ways we could not understand them using positivist methods of analysis.

The adoption of a mixed methods approach meant analyzing the same GSS data—referring both to the discussion transcripts and to the organizational context—using both positivist and interpretive methods. In this paper, we both utilize and evaluate the positivist and interpretive approaches to GSS analysis. Printed transcripts of four GSS discussions were analyzed first using a positivist approach. This approach was narrower in that the scope of analysis was limited to the GSS sessions themselves and incorporated only a limited amount of contextual data.[3] The same transcripts were then analyzed using an interpretive approach. This approach was broader in that the scope of analysis included contextual data as well. The objective in conducting this dual analysis was to contrast the results of positivist and interpretive analyses of GSS use in the field in order to learn how our understanding of the information generated in the GSS sessions might be enhanced by the addition of interpretive techniques. We posed the following research question about the use of these alternative methods:

> Does an interpretive analysis of GSS use result in a different understanding of the GSS discussions than that provided by a positivist analysis?

In the next section, we present the context within which the GSS was used: a university-wide discussion of gender equity on campus. (We sought an emotionally charged setting in order to explore the potential benefits of anonymous computer-mediated discussions.) We then conduct two analyses of the discussion transcripts—first positivist then interpretive—and present the respective results. We consider the differences in the findings that result from use of these two different methods. The contribution of this research is two-fold. First, contrasting analyses of the same data facilitates critical analysis of each research method. Second, by adding the interpretive layer, our analysis provides evaluative insight into GSS use that is not captured easily by positivist methods of GSS analysis.

The Context: Coping with Gender Equity at State University

When State University,[4] a small, public university in the American southwest, was created in the early 1990s, the topics of gender equity and affirmative action loomed large on the political horizon of academia and particularly within this state and this state university system. As a result, one of the goals of this new university was to start with a clean slate and to build into the university an egalitarian mentality that would pervade its programs and its organizational structure and culture. Indeed, this stated goal and the expected opportunities it would produce were used in recruiting new faculty, staff, and administrators.

Despite the excitement of building a new university, literally from the ground up, several sources of conflict quickly emerged. First, the President pursued a hiring strategy weighted heavily toward administrators and staff. In response to grumbling from founding faculty members that the university was starting off rather top heavy, the President explained that whether a university had 2,000 or 20,000 students, it needed the same administrative infrastructure: a President, Vice President, Vice President/Provost, academic Deans, a Registrar, a Director of Financial Aid, librarians, janitors, a facilities manager, campus police, and so on.

A second source of conflict involved pay equity. Since the university system of which this campus was a part was heavily unionized, lower level staff members with seniority were generally well paid relative to their counterparts at other universities across the country. Faculty members (especially at senior levels), on the other hand, were generally not well paid when compared to their counterparts at other universities across the country, particularly considering the cost of living in the region. In addition, the University administration promoted an organizational culture of egalitarianism—some called it socialism—across levels and functions within the University; this received mixed support. For example, staff support and secretarial personnel were invited to participate in Dean searches, with voice and vote equal to that of tenured professors, resulting in one situation in which a cadre of secretaries was able to oust a Dean candidate from consideration in direct opposition to the tenured faculty.

Woven into this pay equity milieu was another issue of imbalance. Because faculty in business, engineering, and computer science commanded higher salaries in the marketplace, they were on a higher pay scale than other faculty in this university system. All universities within this state system follow a rigid pay scale of grades and steps for faculty, staff, and administrators, much like the federal government. This situation caused ill feelings among arts, sciences, and education faculty that were vocalized frequently in informal settings as well as in formal settings such as the Academic Senate and Administrative Council meetings. This resentment permeated other areas of the University such as the Provost's Office, which was heavily influenced by arts and sciences faculty members. When the infrastructure and rules to govern revenue-generating programs such as executive education were being established, business school faculty viewed the policy as an impediment. The Provost's position was that all programs would be viewed as equally valuable, would all charge participants the same fees, and would pay all participating faculty members the same rate. To some, this policy seemed just and egalitarian. To others, the idea of restricting what the business school could charge participants and pay participating faculty for their external programs seemed to fly in the face of free market principles and doomed the external programs to failure.

One final source of conflict was the heavy workload imposed upon employees as the University was being established. New courses and programs, new services, new buildings, and hiring new faculty all placed a heavy service load on people. Burnout was in evidence and junior faculty members approaching the tenure decision worried that the service demands would not be sufficiently taken into account come tenure time.

This description of State University provides the organizational and historical context within which issues of gender equity arose. While State University was full of opportunity and excitement, it also contained underlying and growing tensions among a variety of constituencies: administration-faculty, faculty-staff, and business/technology-arts/sciences faculty. In addition, within a few years of its inception, gender issues were already a controversial topic at State University. The primary evidence to support the claim of inequity was the absence of women in senior administrative positions. The President, Vice President of Academics and Provost, Vice President of Business Affairs, and the three deans were men. Only a small number of mid-level directors were women. The event that escalated the gender issue and precipitated this research was the resignation of a female dean after serving only 18 months in the position. Following a negative ad-hoc personnel evaluation from the faculty, an acting dean replaced her and soon after a permanent dean, both of whom were men. Female employees on campus claimed that such behavior reinforced their perception of a hostile workplace with a glass ceiling for women.

In this increasingly divisive environment, many members of the campus community wanted these issues to be addressed. The process began when the faculty, through the Academic Senate, called for efforts to address perceived gender inequities on campus. A four-person team of two faculty and two administrators subsequently attended a workshop on gender equity. Upon their return to campus, this team began to explore innovative ways to collect information about gender inequities. They wanted to foster dialogue that would help to uncover problems and possible misperceptions and that might lead to concrete steps that could be taken to resolve these inequities. Because the team members sought a forum in which these employees could openly and honestly discuss gender equity on campus, they decided to hold a series of small group discussions on the topic. The groups would be small enough to promote discussion and each group would have a mix of faculty, staff, and administrators. In order to be fair, to build commitment to the process, and to enrich these discussions, participation from a diverse set of as many employees as possible was needed. At the same time, the team was aware that discussions might be inhibited because those whose input was especially desired were those most vulnerable to repercussions for providing it.

The solution to this dilemma was to call upon a faculty member—the second author—who had been conducting research on computer-supported collaborative work and the use of GSS software for anonymous brainstorming and voting. The team hoped that people would be more willing to participate and be more open when they did so if the identity of the person making a comment were not known to other participants. Consequently, with the blessing of the President's Cabinet, a member of the team worked with this faculty member to develop a plan for using GSS software to support the discussions of gender equity. All University employees were invited by e-mail to participate in one of four anonymous, computer-based discussions of gender equity on campus. The goal of these discussions, as formulated and agreed upon by the team, the faculty member with GSS expertise, and the President, was to bring together employees from various levels, attempt to raise their awareness of gender equity issues within their organization, and enable them to generate alternatives for managing these inequities. The sessions were anonymous in that identities were not divulged or linked to comments. However, insofar as all participants sat in the same room and knew each other, the sessions were not "anonymous." While it is typical with GSS use to spend several hours per session in order to delve deeply into issues with intact work groups, the University administration was concerned that lengthy sessions might inhibit widespread participation. Therefore, it decided that total session times should be kept to a minimum: participants were asked to allow one and a half-hours for this endeavor.

The four GSS sessions were conducted in a computer-based classroom with 30 personal computers recessed into special desks. The room contained five rows with six desks/computers in each; the rows of desks ran across the room so that each participant was facing the front. A facilitator's computer sat on top of a desk in the front of the room off to one side. The room contained a large projection screen on the front wall and a ceiling-mounted, color projection system. The second author facilitated all four of the GSS sessions.

Each of the sessions began with the facilitator giving a brief introduction to the GSS[5] followed by a 20 minute anonymous, interactive brainstorming phase using a tool which sent to each participant's computer monitor the seed question:[6]

Do you believe that this university is a place where both genders receive equal treatment? Why or why not?

Participants saw the seed question at the top of their computer screens and had two windows available to them: one for entering their own comments and another for viewing the comments that had been entered by other participants. Participants could at any time input ideas, questions, suggestions, clarifications, arguments, or any other relevant thoughts, and they could scroll through this list of all comments and reply directly to someone else's comment. The first session was followed by another 20 minute brainstorming phase using the same software tool, but with the new seed question:

What should be done to insure that both genders receive equal treatment at this university?

While participating in this second brainstorming process, the participants could at any time revisit the previous brainstorming topic.

After this second brainstorming phase, the participants spent 10 minutes individually, anonymously ranking a list of approximately eight alternative solutions gleaned from the second brainstorming session. The facilitator worked with the primary contact person from the original gender study team to glean these solutions from the list generated during the second brainstorming phase.[7] For this ranking, the participants were asked to use the following questions as criteria:

Does the solution address the problem?

Is the solution feasible?

Will the solution have a high degree of potential success?

The GSS software quickly tallied the rankings and produced the group's average ranking for each item, which was projected on the large screen and reviewed with the group. The participants then completed a post-session questionnaire and the session was ended.

Out of approximately 400 University employees, 40 people participated in the four sessions. There were seven participants in the first session, 14 in the second, nine in the third, and 10 in the fourth session. A total of 30 participants identified themselves on a post-session questionnaire as being female. Two participants chose not to identify their genders on the questionnaires. Despite persistent lobbying from the session organizers, no male faculty members participated. The men who did participate in the sessions came from the ranks of university staff and administration. They came in response to a presidential request sent via e-mail. The rest of the men indicated that either they didn't think gender equity was a real problem or else they didn't think it was their problem. Some men simply failed to respond to the calls for participation. Others indicated that they were too busy to participate. In subsequent discussions with male faculty members, we learned that some men did not really believe there were gender inequities to the extent that female employees claimed. Those who acknowledged that there were gender equity issues thought that this was a matter for female employees to take up with the administration (see Appendix A for further demographic details).

Despite the intentions motivating these GSS sessions, nothing was ever done with the information collected. A University-wide committee comprised of faculty, staff, and administrators charged with exploring diversity on campus planned to have an organized, full day activity during which the data from the GSS sessions and other data could be presented. This event never occurred. Members of that committee explained that they ran out of money allocated for such activities. A daylong workshop on diversity, affirmative action, gender, and other issues was announced for some later, unspecified date. In addition, the individuals charged with launching a campus-wide assessment of diversity and gender equity issues never got the project going. Some of these individuals kept trying to design a "perfect" research instrument tailored to State University rather than use a less-than-perfect or an existing instrument. To date, there is no plan to conduct a comprehensive survey of the campus climate on these issues.

While there was no direct follow-up from the GSS discussions of gender equity, there was some activity on campus related to gender equity. Not long after these computer-based discussions, the faculty called for an ad-hoc personnel evaluation of one of the senior administrators in which the administrator was criticized for gender inequities. When this person subsequently left the University, a committee comprised of faculty, staff, and administrators charged with recommending an internal, interim candidate produced a short list of four individuals—three white males and one African-American female—from which the female candidate was chosen. Some people on campus believed that this choice was partly a result of calls to evaluate the performance of the other top administrators, which would include their record on gender equity. Informal discussions with members of the University community revealed the perception that the GSS sessions contributed to this subsequent change in the gender composition of the top administration by helping to raise awareness about gender inequities. People described the GSS sessions as the spark that raised people's awareness of the problem and led to a more proactive effort to fill the position with a woman. However, the search for a permanent candidate to fill the position resulted in a white male being offered the job.

The events described in this case suggest two reasons for triangulating our analysis of the data. First, given the highly charged nature of the topic, we suspected that a positivist approach with structured content coding focused on idea generation and using a priori categories might not capture all of the interesting and important aspects of the computer-mediated discussions that occurred in the sessions (even though surfacing ideas and opinions was a goal of the sessions). Second, a structured content analysis of the GSS sessions, alone, without some wider analysis of the context—what led up to and what happened subsequent to the sessions—would not tell the entire story. As this GSS analysis proceeded, it seemed increasingly evident that while a positivist analysis would be helpful, it would not be sufficient. Therefore, in addition to conducting an analysis in the positivist paradigm, using a more conventional GSS research methodology to understand the discussions, we expected to gain additional insight from conducting an analysis from the interpretive paradigm as well. For these reasons, we conducted an interpretive analysis of the GSS session transcripts in addition to the more positivist analysis.

Analysis of GSS Discussions

The case detailed how the authors gathered the data used in the analysis. The following sections address the analysis and interpretation and comparison of insights provided by positivist and interpretive analyses of this same data. First we present the positivist analysis of the GSS sessions, which includes a structured content coding of the session transcripts and an analysis of the results of this coding. We then present an interpretive analysis of the same transcripts borrowing from several interpretive traditions. Each of these sections is organized to provide an overview of the methodology before applying the methods to this case situation. The comparison, critique, and interpretations are presented after both of the analysis approaches are covered.

Positivist Approach to GSS Analysis

The underlying assumption of GSS use is that the technology can be useful in helping people to work together to solve problems and make decisions. The technology is typically used to support idea generation (referred to as divergence and/or conveyance) followed by idea evaluation and coming to consensus around one or a small set of selected solutions to the problem at hand (referred to as convergence; Niederman and DeSanctis 1995). The research then measures idea generation, evaluation, and consensus in quantifiable ways. Since the purpose of the technology is to perform a specific task, it is assumed that people will be task-oriented when using the technology and that the measurement techniques used will capture this.

One common approach used to gauge the effectiveness of GSS sessions is content coding to determine whether or not the sessions were effective in enabling the participants to achieve their goals. In this case, the goal was to bring together employees from various levels of the organization, attempt to raise their awareness of gender equity issues within their organization, and enable them to generate alternatives for managing these inequities. A conventional approach in conducting structured content coding and quantitative analysis of the session transcripts is to borrow an existing, tested content coding scheme from the research literature. Such an instrument enables the researchers to quantify the types and total number of comments that were generated in the brainstorming phases of each session and the degree of consensus around one or more of the solutions generated.

Since the comment categories used in the coding already exist, the job of the evaluators performing the content coding is essentially to parse and read each comment on each transcript and place it in the pre-existing category of best fit. There are typically strict definitions for comment categories and rules for determining when to place a comment in a category. The a priori analytical lens used for coding is idea generation, evaluation, and consensus; the results are expected to lend themselves nicely to quantitative analysis of group and member behavior. In addition, the researcher takes an "outsider's view" of the discussion. The focus is not on the meaning that participants assign to comments but rather on the type of comment made, such as a new idea versus supporting an idea already given. Comments from the session transcripts are taken at face value, for the most part, ignoring context when analyzing them. Indeed, a coding scheme would be most useful if a neutral, third party not involved with the use of the GSS and unfamiliar with the context of the participants could use the coding scheme to easily code the transcripts from the GSS sessions.

Positivist GSS Methods

The transcripts from each of the brainstorming sessions were content coded using the content coding categories, coding process, and corresponding constructs and measures from Connolly et al. (1990). These methods have been shown to be reliable and valid and have been used in Jessup et al. (1990), Jessup and Tansik (1991), Wilson and Jessup (1995), and many similar GSS experiments. This scheme is a classic example of a positivist approach and is ideal for determining how many unique ideas have been generated with a GSS. Each comment was read and placed into one of the preexisting categories of best fit. Redundant and/or infeasible ideas were not counted in the category labeled "Unique Ideas" and any comments that did not easily fit into any comment categories were placed in the category labeled "Off Topic." The complete list of comment categories is listed in Table 1.

Tot Comments – Total number of comments generated.

Tot Ideas – Total number of ideas generated.

Unique Ideas – Total number of ideas less redundant ideas.

Supp Remark – Expresses support for a proposal without adding evidence or argument.

Supp Argument – Supports a proposal and gives evidence or argument.

Sol Clar – Adds detail or new features to a solution.

Prob Clar – Adds detail or new features to problem statement.

Crit Remark – Expresses opposition to proposal without adding evidence or argument.

Crit Argument – Opposes a proposal and offers evidence or argument.

Ques Sol – Requests clarification of a proposed solution.

Ques Prob – Requests clarification of problem specification or solution criteria.

Computer – Remark about the computer network or its operation.

Group – Remark about the interpersonal processes of the group.

Off Topic – Remarks that are "off the topic" and do not fit into the existing categories.

Uncodable – Uncodable text.

Table 1. Content coding categories (comment categories from connolly et al. 1990).

Discussion of the Results of Positivist Analysis of GSS

Table 2 shows how many of each type of comment were generated during the first and second brainstorming phases within each of four GSS sessions. An average of 53.5 unique contributions were made in each of the eight brainstorming phases. The brevity of the brainstorming phases (only 20 minutes) combined with an average of 10 participants in each session indicates that the first goal of a high degree of participation was achieved. Such data suggests that in addition to reading other people's ideas and opinions, each person was able to contribute, on average, between 10 and 11 ideas and/or opinions during the 40 minutes of brainstorming.

Another measure of effectiveness in achieving participation was redundancy. This measured the extent to which ideas generated during a session had already been generated by others. A high degree of redundancy would suggest that participants were not reading and/or understanding each other's comments effectively. We interpret the low degree of redundancy (approximately 6.75% of all ideas generated were redundant within sessions) as an indication that participants were reading and understanding each other's comments.

To measure effectiveness in achieving the second goal of generating ideas about gender equity, we analyzed the total number of ideas (and the total number of ideas minus any redundancies) generated during the brainstorming phases of the sessions. An average of 10.38 total ideas and 9.25 unique ideas were generated during each of the eight brainstorming phases. While the first brainstorming phase of each GSS session was problem diagnosis, the second was problem solving. When focusing on the number of ideas generated during the second half of each GSS session, we found that in each of these problem solving, brainstorming phases an average of 17 ideas and 14.75 unique ideas were generated. Within a relatively short amount of time in each of these sessions, an average of nearly 15 workable, unique ideas were generated for insuring gender equity, with a total of 83 ideas generated across all sessions. This data suggests that the sessions were effective in achieving the goal of surfacing ideas about gender inequity.

We also analyzed the frequency of comments in other comment categories to measure the effectiveness of the GSS sessions in idea generation.

	Sess 1, Q1	Sess 1, Q2	Sess 2, Q1	Sess 2, Q2	Sess 3, Q1	Sess 3, Q2	Sess 4, Q1	Sess 4, Q2	Cat Average
Tot Comments	45	45	69	70	41	44	50	64	53.50
Tot Ideas	3	11	3	20	3	23	6	14	10.38
Unique Ideas	3	9	3	16	3	21	6	13	9.25
Supp Remark	4	2	1	5	0	2	2	2	2.25
Supp Argument	0	4	3	5	1	10	4	6	4.13
Sol Clar	0	12	0	4	4	3	2	17	5.25
Prob Clar	25	5	39	14	21	1	19	3	15.88
Crit Remark	2	1	0	0	1	0	3	0	0.88
Crit Argument	2	8	8	7	1	2	3	7	4.75
Ques Sol	0	2	0	7	3	1	1	13	3.38
Ques Prob	8	0	15	7	7	1	10	0	6.00
Computer	1	0	0	0	0	0	0	0	0.13
Group	0	0	0	0	0	0	0	0	0.00
Off Topic	0	0	0	0	0	1	0	1	0.25
Uncodable	0	0	0	1	0	0	0	1	0.25

Table 2 Raw data from content coding (comment totals by session and brainstorming phase).

For example, the relatively low frequency of supportive remarks and supportive arguments, and the relatively high frequency of questions about the problem and problem clarifications, particularly for the problem diagnosis brainstorming phases, suggests that during these sessions the participants were probing and challenging each other's ideas and opinions. It is also important to note that while the participants were probing and engaged in critical thinking, they were not particularly critical in a negative sense. Indeed, blatantly negative remarks were relatively low. Similarly, the frequency of critical arguments was significantly higher than critical remarks, and the frequency of supportive arguments was significantly higher than supportive remarks, suggesting that the comments (whether critical or supportive) tended to be substantial and thoughtful. In other words, it was much more probable to see a comment in which the participant explained why s/he agreed or disagreed with another person than it was to see a comment in which a person simply agreed or disagreed without providing any supporting information.

With respect to the third goal of generating alternatives for managing gender inequities there also appeared to be a fair degree of consensus among the participants—expressed in their rankings at the end of each session—as to alternative courses of actions. As described above, after the second brainstorming phase, each individual spent 10 minutes anonymously ranking a list of approximately eight alternative solutions from the second brainstorming session using as criteria the questions: Does the solution address the problem? Is the solution feasible? Will the solution have a high degree of potential success? Participants saw on their computer screens an unordered list of the alternatives to be ranked, they each ranked their own lists independently and anonymously, and they then submitted their rankings electronically. The GSS software quickly tallied the rankings and produced the group's average ranking for each item, which was projected on the large screen. Table 3 shows the raw votes for the highest ranked alternative solution within each of the four sessions. For example, the first column reveals that for the highest ranked alternative solution (enabling better mentoring for employees) within the first session, two people ranked this as the best alternative, one person ranked it as the second-best alternative, three people ranked it third, and one person ranked it sixth.

In the first session, seven people ranked six alternatives. Their highest ranked alternative was "enabling better mentoring for employees," with an average ranking for that alternative of 2.71 and a standard deviation of 1.58. In the second session, 14 people ranked eight alternatives. Their highest ranked alternative was "creating equity across job classifications and pay categories," with an average ranking for that alternative of 2.64 and a standard deviation of 2.02. In the third session, eight people ranked four alternatives. Their highest ranked alternative was "conducting an equity survey and then analyzing the data closely," with an average ranking for that alternative of 2.38 and a standard deviation of 0.86. In the fourth session, 10 people ranked nine alternatives. Their highest ranked alternative was "promoting mutual respect for all employees," with an average ranking for that alternative of 2.8 and a standard deviation of 2.71. In two of the sessions, "do nothing" was ranked as one of the alternatives and in both cases was nearly unanimously ranked at the bottom of everyone's list.

An overall indication of the groups' effectiveness in these computer-mediated discussions is the low frequency of comments that were "off the topic," uncodable, about the computer system, or about the group. On average, approximately 1.17% of the comments generated in each brainstorming phase were in these four comment categories combined. Therefore, it is safe to say that, for the most part, the people participating in these sessions were "on task."[8]

The results of the positivist analysis can be summarized as follows. Both the degree of participation and participants' engagement appeared to be high. There was a fair degree of consensus among the participants about what to do and not to do. While participants were probing and challenging each other's ideas and opinions, conflict appeared to be low. Participants were not critical in a negative sense; comments (whether critical or supportive) tended to be substantial and thoughtful. The groups were effective in achieving the goal of surfacing many workable ideas about gender inequity. The information they exchanged was for the most part related to issues and possible solutions.

Session 1	Session 2	Session 3	Session 4
1	1	1	1
1	1	2	1
2	1	2	1
3	1	2	1
3	1	2	1
3	1	3	2
6	2	3	2
Mean = 2.7	2	4	3
SD = 1.6	2	Mean = 2.4	7
—	4	SD = 0.9	9
—	4	—	Mean = 2.8
—	4	—	SD = 2.7
—	5	—	—
—	8	—	—
—	Mean = 2.6	—	—
—	SD = 2.0	—	—

Table 3 Raw votes for highest ranked alternative within each session (highest ranked alternative was ranked #1).

Stripping participants' identities from comments provided anonymity. They appeared to be reading and understanding each other's comments. Finally, the groups were useful insofar as their comments were very much on task.

Taken together, these results suggest that the sessions were effective in helping individuals to surface ideas for better managing gender equity within their institution. However, these results do not tell us about any attitudinal changes that might have occurred because of these discussions. The nature of the comments and replies, particularly in the areas of questions and clarifications, suggests that people were interacting with and informing each other. We cannot directly ascertain from this analysis, however, whether their awareness of gender equity issues was altered. We can only surmise that people were reading and understanding the ideas and opinions of others.

Interpretive Approach to GSS Analysis

While the positivist approach to analysis of GSS data is guided by a priori coding categories, the interpretive approach used in this study developed meaningful categories in grounded fashion. In addition, whereas the previous analysis provides a quantitative representation of the groups' interactions, the interpretive approach produces a qualitative rendering of group behavior. The objective of interpretive research is to piece together people's words, observations, and documents into a coherent picture expressed through the voices of the participants.

In choosing interpretive methods, the researcher is acknowledging that access to the world of the people being studied comes through social constructions such as language, consciousness, and shared meanings. We learn about the groups being studied by inductively exploring their behavior and communication in context. We engaged in this endeavor with no a priori lens regarding the information we would obtain. Unlike the positivist GSS analysis, participants' comments are not focused on convergence and decision making. Rather, we allowed the relevant information to emerge in grounded fashion through the iterative process of examination, connection to their world, and reexamination. This approach adopts an "insider's view" of the participants, their motivations, and their interactions by interpreting their voices within both the immediate and the larger organizational contexts.

Interpretive Methods

In conducting our interpretive study of these GSS sessions, we consciously drew from several different interpretive traditions including ethnography, hermeneutics, and grounded theory. Doing so is consistent with the published literature of interpretive research. Walsham and Sahay (1999) used ethnographic criteria to assess the quality of their research even though their work is not an ethnography, while Geertz (1973), an ethnographer, wrote about hermeneutics.

In contrast with the positivist analysis presented in the previous section, the scope of the interpretive analysis presented here is broadened to make significant use of the contextual data found in the organizational setting. By incorporating the organizational context as well as the context of use into the interpretation of the computer-mediated discussions, our analysis can be viewed as an interpretive case study of an organization's use of computer-mediated communication tools.

A broad criterion for our interpretive analysis is that after having read our work, an outsider would be able to read the transcripts and understand the logic of the comments within the context of this particular setting. We hope to satisfy this criterion by walking the reader through our development of our interpretation, involving "breakdowns" and "absurdities" and their resolution. Geertz expresses this as the reduction of puzzlement and the sorting out of local meanings. Open coding was used to sort out these local meanings.

The use of open coding to develop themes and meanings is a well-recognized technique employed in ethnography[9] as well as in the hermeneutic analysis[10] of text. According to Boland (1991, p. 439):

> Hermeneutics is the study of interpretation, especially the process of coming to understand a text. Hermeneutics emerged as a concern with interpreting ancient religious texts and has evolved to address the general problem of how we give meaning to what is unfamiliar and alien.
>
> Searching for the meaning of an aberrant passage occurs by using the hermeneutic circle. The reader seeks alternative understandings by cross-referencing the passage in question with other passages. In the process of reaching understanding, the reader goes back and readjusts previous understandings. Hermeneutic methodology has been employed in a variety of IS research settings (Boland 1985, 1991; Davis et al. 1992; Lee 1994; Myers 1994; Rathswohl 1991).

When the interpretive research includes open coding, the researcher approaches the data without an a priori framework to shape the understanding of the information. Instead, s/he allows the interpretive lens to evolve through the iterative analysis of the information within its context. This process is part of the grounded theory approach to qualitative analysis developed by

Glaser and Strauss (1967). Using this approach, the researcher engages with the data without a preconceived commitment to a particular line of thinking. The essential features of open coding are (1) the inductive development of provisional categories; (2) ongoing testing of categories through conceptual analysis and comparison of categories with data that is already coded; and (3) the altering of existing categories as other ones are created or eliminated (Strauss 1987, pp. 11–13). Open coding requires considerable flexibility by the researcher who must let go of initial control over the categories and be willing to adjust them as the analysis progresses.

The more traditional method of analyzing the discourse of GSS sessions involves imposing an extant analytical framework on the data. This framework is used to guide the data collection and analysis and to focus the researcher's attention in on what is "relevant" to the task at hand. In contrast, the interpretive approach that we employed called for us to begin with a blank slate and let the coding categories emerge as our interpretive understanding and engagement with the text progressed. In this way, we let the data "speak" to us. The way in which we allowed the interpretive lens to evolve through the use of open coding is illustrated in the way in which these categories evolved. We began with inductive development of provisional categories, engaged in ongoing testing of categories and comparison of new categories with data that was already coded, and subsequently altered existing categories as others were created or eliminated. The categories and subcategories that emerged from this open coding process are shown in Table 4.

Interpreting Information Types: Having identified the interpretive process whereby the three types of information emerged, we can now take a closer look at the groups' discussions. By tracing the way open coding produced these categories of information from the GSS sessions, we can see how the interpretive understanding of the groups' discussions evolved. From our exploration of the discussion transcripts, we learned that, while participating in the computer-mediated discussions, group members operated on three different information levels. One was exchanging cognitive content, a second level was expressing emotions, and the third was evidencing consciousness change.

In our interpretive process, we use Agar's (1986) language: the meaning of *strips* is interpreted through the *resolution* of *breakdowns* that occurs by revising one's *schema*—the world view held by the researcher. The transcripts of each GSS discussion session constituted the recorded information that we analyzed. These discussions were subsequently segmented, categorized, coded, and revised in order to make sense of them within the participants' own understandings. Categorizing involved segmenting commentary into a meaningful unit of discourse (i.e., either a single statement or several people's comments about a single idea) called a strip. A strip can be an observable act, an interview, an experiment, a document, a comment, or any other bounded phenomenon against which the researcher tests his or her understanding. In this study, all strips were text segments, either a single comment or a collection of comments made by different people

Cognitive Information	Information about the content of the communication
Affective Information	Information about the emotions of the participants
Talking about feelings	Discussing people's feelings
Giving voice to feelings	Expressing one's feelings
Behavioral Information	Action-oriented information
Talking about behavior	Discussing actions
Consciousness raising	Expressing a personal change in consciousness or attempting to change another participant's consciousness

Table 4 Results of the open coding: The information exchanges of the GSS sessions.

about a single thought. An example of a strip made up of two comments is the following:

> P1:[11] We need constant reminders that we are aware that there are historical problems and that we are not satisfied with the status quo. Dwelling on the past will not help (although I am NOT suggesting that we ignore the past).
>
> P2: Yes, I think some of us are trying. Unfortunately, those who need to be cognizant of gender issues will not, and probably truly believe they don't need to make the effort.

An example of a strip made up of a single comment is the following:

> The problem with this solution is that until the men on campus believe that there is inequality, anything done by "all women for the improvement of women" will be seen as worthless. There are only a few male faculty members both junior and senior that believe in equality and equal opportunity. It is a sad affair, but one that is promulgated in Higher Education and Academia in general. Let's be real. The world's perspective has to change, before [State] University personnel can truly make a difference. We get our employees from the world outside these walls.

After the transcripts were categorized into strips, each strip was coded according to its cognitive content. This was a process of reading the text strips and assigning provisional labels with respect to the content of the exchange. The labels assigned to strips summarized the theme of the strip as shown in the following examples:

> Gender discrimination is a systemic, societal problem.
>
> There is resistance to change.
>
> There are differing views in the groups about men's vs. women's ability to have power.
>
> Women are not mentored.
>
> Formal mechanisms currently exist to deal with diversity issues.

Sometimes the information that was shared was a direct response to a seed question and other times it was in response to another respondent. For example, in response to the seed question about both men and women receiving equal treatment in the University, the following types of information were shared.[12]

> People come in with baggage they've accumulated from sexist socialization and institutions.

****

> One of our problems that, "We have always done it this way before," is compounded by the notion that, "We have always done it MY WAY" at individual or campus sites that we don't explain or discover among our colleagues' backgrounds.

****

> I don't believe that women receive the same type of mentoring or support necessary to solve problems or advance their careers.

When the seed question asking what should be done to ensure equal treatment was given, the following types of responses were given.

> It would be fascinating to see how many individuals already feel they have "paid their dues" only to watch individuals from outside the campus community come on board and take the better positions. It would seem we need a situation that calls for an investment from the employee AND the University.

****

> P1: We need a mechanism for evaluating perceptions of equity on an ongoing basis.

****

> P2: Yes, perceptions seem to rarely match reality, especially when it is beneficial to see things in a given way.

****

> There is only one salary schedule but faculty can be placed on a step or at a level based on some pretty nebulous variables. Also, managers can be placed anywhere in the

range. When prior salary is used as a jumping off place for salary placement, women suffer because they usually come from a position of underpayment.

In addition to responses to the seed questions, other examples of exchanging information content occurred when respondents shared information in response to something said by another respondent. For example, in response to two individuals discussing the pervasiveness of sexism in American society and whether anything can really be done about it, one person offered additional information:

> [T]he point is that there are people who benefit from the sexist status quo and will defend it, that there are people who are invested in sexism.

During a discussion of affirmative action and quotas, several respondents had reacted negatively to the notion of targeting a particular group such as females in a search, suggesting that this would be a departure from common practice. In response, one participant commented:

> Limiting searches to the male gender has been going on for a fair number of years. Hiring another man into an executive position will provide absolute proof that this university cares nothing about women. It's pretty evident now, but one always holds out hope for justice.

As the coding proceeded, we began to recognize that not all of the interactions could be categorized according to the cognitive content of the statement. We noted that some segments were not instancing an existing content category or even suggesting a new one, for that matter. Instead, we recognized that there was something substantially different in some of these segments. We had encountered an anomaly. Both ethnographic and hermeneutic methods make use of the anomaly as the vehicle for focusing attention and gaining better understanding of the information in context. Attention is focused on a strip that appears to be a contradiction or somehow does not make sense to the researcher. This is what Agar calls a *breakdown*. It was through breakdowns that the categorization of types of information emerged in our interpretation. The first of these occurred with the following strip:

Strip 1:[13]
P1: We have no female representation at the executive level and currently we have no female college deans. Without such representation at this level, how can women receive equal treatment?

P2: What do you call [name of a female Director]?

P3: I call this woman executive an excellent leader. I also call the usage of her in this response an example of the exact type of tokenism practiced by people unaware of gender inequities. One woman in a mid-level power position does not equate with gender equality or equity on a campus. She will be wielded against other women as an example of the falsity of women's complaints about lack of female leadership. One woman in power is not enough. One white woman in power is not enough. One lesbian woman in power is not enough. One woman of color in power is not enough.

This breakdown was resolved by revising the *schema* to make sense of such strips. The schema that had to be revised was the assumption that the information being coded was limited to *cognitive* content. We needed to revise our interpretive stance to acknowledge that sharing cognitive information was not the only type of information exchange occurring in these sessions. This recognition, then, required us to go back to previously coded strips in search of other instances in which something other than exchanging cognitive content was the essential meaning of the strip. Another example is:

P1: Does the structure of the place offer equal treatment opportunity? I think so. Salary structures are essentially equal. Search committees look "equally" for both genders.

P2: Are we in the same institution?

In the language of hermeneutics, this strip is an *absurdity* (Davis et al. 1992, p. 302). Of course the second speaker knows that P1 works at the same university. However, by examining this exchange in the context of the other exchanges occurring before it and in the context of the GSS sessions themselves, P2's reply can be interpreted as a sarcastic response. *Resolution* of this breakdown occurred when the schema was revised to include a second category: emotional information.

As the coding progressed, we noted that this new categorization of emotional information also required adjustment. The resolution that led us to establish the category of emotional information led, in turn, to another breakdown. Not all affective information was the *expression* of emotion; sometimes the speaker was only *talking about* emotion, as the following strip indicates:

> Strip 2:[14]
> I think the first thing that would need to be done is an attempt to eliminate whatever fear this question generates. Affirmative action issues always seem to strike fear in everyone's hearts—at least that's equal!!

This breakdown was resolved by expanding the category of emotional information into two subcategories: "Talking about feelings" and "Giving voice to feelings." In resolving this breakdown we were distinguishing between *talk* and *behavior*. An instance of the former is when a respondent refers to feelings that people might have about gender inequity. For example, respondents indicated that there is considerable reluctance to speak openly about this topic. Other feelings that were discussed include the emotional reaction that people have to the power dynamics and the need for people to really listen to and take seriously each other's feelings about gender inequity.

An example of the latter—actually expressing one's own feelings—is the exchange about female representation at the executive level. One respondent gives a strong emotional reaction, a tirade almost, when a counter-example is given to the claim that there were no women in executive positions at the University:

> One woman in power is not enough. One white woman in power is not enough. One lesbian woman in power is not enough. One woman of color in power is not enough.

The speaker is expressing her/his considerable frustration with the use of a single instance being used to refute the claim of an entrenched pattern. This person is expressing an emotion—frustration—with respect to what s/he perceives to be tokenism. The ability to distinguish between talk and behavior with respect to this category of information enabled us to gain deeper insights into the interpersonal dynamics of these discussions. Exchanges such as this were clues that the GSS sessions were not always calm, logical discussions of issues and solution alternatives. Indeed, when the first author read the above exchange for the first time, she heard her own internal voice rise in pitch and volume as she connected with the emotion that was being expressed. She had heard these stories before in her research on gender.[15]

Once we adjusted our schemes to include affective information, we relied upon context and language cues in order to uncover and interpret the emotional meaning of the group members' discourse. The emotional messages were signaled in several ways. One was the use of writing conventions. For example, in response to the seed question, "Do you believe this university is a place where both genders receive equal treatment?" came the following reply:

> Absolutely NOT!! Not only are there differences in pay for the same job and qualifications, but also, women's ideas are generally dismissed and demeaning comments are made to females that would never be made to men.

Here the use of grammatical devices such as capitalization and punctuation are indicative of the strong negative feeling being communicated.

A second way of interpreting expressions of emotion was by examining a statement in relation to what surrounded it, as when the participant inquired:

> Are we in the same institution?

This response can only be understood as an emotional—specifically, an incredulous—reaction when it is viewed against the backdrop of the preceding comment. Two individuals are discussing the role of institutional structures in offering equal opportunity to both genders. The first respondent thinks that the structure of the university does offer equal treatment and opportunity to both genders. One can almost hear the sarcasm in the second person's voice as s/he wonders whether they work at the same place.

A final way that emotion was expressed was simply through the meaning of the words themselves. In the excerpt below, the final speaker is expressing skepticism about the recommendation for ensuring fairness.

> P1: *We need to redouble our commitment.* Winding up three important searches just now....What is the gender picture of interviewees? Who searched? What happened?
>
> P2: [We need] discussions in considering positions. Talk about the person not the gender. Every few months [make sure] there is a "check" to make sure there is no de facto bias.
>
> P3: Who does the "checking"? Foxes in the henhouse.

The first two comments in this strip set the stage for the cynical remark made in the final entry, but even without these, the cynicism in the third comment is evident in the phrase "foxes in the henhouse."

In some cases the emotional information was expressed through a combination of mechanisms. In the following exchange about diversity and assimilation, the feelings come across from a combination of the context, the language and the writing conventions.

> P1: By keeping the feeling of groups, underrepresented or not, we create a gang-type climate. We are all here because we are Americans and this is our culture NOW. Think of the reasons that many "underrepresented" groups left their country—because those customs brought about societies that did not work for them and they came here to be an American.
>
> P2: Maybe, but you guys think we all want to be white middle class males.
>
> P3: Oh, yuck!!!!!

The use of multiple means to express emotion is akin to face-to-face communication in which the words and nonverbal communication work together to convey an emotional response.

As we proceeded with our coding, another breakdown occurred when we encountered strips that fit neither the cognitive nor the emotional categories.

Strip 3:[16]
P1: It doesn't make sense, to me, to think that we could possibly be an oasis of equity in a society where social inequities are so deeply institutionalized; we are also part of the larger [university] system and as such, must be cognizant of the larger system's lengthy history of sexism—in every aspect.

P2: Yes, but we can try. Why give up and say that the "picture" is just too large to deal with?

The response made by P2 in this exchange contains something other than cognitive content or emotional expression. This breakdown was resolved when we recognized that a third type of information was being expressed: behavioral information. This was information connected to action. In this exchange, P2 is attempting to rouse people into action; s/he is sounding a "call to arms."

However, as with emotional information, the development of this category subsequently led to another breakdown, which we resolved as we adjusted our interpretive schema, yet again, to recognize that strips contained information about behavior in two ways. Sometimes the people simply talked about the behavior of others: what should be done to change the gender inequity situation at the university or in society, as Strip 4 shows:

Strip 4:[17]
I have talked to female faculty members who feel they weren't hired at a fair salary. They didn't negotiate as ruthlessly as others. Maybe women need to network on how to negotiate like the guys.

However, at other times participants were more active. They were expressing some altering of their consciousness about this topic—a reinforcement of an existing view or a change in consciousness—or an attempt to alter another's consciousness. It is in this sense of consciousness change that the respondents are evidencing some sort of behavioral change.

In resolving these breakdowns, we created two subcategories for behavioral information: information about behaviors (talk) and information signaling consciousness change (action). Examples of simply talking about behaviors that do or should occur are comments describing behaviors participants say ought to occur in order to respond to instances of gender inequity. The behavior changes that they recommended range from women changing to the organization changing to society changing.

The evidence of a consciousness change was deduced from what the person said or how she or he said it. Examples are conveying a tone of surprise or a spirit of activism. As with expressions of feeling, expressions of consciousness change occurred in several ways. The most discernible indicator of consciousness change was the use of action words such as those used in Strip 3:

> Yes, but we can try. Why give up and say that the "picture" is just too large to deal with?

While the preceding comments serve to reinforce this respondent's call to action, they are not essential for understanding this strip. It stands on its own as a motivational statement to the rest of the participants to engage in activity that will change the inequity at the university. We see that s/he wants to do something to rectify the situation. In another place a strip stands totally on its own. The following comment was not a direct response to previous comments.

> I think we ARE trying, even now. Otherwise, why did we come and engage in this process? It just isn't going to change overnight, but it will change.

The second indicator of consciousness change was a statement observed in context. For example, the final statements in one of the sessions were the following:

> P1: I think it begins with each individual learning that each person, no matter which gender, is capable of achieving and attaining the same goal. This may not be the case in today's society, but if we advocate and teach toward that belief—it may someday come true.

> P2: This is absolutely beautiful. I hope I see this statement again and again. It ought to drive our behavior. It has real promise for us.

If viewed on its own, the second comment in this strip could be classified as simply talking about behavior, but its placement at the end of the session in a discussion about creating awareness suggested to us that some other meaning could be gleaned from the words. At a minimum, P1's comment resonates and serves to reinforce P2's consciousness about this matter. Alternatively, this participant's consciousness about gender equity could have been altered in some way by the discussion. Either way, the statement belongs in a category different from that which contains distanced discussion of other people's behavior.

The final expression of consciousness change is the subtlest and most difficult to discern. While it is akin to the other two forms of expression, it is also different. This type of consciousness change is signaled through the presence of emotion in discussing behavior. Without the emotional component, the words would be classified as talking about behavior rather than expressing consciousness change. Here, a sense of activism emerges from the exchange. It is as though the respondents are saying, "We have to do something!" The discussion is about pay equity across genders.

> P1: [P]aying men and women equally is a requirement to equal treatment. Equality is only word without salary equity.

> P2: This is really hard to believe. Do you have concrete evidence of this happening on our campus?

> P3: Absolutely. The figures are available for all to see. The EOE[18] officer can gather them for you.

> P4: Is it really the case that we have different salary schedules for male and female? That's hard to believe [in this age].[19]

Before leaving this discussion of types of information, it should be noted that multiple coding of strips also occurred. That is, segments were simultaneously placed into more than one category. Consider the following excerpt that talks about change behavior. At the same time it creates awareness, provides cognitive content, and expresses emotion.

> There should be at least one (and of course preferably MORE THAN ONE) female at the top level of administrative decision-making. (That is, a President or a Vice President.) If there are not females present and participating when highest level decisions are being made, gender bias is almost inevitable (even if it is not conscious).

Excerpts such as this are typical of face-to-face communication in which multiple motives inform messages.

The process employed to interpret the meaning of the information exchanges was an iterative one. As we moved through the discussion transcripts, we continuously adjusted our worldview (and expectations) about the nature of the information contained in them. Figure 1 depicts the way in which this method was applied in the interpretation of this text using Agar's (1986, pp. 27–29) method of breakdown resolution. Our initial schema or level of understanding encountered a breakdown when a strip did not conform to the expectations embodied in the schema. Resolution came when we adjusted our schema and revised our knowledge about the meaning of strips contained in the discussion transcripts. But then a new breakdown would occur when a strip challenged this revised knowledge and the process began all over again. This process of schema revision continued until all of the strips were able to fit with the schema that ultimately resulted.

Interpreting the Meaning of the Information: Having used interpretive methods to uncover the types of information that were exchanged in these GSS sessions, we then employed the interpretive process to understand the meaning of the textual transcripts that resulted. In this section, we discuss our use of the hermeneutic circle to develop these interpretations.

In the course of expressing their feelings, we noted that the participants revealed very different perceptions of gender equity at the University. One group of participants used these sessions as an opportunity to express a number of feelings about the position of women at State University: anger, frustration, annoyance, and fear.

> I am always being told that I perform better than expected for a woman.

> **********

> I trust absolutely no one here. I am afraid to discuss the hard reality of the situation here for women with anyone.

> **********

Another group of participants used this as an opportunity to express feelings about male issues:

> Oh, this really bugs me. Why is there WOMEN'S STUDIES and not Men's Studies offered here?????!!!!! I think the change in a man's role in society is often overlooked as women come forward in the workplace.

In response to the question "Do you believe this university is a place where both genders receive equal treatment," they revealed very different perceptions about the status of inequality at the University. The following is a sampling of comments from one session:

> No. Males at distinct disadvantage; in some case actively discriminated against.

> **********

> It seems so; distribution of faculty and administrators seems equal.

> **********

> I think that women in leadership roles are not recognized as such. A person told me that white males ran this university, and after checking, I found out that women were in the majority of management positions on campus. The perception was diametrically opposed to the fact*s*.

This group appeared to resent the focus on women. One participant expressed the viewpoint that focusing on gender or other "special interest groups" is diverting attention from the real interpersonal issues at the university, namely, facultystudent interactions. Another believed that an appropriate response to a group's charge about feeling disadvantaged is to counter with "the facts," implying that these "facts" would show otherwise.

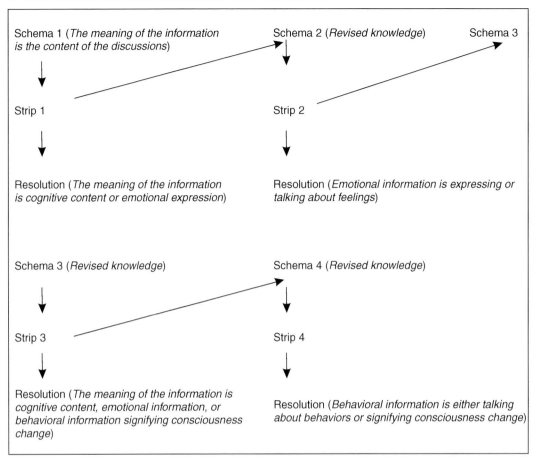

Figure 1 Interpreting the meaning of information exchanges through breakdown resolution.

Observation of these responses suggested that the discussion transcripts contained divergent meanings about gender equity at the university, but this conclusion produced an anomaly. The purpose of people coming together in the GSS sessions, at least as seen in the perspective of conventional positivist GSS research, was to do problem solving: to discuss aspects of the gender equity issue at the university and then to make suggestions about improving the situation. Implied in this motivation is the assumption that there is a common recognition of "the problem." Upon examination of the textual passages, however, we came to a different conclusion. It seemed that there were two points of view: one vehemently consistent with this assumption and another which did not "buy in" to it at all. We sought to resolve this contradiction by returning to the meanings that lay in the larger context. As we did so, we returned to the particular passages with new insights that led us—after a series of iterations—to new understanding about the meaning of the discussion transcripts.

Our first insight, therefore, was that there was an absence of shared consciousness. Participants' comments often seemed to be operating on two different "wave lengths" as though they were carrying on two different—and parallel—discussions. In one dialogue, the understanding of equality ranged from the lofty ideal to the gritty reality:

> P1: Equality transcends "equal pay for equal work" issues. The essence of equality emanates from mutual respect.

P2: Mutual respect is lovely. Status gains more respect. Equal pay for equal work is vital for self-respect and the respect of others because it translates into status. This is a patronizing statement.

Other examples of being on different "wave lengths" come from the differing interpretations of affirmative action and mentoring. With respect to affirmative action, there were two distinct connotations given to the term. One was promoting equality of opportunity for underrepresented groups; the other was rewarding incompetence. With respect to mentoring, the same phenomenon occurred. One connotation was helping members of underrepresented groups to fill positions at all levels of the University hierarchy. Another interpretation linked mentoring for purposes of enhancing upward mobility to a "warm and fuzzy place" that based rewards not on competence but on length of employment.

Dialogues also revealed a low level of understanding about the reality and the language of gender discrimination. These authors find this ironic in view of the fact that the majority of the male participants were in management positions.

P1: I have not witnessed nor experienced what I would consider unfair treatment of either gender. However it is true that the [State] University staff is well over half women. I am not sure why.

P2: Which half of the staff is female? Most support staff is female.

In another exchange, a participant refers to a "pink ghetto" to describe positions generally held by women and which tend to be low paying with little advancement, power, and prestige. This person is contrasting the low paying "pink ghetto" with positions in the trades held by men which receive higher pay. In response, another participant wonders:

What is a "pink ghetto"? It is difficult to understand why trade positions receive a higher salary than secretarial levels considering the advanced technology that "secretaries" have been required to learn in a short amount of time—jack of all trades....

The abbreviated time period for the GSS sessions exacerbated this absence of shared consciousness with which participants embarked upon the discussions. There was not even enough time to negotiate common meanings of relevant terms, define the scope of the conversation, and establish participants' points of view much less to move on to finding an acceptable solution to the issue.

A second insight was that these transcripts did not reflect widespread participation. Our analysis of participation, which was derived from an understanding of the broader context, was the opposite of that which resulted from the positivist analysis. Only 10% of the University community participated. Among those who did participate, there were both gender and professional imbalances. Three-quarters of the participants were women. Faculty members were heavily under represented while administrators were over represented. There were no male faculty members in any of the sessions and only seven women self-identified as "faculty or librarian." On the other hand, there were 15 administrators, several of whom were part of senior administration. To generalize, male administrators/staff and female staff dominated these GSS sessions.

Finally, our growing understanding of the whole led us to question the anonymity that was suggested by the positivist perspective on the GSS sessions. Despite the technical attention to confidentiality, we questioned whether participants believed they were provided with a safe space to talk. While the name of the participant may not have appeared on the screen along with her/his comment, the people sitting in the room together all knew each other. They talked to each other by name before and after the sessions. Given the power relationships in the group makeup—males were primarily administrators, females were primarily staff and faculty—one wonders just how safe a space it really was. Indeed, on the post-session questionnaire 25% of the participants were reluctant to provide some form of identifying information: two participants chose not to reveal gender, five would not indicate ethnicity, and three would not give position titles (see Appendix A). This might also have contributed to the low turnout by female faculty, despite the Faculty Senate call for such a forum. There may have been a selfimposed censorship on the part of the women faculty.

Raising awareness about gender equity issues

1. Uncovering problems and misperceptions
2. Giving women an opportunity to air perceptions about a hostile workplace and a glass ceiling
3. Collecting information about gender inequity

Bringing people from various parts of the University together

4. Garnering widespread participation
5. Providing people with a safe space to talk
6. Fostering dialogue within the University

Generating alternatives for managing gender equity

7. Identifying concrete steps to resolve gender inequity

Table 5 Interpretive perspective on goals of the GSS sessions.

By looking beyond the apparent meaning of the individual textual passages and into the context within which they occurred we developed insights into the whole. We came to understand the university context as one fraught with tension, suspicion, frustration, and incompatible differences in perception. The insertion of the GSS sessions into this setting was, at best, like dropping a pebble into a stream to build a dam; at worst, these sessions were mere public relations. The more we understood about this context, the more we were able to resolve the anomalies in the particular meanings of the individual passages. In this way, we moved back and forth from the larger context to the individual passages until we were confident that the story that was emerging was consistent with both the data in the transcripts and the meaning of the context.

Discussion of the Results of Interpretive Analysis of GSS

What guided the interpretation of the information—and the criteria used to evaluate the effectiveness of these computer-mediated discussions—were the goals for the GSS sessions that were presented in the context description section of this paper: raising awareness about gender equity issues, bringing people from various parts of the University together, and generating alternatives for managing gender equity. Whereas the positivist type of GSS analysis used the goals at this level of detail, the interpretive analysis went deeper into the case description to produce greater refinement of the GSS session goals. These are presented in Table 5.

The evidence from the interpretive analysis of the groups' discussions suggests that the first and third goals of the GSS sessions—raising awareness about gender inequity issues and generating alternatives for managing gender inequity—were addressed if not fully achieved. The interpretive analysis, however, casts doubt upon the achievement of the second goal: bringing disparate groups together to facilitate communication.

With respect to the first goal of exchanging information in order to raise awareness about gender equity issues, the interpretive analysis shows that participants exchanged information on three levels, thereby going beyond the original intention of the discussions. The intention of these GSS sessions—based upon the assumptions of conventional GSS use—was, first, to discuss whether participants believed the University is a place where both genders receive equal treatment. It was, then, intended that they would discuss what should be done to ensure that both genders receive equal treatment. The intention was to have a focused discussion of two specific questions after which some concrete alternatives would result. However, our interpretive analysis shows that the participants in these particular GSS sessions went beyond these expectations. They exchanged other types of information in addition to responses about the two seed questions:

> Do you believe that this university is a place where both genders receive equal treatment? Why or why not?
>
> What should be done to insure that both genders receive equal treatment at this university?

While the results of the interpretive analysis show that the GSS sessions were effective in achieving the stated goals of raising the issues and generating solution alternatives, they show that the second goal was

	Positivist	Interpretive
Unit of Analysis	The meeting	The meeting in its organizational and historical context
Participation level	High – number of comments relative to number of participants in sessions	Low – imbalanced representation of key University constituencies in GSS sessions
Participant's engagement	High – critical thinking evidenced by frequency of question/problem clarification	High – engagement shown through emotionally charged responses
Consensus	High – consistency in ranking of alternative solution scenarios generated from second brainstorming session	Some – widely varying perceptions of extent and reality of problem that were never resolved
Conflict	Low – frequency of explicitly negative remark	High – number of impassioned, emotional reactions to each other. Expression of opposite world views on key topics (such as affirmative action). Use of sarcastic, ironic language.
Information type	Cognitive	Cognitive
	Behavioral (talking about)	Emotional (talking about and showing)
	Expected	Behavioral (talking about and showing)
		Unexpected
		Contextual
Anonymity	High – participant's identities stripped from comments	Low – people who knew each other well sat together in the same room
Shared consciousness	High – participants were successfully reading and understanding each other's comments	Low – major differences in perceptions and opinions, "ships passing in the night"
Usefulness	Yes – Useful, "on task" solution scenarios generated	Partial – questionable meaningful interaction, failure to achieve shared consciousness, low participation, nothing ever done with session information

Table 6 Comparison of findings using positivist and interpretive analyses of GSS sessions.

achieved only minimally, at best. This goal was to bring people from various parts of the university together in order to facilitate meaningful communication. Instead, the interpretive analysis revealed an absence of shared consciousness. Contextual factors that contributed to this were uneven participation by gender and status in the university, absence of real anonymity during the discussions, and insufficient internal motivation to participate in meaningful dialogue about this topic.

With respect to the third goal of generating alternatives for managing gender equity in order to encourage action, participants gave evidence of behavioral information exchange in the form of consciousness change which went beyond *talking about* what needs to be done. Once again, however, the viewpoints were at opposite extremes. While numerous concrete suggestions for change were made, others expressed the viewpoint that gender equity is not an important and actionable issue.

I think we do enough already. I don't think that we could achieve a substantial increase in equity (and we could probably not agree on it if we did) without a lot of cost—time, yet another committee.... Better ways to spend our energy.

There were three overall findings that resulted from the interpretive analysis of the GSS transcripts. First, the GSS session participants exchanged three types of information: cognitive (i.e., content about the topic in question); emotional (discussing and expressing emotions); and behavioral (discussing the need for new behaviors and indicating personal consciousness change). The second finding is that these computer-mediated discussions were emotionally charged events. Participants gave impassioned, emotional reactions to each other and to the seed questions. Finally, the participants exhibited widely diverging worldviews about the problem at hand. There was no consensus that a gender equity problem even existed much less about the extent of it. These findings and a comparison with those of the positivist analysis are summarized in Table 6.

As Table 6 shows, the results of these two analyses paint very different pictures of the GSS sessions. The positivist analysis suggests that the GSS sessions were effective in helping these people to achieve their goals. There appeared to be an effective exchange of information. The technology appeared to have facilitated communication by helping the people involved to generate useful ideas to solve their problems. Finally, there was some consensus around possible solutions. Thus, we can conclude that the sessions appeared to have encouraged action.

The interpretive analysis, on the other hand, suggests that the discussions were, at best, only partially effective in helping the University to achieve its goals. There appeared to be an effective exchange of information associated with problem identification and solution scenarios. However, the sessions did not help to achieve a significant goal that was motivating these sessions: facilitating real communication among the participants. An interpretive analysis shows that the GSS sessions were most successful in exchanging concrete information or perceptions about gender equity, or in making concrete suggestions for action. In the course of sharing these facts and perceptions, however, wide divisions were made evident. Participants were not able to overcome the absence of a shared consciousness. The failure lies not in the technology, however. As exemplified in the time constraints imposed upon the process, factors that reside in the organizational context were responsible for the failure of the sessions to facilitate real communication.

The results of this interpretive analysis reinforce the point that GSS, like all information systems, are socio-technical systems. As such, technological characteristics alone will not ensure their successful use. Through the interpretive lens, we learned that the GSS was most effective in addressing the narrower goals of information generation and solution identification. Where this GSS project was least successful was in addressing the broader goals that were more connected to the organizational context. Because of features in the context—fear of reprisals, disconnect from the issue, lack of real anonymity, time devoted to the discussions, absence of shared consciousness—these computer-mediated sessions fell short of the goals they were, perhaps unrealistically, expected to achieve.

CRITERIA FOR EVALUATING INTERPRETIVE FINDINGS

We derive, from different interpretive research traditions, four criteria for evaluating our findings. They are triangulation, authenticity, breakdown resolution, and replication. We describe each of these criteria below and then show how we employed them in our study.

Triangulation Triangulation—the use of multiple sources, methods and investigators to provide corroborating evidence—is commonly used in a variety of qualitative methodologies to show that there is evidence other than the researcher's own interpretation to support the discovery (Creswell 1998; Fetterman 1998; Miles and Huberman 1994; Silverman 1993; Yin 1989). The objective of triangulation in our study is to find information from other sources to corroborate our findings that are based upon the interpretation of textual materials (the transcripts of the GSS sessions). The two sources of information that were used in this study were participant observation and member checking. The second author carried out participant observation in order to enable comparison of the interpretive findings with observations about the organization before, during, and subsequent to the GSS sessions. The second author was a member of this organization before and after the GSS sessions and was a cofacilitator for each of the sessions. On many occasions, such first-hand experience within this organization was used as a barometer with which to compare, challenge, and confirm interpretations that the first author was drawing from the analysis of the session transcripts.

The other source of information employed for purposes of triangulation was member checking. Member checking is a method of establishing the credibility of the findings in which the researcher checks her/his interpretations with representatives of the people being studied (Cresswell 1998; Ely et al. 1991; Lincoln and Guba 1985; Miles and Huberman 1994; Silverman 1993; Trauth 1997). In member checking, the researcher solicits the *inside* perspective on the credibility of the findings by reviewing the data, analyses, and interpretations with the participants. There was one key informant, in particular, who provided valuable feedback on interpretations. He is one of the staff members working within the library at State University.[20] He was heavily involved in all the significant events in the case study. In addition to his input, member checking occurred after the analysis of the transcripts and a complete draft of the manuscript had been written. Employees at State University were asked to review and validate portions of the analysis and manuscript as they were being developed. For example, on several occasions, the second author asked colleagues at State University to review interpretations that dealt with the subsequent effects of the GSS sessions on people's behaviors and decisions.

Once we had probed a little deeper into the context surrounding the computer-mediated sessions, we became more sensitized to the multiple layers within the computer-mediated conversations. We could more easily perceive the dynamics within the physical settings of the sessions and the broader, emotionally charged context within the organization. In short, the triangulation of case study analysis, interpretive transcript analysis, and member checking caused us to rethink and change our initial impressions derived from the GSS analysis about the degree of anonymity during the sessions and the subsequent effects that use of the GSS had.

Authenticity An evaluative criterion that reaches across the spectrum of interpretive research is that the account must *make sense* or *ring true* to the reader (Geertz 1973; Miles and Huberman 1994; Sanday 1979). This is expressed as the *persuasiveness* (Reissman 1993) or *authenticity* (Golden-Biddle and Locke 1993) of the narrative analysis. An authentic account is one that is perceived by the reader as genuine and conveys the researchers' understanding of the members' world. The evocative quality of the narrative, including the use of quotations and rich, detailed description, is indicative of the researchers' connection to the people being studied. Walsham and Sahay (1999) have used this criterion to evaluate their interpretive information systems research.

In this research, authenticity refers to the interpretive rendering of both the discussion transcripts and the context from which they arose. By providing rich detail about the organization, the participants, the relevant perceptions, actions, and events, and the relevant issues, and by describing the local and broader contexts within which the research took place, we helped the reader to better sense the meaning of this context. In the second section of this paper, we provided an overview of the case study. In the third section of the paper, we made use of excerpts from the transcripts to illustrate the emerging interpretations.

We also endeavored to produce an authentic account by revealing the two authors' connection to the context and the transcripts. We revealed that the second author worked at the institution and was physically present for all four GSS sessions. What he brought to the research was his own experience of the topic of gender inequity from participant observation in the context. We also revealed that the first author is a female university professor who has also done research on gender. We noted how her past experiences with gender issues came to light as she engaged in the interpretive process. In different ways, then, the authors were able to use their backgrounds in order to establish a connection to the case study. According to the principle of interaction between the researcher(s) and the subjects (Klein and Myers 1999, p. 10), the "facts are produced as part and parcel of the social interaction of the researchers with the participants." In our case, the authenticity of our interpretation is due, in part, to the way in which the authors interacted with the text and the context, and the way in which we deliberately shared the process of developing our interpretation openly with the readers, rather than simply presenting it as a finished product to them.

Breakdown Resolution or Hermeneutic Circle
What is necessary but not sufficient for reliability[21] of interpretive research is that detailed documentation of procedures be provided (Kirk and Miller 1986; Yin 1989). It is also necessary to employ methods that can demonstrate how the interpretation is consistent with the data. This occurs in interpretive research when the reader, after having read the researcher's account of the process, would be able to see how the interpretation is meaningful rather than simply made up. This is done by walking the reader through the process of developing the interpretations.

We accomplished this through the resolution of breakdowns, to use the words of ethnography (Agar 1986), or through the hermeneutic circle, to use the words of hermeneutics. Both terms characterize interpretation as an iterative process of examining the particular in relation to the greater whole and revising meanings as these iterations progress. When an anomaly or breakdown in understanding occurs, the individual strip is revisited with respect to the schema or "spirit of the whole," the one guiding idea that governs the text (Ormiston and Schrift 1990, p. 12). Through this dialectic process of reexamining strips and readjusting our schemas, we moved toward improved understanding of the whole text.

The following examples serve to show how understanding their relationship to the whole strengthened the interpretation of parts. Breakdown/strip-reformulation was explained and depicted in Figure 1 to show how interpreting the meaning of the information resulted in the understanding that several different types of information were being exchanged in the GSS discussions. As part of the verification of our work, we analyzed the texts in an iterative process, invoking the hermeneutic circle to verify that our interpretations had, in fact, uncovered and resolved as many anomalies as could be identified in the texts. An example of one of these anomalies was the absence of shared consciousness, as discussed earlier in the section on breakdown resolution. We believe that a robust approach to breakdown analysis, such as that offered through the hermeneutic circle, and subsequent reformulation of schemas increased the validity of our interpretations and conclusions.

Replication A method used in case study research to support validity[22] is replication. Through replication across multiple cases, the findings are shown to be generalizable beyond the immediate case (Yin 1989, pp. 43–44). The interpretation that yielded the three findings[23] from this study was obtained by pooling the four transcripts and interpreting them as a single document. Thus, at the end of the interpretive analysis we did not know whether or not each of the four sessions instanced all three of the findings. Therefore, in order to check the validity of our findings, we revisited the transcripts of each GSS session looking for evidence of each of the findings in each set of transcripts. In this second interpretation of the transcripts, we were considering each of the four sessions to be "replications" of our initial, pooled case.

By replicating, in the individual sessions, the analyses that derived from the analysis of the pooled transcripts, we were able to test our emerging interpretations. In the case of the first two findings (information sharing and emotion expression), the replication exercise was consistent with our initial interpretation. But the case of the third finding (worldview about the problem) was different. Coding of the pooled transcripts yielded the initial interpretation that participants exhibited a changed consciousness about the problem as a result of the GSS sessions. However, the replication caused us to revise that interpretation, highlighting a breakdown in schema that had not surfaced when aggregating the data and removing it from the context of the individual sessions. The literal replication using the individual GSS sessions became part of the "iterative" process common to interpretive research and specifically to breakdown resolution.

Whereas breakdown resolution is a process-oriented way to show how we developed our

	Positivist	Interpretive
Goal	Efficient conveyance of ideas and convergence on a solution	Understanding the meaning of the information exchanges of a computer-mediated group
Analysis	Quantitative	Qualiltative
Assumptions	That technology would help participants to generate and evaluate useful ideas	No explicit a priori assumptions as to what meaning would arise from the transcripts
Coding	Established, pre-tested, a priori categories applied to transcripts	Categories developed in grounded fashion through open coding
Decision perspective	Focus on considering alternative solution scenarios	Focus on better understanding of the problems and issues
Viewpoint	Outsider's: what text the participants produced	Insider's: the meaning of the participants' text
Coding assumptions	Text has static meaning	Meaning of text is dependent upon the context

Table 7 Comparison of GSS analysis methods.

interpretations, replication is a post-hoc way to show how we evaluated and verified our interpretations.

Discussion and Implications

Having conducted these two analyses of the GSS sessions we can now return to the research question motivating this study:

> Does an interpretive analysis of GSS use result in a different understanding of the GSS discussions than that provided by a positivist analysis?

To answer this question about the value added by using an interpretive lens in addition to a positivist GSS lens to analyze the session transcripts, we begin by comparing the methods used and the results that were obtained from each approach. A comparison of the methods is summarized in Table 7.

When comparing these two approaches, it is important to highlight the difference in the goal of each method. The goal of the traditional way that researchers and practitioners have used GSS is to efficiently produce conveyance of ideas and then convergence on a solution. This goal, in turn, shapes the collection and interpretation of the information generated during GSS sessions. Information that is useful in this analysis is that which gives evidence to conveyance of ideas, convergence on a limited set of workable ideas, and consensus of viewpoints. Participant comments that are not directly related to this specific goal are deemed not on task and, hence, are excluded from analysis and interpretation.

In contrast, the interpretive analysis had no such a priori goal and, therefore, no screening mechanism for "extraneous" information. The goal of the interpretive approach to GSS use was to understand all of the information exchanges of the computer-mediated groups. It was through immersing ourselves in the world of the participants through open coding of the transcripts that the information categories used for coding and interpreting the discussions emerged. While the conventional analysis placed the focus of the groups' attention on idea generation and evaluation, and on consensus about an action plan, the interpretive analysis placed the focus on better understanding of the problems and issues without regard to the development of an action plan.

One important distinction between positivist and interpretive understandings of the GSS sessions is the point of view taken during analysis of the transcripts. With the former approach, the perspective of the researcher is an "outsider looking in" on the group. With the latter approach,

Dimension	Additional Information
Participation level	Key University constituencies were absent from the meetings
Participant's engagement	Emotionally charged exchanges
Consensus	There was a lack of consensus about the perception and extent of the problem
Conflict	No additional information

Information type	Several types of information were exchanged:
	   emotional
	   behavioral
	   unexpected
	   contextual

Anonymity	No additional information
Redundancy	There was an absence of shared consciousness
Usefulness	No additional information

Table 8 Additional information acquired from interpretive analysis of GSS sessions.

however, the researcher's perspective is that of one who is "inside" the group, observing and interpreting what is happening. While the conventional analysis documents and quantifies *that* people communicated, the interpretive analysis seeks to understand the meaning of *what* people communicated. The difference between the two approaches is evident in the richness of the information that is captured. Whereas the conventional approach would disallow certain topics as not being on task, everything was on task when viewed through the interpretive lens.

Consistent with the differences in methodological approach are the different results that emerged from these two analyses. In order to understand why these differences in results have occurred as well as to probe the contribution of the interpretive lens, we now take a closer look at the interpretive results. We define the contribution of the interpretive analysis as (1) developing different conclusions from the same evidence (see Table 6) and (2) acquiring additional information to that which resulted from the positivist analysis. This additional information is shown in Table 8.

On seven of the eight dimensions listed, different interpretations of the same evidence resulted. Whereas the participation level of those in attendance at the GSS sessions seemed to be high, the broader lens of the interpretive analysis, which took into account the organizational context within which the discussion sessions occurred, shows that participation was low relative to the key University constituencies. While both analyses showed that there was high participant engagement in the sessions, they did so for different reasons. The conventional analysis pointed to a low frequency of clarification requests to show that people were engaged in critical thinking; the interpretive analysis pointed to the way in which people interacted: with emotionally charged responses. The positivist GSS analysis used the rankings of solution scenarios to conclude that there was consensus within the sessions. The interpretive analysis found that people entered the sessions with widely varying perceptions about whether there was a problem and, if so, its extent. As the interpretation of the transcripts ended, there was only limited movement toward changed consciousness about the issues.

The conflict dimension, perhaps more than others do, points to the richness of the information that resulted from interpretive analyses. The low frequency of explicitly critical comments in sessions is used by the conventional approach as evidence of a low level of conflict in the sessions. The interpretive analysis, by turning to such literary devices as sarcasm, irony, and grammar along with recognition of several types of information including emotional expression, concluded that there were was an atmosphere of conflict regarding worldviews, feelings, and reactions to each other.

The overall conclusion drawn from the positivist analysis is that the sessions were useful. They showed consensus around viable solution scenarios and were generally on task with their comments. The interpretive conclusion, however, is that the sessions were only partially useful. This analysis wonders how much meaningful interaction really occurred. The absence of a shared consciousness on the part of entering participants was not significantly altered in these deliberately abbreviated discussions. Finally, nothing was ever done with the information generated from these sessions.

In addition to the different understandings that resulted on these dimensions, Table 8 also shows the additional information that the interpretive analysis provides. First, incorporating the wider context of the case into the analysis of the GSS discussions shows that key University constituencies were absent in these discussions. This information is important in determining the representativeness of the results of the discussions. Second, whereas there was consensus around the solution alternatives that surfaced in the second brainstorming session, to begin with, there was no consensus about the nature of the problem. Despite buy-in to a solution, in theory, a participant's failure to consider gender inequity to be a significant issue in the first place will influence her/his motivation to enact such a solution. Probably the most significant contribution of new information occurred for the dimension of information type. Emotional, behavioral, unexpected, and contextual information all helped to enrich the understanding of the information exchanges in these computer-mediated discussions. Finally, the absence of a shared consciousness with which participants entered the sessions and which was only partly diminished by the end is useful in trying to understand why nothing ever resulted from these sessions. To the extent that senior administrators did not perceive a significant gender inequity issue to exist—and only attended the sessions because of a presidential directive—there would be low motivation to take action.

For these reasons, then, our research question is answered in the affirmative. The interpretive analysis of the GSS sessions did, in fact, provide different information from that which resulted from the positivist analysis of the same transcripts. Further, the understanding of the information exchanged within the GSS sessions was enhanced by the new information that only the interpretive analysis provided. The additional triangulation, which was achieved by this dual analysis of the same data, can give the researcher greater confidence in the results. In this research, triangulation verified the value of the GSS sessions. For example, the interpretive analysis revealed the presence of both passion and the absence of shared consciousness during the sessions, evidence that the participants spoke openly and honestly about their feelings and biases. Thus, the addition of this type of information helps to strengthen one's confidence in the issues that were raised in the discussions.

The other benefit of adding the interpretive lens—producing new insights into GSS—resulted from the hermeneutic analysis of both the transcripts and the organizational context. This analysis documented the absence of a shared consciousness and consensus about the issue of gender equity. What emerged from the interpretive analysis of GSS use in a highly politicized and volatile setting is that the technology can facilitate the process of laying the issues out for consideration. The "anonymity" of the discussions—partial as it was—nevertheless facilitated the expression of people's thoughts and feelings on this emotionally charged topic. Thus, while awareness in the sense of changed consciousness may not have occurred, tangible issues were made available for management's consideration. In addition, concrete steps for addressing gender inequity were raised. Again, while there was not common agreement about the nature and extent of the problem, State University's management was nevertheless provided with employee's suggestions about improvements. The insight into the use of GSS in this setting is that positivist analysis of the discussion data, alone, did not reveal for management all of the important information that was present in the sessions. While use of the GSS revealed information about issue and solution identification, management would have been left without information about the feelings of the participants on the issues or their motivation to enact the proposed solutions.

This insight suggests that different methods of GSS analysis might be appropriate for different circumstances. An interpretive analysis seems particularly suited to GSS sessions with greater uncertainty about the type of information that will be exchanged. This might occur when the problem is incompletely understood, when the problem is emotionally charged, or when the organizational context is highly politicized. A positivist analysis of GSS sessions as conducted here, on the other hand, seems best suited to documenting the communication characteristics of a group that is moving toward convergence about a decision.

Lee (1991) suggests that an interpretive study could be useful for indicating reformulated or new variables for use in subsequent positivist studies. The results of this study are consistent with this suggestion. The kind of information that only the interpretive analysis produced, such as feelings, attitudes, and consciousness about the issue, could become the seed questions for future GSS sessions. For example, a follow-up positivist study at State University or a positivist replication of this study at another university could use the coding categories that resulted from the interpretive analysis.

This paper shows the way in which two different stories can be told from the same set of facts. In doing so, it contributes new insights into our understanding about the choice of methods in IS research. The story portrayed by the positivist analysis of the GSS sessions is about four groups of university colleagues who came together for a brief period of time to generate alternative solutions for addressing a highly threatening topic: gender equity. The story told in the interpretive analysis is about a nonrepresentative group of people from a highly contentious university setting and with questionable motivation embarking upon a computer-mediated discussion of gender inequity with perhaps unrealistic expectations.

By focusing on the critical role of research methodology, this study makes a contribution to our understanding of IS, in general, and GSS, in particular. Table 6 illustrated how preconceived rules and instruments for coding utilized in positivist research might (perhaps, incorrectly) conclude that participation level is high, consensus is high, conflict is low, anonymity is high, and usefulness is high, even though there may be convincing pieces of evidence (not fitting the predetermined framework of the positivist researcher) that suggests otherwise. Further, because the context of use is not typically incorporated into the analysis of the transcripts, rich contextual information that is part of the hermeneutic analysis is not taken into account. This study shows how an interpretive analysis can complement the positivist understanding of GSS use.

In a broader sense, this study contributes to our growing understanding of the application of interpretive research methods to IS problems. While interpretive methods have been used to study electronic mail and on-line discussion forums, as noted earlier, they have rarely been used to study computer-mediation in same-time, same-place contexts as was done here. As we move into the world of virtual organizations and electronic commerce, one fruitful line of research would be to extend to virtual groups what has been learned about interpretive analysis of computer-mediated discussions from this study. We believe that studying virtual groups in this way could present new and interesting challenges to both positivist and interpretive traditions and would be valuable to both managers and researchers in understanding the information exchanged in the physical workplace and the emerging virtual workplace.

NOTES

1 Allen Lee was the accepting senior editor for this paper.
2 The characterization of positivism provided in this paper is not a characterization of logical positivism as discussed in the philosophy of science literature (Hempel 1966; Kolakowski 1968) but of positivism-in-practice in the GSS arena.
3 This is not to suggest that all or most of GSS research is necessarily positivistic, narrow, or focused on individual and group phenomenon to the exclusion of organizational context. Indeed, the review of GSS literature provided above attests to the diversity in GSS research. Nevertheless, the positivist approach taken in this research is consistent with the significant body of GSS research that is conducted in similar fashion and was cited above. More important, this approach was selected because it demonstrates how, within the GSS context, one useful research approach and a set of corresponding methods (i.e., positivist) can be complimented with another, different, useful research approach and set of corresponding methods (i.e., interpretive).
4 To preserve anonymity, the actual name of the institution is not used here.
5 OptionLink, by Option Technologies.
6 Seed questions were formulated and agreed upon prior to the sessions by the four-person team and the faculty member with GSS expertise.
7 These two people basically monitored each of the solution-oriented brainstorming sessions and dynamically built-separate lists of the primary, workable ideas for each session. Then, during the brief breaks just after the brainstorming and before the ranking, they edited this list of alternatives and entered it to be used as the basis of the subsequent rankings within each session.

8. By on task, we mean to say that the participants were focused on the task at hand.
9. See Trauth (1997) for further discussion of open coding in IS research and Orlikowski (1993), Trauth (1995, 1996, forthcoming), and Urquhart (1997) for some applications of this method.
10. For an application see Davis et al. (1992).
11. The labels P1, P2, etc. are used to differentiate the participants in a strip.
12. Asterisks are used to separate distinct strips.
13. This is the "Strip 1" to which reference is made in Figure 1.
14. This is the "Strip 2" to which reference is made in Figure 1.
15. See Trauth (1995), Kwan et al. (1985), and Mitroff et al. (1977).
16. This is the "Strip 3" to which reference is made in Figure 1.
17. This is the "Strip 4" to which reference is made in Figure 1.
18. EOE stands for Equal Opportunity Employment.
19. Since the identity of the contributors is not known, it is not possible to say whether this exchange involved two, three, or four different individuals.
20. As noted above, staff members such as this gentleman are considered to be the equivalent of faculty members within the University.
21. Whereas reliability of positivist research is confirmatory—achieving the same results across repeated "experiments" (i.e., all research of a hypothesis testing nature)—the objective of reliability in interpretive research considers the extent to which the observational procedure yields consistent findings.
22. The objective of validity in interpretive research is not to verify a *correct answer* but rather to convince the reader that a *believable story* is being told.
23. These findings are (1) that GSS participants exchanged three types of information: cognitive, emotional, and behavioral; (2) that the GSS sessions were emotionally charged events; and (3) that the participants exhibited widely diverging world views about the problem at hand.

ACKNOWLEDGMENTS

We would like to acknowledge the helpful suggestions of the senior editor, the associate editor, and the reviewers in helping us to shape and improve this manuscript as it evolved.

REFERENCES

Agar, M. H. *Speaking of Ethnography*, Sage Publications, Newbury Park, CA, 1986.

Avison, D. E. and Myers, M. D. "Information Systems and Anthropology: An Anthropological Perspective on IT and Organizational Culture," *Information Technology and People* (8:3), 1995, pp. 43–56.

Boland, R. J. Jr. "Information System Use as a Hermeneutic Process," in *Information Systems Research: Contemporary Approaches and Emergent Traditions*, H.-E. Nissen, H. K. Klein and R. A. Hirschheim (eds.), North-Holland, Amsterdam, 1991, pp. 439–58.

Boland, R. "Phenomenology: A Preferred Approach to Research in Information Systems," in *Research Methods in Information Systems*, E. Mumford, R. A. Hirschheim, G. Fitzgerald and T. WoodHarper (eds.), North-Holland, Amsterdam, 1985, pp. 193–201.

Chidambaram, L. "Relational Development in Computer-Supported Groups," *MIS Quarterly* (20:2), 1996, pp. 143–66.

Chidambaram, L. and Bostrom, R. P. "Evolution of Group Performance Over Time: A Repeated Measures Study of GDSS Effects," *Journal of Organizational Computing* (3:4), 1993, pp. 443–69.

Chidambaram, L., Bostrom, R. P. and Wynne, B. E. "A Longitudinal Study of the Impact of Group Decision Support Systems on Group Development," *Journal of Management Information Systems* (7:3), 1990/91, pp. 7–25.

Connolly, T., Jessup, L. M. and Valacich, J. "Idea Generation Using a GDSS: Effects of Anonymity and Evaluative Tone," *Management Science* (36:6), 1990, pp. 689–703.

Creswell, J. W. *Qualitative Inquiry and Research Design: Choosing among Five Traditions*, Sage, Thousand Oaks, CA, 1998.

Davidson, E. J. "Examining Project History Narratives: An Analytic Approach," in *Information Systems and Qualitative Research*, A. S. Lee, J. Liebenau and J. I. DeGross (eds.), Chapman & Hall, London, 1997, pp. 123–48.

Davies, L. J. "Researching the Organisational Cultural Contexts of Information Systems Strategy: A Case Study of the British Army," in *Information Systems Research: Contemporary Approaches and Emergent Traditions*, H.-E Nissen, H. K. Klein and R. Hirschheim (eds.), North-Holland, Amsterdam, 1991, pp. 145–67.

Davies, L. J. and Nielsen, S. "An Ethnographic Study of Configuration Management and Documentation Practices in an Information Technology Center," in *The Impact of Computer Supported Technology on Information Systems Development*, K. E. Kendall, K. Lyytinen and J. I.

DeGross (eds.), North-Holland, Amsterdam, 1992, pp. 179–92.

Davis, G. B., Lee, A. S., Nickles, K. R., Chatterjee, S., Hartung, R. and Wu, Y. "Diagnosis of an Information System Failure: A Framework and Interpretive Process," *Information and Management* (23), 1992, pp. 293–318.

Dennis, A. R., Heminger, A. R., Nunamaker, J. F. Jr. and Vogel, D. "Bringing Automated Support to Large Groups: The Burr-Brown Experience," *Information and Management* (18:3), 1990, pp. 111–21.

Dennis, A. R. and Valacich, J. S. "Computer Brainstorms: More Heads are Better Than One," *Journal of Applied Psychology* (78:4), 1993, pp. 531–7.

Dennis, A. R. and Valacich, J. S. "Group, Subgroup and Nominal Group Idea Generation: New Rules for a New Media?," *Journal of Management* (20:4), 1994, pp. 723–36.

Dennis, A. R., Valacich, J. S., Connolly, T. and Wynne, B. E. "Process Structuring in Electronic Brainstorming." *Information Systems Research* 7(2), 1996, pp. 268–77.

Dennis, A. R., Valacich, J. S. and Nunamaker J. F. Jr. "An Experimental Investigation of the Effects of Group Size in an Electronic Meeting Environment," *IEEE Transactions on Systems, Man and Cybernetics* (20:5), 1990, pp. 1049–57.

DeSanctis, G. "Shifting Foundations in Group Support System Research.," in *Group Support Systems: New Perspectives*, L. M. Jessup and J. S. Valacich (eds.), Macmillan Publishing Company, New York, 1993, pp. 97–111.

DeSanctis, G., Dickson, G. W., Jackson, B. and Poole, M. S. "Using Computing in the Face-to-Face Meeting: Some Initial Observations from the Texaco-Minnesota Project," presented at the *Annual Meeting of the Academy of Management*, Miami, FL, August 10–14, 1991.

DeSanctis, G. and Gallupe, B. "Group Decision Support Systems: A New Frontier," *Database* (16:2), 1985, pp. 2,10.

DeSanctis, G. and Gallupe, R. B. "A Foundation for the Study of Group Decision Support Systems," *Management Science* (33:5), 1987, pp. 589–609.

DeSanctis, G. and Poole, M. S. "Capturing the Complexity in Advanced Technology Use: Adaptive Structuration Theory," *Organization Science* (5:2), 1994, pp. 121–47.

DeSanctis, G., Poole, M. S., Dickson, G. W. and Jackson, B. M. "An Interpretive Analysis of Team Use of Group Technologies," *Journal of Organizational Computing* (3:1), 1993, pp. 1–29.

DeSanctis, G., Poole, M. S., Lewis, H. and Desharnais, G. "Using Computing in Quality Team Meetings: Some Initial Observations from the IRS-Minnesota Project," *Journal of Management Information Systems* (8:3), 1992, pp. 7–26.

Ely, M., Anzul, M., Friedman, T., Garner, D. and Steinmetz, A. M. *Doing Qualitative Research: Circles Within Circles*, The Farmer Press, New York, 1991.

Fetterman, D. M. *Ethnography: Step by Step* (2nd ed.), Sage Publications, Thousand Oaks, CA, 1998.

Fulk, J. and DeSanctis, G. "Electronic Communication and Changing Organizational Forms," *Organization Science* (6:4), 1995, pp. 1–13.

Gallupe, R. B., Dennis, A. R., Cooper, W. H., Valacich, J. S., Nunamaker J. F. Jr. and Bastianutti, L. "Electronic Brainstorming and Group Size," *Academy of Management Journal* (35:2), 1992, pp. 350–69.

Gallupe, R. B., DeSanctis, G. and Dickson, G. W. "Computer-Based Support for Group Problem-Finding: An Experimental Investigation," *MIS Quarterly* (12:2), 1988, pp. 277–96.

Geertz, C. *The Interpretation of Cultures*, Basic Books, Inc., New York, 1973.

Glaser, B. and Strauss, A. *The Discovery of Grounded Theory*, Aldine Publishing Co., Chicago, 1967.

Golden-Biddle, K. and Locke, K. "Appealing Work: An Investigation of How Ethnographic Texts Convince," *Organization Science* (4:4), 1993, pp. 595–616.

Greene, J. C., Caracelli, V. J. and Graham, W. F. "Towards a Conceptual Framework for Mixed-Methods Evaluation Design," *Educational Evaluation and Policy Analysis* (11), 1989, pp. 255–74.

Harvey, L. "A Discourse on Ethnography," in *Information Systems and Qualitative Research*, A. S. Lee, J. Liebenau and J. I. DeGross (eds.), Chapman & Hall, London, 1997, pp. 207–24.

Harvey, L. and Myers, M. D. "Scholarship and Practice: The Contribution of Ethnographic Research Methods to Bridging the Gap," *Information Technology and People* (8:3), 1995, pp. 13–27.

Hempel, C. G. *Philosophy of Natural Science*, Prentice-Hall, Englewood Cliffs, NJ, 1966.

Hughes, J. A., Randall, D. and Shapiro, D. "Faltering from Ethnography to Design," *ACM*

Conference on Computer-Supported Cooperative Work, ACM Press, New York, 1992, pp. 115–23.

Jessup, L. M. "Group Decision Support Systems: A Need for Behavioral Research," *International Journal of Small Group Research* (3:2), 1987, pp. 139–58.

Jessup, L. M., Connolly, T. and Galegher, J. "The Effects of Anonymity on GDSS Group Process with an Idea-Generating Task," *MIS Quarterly* (14:3), 1990, pp. 312–21.

Jessup, L. M., Egbert, J. L. and Connolly, T. "Understanding Computer-Supported Group Work: The Effects of Interaction Frequency on Group Process and Outcome," *Journal of Research on Computing in Education* (28:2), 1996, pp. 190–208.

Jessup, L. M. and Tansik, D. A. "Group Decision Making in an Automated Environment: The Effects of Anonymity and Proximity with a Group Decision Support System," *Decision Sciences* (22:2), 1991, pp. 266–79.

Jessup, L. M. and Valacich, J. S. *Group Support Systems: New Perspectives*, Macmillan Publishing Company, New York, 1993.

Jick, T. D. "Mixing Qualitative and Quantitative Methods: Triangulation in Action," *Administrative Science Quarterly* (24), December 1979, pp. 602–11.

Kaplan, B. and Duchon, D. "Combining Qualitative and Quantitative Methods in Information Systems Research: A Case Study," *MIS Quarterly* (4), 1988, pp. 571–86.

Kaplan, B. and Maxwell, J. A. "Qualitative Research Methods for Evaluating Computer Information Systems," in *Evaluating Health Care Information Systems: Methods and Applications*, J. G. Anderson, C. E. Aydin and S. J. Jay (eds.), Sage, Thousand Oaks, CA, 1994, pp. 45–68.

Kirk, J. and Miller, M. L. *Reliability and Validity in Qualitative Research*, Sage, Beverly Hills, CA, 1986.

Klein, H. K. and Myers, M. D. "A Set of Principles for Conducting and Evaluating Interpretive Field Studies in Information Systems," *MIS Quarterly* (23:1), 1999, pp. 67–93.

Kolakowski, L. *The Alienation of Reason: A History of Positivist Thought* (1st ed.), N. Guterman (trans.), Doubleday, Garden City, NY, 1968.

Kwan, S. K., Trauth, E. M. and Driehaus, K. C. "Gender Differences and Computing: Students' Assessment of Societal Influences," *Education and Computing* (1:3), September 1985, pp. 187–194.

Lee, A. S. "Electronic Mail as a Medium for Rich Communication: An Empirical Investigation Using Hermeneutic Interpretation," *MIS Quarterly* (18:2), 1994, pp. 143–57.

Lee, A. S. "Integrating Positivist and Interpretive Approaches to Organizational Research," *Organization Science* (2:4), 1991, pp. 342–65.

Lincoln, Y. and Guba, E. *Naturalistic Inquiry*, Sage Publications, Beverly Hills, CA, 1985.

Miles, M. B. and Huberman, A. M. *Qualitative Data Analysis: An Expanded Sourcebook* (2nd ed.), Sage Publications, Thousand Oaks, CA, 1994.

Mitroff, I. I., Jacob, T. and Trauth Moore, E. "On the Shoulders of the Spouses of Scientists," *Social Studies of Science* (7), 1977, pp. 303–27.

Myers, M. "Critical Ethnography in Information Systems," in *Information Systems and Qualitative Research*, A. S. Lee, J. Liebenau and J. I. DeGross (eds.), Chapman & Hall, London, 1997, pp. 276–300.

Myers, M. D. "A Disaster for Everyone to See: An Interpretive Analysis of a Failed IS Project," *Accounting, Management and Information Technologies* (4:4), 1994, pp. 185–201.

Niederman, F. and DeSanctis, G. "The Impact of a Structured-Argument Approach on Group Problem Formulation," *Decision Sciences* (26:4), 1995, pp. 451–74.

Nunamaker, J. F. Jr., Vogel, D., Heminger, A., Martz, B., Grohowski, R. and McGoff, C. "Experiences at IBM with Group Support Systems: A Field Study," *Decision Support Systems* (5:2), 1989, pp. 183–96.

Orlikowski, W. J. "CASE Tools as Organizational Change: Investigating Incremental and Radical Changes in Systems Development," *MIS Quarterly* (17:3), 1993, pp. 309–40.

Orlikowski, W. J. "Integrated Information Environment or Matrix of Control? The Contradictory Implications of Information Technology," *Accounting, Management and Information Technologies* (1:1), 1991, pp. 9–42.

Orlikowski, W. J. "Improvising Organizational Transformation Over Time: A Situated Change Perspective," *Information Systems Research* (7:1), 1996, pp. 63–92

Orlikowski, W. J., Yates, J., Okamura, K. and Fujimoto, M. "Shaping Electronic Communication: The Metastructuring of Technology in the Context of Use," *Organization Science* (6:4), 1995, pp. 423–44.

Ormiston, G. L. and Schrift, A. D. "Editor's Introduction," in *The Hermeneutic Tradition: From Ast* to *Ricoeur*, G. L. Ormiston and A. D. Schrift (eds.), State University of New York Press, Albany, NY, 1990, pp. 1–35.

Prasad, P. "Systems of Meaning: Ethnography as a Methodology for the Study of Information Technologies," in *Information Systems and Qualitative Research*, A. S. Lee, J. Liebenau and J. I. DeGross (eds.), Chapman & Hall, London, 1997, pp. 101–18.

Preston, A. M. "The 'Problem' in and of Management Information Systems," *Accounting, Management and Information Technologies* (1:1), 1991, pp. 43–69.

Rathswohl, E. J. "Applying Don Ihde's Phenomenology of Instrumentation as a Framework for Designing Research in Information Science," in *Information Systems Research: Contemporary Approaches and Emergent Traditions.* H.-E. Nissen, H. K. Klein and R. Hirschheim (eds.), North-Holland, Amsterdam, 1991, pp. 131–44.

Rebstock Williams, S. and Wilson, R. L. "Group Support Systems, Power and Influence in an Organization: A Field Study," *Decision Sciences* (28:4), 1997, pp. 911–37.

Reissman, C. K. *Narrative Analysis*, Sage, Thousand Oaks, CA, 1993.

Sanday, P. R. "The Ethnographic Paradigm(s)," *Administrative Science Quarterly*, (24:4), 1979, pp. 527–38.

Silverman, D. *Interpreting Qualitative Data: Methods for Analyzing Talk, Text and Interaction*, Sage Publications, Thousand Oaks, CA, 1993.

Simonsen, J. and Finn, K. "Using Ethnography in Contextual Design," *Communications of the ACM.* (40:7), 1997, pp. 82–8.

Strauss, A. *Qualitative Analysis for Social Scientists*, Cambridge University Press, New York, 1987.

Trauth, E. M. "Achieving the Research Goal with Qualitative Methods: Lessons Learned Along the Way," in *Information Systems and Qualitative Research*, A. S. Lee, J. Liebenau and J. I. DeGross (eds.), Chapman & Hall, London, 1997, pp. 225–45.

Trauth, E. M. *The Culture of an Information Economy: Influences and Impacts in the Republic of Ireland*, Kluwer Academic Publishers, Dordrecht, The Netherlands, forthcoming.

Trauth, E. M. "Impact of an Imported IT Sector: Lessons from Ireland," in *Information Technology Development and Policy: Theoretical Perspectives and Practical Challenges*, E. M. Roche and M. J. Blaine (eds.), Avebury Publishing Ltd., Aldershot, UK, 1996, pp. 245–61.

Trauth, E. M. "Women in Ireland's Information Industry: Voices from Inside," *Eire-Ireland* (30:3), 1995, pp. 133–50.

Trauth, E. M., Derksen, F. E. J. M. and Mevissen, H. M. J. "The Influence of Societal Factors on the Diffusion of Electronic Data Interchange in The Netherlands," in *Human, Organizational, and Societal Dimensions of Information Systems Development*, D. Avison, J. E. Kendall and J. I. DeGross (eds.), North-Holland, Amsterdam, 1993, pp. 323–35.

Trauth, E. M. and O'Connor, B. "A Study of the Interaction Between Information Technology and Society: An Illustration of Combined Qualitative Research Methods," in *Information Systems Research: Contemporary Approaches and Emergent Traditions.* H.-E. Nissen, H. K. Klein and R. Hirschheim (eds.), North-Holland, Amsterdam, 1991, pp. 131–44.

Urquhart, C. "Exploring Analyst-Client Communication: Using Grounded Theory Techniques to Investigate Interaction in Informal Requirements Gathering," in *Information Systems and Qualitative Research*, A. S. Lee, J. Liebenau and J. I. DeGross (eds.), Chapman & Hall, London, 1997, pp. 149–81.

Valacich, J. S., Dennis, A. R. and Connolly, T. "Idea Generation in Computer-Based Groups: A New Ending to an Old Story," *Organizational Behavior and Human Decision Processes* (57), 1994, pp. 448–67.

Valacich, J. S., Dennis, A. R. and Nunamaker J. F. Jr. "Group Size and Anonymity Effects on Computer-Mediated Idea Generation," *Small Group Research* (2:1), 1992, pp. 49–73.

Valacich, J. S. and Schwenk, C. "Devil's Advocacy and Dialectical Inquiry Effects on Group Decision Making Using Computer-Mediated Versus Verbal Communication," *Organizational Behavior and Human Decision Processes* (63:2), 1995, pp. 158–73.

Valacich, J. S., Wheeler, B. C., Mennecke, B. E. and Wachter, R. "The Effects of Numerical and Logical Group Size on Computer-mediated Idea Generation," *Organizational Behavior and Human Decision Processes* (62:3), 1995, pp. 318–29.

Walsham, G. "The Emergence of Interpretivism in IS Research," *Information Systems Research* (6:4), 1995, pp. 376–94.

Walsham, G. and Sahay, S. "GIS for District-Level Administration in India: Problems and Opportunities," *MIS Quarterly* (23:1), 1999, pp. 39–65.

Watson, R. T., DeSanctis, G. and Poole, M. S. "Using a GDSS to Facilitate Group Consensus: Some Intended and Unintended Consequences," *MIS Quarterly* (12:3), 1988, pp. 463–78.

Wheeler, B. and Valacich, J. S. "Facilitation, GSS, and Training as Sources of Process Restrictiveness and Guidance for Structured Group Decision Making: An Empirical Assessment," *Information Systems Research* (7:4), 1996, pp. 429–50.

Wilson, J. and Jessup, L. M. "A Field Experiment of GSS Anonymity and Group Member Status," *Proceedings of the Twenty-eighth Hawaii International Conference on System Sciences*, IEEE Computer Society Press, Los Alamitos, CA, January 1995.

Yin, R. K. *Case Study Research: Design and Methods*, Sage Publications, Newbury Park, CA, 1989.

Zigurs, I., Poole, S. and DeSanctis, G. L. "A Study of Influence in Computer-Mediated Group Decision Making," *MIS Quarterly* (12:4), pp. 625–44.

ABOUT THE AUTHORS

Eileen M. Trauth is an associate professor of Management Information Systems in the College of Business Administration at Northeastern University. Her teaching and research interests center around the societal, organizational and educational impacts of information technology. Dr Trauth has published recently in the areas of global informatics, information policy, IS education and qualitative research methods. She is on the editorial boards of two international journals and serves as a reviewer for several other journals. Dr Trauth is the author of *Information Literacy: An Introduction to Information Systems* and *The Culture of an Information Economy: Influences and Impacts in the Republic of Ireland*, which documents her ethnographic study of that country's information economy. She has taught and conducted research in several countries and was a Fulbright Scholar in Ireland. Dr Trauth is a member of the Association for Computing Machinery, the Society for Information Management, the International Federation for Information Processing, and the Information Resources Management Association. Dr Trauth received her Ph.D. in Information Science from the University of Pittsburgh.

Leonard M. Jessup is an associate professor of Information Systems and Chairperson of the Technology Committee for the Kelley School of Business at Indiana University. Professor Jessup received his B.A. in Information and Communication Studies in 1983, and his M.B.A. in 1985, from California State University, Chico, where he was voted Outstanding MBA Student. He received his Ph.D. in Organizational Behavior and Management Information Systems from the University of Arizona in 1989. He is a member of the Association for Information Systems, an associate editor for the *MIS Quarterly*, and a member of the Editorial Board for *Small Group Research*. He teaches in various areas of Management and Management Information Systems and has published, presented, and consulted on electronic commerce, computer-supported collaborative work, computer-assisted learning, and related topics. With Joseph S. Valacich, he coedited the book *Group Support Systems: New Perspectives*, Macmillan Publishing Company, and has cowritten the book *Information Systems Foundations*, QUE Education and Training. With his wife, Joy L. Egbert, he won Zenith Data System's annual Masters of Innovation award.

Appendix A

Further Demographics on Session Participants

Three-quarters of the participants reported on the University's ethnicity questionnaire used at the end of the sessions that they were White; the remainder labeled themselves as African American, Hispanic, Native American, and Mexican American. Despite the confidentiality provided by the GSS software, five participants chose not to identify ethnicity on the questionnaires. When we inquired about this, one participant, who was a member of a small ethnic group on campus, expressed fear that people might be able to identify him/her from answers to demographic questions and be able to identify his/her comments. S/he feared potential repercussions.

Seven participants identified themselves on the questionnaire as being either faculty or librarians, 15 participants identified themselves as staff, and 15 identified themselves as administrators. At this university, faculty and librarians are considered equal and, as a result, were grouped together on the university's survey instrument. Three participants chose not to identify their job titles on the questionnaire. The senior administration was well represented in the sessions: the President, vice presidents, and directors all attended the sessions. Thus, the group composition in each session spanned the organizational hierarchy, with participants in each session interacting directly with people both at their own and at other levels. In most cases, faculty/staff participants were in sessions with the supervisors and/or administrators for whom they worked. Because the university was so small, employees knew each other well. Participants in the sessions knew each other, referred to each other by name before and after the GSS sessions, and could identify the participants with organizational power and authority.

Participation in Groupware-mediated Communities of Practice: A Socio-political Analysis of Knowledge Working

N. Hayes and G. Walsham

Abstract

This paper adopts a communities of practice approach to examine how the introduction of a groupware application in a UK pharmaceuticals company enabled and constrained knowledge working. We will refine the analysis by distinguishing between participation that is undertaken in what is referred to as political enclaves, and participation that takes place in safe enclaves. We will discuss how the deliberate intervention of some employees moderated some of politicising, and facilitated increased participation. The paper concludes by suggesting ways in which existing theoretical conceptualisations of information systems and knowledge work may be expanded to consider socio-political issues in more depth. © 2001 Elsevier Science Ltd. All rights reserved.

Keywords: groupware; knowledge management; communities of practice; participation; politics; norms; case study; safe and political enclaves

1. Introduction

Commentators on contemporary themes of organising have suggested that organisations have increasingly become dependent on the exercise of specialist resources and on workers that ply their trade through their cognitive abilities and their specialist knowledge (Blackler 1993, 1995). Reich (1991) terms these "symbolic analytical workers", whose intellective abilities are varied, difficult to duplicate and who frequently command high rewards. Several writers argue that many "expert dependent" organisations are increasingly going through a shift in becoming communication intensive (Barley, 1996; Blackler, Crump, and McDonald, 1998). This shift, in part, is attributed to the emergence of ubiquitous, low cost distributed technology (Blackler, Crump, and McDonald, 1997; Ruhleder, 1995). Zuboff (1996) concurs with this view, by noting that as technologies, such as groupware systems, become more ubiquitous, they are: "fully imbuing tasks of every sort and providing ever more powerful opportunities for the kind of learning that translates into value creation". Indeed, some scholars view knowledge work as being inseparable from the development of contemporary technologies (Knights, Murray, and Willmott, 1993; Ruhleder, 1995; Star and Ruhleder, 1994).

One important approach to the study of knowledge working is the communities of practice approach to learning, which argues that we can not separate knowledge from practice (Brown and Duguid, 1991; Blackler, 1995; Lave and Wenger,

1991). Dominant accounts in the knowledge work literature portray knowledge as being an entity that can be possessed and traded (Bell 1973, 1978; Nonaka and Takeuchi, 1994). In contrast to this Brown and Duguid (1991) explain that the community of practice approach: "draws attention away from abstract knowledge and cranial processes and situates it in the practices and communities in which knowledge takes on significance". They suggest that what is learned is highly dependent on the context that the learning takes place in. Writers from this tradition suggest that this requires looking at the actual practice of work, which consists of a myriad of fine-grained improvisations that are unnoticed in any formal mapping of work tasks. Consequently, rather than referring to "knowledge", "knowledge work" or "knowledge management", Blackler et al. (1998) argue that we should move to a study of "knowing" or "knowledge working". These are not merely semantic differences, but represent the view that knowledge working is an active process, where employees from different domains of expertise engage in collaborative endeavours as they seek to utilise their different histories and experiences.

A limited number of studies have considered the role of information technology from a communities of practice perspective. Boland and Tenkasi (1995) drew on these foundations to theorise different opportunities that communication technologies present to knowledge working. They argued that communication forums may provide a mechanism that will allow community identity to be strengthened, whilst also helping specialised knowledge workers to make sense of other community perspectives (Boland, Tenkasi, and Te'eni, 1994). Brown (1998) provided an insightful account of the use of the Internet to support knowledge working and argued that a reliance on technology as a means of transferring knowledge is insufficient. Instead he contended that abstractions recorded and shared on the Internet need to be considered as being inseparable from their own historical and social locations of practice. He suggested that the challenge lies in: "getting the formal organisation and the informal one to work together wherever possible". Hayes (2001) considered the opportunities and limitations that surrounded the use of groupware to support knowledge working within and between temporal and spatial boundaries. He further considered how the improvised activities that some employees in a UK company undertook, provided opportunities to work around some of the temporal and spatial limitations the groupware technology presented.

Collectively, studies such as these, together with Lave and Wenger's (1991) analysis, suggest that a knowledge–practice separation is unsound, both in theory and in practice. Further to these studies of work-a-day practice, other accounts have emerged which view power/knowledge as also being inseparable. Critical writers such as Alvesson (1993) suggest that the notion of knowledge intensive firms is no more than an institutionalised myth, which seeks to ensure employee conformity with the institutionalised expectations of their environments (Willmott, 1995). Within the communities of practice school, Lave (1993) points out that claims to decontextualised knowledge are often no more than a "power play" as experts stake out their claims to be respected against the rival claims of other interested groups. Blackler et al. (1998) also point out that the relationships between culture, knowledge and power are impossible to disentangle.

Other relational writers have begun to explore issues surrounding power/knowledge and information communication technologies (ICTs). Hayes and Walsham (2000b) discussed how the political and normative context influenced the nature of discussions and interactions in differing electronic communication forums. Newell, Scarborough, Swan, and Hislop (2000) found that the introduction of communication forums reinforced the "powerful centrifugal forces operating as the strategic development of the firm". They suggested that this was prohibitive to knowledge working. McKinlay (2000) suggests that though databases can capture "knowledge bytes", they can not appropriate the ephemeral social processes that constitute actual practice. He argues that where surveillance and control is extended into the latter area, then this will "fatally compromise knowledge management as a project".

Emerging from this brief literature review are two key observations. First, there have been few interpretive studies that have explicitly considered technology and knowledge working from a communities of practice perspective. Second, these accounts have not normally considered how the work-a-day use of IT shapes and is shaped by the political and normative context. This paper will consider these two underdeveloped themes by exploring how the political and normative context influences the nature of predominantly (but not exclusively) groupware-mediated participation within the UK selling division of multinational pharmaceuticals company. Compound UK (the selling division) introduced Lotus Notes, a leading groupware product, to assist employees to work more qualitatively within and between functions. Employees used Lotus Notes to share views and perspectives with members of their own and other functions. Specifically, this paper will consider how the political and normative context largely constrained the nature of participation within and between functions that surrounded the use of Notes, and how, through their actual practices, some employees brought about changes to the political and normative context, which presented opportunities for improved participation. To do this, we will draw on Lave and Wenger's (1991) communities of practice approach to knowledge working, and Hayes and Walsham's (2000b) conceptual distinction between safe and political enclaves. Finally, we will critique and extend several of Lave and Wenger's (1991) concepts within their theorisation of situated learning, and specifically in relation to information technology.

The following section will develop the theoretical basis of the paper. A brief description of the methodology of the field study will then be given. Section 4 will introduce Compound UK. Section 5 will then discuss how the visibility that was made possible through Lotus Notes influenced the political character of interaction between employees located in the same and in separate disciplines. The subsequent section will analyse how some shared forums were considered to be safe, and further to this, will examine how the deliberate intervention of some employees moderated some of the limitations to effective knowledge working. The final section will discuss the implications and conclusions arising from this study, and reflect upon the theoretical basis of the paper.

2. Conceptualising the Role of Groupware Technologies in the knowledge Work Process

The communities of practice school is largely attributed to the work of Brown and Duguid (1991) and Lave and Wenger (1991). Table 1 lists the key

Concept	Description
Learning as being situated	Learning as an integral and inseparable aspect of social practice
Full participation	Full participation allows newcomers to acquire the ability to behave as community members
Peripheral participation	Allows access to observe and participate in a community's practices
Legitimate participation	Participation is viewed as being normatively permissible

Table 1 Key aspets of legitimate peripheral participation.

aspects of Lave and Wenger's approach to conceptualising knowing and learning. Firstly, *learning is viewed as being a situated* and improvised activity, which as Brown (1998) argues: "is profoundly connected to the conditions in which it is learned". Lave and Wenger (1991:31) view learning to be an integral and inseparable aspect of social practice. Based on this situated view of learning, Lave and Wenger developed the theoretical construct of Legitimate Peripheral Participation.

Opportunities for learning from the viewpoint of legitimate peripheral participation arise when learners or newcomers are given access to *fully participate* in the sociocultural practices performed by particular communities. By fully participating, learners acquire the embodied ability to behave as community members. However, Lave and Wenger (1991:36) warn that newcomers need to see the value of them becoming full practitioners in order for them to participate. They do not view full participation as being ever a complete and closed domain of knowledge that newcomers may acquire, instead suggesting that it is intended to do justice to the "diversity of relations involved in varying forms of community membership". When full participation comes about, Lave and Wenger (1991:122) argue that: "the *person* has been correspondingly transformed into a practitioner, a newcomer becoming an old-timer, whose changing knowledge, skill, and discourse are part of a developing identity —in short, a member of a community of practice".

Peripheral participation is concerned with connecting to the sources of understanding through greater access being gained by the practices of members of a community. Having access to the periphery of a community of practice allows newcomers to observe and participate in the practices of that community which is vital if they are to become insiders and full participants. Several authors from the communities of practice tradition explain that technological support is available (such as e-mail and groupware) which allows learners opportunities to observe and participate, often across spatial and temporal divides (Boland and Tenkasi, 1995; Brown and Duguid, 1991; Hayes, 2001).

The final component of Lave and Wenger's conceptualisation of learning relates to *legitimate participation*. Legitimate participation refers to whether access to participation in community groups is normatively allowed, limited or prohibited. They suggest that due to the power relations within specific contexts, learners may either be permitted or kept from observing and participating intensively in a community's practices (Lave and Wenger, 1991:36). They argue that the former is empowering while the latter is disempowering to learning.

Though Lave and Wenger focus on participation in communities of practice, their conception of learning does not explicitly consider how the political and normative context influences learning. To further develop the analysis of how participation in communities of practice shapes and is shaped by the normative context in which learning takes place, this paper draws on Hayes and Walsham's (2000b) concepts of political and safe enclaves which have been specifically developed in relation to knowledge working. Their concepts derive from Goffman's (1959) influential work on the dramaturgical analysis of social action, which sought to understand everyday social intercourse in terms of the crafting of theatrical performances. Goffman suggested that within establishments, during periods of co-presence: "In their capacity as performers, individuals will be concerned with maintaining the impression that they are living up to the many standards by which they and their products are judged".

Hayes and Walshham (2000b) develop the concept of political enclaves as akin to what Goffman (1959) terms "front regions", which refers to that part of the performance that is visible to an audience. Political enclaves in this study are portrayed as being used as a resource by all politically orientated actors who seek to further their own agenda. Safe enclaves are akin to what Goffman terms "back regions" which is where the performance of a routine is prepared. Access to these regions is controlled in order to prevent the audience from seeing back stage, and to prevent an outsider from coming into a performance that is not addressed to them. Hayes and Walsham (2000b) suggest that in safe enclaves, employees feel able to express their own underlying views of an activity, and are open to discussion and reflection surrounding on-going activities and events. However, they warn that the distinction between safe and political enclaves refers to the character of the use of the shared databases and other encounters between employees, but that safe enclaves are political in so far as they are shared social spaces. For example, opting out of political enclaves in preference to safer ones is itself a political act.

To summarise our theoretical perspective, then, we are extending Lave and Wenger's theory of participation in and between communities of practice by considering explicitly the socio-political context through the concepts of safe and political enclaves. However, although we are building on Lave and Wenger's work as a theoretical basis, in the case material that follows we will not use all of their terms as summarised in Table 1, since we believe that they are problematical in a number of respects. We will reflect on this theoretical point in the final section of the paper.

3. Methodology

The primary research in Compound UK was carried out in two phases over a two and a half-year period. The first phase of fieldwork was undertaken between October 1995 and February 1996. During this phase, 33 in-depth interviews were carried out, lasting from 1 to 3 h each, as well as considerable informal interaction. Between February 1997 and May 1997, 21 follow up interviews were conducted in order to try to understand the use of co-operative systems over time, and to provide the longitudinal element that is seen as highly desirable in tracing changing attitudes and actions over time.

In the first phase, interviews were relatively structured. They sought to glean the changing perceptions and work practices that had emerged as a result of the introduction of Lotus Notes. As an increased awareness about both Compound UK's operations and a provisional understanding of the emerging issues was gathered, the interview questions were not followed as rigidly as they were at the outset of the research. In the second phase of interviews, any emerging themes and issues that remained unresolved from phase one were pursued in a more unstructured way. This flexibility allowed for the modification of the research design in the light of emergent or unanticipated analytical problems thrown up by the context or the data (Layder, 1993). In addition, interviews were confidential and conducted in private. The initial part of the interview would be spent explaining the identity and purpose of the researcher(s), and reassuring interviewees that no attribution would be given to their views in any subsequent discussion or reports. Relatedly, detailed field notes were preferred to the use of a tape recorder, as it was thought that tape recording would have led to less candid responses.

Interviews were supplemented by social interactions in the cafeteria and during drinks in the evening with employees. The aim of this social interaction was to gain a feel for what it is like for the people in the situation being studied. In particular, by undertaking these social interactions at every possible moment, it allowed a way to compensate for any limitations arising from undertaking interviews. These extensive interactions were intended to further reveal a "rich under life" that is usually seen as being masked to quantitative researchers (Geertz, 1973), and to those qualitative researchers that place their emphasis on solely undertaking interviews.

4. Compound UK

Compound UK (a pseudonym) is the UK selling division of Compound Pharmaceuticals International. It has two key roles. First it is concerned primarily with selling products to hospitals and general medical practices within the UK. Secondly, it undertakes clinical trials of the new drugs that have been developed, and require statutory testing of their safety. Consequently, medical experts liaise with participating doctors and patients in Great Britain and Northern Ireland to test the efficacy and effectiveness of new drug development. The selling division had undergone considerable change in response to reforms in the UK health care sector, as outlined below.

During the late 1980s there was an acceleration in the reform of the UK National Health Service (NHS) (Connah and Pearson, 1991). Government controls on public expenditure meant that for a number of years the health care budget had not kept pace with inflation, and thus had reduced in real terms. To try and achieve cost savings, the NHS attempted to mirror market principles by introducing an internal market place (Flynn and Williams, 1997; Robinson and Le Grand, 1994). This led the NHS to rethink the purchase of pharmaceuticals products; the criteria for purchasing such products no longer concentrated solely on their efficacy, but also on their cost and efficiency.

The introduction of these market reforms split the health care sector between primary care and specialist care. The primary care sector covers general medical practices, while the specialist care sector covers hospital markets. Since the reforms, many primary care doctors have become fundholders. They have budgetary responsibilities for drugs, hospital referrals, staff as well as for their fixed costs. Hospitals are also more autonomous from the Department of Health, and are responsible for their own budgets. As a consequence, specialist care doctors are part of a large group of decision-makers, including managers and accountants. From the point of view of Compound UK, at the time of the research study, not only had the criteria for purchasing pharmaceuticals products added cost savings to each drug's efficacy, sales situations had also become far more complex, since client groups such as hospitals now included a wider range of actors in their purchasing decisions.

An outline organisational structure for Compound UK is shown in Fig. 1. Much of this paper focuses on the commercial function. In 1993, this was restructured into eight regions. It was thought that this would provide each region with considerable autonomy to plan and respond to their own locality, and thus to make the organisation more responsive to the new market place. In 1996, the commercial function consisted of its director, Tom Saunders, eight regional managers, twelve area managers, and around 150 sales representatives (reps). All members of the commercial function, apart from the director,

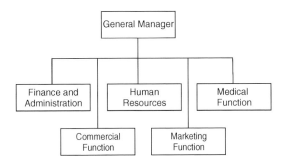

Figure 1 Organisational structure of Compound UK.

worked from their own homes, while employees working in other departments were located at the head office, Compound Square.

The other functions shown in Fig. 1 remained largely unchanged from the ways they were organised prior to the restructuring of the commercial function, apart from the clinical and medical function. Staff in this latter function consisted of the medical director, medical advisors and health economists. They undertook the UK component of Compound Pharmaceuticals International's clinical trials that are part of an international clinical trials programme. In addition to these traditional responsibilities, after the restructuring the medical staff also started to work more closely with marketers and the commercial function, to help assist in developing a broader awareness of medical issues. They also visited doctors with reps to assist in the selling process.

As part of the response to the environmental changes in Compound UK, senior management felt that the selling division could become more competitive by encouraging employees to draw on all areas of the organisation to work and share information and knowledge. Lotus Notes was seen as a software package that could assist with sharing information and improving group working. Notes was introduced about six months after the restructuring in 1993 to facilitate knowledge sharing across several boundaries. First and foremost, Notes was introduced to bridge the functional boundaries in Compound UK. The intention was that Notes would allow employees from different professional backgrounds, who had previously worked in isolation from each other, to be able to share information, views and experiences with each other regardless of their disciplinary background. Secondly, and vital for this multidisciplinary initiative to be possible, Notes was introduced to bridge the spatial boundary that existed between the field force and those located in the Compound Square, and to cross the temporal boundaries between the field force and Compound Square. Often, reps were unavailable during the day as they were travelling to visit doctors and other stakeholders, while employees located in Compound Square would typically work between nine and five. Notes allowed reps, for example, to review and respond to their electronic mail before they left home in the morning or when they returned in the evening.

In addition to the electronic mail (e-mail) facility, there were three main uses of Notes after its introduction in Compound UK, as outlined in Table 2. First, databases to support the co-operative activities involved in strategic selling were created. These databases allowed employees to input their views and information in a structured way with the aim of bringing together the employees' shared knowledge so that they might contribute to a successful sale. Issues were recorded on electronic strategic selling sheets. The strategic selling database contained all the current and previous strategic selling sheets that related to a particular doctor, general practice or hospital.

A further use of Notes was the provision of a wide variety of discussion databases that focused

Lotus notes component	Intended opportunity for participation
E-mail	To enable one-to-one communication between individuals
Strategic selling	To enable employees in different functions to input their views and information in a structured way with the aim of bringing together the employees' shared knowledge so that they might contribute to a successful sale
Discussion databases	To enable employees to review the thread of discussions that had emerged on a particular issue
Contact recording	To enable employees to record and review the views, interests and requirements of particular doctors

Table 2 Key uses of Lotus notes.

on issues, products or a particular role. The database allowed employees to review the thread of discussions that had emerged on a particular issue. The final and most prevalent use of Notes was the contact recording database. This database provided a shared resource for employees to record details of customers. This is a widely used practice in selling companies, and had been present in various paper and electronic forms in Compound UK since the early 1970s. The contact recording database allowed employees to record the views, interests and requirements of particular doctors. This information was then available to any member of the organisation to review and expand before and after they contacted a particular doctor.

5. Political Enclaves

This section draws on the concept of political enclaves, and the case material, to examine how the visibility that Notes provided was implicated in limiting participation within and between geographic, spatial and specifically functional boundaries. The first Section 5.1 will examine how the transparency of the contact recording and strategic selling databases provided a resource for senior managers to co-ordinate and control the participation of employees, as had happened with the previous contact recording system. The second Section 5.2 will consider how the transparency also allowed more junior employees to politicise. Politicising took the form of employees drawing on the forum that Notes provided them, to portray themselves in a way that they thought senior managers would approve of. The final Section 5.3 will discuss how and why some employees felt normatively bound to participate in some databases, while free to choose if they would participate or not in others.

5.1. Extended capacity to monitor and control through the contact recording and strategic selling databases

As with Jaguar, the previous contact recording system, Lotus Notes provided the facility to monitor the number of contacts that reps made with doctors. However, Jaguar had merely provided a facility to input the number of visits individual reps had made to specific doctors. In contrast, Notes provided far more scope to record and share details about each visit a rep or other employee made to a particular doctor or other stakeholder. Not long after Notes was introduced, the commercial director instructed the Notes developer to devise what were referred to by most employees as a "league table" which could indicate centrally how many contact records and strategic selling sheets had been completed by each sales representative. The league table was a simple two by two matrix. On the left-hand side it listed the rep's name, and on the right the number of contacts the rep had recorded within a specific time scale (usually a month). Reps would be listed in descending order, with the highest number of contacts at the top, and the lowest at the bottom. Once the contact records had been collated centrally in the form of contact recording league tables, Robert Cross, the strategic selling manager, would send out electronic messages to field force managers to inform them about their reps' position in the league tables, and the number of strategic selling sheets that each rep had completed. Area managers felt obliged to inform some members of their sales team that their low contact recording rates had been noticed centrally. The creation of the league table confirmed many of the reps' suspicions about the intention behind the electronic contact recording system, namely to increase central surveillance and control.

Contact recording on Lotus Notes not only allowed reps to represent the number of calls they made, it also had the scope to record employees' observations and comments. Some senior managers would regularly review not only the league tables, but also the detailed comments and observations that reps had recorded, so that they could gain an insight into how their products were faring. This meant that with reps recording not only the number of visits, but also the content of their visits, the storage capacity of Lotus Notes expanded the degree of authoritative resources that senior managers could draw on to sanction reps from that which existed with Jaguar, the previous contact recording system.

Many of the more cynical reps were negative about the use senior managers made of the details stored on Lotus Notes, as is exemplified by one primary care rep:

> The computer allows them a handle on what we do. How would they have known that I

was sat in a hotel meeting you before they introduced Notes? Now I will put it into my contact recording database.

The introduction of the strategic selling database similarly extended the ability for senior managers to exercise tight control over sales reps. All employees, but most notably the sales reps, were required to record details on the shared database, to structure their visits, and to draw on the expertise of other employees. Strategic selling sheets represented a simplification of the selling process; they provided a way to visibly structure the complex work of the reps. By recording details on the strategic selling database, reps could prove to senior management that they were planning their more difficult sales in a way that senior management advocated. As one area manager mentioned:

> Robert Cross looks randomly at everyone's strategic selling sheets to check up on the quality of them. Robert will send me a note saying that the reps use of the strategic selling sheets was either appropriate or not appropriate! Sometimes he will write more detailed comments.

Prior to the introduction of Notes, there was more reliance on area managers to supervise and control much of the reps' work due to the limited awareness that those in Compound Square had of a rep's day-to-day activities. The only direct information that senior managers could draw on to monitor the reps' activities were the contact rates recorded on Jaguar. However, Notes allowed senior management in Compound Square to co-ordinate more of the reps' day-to-day activities. Consequently, the authority of senior management was increased and the autonomy of the field force management was reduced.

5.2. The career reward structure

Some reps welcomed the reintroduction of contact recording and the extended surveillance capacity, and harnessed this increased visibility for their own individual purposes. In relation to the league table, many reps were competitive with each other about the league position they attained. They were aware that senior managers reviewed the overall rankings, and saw this as a means to be noticed by them in the hope that this would further their career aspirations. They would record all the doctors, managers, accountants and even nurses that they met in general practices and hospitals. This went contrary to the espoused ethos that arose in tandem with the restructuring and the introduction of Lotus Notes that emphasised only recording "relevant" calls. Many reps that were not career orientated were quite resentful of some other sales reps and of those working in the Compound Square, who they saw as using the Notes to try to further their careers. Non-careerist members of the field force were generally those that had been a sales representative or area manager for some time, and saw themselves as professional sales people. In contrast, ambitious reps tended to be younger, inexperienced and saw being a sales rep as a stepping stone to gaining a job in the head office of Compound UK or Compound International (the parent company).

Professional reps were generally favourable to the move away from only recording the quantity of calls, and saw this approach to working as being appropriate to the new health care environment. However, they feared that as a result of the activities of the ambitious reps, senior management might revert back to explicitly directing reps to record high contact rates. As one experienced rep noted:

> Younger people just see reping as temporary, as a way to pass through to the head office. They think they can put up with it for a few years, and are quite happy to input loads of contacts, and work flat out; this has an impact on the long term reps, who have to keep up with them.

Many ambitious reps would seek to gain favour with Robert Cross, Tom Saunders and Martin Garratt (the general manager), by working on a considerable number of strategic selling sheets at any moment in time. As with the contact recording database, many of the reps who were not seeking promotion felt frustrated with their peers, as they saw them legitimating the extensive use of strategic selling, regardless of whether they had any complex sales accounts in their territory at that moment in time or not.

This was particularly the case for primary care reps, who did not see the benefit of having too many strategic selling sheets open at any one moment in time. Primary care was not seen to consist of as many stakeholders as specialist care and, in effect, primary care reps only had one or two general practices that could be classified as complex sales situations within their territory. Thus, with many primary care reps keeping active a considerable number of strategic selling sheets, experienced reps resented this, feeling that this was being done to gain favour with the strategic selling manager who reviewed how well or how poorly the rep had used the strategic selling database. As one rep mentioned:

> There are not that many complex sales situations in my area, and I only need to keep active a few strategic selling accounts at a time, unlike some of the shining stars!

The increased transparency of both senior managers and more junior employees that Notes provided influenced the nature of participation within Compound UK. Senior managers were motivated to use Notes so that they could monitor and control the activities of junior employees, particularly those that worked from their own homes throughout the UK. Ambitious employees were motivated to participate in the different components of Notes so as to indicate to senior managers that they were working hard and in the ways that they advocated. Due to the controlling and the politicising forms of participation undertaken by many employees, the use of Notes merely reinforced the dominance of the senior management perspective. This hindered expansive dialogue and participation between members located in both the same and different communities.

5.3. Participation and non-participation in the discussion databases

The use made of the discussion databases that had a national and cross-functional audience was similar to the contact recording and strategic selling databases. Politically orientated employees were aware that the national discussion databases were reviewed and contributed to by many senior managers, including the commercial director and the general manager. Those employees who wanted their comments to be noticed by senior managers felt that contributing to these databases would be beneficial to their career development. As one rep mentioned:

> There is a political element definitely. People hijack databases to make political statements. They want to be seen and heard. The national discussion databases are where the political animals are to be found.

As a result of the politicising, many employees that were not seeking career progression did not make use of the national discussion databases. They did not see any benefit from participating in discussions that were dominated by what one rep described as: "ambitious people competing with each other about who could shout the loudest". In addition, due to them feeling normatively obliged to include themselves in a minimal use of the contact recording and strategic selling databases, they had little time or inclination to pursue the national discussion databases. Also, many employees were fearful that any non-consensus forming contributions that they made to the cross-community discussion databases might be misunderstood or seen as irrelevant. By not contributing, or only contributing in a limited way, this meant that the nature of debate merely reaffirmed the legitimacy of the views of senior management and the work practices of ambitious reps. Consequently, the dominance of these consensus-forming views recorded on Notes did not assist others with different expertise to recognise and accept how their own community's views and conventions differed to those recorded by members of other functions.

6. Safe Enclaves

This section will contrast the limitations reported in the previous section with the positive opportunities that the use of Notes and other forums presented to knowledge working. The first Section 6.1 will explore how several databases were seen as being non-political, and how the participation in these communication forums was deemed to be optional. The second Section 6.2 will discuss how the deliberate intervention of some employees moderated some of the limitations to effective knowledge working arising from the visibility and politicisation of activities reported earlier.

6.1. Non-career oriented participation

Though there were no company-wide databases that were deemed safe, some non-careerist employees did optionally include themselves in the functional or regional specific shared databases that were viewed by, and contributed to, exclusively by members of their own community. For example, several of the regional and role specific databases were used extensively to express and share views and comments between colleagues in the same function. As one rep described:

> The regional databases are not the same (as the national ones). They are about local and regional issues, and are more concerned with sharing than anything else.

On community-specific safe enclaves, they would discuss the ways that they had approached sales situations and other areas of interest that they shared. In addition, those employees that had experience of working with members of other functions would discuss and provide advice to less experienced members of their own community, on how they could best interact with members of other functions. Thus, unlike the national discussion databases, these community specific databases were perceived to be safe to express underlying views, and allowed for some discussion and reflection to occur about how they went about their work. In this sense, the region specific safe enclaves allowed reps the most comprehensive and in-depth forms of participation associated with Notes.

However, the participation that regional databases afforded was not harnessed uniformly. Some ambitious reps optionally excluded themselves from the use of the regional or role specific databases, as they did not see them as being beneficial to their careers and, as such, saw their time being better spent making use of the national rather than the function or role specific databases. In addition, due to the resentment of these more experienced reps for feeling normatively bound to use the non-optional political enclaves, such as the contact recording databases, and the additional workload this presented, many did not use these safe enclaves extensively.

6.2. Mediating the political and normative context

Though the pervasiveness of political enclaves was largely restrictive to engendering widespread participation in community wide enclaves, the actions of some area managers and medical staff assisted in working around these limitations. These strategies will be outlined and discussed in four subsections below. It is important to note that the activities discussed below could be regarded as the exception rather than the rule.

6.2.1. Managing competing agendas

Some area managers and experienced reps particularly welcomed the new emphasis on sharing perspectives within and across professional boundaries, as they felt that the increasing complexity of the UK health care sector required an approach that focused on detailed and cross-functional selling approaches. As we have seen earlier, they did not view the recording of considerable numbers of contacts, particularly in primary care, as being relevant to the current health care climate. As a way of trying to help their reps deal with this issue, many area managers suggested to their primary care reps that they need only complete enough contact records, and keep a satisfactory amount of strategic selling sheets active, so as to not attract the negative attention of senior managers in Compound Square. One area manager discussed this as follows:

> Primary care is not complex, and all our products are old and have been in place for a while. We tell them just to keep the contacts and strategic selling sheets ticking over so they are not noticed by head office, why should they waste their time on the computer, when they should be out selling drugs? In specialist care we do encourage them to have a few more strategic selling sheets, as it is more useful in hospitals.

As the quotation above indicates, area managers saw strategic selling sheets as being more relevant to specialist care, but even within this sector, would suggest to their reps that they need only complete as many strategic selling sheets as there were complex sales.

The safe-guarding activities that several area managers undertook mitigated some of the workload effects on reps that arose from them feeling compelled to make at least a minimal use of political enclaves. This provided some additional scope for experienced reps to be able to undertake the qualitative forms of working they favoured with colleagues within their own and other communities. These safe-guarding activities indicated to such reps that it was legitimate to use Notes to exchange subjective viewpoints which are vital to learning from a communities of practice perspective (Brown and Duguid, 1991).

6.2.2. Evading the strictures of tight control

Some experienced reps found a somewhat subversive way of representing to senior managers the futility they associated with the league tables. Some reps refused to input any contacts that were not directly beneficial to their work, others would go back to a contact record and only input a line, or even a "full stop". For quite some time, inputting the bare bones of a contact would satisfy the strategic selling manager, who monitored the rep's use of Notes, as it would register that a new contact had been made. This issue was exemplified by one area manager who explained:

> The numbers that they generate in the league tables do not reflect what level of detail someone has put in. You can enter a comma onto one contact record and nothing else and this will be judged to be active and recorded next to your name, but if they want to move back to number generating then what else do they expect?

It was only towards the end of the research study that the strategic selling manager became aware of these activities. When the strategic selling manager did notice, he adopted a slightly different strategy, only contacting a rep's area manager if the number of contact records and strategic selling sheets were "significantly low", or if they were not completed in enough detail. It was then left to the field force managers to act upon this. Quite often, area managers would not inform their reps of the strategic selling manager's observations, as they thought it might be de-motivating.

When area managers were eventually questioned by the strategic selling manager about the limited use many of their reps made of the contact recording and strategic selling databases, they would explain their reps' predicament, and the sense of futility they associated with strategic selling and particularly contact recording. The subversive nature of the reps' activities and the support they received from their area managers contributed to some destabilisation of senior managers' views about how employees should go about their work. This contributed to non-careerist reps feeling less under pressure to include themselves in the contact recording and strategic selling databases than they had previously done. Coupled with the extensive debate that these subversive activities initiated about which activities should be monitored, this allowed more scope for reps who were seeking genuine forms of collaboration to develop in-depth relationships with members of their own and other functions who shared similar wishes for participation. Though these instances were few, they did present positive opportunities for experts to be able to understand and develop more coherent perspectives of their own communities.

6.2.3. Developing and maintaining in-depth relationships with members from other functions by alternative means.

Some employees actively pursued alternative approaches to the development and maintenance of relationships with other experts within and between functions after the introduction of Notes. Area managers had been one of the few groups of employees who had worked across functional boundaries prior to the introduction of Notes, and they maintained this role after its introduction. They drew on some of the opportunities that the different components of Notes provided for bridging the temporal, spatial and functional divides, while in addition they continued to meet and talk to other experts at regional meetings, and at meetings in Compound Square. They also reinforced these relationships with regular telephone calls. This stood them in good stead to work cross-functionally themselves, and to encourage and assist others to engage in work that crossed professional boundaries after the restructuring.

Some employees working in Compound Square also tempered their reliance on the use of Notes to work between professional groups that were separated in space and often time. Most notable amongst these were medical staff working in the clinical and medical function who previously had little experience of working with members of other functions prior to the introduction of Notes. However, after its introduction, many such medics were expected to work with other employees on complex sales situations and the development of new products. James Black, a senior medical advisor, took this on board more than most and supported this new ethos wholeheartedly. Black consciously harnessed every opportunity he could to meet with other employees, rather than relying on the use of Notes to work across spatial, temporal and professional boundaries, and saw this as a way to be supportive of generating familiarity and understanding between experts working across different functions. He had the view that, by developing such relationships, they could move away from the politicising that he saw as pervading the use of Notes, to genuine working across functional boundaries.

One example of how Black's activities brought about changes in ways of working arose from his involvement during induction and training courses for new reps. Reps joining Compound UK underwent a six week training course in Compound Square before they were allocated their own sales territories and went out on the road themselves. Black attended many of these training sessions, and he viewed them as an opportunity to introduce himself, outline what he did, what he required, and to socialise formally and informally with reps over lunch and in the evenings. He attended as many regional meetings as possible, as well as the national reps conferences that took place bi-annually. Black, and other medical staff with similar views, accompanied reps on their visits to doctors as often as possible. As Black mentioned:

> I make it my business to meet reps on their training courses and at regional meetings. I do this so they feel confident enough to call me if they need me. I also try to go out and meet doctors with reps whenever possible.

This was very much welcomed by the reps. It made them comfortable to call Black and other medics to ask for assistance, as well as providing medics with a deeper appreciation as to what was involved in being a rep, and vice-versa. Black explained how, as reps became more familiar with him, he had noticed that they would increasingly send him e-mails, complete a considerable amount of detail on the strategic selling and contact recording databases, and even after a particular account had been won or lost, would e-mail him and ask for advice. It opened up a forum of discussion surrounding the assumptions and perspectives of experts from different functions. Further to this, Black explained how he would get the people he worked with, to see if they were careerist or were genuine in their intentions for interaction. He was aware that some ambitious people wanted to form a relationship with him as they knew he was senior, and thought their careers might benefit from an association with him.

Within the Compound Square, Black also adopted strategies to encourage cross-functional working. He was critical of how, after the introduction of Notes, many employees had "hidden behind their screens" which had resulted in a reduction in the number of face-to-face meetings in Compound Square. He explained this vividly:

> They can have sent a message to someone in the morning, have sat next to them at lunch not knowing who they are, and then send them another note in the afternoon asking if they could act on their note!

He made a conscious decision to try to change his strategy in dealing with these issues. If people from Compound Square sent him a lengthy e-mail, he would send a reply back to other experts saying, "call in and talk to me about it". Or, if someone had made an interesting comment on the discussion databases, he would arrange a meeting to talk this through. If they worked in another geographic location, and a meeting would have been difficult to organise, Black would discuss the issue at length with them on the telephone. People such as Black were aware that though Notes allowed access to members located in different functions in terms of sharing abstract information, they also needed access to observing the manner of how competent practitioners in other communities go about their business (Brown and Duguid, 1991).

This expansion of safe enclaves' to incorporate not only electronic discussion forums, but also face-to-face, and telephone mediated communication, overcame some of the restrictions that the pervasiveness of political enclaves presented. In Brown and Duguid's (1991) terms, by recognising the importance of adopting strategies to develop in-depth relationships they recognised that: "learning, understanding, and interpretation involve a great deal that is not explicit or explicable, developed and framed in a crucially communal context". In this sense mediators saw that members of other communities not only constructed their understandings from their use of Notes, but this also had to be supplemented, or on some occasions replaced by the development of in-depth social relations between the people involved. By encouraging access to other communities in this way, employees were better able to take into account conventions in their work practices.

6.2.4. Rewarding collaboration

Although the broad reward structure remained unchanged after the introduction of Notes, some managers adopted strategies to try and reward effective collaboration. Black, for example, was adamant that those senior employees who advocated the development of in-depth relationships should actively encourage and motivate others to engage in genuine collaboration. He was conscious that, by encouraging people in this way, it would make them feel that their in-depth qualitative forms of working were appreciated. He thought this was an important contrast to those who gained recognition for optionally including themselves in political enclaves.

Several area managers similarly went to great lengths to encourage and motivate their sales representatives to engage in collaboration within and between functions. However, they also provided a further more tangible reward for those reps that worked in these qualitative ways. Thirty percent of the reps salaries were awarded at the discretion of their area managers. This discretionary amount was traditionally allocated depending on whether reps worked weekends and evenings, if they surpassed sales and contact targets, and on any extra work they did for the region. However some area managers used this discretionary thirty-percent to reward those reps they saw as working qualitatively within and between functional boundaries. For example, many of the more experienced sales reps often undertook additional responsibilities to their basic selling activities. These reps coached and trained less experienced staff, and they were rewarded for these activities by their area manager from their discretionary thirty-percent. In addition, some area managers took into account the reps' detailed use of the different shared databases, and those who worked cross-functionally with other experts, within this discretionary proportion of their salaries.

Not all area managers would do this, and even those that did were aware that it was relatively insignificant in relation to the other 70%. This latter proportion included a basic salary, as well as taking into account the sales volume that reps had brought in. As one rep in the North West of England said:

> The reward structure does not offer any incentives for us to use the technology. The reward structure needs changing to take into account contact recording. At the moment we do get some reward for putting in the contacts from our area manager's assessment, but this is unclear and minimal.

Despite such negative views, some experienced area managers did succeed in using the reward structure to encourage reps to feel that it was legitimate for them to engage in working collaboratively with members of other functions.

7. Implications and Conclusions

This final section considers the implications and conclusions arising from this study. The first subsection considers the conclusions in relation to the theoretical conceptualisations of safe and political enclaves in the context of groupware technology. Following this we outline four generic categories of participation that may help sensitise analysis in future studies. We then reflect on the limitations of Lave and Wenger's (1991) theorisation of learning. Finally we call for an expansion of existing theoretical conceptualisations of knowledge work to incorporate socio-political issues in more depth.

7.1. Safe/Political Enclaves

A first implication arising from this study relates to the theoretical constructs of safe and political enclaves (Hayes and Walsham, 2000b). This study has confirmed the applicability of these concepts to analysing communication and participation on shared electronic forums, while also extending earlier accounts by suggesting that it is important to consider all forms of participation, rather than focusing solely on computer mediated participation.

In relation to the specific issues emerging from the micro-level empirical study of Compound UK, the character of interaction varied from enclave to enclave, some having a more political air than others. Political enclaves were distinguished as being shared social spaces, which were characterised by participation resembling a "public facade". This study indicated how the career reward structure, the surveillance activities senior managers undertook, and the minimal or non-use that non-ambitious employees made cumulated in the consensus forming character of interaction within and between functional boundaries. However, Brown and Duguid (1991) suggest that, in knowledge work, communities-of-practice must be: "allowed some latitude to shake themselves free of received wisdom", and the univocality of expressed perspectives runs contrary to this. Furthermore, as a result of the normative discipline and regulation that arose from the threat of continuous observation which Lotus Notes provided, we suggest that groupware technologies may well be deeply involved in the homogenisation of perspectives, reflecting the views expressed by dominant groups (Hayes and Walsham, 2000a; Knights and Murray, 1994).

In this paper, we have extended the concept of safe enclaves to include not only computer-mediated participation, but also the many face-to-face and telephone mediated forms of participation that largely went unnoticed in Compound UK. These mitigating activities were central to generating opportunities for participation, and the establishment of safe enclaves. Safe enclaves were characterised as being shared electronic and non electronic social spaces that allowed for underlying views to be expressed, and for discussion and reflection to take place on the different ways of participating within and between communities. However, as we have seen, many aspiring reps optionally excluded themselves from the use of community specific enclaves, as they viewed their time as being better spent using the community wide political enclaves for their own individualist agendas. This limited the extent and nature of participation, and resulted in the irony that electronic safe enclaves, which presented opportunities for widespread participation, were used in a limited way, while the political enclaves, which limited the nature and extent of participation, were used extensively. In safe electronic enclaves, which can be characterised as allowing more intensive participation, participation was optional and not associated with the surveillance criteria of senior managers, nor with the career reward structure. Thus, one central challenge for the establishment of safe enclaves is to recognise that aligning IT-enabled participation to competitive based reward systems, reflecting both financial and career aspirations, is likely to lead to political rather than genuine forms of participation.

As Table 3 summarises, mediating activities to maintain and establish safe enclaves started with some employees being aware of the constraining nature of political enclaves, and reflexively adopting strategies to move more of their interactions with other experts into safe enclaves outside the restricted bounds of technology. Further to this, the implicit and explicit support of these managers for their subordinates' subversive use of the political enclaves, to indicate the futility they associated with working in this way, was central to mitigating some of the negative effects of the pervasiveness of political enclaves. These activities, such as the moving of interactions outside the technology domain, initiated the establishment and maintenance of safe shared social spaces, which brought about some change in the character of interaction within and between functional boundaries. Those employees that engaged in these mediating activities found ways to "theorise" about the constraining nature of political enclaves, and endeavoured to break up even the most enduring routines which constituted the political and what was deemed to be the legitimate character of many of the enclaves. These mitigators recognised that learning is fostered by allowing legitimate access to all social relations in a target community of practice, rather than merely relying on electronic mediated participation. They further recognised the negative impact that the individualistic reward structure may have on participation. We suggest that mediating activities of the nature of those

Mitigating activities that assisted in the creation and establishment of safe enclaves
Aware of the constraining nature of the political and normative context
Moving interactions into safe enclaves outside the technology domain
Reducing the individualistic components of the reward structure
Acknowledging that mediating activities will always be possible

Table 3 Maintaining and establishing safe enclaves.

reported in this paper will always be possible to some extent, and that actors will always have opportunities to change the political and normative context, and in doing so, present opportunities for in-depth participation to take place (Hayes, 2000).

7.2. Different motivations and intentions for participation

The previous subsection discussed safe and political enclaves as contexts for particular forms of participation within and between groups. In this second subsection we will outline four different forms of participation that may occur in the establishment and maintenance of communities of practice that intersect spatial, temporal and professional boundaries. We contend that it is important to recognise the motivations that participants bring with them when participating within and between communities.

7.2.1. Surveillance and control motivations for participation

One dominant form of participation that emerged from this study was that of surveillance and control. Senior managers harnessed the increased transparency that Notes provided to survey and control the participation of employees. Due to the more expansive scope of Notes, as opposed to the previous system, this extended the control that they previously exercised. Senior managers harnessed the visibility to ensure that employees were working hard and in the ways that they advocated. Surveillance and control motivations for participation had significant implications for learning. They had the impact of influencing the consensus forming and careerist nature of participation across functional boundaries. Thus when large amounts of participation are transparent as a consequence of the technology, and senior managers harness this to reaffirm their position in the power relations in an organisation, then expansive participation is unlikely to arise. A final note worth making is how and who may instigate the motivation for surveillance and control. In Compound UK, though most employees viewed the commercial director, the strategic selling manager and the general manager as the perpetrators, it is important to note that they were also under pressure from the international company to provide figures on contacts. This indicates the tendency for those most removed to want to control their organisation based on abstract representations of work activity conducted at a distance. Though countering this form of participation is difficult, bearing in mind the positions of those concerned, it negatively influences the participation of all those being surveyed.

7.2.2. Career enhancing motivations for participation

The individualistic and career enhancing motivations that lay behind much of the participation in Compound UK is a second generic category of participation. These motivations arose as a consequence of the reproduction of the financial and career reward structure that had existed prior to the introduction of Notes and the introduction of the new ways of working. Limitations such as these have been highlighted by Orlikowski's (1993) paper on collaboration and groupware, but have not been previously considered in relation to the role of IT in knowledge working. Even though this individualistic form of participation was optional, many felt normatively constrained to work in this way so as not to either stand out or jeopardise their careers. Career orientated motivations for participation were exacerbated, as staff could harness the visibility Notes provided to allow them direct access to the electronic spaces used by those that determined career progression. Thus it is important to consider individualistic and careerist motivations when examining participation and learning within and between functional boundaries.

7.2.3. Non-careerist motivations for participation

Those that participated in safe enclaves did so for non-careerist intentions. They saw the value of participating in the databases that were not used by senior managers or ambitious reps, as they assisted in developing a sense of community membership. Genuine participation only arose when the use of the technology did not mirror the career or financial reward structure, or the surveillance or control activities of senior managers. Furthermore, even if the pervasiveness of the controlling and aspiring motivations were reduced, it is not clear that participation would be any more intensive. Perhaps those whose participation was motivated by controlling or career oriented reasons would not see the worth in more genuine forms of participation?

7.2.4. Motivations that mitigators have for participation

This final category of participation was vital to work around the constraints that the political and normative context presented to knowledge working. Those involved in mitigating some of the constraints to widespread participation tended to be medical experts or field force managers who had limited aspirations to improve their career prospects in the organisation. They indicated the limitations to participation when there was a reliance solely on the use of technology to work within and particularly between functions.

Through these generic categories of participation we have provided a grounded description of nuanced participation in this context. We are not suggesting that these forms of participation are applicable in every context, nor that there are only four motivations for participation. Instead, we suggest that it is vital to explore the motivations for participating so as to develop existing accounts of participation from a communities of practice perspective.

7.3. Extending the communities of practice theorisation of learning

This third subsection will reflect upon, complexify and extend several of the key concepts within Lave and Wenger's (1991) theorisation of situated learning and specifically their concepts outlined in Table 1. We will consider the applicability of these concepts to knowledge work initiatives in contemporary organisations and how information technology is implicated in this.

7.3.1. Peripheral participation

This study has concurred with those authors who suggest that IT, such as groupware systems, provides opportunities for peripheral participation due to the opportunities it provides for employees to observe and collaborate with others across spatial, temporal and, importantly, professional boundaries (Boland and Tenkasi, 1995; Brown, 1998; Hayes, 2001). However, we have specific concerns surrounding this concept. First we found that peripheral participation is more heterogeneous than Lave and Wenger (1991) and subsequently Brown and Duguid (1991) suggested. In this study we found that though Notes allowed employees equal access to the periphery of members of other professional groups on the various shared discussion forums, staff chose to identify with and form two separate safe and political communities of practice that matched their differing motivations for participation. Thus we suggest that peripheral access to communities does not imply that a singular homogenous community will necessarily form. Instead, peripheral access to different groups is likely to lead to the formation of differing communities that reflect the overlapping motivations and normative assumptions of participants.

We further suggest that the transparency the technology provided was deeply implicated in the formation of the two distinct communities, as it allowed for an increased awareness of the differences in the dynamics of practice across functional, temporal and spatial boundaries such as the controlling, aspiring, non-careerist or mitigating motivations. This led to employees responding to this transparency by associating themselves with a community that best matched their motivations for participation.

A further reflection arising from our study relates to the applicability of Lave and Wenger's (1991) apprenticeship/master distinction to contemporary knowledge work initiatives such as the one described in Compound UK. Working across functional boundaries within Compound UK did not reflect a master/apprentice relationship. It was not about becoming an expert in all business areas in which there were interactions, rather it involved becoming proficient enough to be able to appreciate and synergistically utilise the perspectives of another community in their day-to-day activities (Boland and Tenkasi, 1995). Thus we suggest that the notion of an apprentice/master relationship may be relevant to traditional and well established areas of specialism (such as within the medical or sales function), but is not relevant to many knowledge work contexts, where the aim is to become competent rather than expert. Similarly, though the newcomer/old timer distinction expressed by Brown and Duguid (1991) diffuses some of these concerns, it is not explicit in their accounts whether this distinction is applicable to the nature of knowledge work initiatives undertaken in contemporary organisations such as the one described in Compound UK.

7.3.2. Full participation

In addition to raising the above concerns about peripheral participation, we also wish to comment on Lave and Wenger's (1991) conceptualisation of moving from peripheral to full participation. First we suggest that they place too much emphasis on newcomers moving from peripheral to full participation modifying their social practices to replicate those taken by the masters, or more experienced. Our study indicated that the process of moving from peripheral to full (or fuller) participation required changes in the work practices of all employees, and not just the newcomers. Those more experienced, such as Black and several field force managers, adopted new strategies that assisted newcomers to be able to learn from them. They also indicated how IT might simultaneously assist or hinder this movement from peripheral to full participation, through their careful establishment of electronic and non-electronic safe enclaves. This was particularly challenging when working between functional boundaries, as opposed to working with less experienced employees from the same function. Thus we suggest, that future studies could usefully develop accounts of the practices of those more experienced community members in this regard.

A connected concern arising from our study relates to the image Lave and Wenger (1991) conjure of a relatively homogeneous professional group of "Full Participants". We suggest that this image downplays the complexities of working in communities with diverse participants. Our study indicated that moving towards fuller community participation involves constant renegotiation and fine-grained improvisations and work arounds, perhaps best demonstrated through the mitigating activities of people such as Black, which are fundamental to the maintenance and recreation of community identity. We suggest that if Lave and Wenger's (1991) concepts are to be better applicable to the diversity of participants and contexts within contemporary organisations, then issues such as improvisation and work arounds should be central to a theorisation of situated learning (Hayes, 2000).

7.3.3. Legitimate participation

Perhaps the most striking omission in Lave and Wenger's concept of legitimate peripheral participation; is the lack of provision of any conceptual lens for making sense of how legitimate practice is established and sustained in specific contexts. They do note that learners may either be permitted or kept from observing and participating intensively in a community's practices (Lave and Wenger, 1991:36). However, with reference to Goffman (1959), we have illustrated through the analysis of the performances that staff exhibited that the emergence of communities does not necessarily follow the learner to full participant categorical form. Instead, communities are likely to emerge in relation to the differing motivations that underpin the varying forms of participation. Behaving as a community member involves participating in its identity formation — being encultured in relation to a particular community's practices (Brown, 1998). How community practices are shaped, and consequently how identities are formulated and sustained, is influenced by the political and normative context, as we have shown in our case study, and as such will reflect the ongoing power struggles in the organisation.

7.3.4. Situated learning

To take stock of these extensions and criticisms of Lave and Wenger's (1991) concept of situated learning and information technology, we contend that though Lave and Wenger's (1991) theory of learning presents useful insights, it needs to be complexified to provide a deeper theoretical analysis of how different forms of participation arise with regard to power relations in specific contexts. Accounts to date have merely mentioned how issues surrounding problems of power, access and transparency may be important, yet have offered little analysis of how learning may be shaped and is shaped by the socio-political context. Further to this we suggest that the shift from peripheral to full participation, which involves both electronic and non-electronic forms of participation, is more complex than currently presented within Lave and Wenger's (1991) theory of learning. We suggest that future theoretical developments of these concepts need to recognise this if they are to be more applicable to the vast array of knowledge work initiatives in contemporary organisations.

7.3.5. Expanding theoretical conceptions of knowledge work

A concluding theme to emerge from this study concerns the need for theoretical conceptualisations of knowledge working to be expanded. Knowledge work has received considerable attention over recent years from practitioners and academics alike. However it is surprising that few existing relational theoretical conceptualisations of knowledge and learning consider the political and normative context in any depth. For example Blackler et al's (1998) activity theoretical analysis does not develop any detailed analysis of how learning is shaped by the political motivations of participants. Boland and Tenkasi's (1995) theorising on communities of knowing and IT suggests that cultural and political issues may be worthy of consideration but does not enlarge on this. Lave and Wenger (1991) themselves implicitly recognise the centrality of socio-political issues within their concept of legitimate participation, yet offer no conceptualisation of how to make sense of how legitimacy shapes and is shaped by the socio-political context. This paper has started to address this important omission by drawing on and extending the concepts of safe and political enclaves to explore the nuances that pervade participation within and between communities of practice, especially in relation to the use of groupware technologies.

REFERENCES

Alvesson, M. (1993). Organization as rhetoric: knowledge-intensive firms and the struggle with ambiguity. *Journal of Management Studies, 30* (6), 997–1015.

Barley, S. (1996). Technicians in the workplace: ethnographic evidence for bringing work into organization studies. *Administrative Science Quarterly, 41* (1), 146–162.

Bell, D. (1978). *The cultural contradictions of capitalism.* London: Heinemann.

Bell, D. (1973). *The coming of post-industrial society.* New York: Basic Books.

Blackler, F. (1995). Knowledge, knowledge work and organisations, an overview and interpretation. *Organization Studies, 16* (6), 1021–1046.

Blackler, F. (1993). Knowledge and the theory of organisations: organisations as activity systems and the reframing of management. *Journal of Management Studies, 30* (6), 863–884.

Blackler, F., Crump, N. and McDonald, S. (1998). Knowledge, organisations and competition. In G. Kroght, J. Roos. and D. Kleine (Eds.), *Knowing in firms: understanding, managing and measuring knowledge.* London: Sage.

Blackler, F., Crump, N. and McDonald, S. (1997). Crossing boundaries: some problems of achieving expansive learning in a high technology organisation. Working Paper for the *EIASM Conference On "Organising in a Multi-Voiced World".* Leuven, Belgium, June 4–6.

Boland, R., Tenkasi, R. V. and Te'eni, D. (1994). Designing information technology to support distributed cognition. *Organization Science, 5* (3), 456–475.

Boland, R. and Tenkasi, R. V. (1995). Perspective making and perspective taking in communities of knowing. *Organization Science, 6* (4), 350–372.

Brown, J. S. (1998). Internet technology in support of the concept of "Communities-of-Practice": the case of xerox. *Accounting, Management and Information Technologies, 8* (4), 227–236.

Brown, J. S. and Duguid, P. (1991). Organisational learning and communities of practice: towards a unified view of working, learning and innovation. *Organization Science, 2* (1), 40–57.

Connah, B. and Pearson, R. (1991). *NHS handbook* (7th ed.). London: Macmillan Press Ltd.

Flynn, R. and Williams, G. (1997). *Contracting for health: quasi-markets and the national health services.* Oxford: University Press.

Geertz, C. (1973). *The interpretation of cultures.* New York: Basic Books.

Goffman, E. (1959). *The presentation of self in everyday life.* Garden City: Doubleday Anchor.

Hayes, N. (2000). Work-arounds and boundary crossing in a high tech optronics company: the role of co-operative work-flow technologies. *Computer Supported Co-operative Work: An International Journal, 9* (3/4), 435–455.

Hayes, N. (2001). Boundless and bounded interactions in the knowledge work process: the role of groupware technologies. *Information and Organization, 11* (2), 79–101.

Hayes, N. and Walsham, G. (2000a). Competing interpretations of computer supported co-operative work. *Organization, 7* (1), 49–67.

Hayes, N. and Walsham, G. (2000b). Safe enclaves, political enclaves and knowledge working. In C. Prichard, R. Hull, M. Churner. and H. Willmott (Eds.), *Managing knowledge: critical investigations of work and learning* (pp. 69–87). London: Macmillan.

Knights, D. and Murray, F. (1994). *Managers divided: organisation politics and information technology management.* Chichester: Wiley.

Knights, D., Murray, F. and Willmott, H. (1993). Networking as knowledge work: a study of strategic inter-organisational development in the financial services industry. *Journal of Management Studies, 30* (6), 975–995.

Lave, J. (1993). The practice of learning. In S. Chaiklin and J. Lave, *Understanding practice: perspectives on activity and context.* Cambridge: Cambridge University Press.

Lave, J. and Wenger, E. (1991). *Situated learning: legitimate peripheral participation.* Cambridge: Cambridge University Press.

Layder, D. (1993). *New strategies in social research.* Cambridge: Polity Press.

McKinlay, A. (2000). The bearable lightness of control: organisational reflexivity and the politics of knowledge management. In C. Prichard, R. Hull, M. Churner and H. Willmott (Eds.), *Managing knowledge; critical investigations of work and learning.* London: Macmillan.

Newell, S., Scarborough, H., Swan, J. and Hislop, D. (2000). Intranets and knowledge management: decentred technologies and the limits of technological discourse. In C. Prichard, R. Hull, M. Churner and H. Willmott (Eds.), *Managing knowledge; critical investigations of work and learning.* London: Macmillan.

Nonaka, L. and Takeuchi, H. (1994). *The knowledge creating company: how Japanese companies create the dynamics of innovation.* Oxford: Oxford University Press.

Orlikowski, W. (1993). Learning from notes: organisational issues in groupware implementation. *The Information Society, 9,* 237–250.

Reich, R. (1991). *The work of nations: preparing ourselves for 21st-century capitalism.* London: Simon and Schuster.

Robinson, R. and Le Grand, B. (1994). Evaluating the NHS Reforms. Kings Fund Institute, Policy Journals.

Ruhleder, K. (1995). Computerisation and changes to infrastructures for knowledge work. *The Information Society, 11* (2), 131–144.

Star, S. L. and Ruhleder, K. (1994). Steps towards an ecology of infrastructure. In *Proceedings from CSCW'94. Chapel Hill, NC, USA.* New York: ACM Press.

Willmott, H. (1995). The odd couple?: re-engineering business processes; managing human relations. *New Technology, Work and Employment, 10* (2), 89–98.

Zuhoff, S. (1996). The Emperor's new information economy. In W. Orlikowski, G. Walsham, M. R. Jones and J. L. Degross (Eds.), *Information technology and changes in organizational work.* London: Chapman and Hall.

PART TWO

The impact of IT on organizations

INTRODUCTION TO PART TWO

The second section of the book still focuses on information systems within organizations, but on a broader scale than the group focus of Part One. Part Two deals with the impact of IT on organizations as a whole – the second ring in Figure 2.1. Willcocks and Lester (1999: 1) in their book entitled *Beyond the IT Productivity Paradox*, open by stating: "The proposition is simultaneously shocking and attractive – that despite the massive accumulated and rising investments in information technology (IT), on the whole these have not contributed to significant rises in productivity". The original bases of the productivity paradox claim, which originated from the work of economists in the 1980s, was that investments in technology did not relate to increases in productivity, despite the trillions of dollars involved. IT was claimed by some to be a 'sink hole' that was simply a drain on organizational resources that did not contribute to performance in any measurable way. Over time, the analyses have become more sophisticated and it is recognized that simple aggregate data linking IT spending and national or industry-level productivity is not very helpful because of the many confounding variables that will impact macroeconomic activity.

Research has, therefore, focused at the level of the firm to seek to establish the relationship between spending on IT and performance and productivity. Even at this level, however, the link between IT spending and productivity is unlikely to be straightforward. For example, IT might influence organizational performance through allowing firms to better serve customer needs or improving working conditions for staff (Baily and Gordon, 1988). This may not, especially in the short-term, lead to direct improvements in productivity. And, perhaps most fundamentally, it is important to recognize that individual firms will be more or less successful in exploiting the potential of IT for enhancing performance or productivity. As Robey and Boudreau (1999) argue, the idea that IT 'drives', 'forces' or even 'enables' organizational change is relatively simplistic and is unsupported by empirical evidence. Instead, it is important to explore the complex relationship of reciprocal causality between IT and organization (e.g., Walsham, 1993; Orlikowski, 2000). Given this reciprocal causality, the outcomes between IT spending and productivity and performance are inevitably emergent and difficult to predict in advance, suggesting the need for different research approaches to capture this dynamic relationship. As you read the four articles selected in this section it is helpful to think about the extent to which the selected methodology of each study has been able to capture these dynamic and emergent interactive processes between technology and organization.

The first article in Part Two is by Brynjolfsson and Hitt (1996). It begins by reviewing previous research on the productivity paradox considering this research within the framework of the economic theory of production. Their main argument is that previous research has suffered from poor data that has not been able to distinguish between changes in productivity that occur as the result of IT and 'random shocks'. They suggest that they have overcome many of the previous data problems through using new 'firm-level data which are more recent, more detailed, and include more companies' (p. 544). More specifically, they use data compiled by International Data Group based on an annual survey of IT managers in the Fortune 500 manufacturing and service companies. The survey asks managers to provide information on various IS expenses and budgets as well as the value and use of computers of

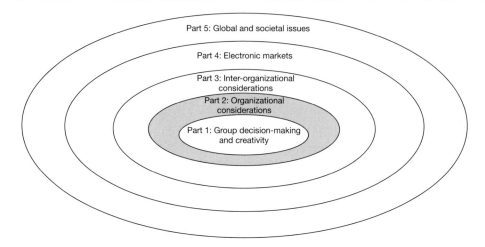

Figure 2.1 Organizational considerations within the broadening focus of Information Systems research considerations.

different kinds. The data from this survey is integrated with the Standard and Poors' Compustat II data, which provides financial and other related information for these same companies, including measures of output, capital investment, expenses and number of employees. From these data sets, they develop measures of computer capital and output, and conclude that IT spending does indeed increase output: "Overall, we found that computers contribute significantly to firm-level output, even after accounting for depreciation, measurement error, and some data limitations" (p. 557). They argue that their results are different to previous studies, and refute the productivity paradox, because the data was collected at a later time period. More specifically, they suggest that the productivity paradox was present during early adoptions of IT but that by the late 1980s and early 1990s, when their data was collected, firms have learnt better to use the adopted IT to enhance performance.

Questions arising from this paper include:

- How useful is the economic theory of the firm as the framework for studying the relationship between IT and productivity, given that it 'black-boxes' the firm?
- This study also 'black-boxes' technology and treats all IS/IT spending as equal. What are the implications of this?
- What are the limitations of the measure of productivity (OUTPUT) used in this study? What about the other measures used (IS STAFF, NON-IS LABOR, EXPENSE, COMPUTER CAPITAL AND NON-COMPUTER CAPITAL)?
- The study assumes a direct relationship between IT and productivity, even though it is recognized that reengineering is also fundamental to accruing the benefits from IT spending. How would it be possible to differentiate between the impact of IT *per se* and the impact of organizational change on performance and productivity?

The Brynjolfsson and Hitt (1996) paper uses commercially available published data to consider the link between IT and performance. Their analysis is entirely quantitative. The next paper in this section by Pinsonneault and Rivard (1998) also explores the credibility of the productivity paradox, but uses a very different method, that includes both qualitative and quantitative data. Qualitative data is gathered from semistructured interviews with middle managers in three different organizations – a bank, a telecommunications company and a utility. Quantitative data is collected with the same respondents, through asking them to log daily activities and IT usage. The premise of the argument in this paper is that it is necessary to look at the interplay between IT and work, specifically managerial work, in order to understand the impact of IT on an organization. The reasoning is that the organizational context will influence

	Brynjolfsson and Hitt (1996)[1]	Pinsonneault and Rivard (1998)[2]	Chan (2000)[3]	Cline and Guynes (2001)[4]
Phenomenon of interest	Measurement of the benefits of IT spending.	The extent to which IT spending yields productivity gains.	How much value IT adds to an organization.	Relationship between IT investment and performance at both firm and industry levels.
Perspective or theory	Critical review of the IT paradox using as a framework the economic theory of production.	The relationship between the nature of managerial work and IT usage.	Critical review of the IT paradox.	No theoretical perspective taken – provides a descriptive case study.
Research method	Analysis of data set on annual IS spending by large US firms.	Semi-structured interviews with middle managers in 3 different organizations.	Comparative review of IT value articles published in four leading N. American journals.	An in-depth case study of the effects of IT investment on related performance returns.
Technology	No specific technology – just general spending on IS.	No specific technology – just general use of IT by middle managers.	No specific technology – just general spending on IT.	No specific technology – investment in different types of technology.
Findings and contributions	The authors conclude that IT spending does indeed increase output. The main contribution of the paper is to explore with better quality data the so-called productivity paradox and demonstrate how overtime firms have learnt better to use the adopted IT to enhance performance so that this paradox is no longer present.	The paper demonstrates that the organizational context influences the way IT is deployed and used. The paper thus contributes to our understanding of the interaction between IT and organization and demonstrates the limitations of using aggregate level data.	The paper finds that quantitative studies, using organizational-level analyses predominate in studies of the organizational impact of IT, even though there are severe limitations with these methods. This contributes to the debate about methods and in particular illustrates the importance of using a variety of measures and levels of analyses in research of this kind.	The paper finds that different types of IT investment have different impacts on performance, so that lumping together all types of IT investment and relating this to overall organizational performance is not likely to be helpful. The study contributes to the deeper exploration of the productivity paradox.

Table 2.1 Comparison of chapters in Part 2.

Notes
1. Brynjolfsson, E. and Hitt, L., (1996) "Paradox lost? Firm-level evidence on the returns on information systems spending". *Management Science*, 42(4), 541–58.
2. Pinsonneault, A. and Rivard, S. (1998) "Information technology and the nature of work: From the productivity paradox to the Icarus paradox?" *MIS Quarterly*, 20(3), 287–311, September.
3. Chan, Y. E. (2000) "IT Value: The great divide between qualitative and quantitative measures". *Journal of Management Information Systems*, 16(4), 225–61.
4. Cline, M. K. and Guynes, C. S. (2001) "The impact of IT investment on enterprise performance: A case study". *Information Systems Management*, 18(4), 70–6, (Fall).

significantly how IT is deployed and used, and therefore its impact. Given this, their focus is on how IT impacts the work process rather than work output, since it is here that IT will have a more direct impact. They review available evidence about the link between IT and managerial work and develop a number of propositions based on this. These propositions are subsequently tested using the empirical data collected. The main conclusion is that the organizational context does indeed effect the way IT is deployed and used so that – "Lumping together firms with different strategic orientations in an aggregate-level study would have blurred the analysis" (p. 304). They find that the pattern of IT usage and the nature of the managerial work depends on the strategic orientation of the organization. They also go further than this and argue that managers use IT to focus on the activities and roles that were identified by senior management as most important. The problem with this, they suggest, is that in evermore focusing, they limit their involvement in other activities. This increasing specialization over time may limit their potential to see the need for change. This, then, is the Icarus paradox – that specializing and focusing in one type of activity may seduce managers into excessive concentration that can eventually lead to their own demise.

Questions arising from this paper include:

- What is the difference between a proposition as used here and a hypothesis, as used by Brynjolfsson and Hitt?
- How well does this study get at the 'blackbox' between IT and productivity that is evident in the Brynjolfsson and Hitt study?
- Would the study have been useful with only the quantitative log data? What do you think the qualitative data add to this quantitative data and analysis? More generally, how useful did you find the log method to be and what did you see as its main strengths and limitations?
- The study discusses the importance of the organizational context and how stable or dynamic it is. Yet, it uses a cross-sectional research design. How appropriate is it to use a cross-sectional design when one of the features you are interested in is change?
- The research includes three different organizations, but the authors do not mention anything about the national context. The Brynjolfsson and Hitt study uses data from US firms, although does not discuss the implications of this. Do you think it is important to situate research in a particular national context and discuss the implications of this?
- The study makes use of 'theory-blind' interviews. This suggests that using 'theory-sensitive' interviewers may cause problems. What are the problems (and advantages) that you see with using theory-blind versus theory-sensitive interviews?
- Given the differences between organizations identified in this study, what are the implications for research that uses a single case? Does it make the single-case simply inappropriate or are there advantages of using a single case that can offset the limitations? Moreover, given the differences between organizations identified here, what are the implications for the type of aggregated data used by Brynjolfsson and Hitt?
- The authors conclude that they have identified the Icarus paradox rather than the productivity paradox. How might you research the validity of their Icarus paradox theory?

The third paper in this section is by Yolande Chan (2000), who again focuses on unpacking the productivity paradox, this time based on a meta-analysis of the methods and findings from previously published studies. The paper begins with a review of the different ways in which researchers have tried to study the productivity paradox and the different explanations given as to why it exists. In doing this Chan contrasts approaches which rely on hard, quantitative data and approaches that include or are based on soft, more qualitative data. She argues for the need for research that incorporates "qualitative, individual and group-level measures" (pp. 228–9). If this is not done, she argues, research falls into the trap of "black-boxing" because only macro-level inputs and outputs can be considered. While she argues for the necessity of these softer analyses of the link between IT and performance, her research is based on the proposition that there has been a trend away from such approaches, so that only hard, organization-level measures of value are included. She identifies the limitations of these hard measures, and while recognizing their contribution, also argues for the need for complementary qualitative analyses. In order to test this proposition, she undertakes a systematic review of the measures used in IT value research in four top North American MIS journals. Having first identified the articles that have a focus on IT benefits (N=38), they are then classified as to the research methods used, whether the measures are quantitative and/or qualitative and whether financial and/or non-financial measures are used.

Chan concludes from her findings that quantitative studies, using organizational-level analyses predominate and that financial measures are most common – "Certainly IT value studies using organization-level analyses appear to be the ones primarily being published in North American journals today" (p. 240) – even though these have not "served researchers particularly well in their search for IT productivity gains" (p. 240). The study also finds a strong link between levels of analysis and research methods used and also a link between the particular journal and the type of research reported, demonstrating the influence of the editorial policy. She concludes that MIS researchers need to recognize the importance of using a variety of measures and levels of analyses in their research, rather than the over-reliance on particular measures and levels that are identified in the paper.

Questions arising from this paper include:

- What kind of research might be designed to answer the questions that Chan feels are inadequately addressed by IT value research right now, which she argues only concentrates on what and not – why, where, when, how, and to whom do these IT investments provide value?
- Which of these questions do the other papers in this section address? Do any of the other papers in this section get inside the 'black box' that Chan is concerned with?
- Chan's paper is based on a systematic literature review? What are the limits of this research approach? How useful do you think this method of research is?
- In this paper the North American bias of the data used is acknowledged but nevertheless the paper was published. Why do you think this bias was allowed through the refereeing system and what impact do you think this bias might have had?
- The paper also only considers MIS literature. Take a look at the references from the Pinsonneault and Rivard paper and suggest what the limitations might be in restricting the literature review on the organizational impact of IT in this way?
- Did reading this paper change the way you interpreted the results from the first two papers in this section? If yes, in what ways?

The final paper in this section is by Cline and Guynes (2001). You will see that the style of this paper is rather different to the others in this section. In particular, while the paper is based on a study which examines "approximately 5.6 million hours of applications development effort on 8700 projects over a 10-year period and relates this investment in IT to organizational performance at the firm and industry levels of analysis" – none of the actual empirical data is presented. This is a point that can be discussed later. The method here, then, is an intensive, archival case study of a major eastern railroad. The main claim of the paper is that it is important to consider the role of the IT investment when evaluating returns since the impact of IT will vary depending upon the objectives managers intended for the IT. So the objectives for introducing IT might be strategic, tactical, transactional or threshold and this will inevitably influence the outcomes of the investment. A list of propositions is suggested that link IT investment to outcomes. The conclusions of the study are that different types of IT investment do indeed have different impacts on performance. For example, strategic IT investment was found to be related to increases in sales volume but not to increased revenues, while tactical investment contributes most to improved profitability.

- While extensive empirical data collection is claimed for this study, none of it is actually presented in the paper. How convinced were you by the argument, given that the data was not presented? How does this compare with your reaction to the other papers in this section?

- The strength of the study is claimed to be that it uses detail rather than aggregate measures of IT investment and performance. So, what do you see as the strengths (and weaknesses) of this study compared to the Brynjolfsson and Hitt study that did use aggregate data?
- The case is presented in a largely descriptive way. There is no particular theoretical framework used for analyzing the case. Is there a role for such descriptive cases and, if yes, what is that role?
- The authors suggest that they are exploring the impact of IT investment at both the firm and industry levels of analyses. Do you think this study provides an analysis at both these levels?
- The case used archival data to reconstruct the level of IT investment over time. What are the limitations and strengths of using archival data rather than following events through longitudinally as they occur?
- Like the Pinsonneault and Rivard paper this paper uses propositions. What is the difference between the propositions here and the propositions in the Pinsonneault and Rivard paper. Which do you find more convincing and why?

REFERENCES

Baily, M. and Gordon, R. (1988) "The productivity slowdown, measurement issues, and the explosion of computer power", in W. Brainard and G. Perry (eds), *Brookings Papers on Economic Activity*, Washington, DC: The Brookings Institute, pp. 347–431.

Orlikowski, W. (2000) "Using technology and constituting structures: A practice lens for studying technology in organizations", *Organization Science*, 11: 404–28.

Robey, D. and Boudreau, M. (1999) "Accounting for the contradictory organizational consequences of information technology: Theoretical directions and methodological implication", *Information Systems Research*, 10: 167–85.

Walsham, G. (1993) *Interpreting Information Systems in Organizations*, Chichester: Wiley.

Willcocks, L. and Lester, S. (1999) *Beyond the IT Productivity Paradox*, Chichester: John Wiley.

FURTHER READING

Cline, M. K. and Guynes, C. S. (2001) "IT investment is strategic to a firm's survival, information strategy", *The Executive's Journal*, 20(3), 2004 (Spring).

Paradox Lost? Firm-level Evidence on the Returns to Information Systems Spending

Erik Brynjolfsson and Lorin Hitt

Sloan School of Management, Massachusetts Institute of Technology, Cambridge, Massachusetts 02139

Abstract

The "productivity paradox" of information systems (IS) is that, despite enormous improvements in the underlying technology, the benefits of IS spending have not been found in aggregate output statistics. One explanation is that IS spending may lead to increases in product quality or variety which tend to be overlooked in the aggregate statistics, even if they increase output at the firm-level. Furthermore, the restructuring and cost-cutting that are often necessary to realize the potential benefits of IS have only recently been undertaken in many firms.

Our study uses new firm-level data on several components of IS spending for 1987–1991. The dataset includes 367 large firms which generated approximately 1.8 trillion dollars in output in 1991. We supplemented the IS data with data on other inputs, output, and price deflators from other sources. As a result, we could assess several econometric models of the contribution of IS to firm-level productivity.

Our results indicate that IS spending has made a substantial and statistically significant contribution to firm output. We find that the gross marginal product (MP) for computer capital averaged 81% for the firms in our sample. We find that the MP for computer capital is at least as large as the marginal product of other types of capital investment and that, dollar for dollar, IS labor spending generates at least as much output as spending on non-IS labor and expenses. Because the models we applied were similar to those that have been previously used to assess the contribution of IS and other factors of production, we attribute the different results to the fact that our data set is more current and larger than others explored. We conclude that the productivity paradox disappeared by 1991, at least in our sample of firms.

Keywords: *Information Technology; Productivity; Production Function; Computers; Software; IS Budgets*)

1. Introduction

Spending on information systems (IS), and in particular information technology (IT) capital, is widely regarded as having enormous potential for reducing costs and enhancing the competitiveness of American firms. Although spending has surged in the past decade, there is surprisingly little formal evidence linking it to higher productivity. Several studies, such as those by Loveman (1994) and by Barua et al. (1991) have been unable to reject the hypothesis that computers add nothing at all to total output, while others estimate that the marginal benefits are less than the marginal costs (Morrison and Berndt 1990).

This "productivity paradox" has alarmed managers and puzzled researchers. American corporations have spent billions of dollars on computers, and many firms have radically restructured their business processes to take advantage of computers. If these investments have not increased the value produced or reduced costs, then management must rethink their IS strategies.

This study considers new evidence and finds sharply different results from previous studies. Our dataset is based on five annual surveys of several hundred large firms for a total of 1121 observations[1] over the period 1987–1991. The firms in our sample generated approximately 1.8 trillion dollars worth of gross output in the United States in 1991, and their value-added of $630 billion accounted for about 13% of the 1991 U.S. gross domestic product of $4.86 trillion (Council of Economic Advisors 1992). Because the identity of each of the participating firms is known, we were able to supplement the IS data with data from several other sources. As a result, we could assess several econometric models of the contribution of IS to firm-level productivity.

Our examination of these data indicates that IS spending has made a substantial and statistically significant contribution to the output of firms. Our point estimates indicate that, dollar for dollar, spending on computer capital created more value than spending on other types of capital. We find that the contribution of IS to output does not vary much across years, although there is weak evidence of a decrease over time. We also find some evidence of differences across various sectors of the economy. Technology strategy also appears to affect returns. For instance, we find that neither firms that relied heavily on mainframes nor firms which emphasized personal computer (PC) usage performed as well as firms that invested in a mix of mainframes and PCs.

In each of the specifications we examine, estimates of the gross marginal product for computers exceeds 50% annually. Considering a 95% confidence interval around our estimates, we can reject the hypothesis that computers add nothing to total output. Furthermore, several of our regressions suggest that the marginal product for computers is significantly higher than the return on investment for other types of capital, although this comparison is dependent on the assumed cost of computer capital. Overall, our findings suggest that for our sample of large firms, the productivity paradox disappeared in the 1987–1991 period.

1.1. Previous Research on IT and Productivity

There is a broad literature on IT value which has been recently reviewed in detail elsewhere (Brynjolfsson 1993, Wilson 1993). Many of these studies examined correlations between IT spending ratios and various performance measures, such as profits or stock returns (Dos Santos et al. 1993, Harris and Katz 1988, Strassmann 1990), and some found that the correlation was either zero or very low, which has led to the conclusion that computer investment has been unproductive. However, in interpreting these findings, it is important to bear in mind that economic theory predicts that in equilibrium, companies that spend more on computers would not, on average, have higher profitability or stock market returns. Managers should be as likely to overspend as to under-spend, so high spending should not necessarily be "better." Where nonzero correlations are found, they should be interpreted as indicating either an *unexpectedly* high or low contribution of information technology, as compared to the performance that was anticipated when the investments were made. Thus, perhaps counter-intuitively, the common finding of zero or weak correlations between the percentage of spending allocated to IT and profitability do not necessarily indicate a low payoff for computers.

To examine the contribution of IT to output, it is helpful to work within the well-defined framework of the economic theory of production. In fact, Alpar and Kim (1990) found that methods based on production theory could yield insights that were not apparent when more loosely constrained statistical analyses were performed. The economic theory of production posits that the output of a firm is related to its inputs via a production function and predicts that each input should make a positive contribution to output. A further prediction of the theory is that the marginal cost of each input should just equal the marginal benefit produced by that input. Hundreds of studies have estimated production functions with various inputs, and the predictions of economic theory have generally been confirmed (see Berndt 1991, especially chapters 3 and 9, for an excellent review of many of these studies).

The productivity paradox of IT is most accurately linked to a subset of studies based on the theory of production which either found no positive correlation overall (Barua et al. 1991, Loveman 1994), or found that benefits fell short of costs (Morrison and Berndt 1990). Using the Management of the Productivity of Information Technology (MPIT) database,[2] Loveman (1994) concluded: "Investments in IT showed no net contribution to total output." While his elasticity estimates ranged from −0.12 to 0.09, most were not statistically distinguishable from zero. Barua et al. (1991) found that computer investments are not significantly correlated with increases in return on assets. Similarly, Morrison and Berndt (1990) examined industry-level data using a production function and found that each dollar spent on "high tech" capital (computers, instruments, and telecommunications equipment) increased measured output by only 80 cents on the margin.

Although previous work provides little econometric evidence that computers improve productivity, Brynjolfsson's (1993) review of the overall literature on this productivity paradox concludes that the "shortfall of evidence is not necessarily evidence of a shortfall." He notes that increases in product variety and quality should properly be counted as part of the value of output, but that the price deflators that the government currently uses to remove the effects of inflation are imperfect. These deflators are computed assuming that quality and other intangible characteristics do not change for most goods. As a result, inflation is overestimated and real output is underestimated by an equivalent amount (because real output is estimated by multiplying nominal output by the price deflator). In addition, as with any new technology, a period of learning, adjustment, and restructuring may be necessary to reap its full benefits, so that early returns may not be representative of the ultimate value of IT. Accordingly, he argues that "mismeasurement" and "lags" are two of four viable explanations (along with "redistribution" and "mismanagement") for the collected findings of earlier studies. This leaves the question of computer productivity open to continuing debate.

1.2. Data Issues

The measurement problem has been exacerbated by weaknesses in available data. Industry-level output statistics have historically been the only data that are available for a broad cross-section of the economy. In a related study using much of the same data as the Morrison and Berndt (1990) study, Berndt and Morrison (1994) conclude, " ... there is a statistically significant negative relationship between productivity growth and the high-tech intensity of the capital." However, they also point out: "it is possible that the negative productivity results are due to measurement problems ... ". Part of the difficulty is that industry-level data do allow us to distinguish firms *within* a particular industry which invest heavily in IT from those with low IT investments. Comparisons can only be made among industries, yet these comparisons can be sensitive to price deflators used, which in turn depend on the assumptions about how much quality improvement has occurred in each industry. Firm-level production functions, on the other hand, will better reflect the "true" outputs of the firm, insofar as the increased sales at each firm can be directly linked to its use of computers and other inputs, and all the firms are subject to the same industry-level price deflator.

On the other hand, a weakness of firm-level data is that it can be painstaking to collect, and therefore, studies with firm level data have historically focused on relatively narrow samples. This has made it difficult to draw generalizable results from these studies. For instance, Weill (1992) found some positive impacts for investments in some categories of IS but not for overall IS spending. However, the 33 strategic business units (SBUs) in his sample from the valve manufacturing industry accounted for less than $2 billion in total sales, and he notes, "The findings of the study have limited external validity" (Weill 1992). By the same token, the Loveman (1994) and Barua et al. (1991) studies were based on data from only 20 firms (60 SBUs) in the 1978–1982 period and derived only rather imprecise estimates of IT's relationship to firm performance.[3]

The imprecision of previous estimates highlights an inherent difficulty of measuring the benefits of IT investment. To better understand the perceived benefits, we conducted a survey of managers to find out the relative importance of reasons for investing in IT (see Brynjolfsson 1994). Our results indicate that the primary reason for IT investment was customer service, followed by cost savings. Close behind were timeliness and quality. In practice, the value of many of the benefits of IT, other than cost savings, are not well captured in aggregate price deflators or output statistics (Baily and Gordon 1988).

Given the weaknesses of existing data, it has been very difficult to distinguish the contribution of IT from random shocks that affect productivity even when sophisticated analytical methods are applied. As Simon (1984) has observed:

> In the physical sciences, when errors of measurement and other noise are found to be of the same order of magnitude as the phenomena under study, the response is not to try to squeeze more information out of the data by statistical means; it is instead to find techniques for observing the phenomena at a higher level of resolution. The corresponding strategy for economics is obvious: to secure new kinds of data at the micro level.

A convincing assessment of IS productivity would ideally employ a sample which included a large share of the economy (as in the Berndt and Morrison studies), but at a level of detail that disaggregated inputs and outputs for individual firms (as in Loveman 1994, Barua et al. 1991, and Weill 1992). Furthermore, because the recent restructuring of many firms may have been essential to realizing the benefits of IS spending, the data should be as current as possible. Lack of such detailed data has hampered previous efforts. While our paper applies essentially the same models as those used in earlier studies, we use new, firm-level data which are more recent, more detailed, and include more companies. We believe this accounts for our sharply different results.

1.3. Theoretical Issues

As discussed above, there are a number of potential explanations for the productivity paradox, including the possibility that it is an artifact of mismeasurement. We consider this possibility in this paper.

More formally, we examine the following hypotheses using a variety of statistical tests:

Hypothesis 1. The output contributions of computer capital and IS staff labor are positive.

Hypothesis 2. The net output contributions of computer capital and IS staff labor are positive after accounting for depreciation and labor expense, respectively.

In our analysis, we build on a long research stream which applies production theory to determine the contributions of various inputs to output. This approach uses economic theory to determine the set of relevant variables and to define the structural relationships among them. The relationship can then be estimated econometrically and compared with the predictions of economic theory. In particular, for any given set of inputs, the maximum amount of output that can be produced, according to the known laws of nature and existing "technology," is determined by a *production function*. As noted by Berndt (1991), various combinations of inputs can be used to produce a given level of output, so a production function can be thought of as pages of a book containing alternative blueprints. This is essentially an engineering definition, but business implications can be drawn by adding an assumption about how firms behave, such as profit maximization or cost minimization. Under either assumption, no inputs will be "wasted," so the only way to increase output for a given production function is to increase at least one input.

The theory of production not only posits a relationship among inputs and output, but also posits that this relationship may vary depending on particular circumstances. Many of these differences can be explicitly modeled by a sufficiently general production function without adding additional variables. For instance, it is common to assume that there are constant returns to scale, but more general models will allow for increasing or decreasing returns to scale. In this way, it is possible to see whether large firms are more or less efficient than smaller firms. Other differences may have to do with the economic environment surrounding the firm and are not directly related to inputs. Such differences are properly modeled as additional "control" variables. Depending on prices and desired levels of output, different firms may choose different combinations of inputs and outputs, but they will all adhere to the set defined by their production function. The neoclassical economic theory of production has been fairly successful empirically, despite the fact that it treats firms as "black boxes" and thus ignores history or details of the internal organization of firms. Of course, in the real world, such factors can make a significant difference, and recent advances in the theory of the firm may enable them to be more rigorously modeled as well.

To operationalize the theory for our sample, we assume that the firms in our sample produce a quantity of OUTPUT (Q) via a production function (F), whose inputs are COMPUTER CAPITAL (C), NONCOMPUTER CAPITAL (K), IS STAFF labor (S), and OTHER LABOR and EXPENSES (L).[4] These inputs comprise the sum total of all spending by the firm and all capitalized investment. Economists historically have not distinguished computer capital from other capital, lumping them together as a single variable. Similarly, previous estimates of production functions have not distinguished IS staff labor from other types of labor and expenses. However, for our purposes, making this distinction will allow us to directly examine hypotheses such as H1 and H2 above. We seek to allow for fairly general types of influences by allowing for any type of environmental factors which affect the business sector (j) in which the company operates and year (t) in which the observation was made.[5] Thus, we can write:

$$Q = F(C, K, S, L; j, t). \quad (1)$$

Output and each of the input variables can be measured in either physical units or dollars. If measured in dollar terms, the results will more closely reflect the ultimate objective of the firm (profits, or revenues less costs). However, this approach requires the deduction of inflation from the different inputs and outputs over time and in different industries. This can be done by multiplying the nominal dollar value of each variable in each year by an associated deflator to get the real dollar values. Unfortunately, as mentioned earlier, this approach will probably underestimate changes in product quality or variety since the deflators are imperfect.

The amount of output that can be produced for a given unit of a given input is often measured as the marginal product (MP) of the input, which can be interpreted as a rate of return. When examining differences in the returns of a factor across firms or time periods, it is important to control for the effects of changes in the other inputs to production. Since the production function identifies both the relevant variables of interest as well as the controls, the standard approach to conducting productivity analyses is to assume that the production function, F, has some functional form, and then estimate its parameters (Berndt 1991, pp. 449–460).

The economic theory of production places certain technical constraints on the choice of functional form, such as quasi-concavity and monotonicity (Varian 1992). In addition, we observe that firms use multiple inputs in production, so the functional form should also include the flexibility to allow continuous adjustment between inputs as the relative prices of inputs change (ruling out a linear form). Perhaps the simplest functional form that relates inputs to outputs and is consistent with these constraints is the Cobb-Douglas specification, variants of which have been used since 1896 (Berndt 1991). This specification is probably the most common functional form used for estimating production functions and remains the standard for studies such as ours, which seek to account for output growth by looking at inputs and other factors.

$$Q = e^{\beta_0} C^{\beta_1} K^{\beta_2} S^{\beta_3} L^{\beta_4}. \quad (2)$$

In this specification, β_1 and β_3 are the output elasticity of COMPUTER CAPITAL and information systems staff (IS STAFF), respectively.[6] If the coefficients $\beta_0 - \beta_4$ sum to 1, then the production function exhibits constant returns to scale. However, increasing or decreasing returns to scale can also be modeled with the above function. The principal restriction implied by the Cobb-Douglas form is that the elasticity of substitution between factors is constrained to be equal to −1. This means that as the relative price of a particular input increases, the amount of the input employed will decrease by a proportionate amount, and the quantities of other inputs will increase to maintain the same level of output. As a result, this formulation is not appropriate for determining whether inputs are substitutes or complements.

The remainder of the paper is organized as follows. In §2, we describe the statistical methodology and data of our study. The results are presented in §3. In §4, we conclude with a discussion of the implications of our results.

2. Methods and Data

2.1. Estimating Procedures

The basic Cobb-Douglas specification is obviously not linear in its parameters. However, by taking logarithms of equation (2) and adding an error term (ε) one can derive an equivalent equation that can be estimated by linear regression. For estimation, we have organized the equations as a *system* of five equations, one for each year:

$$\text{Log}Q_{i,87} = \beta_{87} + \beta_j + \beta_1 \text{Log}C_{i,87} + \beta_2 \text{Log } K_{i,87} \\ + \beta_3 \text{Log } S_{i,87} + \beta_4 \text{Log } L_{i,87} + \varepsilon_{87.} \quad (3a)$$

$$\text{Log}Q_{i,88} = \beta_{88} + \beta_j + \beta_1 \text{Log}C_{i,88} + \beta_2 \text{Log } K_{i,88} \\ + \beta_3 \text{Log } S_{i,88} + \beta_4 \text{Log } L_{i,88} + \varepsilon_{88.} \quad (3b)$$

$$\text{Log}Q_{i,89} = \beta_{89} + \beta_j + \beta_1 \text{Log}C_{i,89} + \beta_2 \text{Log } K_{i,89} \\ + \beta_3 \text{Log } S_{i,89} + \beta_4 \text{Log } L_{i,89} + \varepsilon_{89.} \quad (3c)$$

$$\text{Log}Q_{i,90} = \beta_{90} + \beta_j + \beta_1 \text{Log}C_{i,90} + \beta_2 \text{Log } K_{i,90} \\ + \beta_3 \text{Log } S_{i,90} + \beta_4 \text{Log } L_{i,90} + \varepsilon_{90.} \quad (3d)$$

$$\text{Log}Q_{i,91} = \beta_{91} + \beta_j + \beta_1 \text{Log}C_{i,91} + \beta_2 \text{Log } K_{i,91} \\ + \beta_3 \text{Log } S_{i,91} + \beta_4 \text{Log } L_{i,91} + \varepsilon_{91.} \quad (3e)$$

where Q, C, K, S, L and $\beta_1-\beta_4$ are as before; 87, 88, 89, 90 and 91 index each year; j indexes each sector of the economy; and i indexes each firm in the sample.

Under the assumption that the error terms in each equation are independently and identically distributed, estimating this system of equations is equivalent to pooling the data and estimating the parameters by ordinary least squares (OLS). However, it is likely that the variance of the error term varies across years, and that there is some correlation between the error terms across years. It is therefore possible to get more efficient estimates of the parameters by using the technique of Iterated Seemingly Unrelated Regressions (ISUR).[7]

As equations (3a)–(3e) are written, we have imposed the usual restriction that the parameters are equal across the sample, which allows the most precise estimates of the parameter values. We can also allow some or all of the parameters to vary over time or by firm characteristics, although this additional information is generally obtained at the expense of lowering the precision of the estimates. We will explore some of these alternative specifications in the results section; however, the main results of this paper are based on the system of equations shown in (3a)–(3e).

2.2. Data Sources and Variable Construction

This study employs a unique data set on IS spending by large U.S. firms which was compiled by International Data Group (IDG). The information is collected in an annual survey of IS managers at large firms[8] that has been conducted since 1987. Respondents are asked to provide the market value of central processors (mainframes, minicomputers, supercomputers) used by the firm in the United States, the total central IS budget, the percentage of the IS budget devoted to labor expenses, the number of PCs and terminals in use, and other IT related information.

Since the names of the firms are known and most of them are publicly traded, the IS spending information from the IDG survey could be matched to Standard and Poors' Compustat II[9] to obtain measures of output, capital investment, expenses, number of employees, and industry classification. In addition, these data were also combined with price deflators for output, capital, employment costs, expenses, and IT capital.

There is some discretion as to how the years are matched between the survey and Compustat. The survey is completed at the end of the year for data for the following year. Since we are primarily interested in the value of computer capital stock, and the survey is timed to be completed by the beginning of the new fiscal year, we interpret the survey data as a beginning of period value, which we then match to the end of year data on Compustat (for the previous period). This also allows us to make maximum use of the survey data and is the same approach used by IDG for their reports based on these data (e.g., Maglitta and Sullivan-Trainor 1991).[10]

IDG reports the "market value of central processors" (supercomputers, mainframes and minicomputers) but only the total *number* of "PCs and terminals." Therefore, the variable for COMPUTER CAPITAL was obtained by adding the "market value of central processors" to an estimate of the value of PCs and terminals, which was computed by multiplying the weighted average value for PCs and terminals by the number of PCs and terminals.[11] This approach yields roughly equal values, in aggregate, for central processors ($33.0 Bn) as for PCs and terminals ($30.4 Bn) in 1991. These values were corroborated by a separate survey by International Data Corporation (IDC 1991) which tabulates shipments of computer equipment by category. This aggregate computer capital is then deflated by the computer systems deflator reported in Gordon (1993).

The variables for IS STAFF, NON-IS LABOR AND EXPENSE, and OUTPUT were computed by multiplying the relevant quantity from the IDG survey or Compustat by an appropriate government price deflator. IS STAFF was computed by multiplying the IS Budget figure from the IDC survey by the "percentage of the IS budget devoted to labor expenses . . .", and deflating this figure. NON-IS LABOR AND EXPENSE was computed by deflating total expense and subtracting deflated IS STAFF from this value. Thus, all the expenses of a firm are allocated to either IS STAFF OR NON-IS LABOR AND EXPENSE.

Total capital for each firm was computed from book value of capital stock, adjusted for inflation by assuming that all investment was made at a calculated average age (total depreciation/current depreciation) of the capital stock.[12] From this total capital figure, we subtract the deflated value of COMPUTER CAPITAL to get NON-COMPUTER CAPITAL. Thus, all capital of a firm is allocated to either COMPUTER CAPITAL or NON-COMPUTER CAPITAL. The approach to constructing total capital follows the methods used by other authors who have studied the marginal product of specific production factors using a similar methodology (Hall 1990, Mairesse and Hall 1993).

The firms in this sample are quite large. Their average sales over the sample period were nearly $7.4 billion. However, in many other respects, they are fairly representative of the U.S. economy as a whole. For instance, their computer capital stock averages just over 2% of total sales, or about $155 million, which is roughly consistent with the capital flow tables for the U.S. economy published by the Bureau of Economic Analysis. Similarly, the average IS budget as a share of sales was very close to the figure reported in a distinct survey by CSC/ Index (Quinn et al. 1993). A summary of the sources, construction procedure and deflator for each variable is provided in Table 1, and sample statistics are shown in Tables 2a and 2b.

2.3. Potential Data Problems

There are a number of possible errors in the data, either as a result of errors in source data or inaccuracies introduced by the data construction methods employed. First, the IDG data on IS spending are largely self-reported, and therefore the accuracy depends on the diligence of the respondents. Some data elements require a degree of judgment—particularly the market value of central processors and the total number of PCs and terminals. Also, not all companies responded to the survey, and even those that did respond in one year may not have responded in every other year. This may result in sample selection bias. For instance, high performing firms (or perhaps low performing firms) may have been more interested in participating in the survey.

However, the effect of the potential errors discussed above will probably be small. The information is reasonably consistent from year to year for the same firm, and we have checked the aggregate values against other independent sources. We used a different, independent source (Compustat) for our output measures and for our non-IT variables, eliminating the chance of respondent bias for these measures. We also examined whether the performance of the firms in our sample (as measured by return on equity (ROE) differ from the population of the largest half of Fortune 500 Manufacturing and Fortune 500 Service firms. Our results indicate that there are no statistically significant differences between the ROE of firms in our sample and those that are not (t-statistic = 0.7), which suggests that our sample is not disproportionately comprised of "strong" or "weak" firms. Furthermore, the average size of the firms of our sample is not significantly different from the average size of firms in the top half of the Fortune 500 listings (t-statistic = 0.8). Finally, the response rate of the sample is relatively high at

Series	Source	Construction procedure	Deflator
Computer capital	IDG Survey	"Market Value of Central Processors" converted to constant 1987 dollars, plus the total number of PCs and terminals multiplied by an average value of a PC/terminal, also converted to constant 1987 dollars.	Deflator for Computer Systems (Gordon 1993). Extended through 1991 at a constant rate.
Noncomputer capital	Compustat	Total property, plant, and Equipment Investment converted to constant 1987 dollars. Adjusted for retirements using Winfrey S-3 Table (10-year service life) and aggregated to create capital stock. Computer capital as calculated above was subtracted from this result.	GDP Implicit Deflator for Fixed Investment (Council of Economic Advisors 1992).
IS staff	IDG Survey	Total IS Budget times percentage of IS Budget (by company) devoted to labor expense. Converted to constant 1987 dollars.	Index of Total Compensation Cost (Private Sector) (Council of Economic Advisors 1992).
Non-IS labor and expense	Compustat	Total Labor, Materials and other noninterest expenses converted to constant 1987 dollars. IS labor as calculated above was subtracted from this result.	Producer Price Index for Intermediate Materials, Supplies and Components (Council of Economic Advisors 1992).
Output	Compustat	Total sales converted to constant 1987 dollars.	Industry Specific Deflators from Gross Output and Related Series by Industry, BEA (1977–89) where available (about 80% coverage)—extrapolated for 1991 assuming average inflation rate from previous five years. Otherwise, sector level Producer Price Index for Intermediate Materials Supplies and Components (Gorman 1992).

Table 1 Data sources, construction procedures, and deflators.

over 75%, suggesting that sample selection bias is probably not driving the results.

Second, there are a number of reasons why IS STAFF and COMPUTER CAPITAL may be understated, although by construction these errors do not reduce total capital and total expense for the firm. The survey is restricted to central IS spending in the United States plus PCs and terminals both inside and outside the central department. Some firms may have significant expenditures on information systems outside the central department or outside the United States. In addition, the narrow definitions of IS spending employed in this study may exclude significant costs that could be legitimately counted as COMPUTER CAPITAL such as software and communication networks. Furthermore, by including only the labor portion of IS expenses in IS STAFF as a separate variable (in order to prevent double counting of capital expenditure), other parts of the IS budget are left in the NON-IS LABOR AND EXPENSE category. The effects of these problems

Table 2a Summary statistics.

Sample Statistics — Average over all points (Constant 1987 Dollars)			
	Total$ (Annual Average)	As a % of Output	Per Firm Average
Output	$1,661 Bn	100%	$7.41 Bn
Computer capital (stock)	$34.7 Bn	2.09%	$155 MM
Noncomputer capital (stock)	$1,614 Bn	97.2%	$7.20 Bn
IS Budget (flow)	$27.1 Bn	1.63%	$121 MM
IS staff (flow)	$11.3 Bn	0.68%	$50.4 MM
Non-Is labor and expenses (flow)	$1,384 Bn	83.3%	$6.17 Bn
Avg. number of companies per year	224	224	224
Total observations	1121	1121	1121

Table 2b Sample composition Relative to Fortune 500 population.

Sample Composition Number of firms			
	Fortune 500 Manu-facturing	Fortune 500 Service	Other
1991 Sample Breakdown			
Top Half Fortune 500	157	61	
Lower Half Fortune 500	39	22	
Total	196	83	14
All Fortune 500 Firms in Compustat			
Top Half Fortune 500	240	228	
Lower Half Fortune 500	226	196	n.a.
Total	466	424	

on the final results are discussed in the Results section, especially §3.4.

A third area of potential inaccuracy comes from the price deflators. Numerous authors (Baily and Gordon 1988, Siegel and Griliches 1991) have criticized the current methods employed by the BEA for constructing industry-level price deflators. It has been argued that these methods substantially underestimate quality change or other intangible product improvements. If consumer purchases are in part affected by intangible quality improvements, the use of firm level data should provide a closer approximation to the true output of a firm, because firms which provide quality improvement will have higher sales and can be directly compared to firms in the same industry.

Finally, the measurement of OUTPUT and COMPUTER CAPITAL input in certain service industries appeared particularly troublesome. For financial services, we found that OUTPUT was poorly predicted in our model, presumably because of problems in defining and quantifying the output of financial institutions. In the telecommunications industry, it has been argued (Popkin 1992) that many of the productivity gains have come from very large investments in computer-based telephone switching gear, which is primarily classified as communications equipment and not COMPUTER CAPITAL, although it may be highly correlated with measured computer capital. We therefore excluded all firms in the financial services industries (SIC60–SIC69) and telecommunications (SIC48).[13]

3. Results

3.1. Basic results

The basic estimates for this study are obtained by estimating the system of equations (3a)–(3e) by ISUR (see §2.1). Note that we allow the intercept term to vary across sectors and years.

As reported in column 1 of Table 3, our estimate of β_1 indicates that COMPUTER CAPITAL is correlated with a statistically significant increase in OUTPUT. Specifically, we estimate that the elasticity of output for COMPUTER CAPITAL is 0.0169 when all the other inputs are held constant. Because COMPUTER CAPITAL accounted for an

Parameter	Coefficients	Marginal Product
β_1 (Computer Capital)	0.0169*** (0.00431)	81.0%
β_2 (Non-computer Capital)	0.0608*** (0.00466)	6.26%
β_3 (IS Staff)	0.0178*** (0.00526)	2.62
β_4 (Other Labor and Exp.)	0.883*** (0.00724)	1.07
Dummy Variables	Year*** & Sector***	
R^2 (1991)	97.5%	
N (1991)	293	
N (total)	1121	

Table 3 Base Regressions—Coefficient Estimates and Implied Gross Rates of Return (All parameters (except year dummy) constrained to be equal across years).

*** —$p < 0.01$, ** —$p < 0.05$, * —$p < 0.1$, standard errors in parenthesis

average of 2.09% of the value of output each year, this implies a gross MP (increase in dollar output per dollar of capital stock) for COMPUTER CAPITAL of approximately 81% per year.[14] In other words, an additional dollar of computer capital stock is associated with an increase in output of 81 cents per year on the margin.[15]

The estimate for the output elasticity for IS STAFF was 0.0178, which indicates that each dollar spent here is associated with a marginal increase in OUTPUT of $2.62. The surprisingly high return to information systems labor may reflect systematic differences in human capital,[16] since IS staff are likely to have more education than other workers. The high return may partially explain Krueger's (1991) finding that workers who use computers are paid a wage premium.

The above estimates strongly support hypothesis H1, that the contribution of IT is positive. The t-statistics for our estimates of the elasticity of COMPUTER CAPITAL and IS STAFF are 3.92 and 3.38, respectively, so we can reject the null hypothesis of zero contribution of IT at the 0.001 (two-tailed) confidence level for both. We can also reject the joint hypothesis that they are both equal to zero ($\div^2(2) = 43.9, p < 0.0001$).

To assess H2 (that the contribution of IT is greater than its cost) it is necessary to estimate the cost of COMPUTER CAPITAL and IS STAFF. After these costs are subtracted from the gross benefits reported above, we can then assess whether the remaining "net" benefits are positive. Because IS STAFF is a flow variable, calculating net benefits is straightforward: a dollar of IS STAFF costs one dollar, so the gross marginal product of $2.62 implies a net marginal product of $1.62. For IS STAFF, we can reject the null hypothesis that the returns equal costs in favor of the hypothesis that returns exceed costs at the 0.05 confidence level ($\div^2(1) = 4.4, p < 0.035$).

Assessing H2 for COMPUTER CAPITAL, which is a stock variable, requires that we determine how much of the capital stock is "used up" each year and must be replaced just to return to the level at the beginning of the year. This is done by multiplying the annual depreciation[17] rate for computers by the capital stock in place. According to the Bureau of Economic Analysis, the average service life of "Office, Computing and Accounting Machinery" is seven years (Bureau of Economic Analysis 1987). If a seven-year service life for computer capital is assumed, then the above gross marginal product should be reduced by subtracting just over 14% per year, so that after seven years the capital stock will be fully replaced. This procedure yields a *net* marginal product of 67%. However, a more conservative assumption is that COMPUTER CAPITAL (in particular PCs) could have an average service life as short as three years, which implies that the net rate of return should be reduced by 33%. This would yield a *net* MP estimate of 48%. In either case, we can reject the null hypothesis that the net marginal returns to computers are zero ($p < 0.01$).

However, it should be noted that the full cost of computers involves other considerations than just the decline in value of the asset itself. For instance, calculating a Jorgensonian cost of capital (Christensen and Jorgenson 1969) would also attempt to account for the effects of taxes, adjustment costs, and capital gains or losses, in addition to depreciation costs. On the other hand, firms invest in IT at least partly to move down the learning curve (Brynjolfsson 1993) or create options (Kambil et al. 1993), and these effects may create "assets" offsetting some of the losses to depreciation. The high gross marginal product of COMPUTER CAPITAL suggests that if the total annual cost of COMPUTER CAPITAL were as much as 40%, its net marginal product would be greater than zero by a statistically significant amount.

An alternative approach to assessing H2 is to consider the *opportunity cost* of investing in COMPUTER CAPITAL or IS STAFF. A dollar spent in either of these areas could have generated a gross return of over 6% if it had instead been spent on NONCOMPUTER CAPITAL or a net return of 7% if it were spent on OTHER LABOR AND EXPENSE. In this interpretation, there are only excess returns to COMPUTER CAPITAL or IS STAFF if the returns exceed the return of the respective non-IS component.

As shown in Table 4, we can reject the hypothesis that the *net* MP for COMPUTER CAPITAL is equal to the MP for NONCOMPUTER CAPITAL, conservatively assuming a service life of as little as three years for COMPUTER CAPITAL (and no depreciation for NONCOMPUTER CAPITAL) at the 0.05 confidence level. Similarly, we can reject the hypothesis that IS STAFF generates the same returns as spending on OTHER LABOR AND EXPENSE ($p < 0.05$).

Our confidence in the regression taken as a whole is increased by the fact that the estimated output elasticities for the other, non-IT, factors of production were all positive and each was consistent not only with economic theory (i.e., they imply a real rate of return on non-IT factors of 6%–7%), but also with estimates of other researchers working with similar data (e.g., Hall 1993, Loveman 1994). Furthermore, the elasticities summed to just over one, implying constant or slightly increasing returns to scale overall, which is consistent with the estimates of aggregate production functions by other researchers (Berndt 1991). The R^2 hovered around 99%, indicating that our

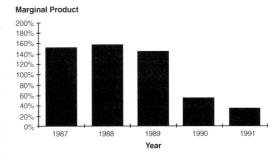

Marginal Product

	1987	1988	1999	1990	1991
Coefficient	.0177**	.0222***	.0239***	.0125**	.0121**
Std. Error	(.00721)	(.00646)	(.00657)	(.00574)	(.00594)
N	135	133	274	286	293

Figure 1 Gross Marginal Product of Computer Capital over Time.

Note

Key: *** - $p<.01$, ** - $p<.05$, * - $p<.1$, standard errors in parenthesis.

independent variables could "explain" most of the variance in output.

3.2. What Factors Affect the Rates of Return for Computers?

The estimates described above were based on the assumption that the parameters did not vary over time, in different sectors, or across different subsamples of firms. Therefore, they should be interpreted only as overall averages. However, by using the multiple equations approach, it is also possible to address questions like: "Has the return to computers been consistently high, or did it vary over time?" and "Have some sectors of the economy had more success in using computers?" We address these questions by allowing the parameters to vary by year or by sector.

Economic theory predicts that managers will increase investments in any inputs that achieve higher than normal returns, and that as investment increases, marginal rates of return eventually fall to normal levels. This pattern is supported by our findings for COMPUTER CAPITAL, which exhibited higher levels of investment (stock increases > 25% per year) and lower returns over time (Figure 1). We find that the rates of return are fairly consistent over the period 1987–1989 and then drop in 1990–1991. We can reject the null hypothesis of equality of returns over time in the full sample ($\div^2(4) =$

Return Difference Tests Computer Capital vs. Other

Capital	Return	χ^2 Statistic	Significance
Gross Return	81%	15.5	$p < 0.001$
Net—7-Year Service Life	67%	10.6	$p < 0.01$
Net—3-Year Service Life	48%	5.5	$p < 0.02$

Table 4 χ^2 Tests for differences in marginal product between computer capital and other capital (return on computer capital greater than return on other capital).

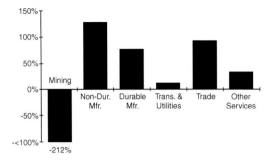

	Mining	Non-Dur. Mfr.	Durable Mfr.	Trans. and Utilities	Trade	Other Service
Coefficient	−.0286	.0122*	.0348***	.00227	.0129	.0153
Std. Error	(.0218)	(.00691)	(.00678)	(.0111)	(.00921)	(.0354)
N (total)	28	414	360	171	123	25

Figure 2 Gross marginal product of computer capital by sector.

Note
Key: *** -p<.01, ** -p<.05, * -p<.1, standard errors in parenthesis.

11.2, $p < 0.02$). Nonetheless, even at the end of the period, the returns to COMPUTER CAPITAL still exceed the returns to NONCOMPUTER CAPITAL. However, these results should be interpreted with caution since they are particularly sensitive to sample changes and time-specific exogenous events such as the 1991 recession.[18]

Roach (1987) has argued that the service sector uses computers much less efficiently than manufacturing and points to aggregate statistics which report higher overall productivity growth for manufacturing than for services. Our data set allows us to reconsider this claim in light of more disaggregated data. The marginal products of COMPUTER CAPITAL across sectors are presented in Figure 2. The marginal product (ignoring the mining sector, which includes only 10 firms and has a large standard error) varies from 10% in transportation and utilities to 127% in nondurable manufacturing. While there have been some suggestions that reorganizing service processes around a "factory" model may help services achieve productivity gains comparable to manufacturing, we cannot confirm that the differences in measured returns are due to fundamental differences, or simply "noise" in the data. Although the returns to computers in durable and nondurable manufacturing are as high or higher than the returns in any other sector, we are unable to reject the hypothesis that these rates of return are the same across most sectors due to the imprecision of the estimates (without mining, $\chi^2(4) = 6.6$, $p < 0.16$).

A second area that can be addressed by our data and method is technology strategy. We have already found that firms with more computer capital will, ceteris paribus, have higher sales than firms with less computer capital, but do the *types* of computer equipment purchased make a

Coefficient Estimates and Marginal Product for Computer Capital Grouping based on mainframes as a percentage of total Computer Capital				
Sample Split	Highest	Middle	Lowest	Statistical Ordering[1]
Elasticity Estimate (β1)	0.0113**	0.0159***	0.0117**	
Standard Error	(0.00500)	(0.00528)	(0.00521)	Med > (High, Low)
Marginal Product (MP$_c$)	49.1%	79.5%	58.2%	(P < 0.03)
Mean % Mainframes	74%	54%	34%	
Group Std. Dev.	9%	5%	8%	

Table 5 Split sample regression results—mainframes as a percentage of total Computer capital.

Note
Key: ***—p < 0.01, **— p < 0.05, *—p < 0.1, standard errors in parenthesis
1 Ordering by χ^2 tests of return difference. P-value shonw represents null hypothesis of equality across groups.

Parameter	OLS Estimates	2SLS Estimates
β_1 (Computer Capital)	0.0284*** (0.00723)	0.0435*** (0.0126)
β_2 (Noncomputer Capital)	0.0489*** (0.00668)	0.0481*** (0.00702)
β_3 (IS Staff)	0.0191*** (0.00795)	0.00727 (0.0116)
β_4 (Non-IS Labor and Exp.)	0.881*** (0.0113)	0.879*** (0.0125)
Dummy Variables	Year*** & sector***	Year*** & Sector***
R^2	98.3%	98.3%
N (total)	702	702

Table 6 Specification Test—Comparison of OLS and Two-Stage Least Squares (All parameters (except year dummy) contrained to be equal across years).

Note:
***—p < 0.01, **—p < 0.05, *—p < 0.1, standard errors are in parentheses
OLS estimates are for sample of same firms as were available for 2SLS regression (n = 702).
Hausman Test Results (instruments are lagged independent variables):
$\chi^2(4) = 6.40$, ($p < 0.17$)—cannot reject exogeneity.

difference? We have data on two categories of equipment: (1) central processors, such as mainframes, and (2) PCs and terminals. For this analysis, we divide the sample into three equal groups based on the ratio of central processor value to PCs and terminals. We find that the rate of return is highest for firms using a more balanced mix of PCs and mainframes (Table 5), and lower for firms at either extreme. One interpretation of this finding is that an IS strategy which relies too heavily on one category of equipment and users will be less effective than a more even-handed approach which allows a better "division of labor."

3.3. Sensitivity Analysis and Possible Biases—Econometric Issues

Our estimates of the return to COMPUTER CAPITAL required that a number of assumptions be made about the econometric specification and the construction of the data set. This section and the following section explore the validity of our assumptions and generally find that the results are robust.

The econometric assumptions required for ISUR to produce unbiased estimates of both the parameters and the standard errors are similar to those for OLS: the error term must be uncorrelated with the regressors (inputs) and homoskedastic in the cross section.[19] ISUR implicitly corrects for serial correlation and heteroskedasticity over time in our formulation, so that additional restrictions on the error structure are not necessary. Nonetheless, we computed single-year OLS estimates both with and without heteroskedasticity-consistent standard errors to test for heteroskedasticity, and plotted the residuals from the basic specification to assess normality. These analyses suggest that neither of these assumptions were violated, although, even if they were coefficient estimates (even for OLS), they would still be unbiased and consistent (but standard errors would be incorrect).

However, the assumption that the error term is uncorrelated with the inputs (orthogonality) is potentially an issue. One way in which this assumption could be violated is if the causality is reversed: instead of increases in purchases of inputs (e.g., computers) leading to higher output, an increase in output could lead to further investment (for example, a firm spends the proceeds from an unexpected increase in demand on more computer equipment). The orthogonality assumption can also be violated if the input variables are measured with error. The direction of bias of the coefficients from measurement error is dependent on both the correlation among the variables as well as the correlation among measurement errors (see Kmenta 1986 for a complete discussion).[20]

Regardless of the source of the error, it is possible to correct for the potential bias using instrumental variables methods, or two-stage least squares (2SLS). We use once-lagged values of variables as instruments, since by definition they cannot be associated with unanticipated shocks in the dependent variable in the following year.[21] Table 6 reports a comparison of pooled OLS estimates with 2SLS estimates and shows that the coefficient estimates are similar although somewhat higher for COMPUTER CAPITAL and lower for IS STAFF. In both cases the standard errors were substantially larger, as is expected when instrumental variables are used. Using a Hausman specification test, we cannot reject the null hypothesis that the error term is uncorrelated with the regressors (see bottom of Table 5 for test statistics), and therefore do not reject our initial specification.

3.4. Sensitivity Analysis and Possible Biases—Data Issues

To further explore the robustness of our results, we examined the impact of the possible data errors discussed in §2.3 that can be tested: (1) error in the valuation of PCs and terminals, (2) errors in the price deflators, and (3) understatement or misclassification of computer capital.

To assess the sensitivity of the results to possible errors in the valuation of PCs and terminals, we recalculated the basic regressions varying the assumed average PC and terminal value from $0 to $6K. Note that as the assumed value of PCs and terminals increases, the increase in COMPUTER CAPITAL will be matched by an equal decrease in NONCOMPUTER CAPITAL, which is calculated as a residual. Interestingly, the return to COMPUTER CAPITAL in the basic regression is not very sensitive to the assumed value of PCs and terminals, ranging from 77% if they are not counted to 59% if PCs and terminals are counted at $6K (peaking at about 85%).

A second contribution to error is the understatement of output due to errors in the price deflators. While it is difficult to directly correct for this problem, we also estimated the basic equations year by year, so that errors in the relative deflators would have no impact on the elasticity estimates. The estimated marginal products ranged from 109% to 197% in the individual year regressions versus 81% when all five years were estimated simultaneously. The standard error on the estimates was significantly higher for all estimates, which can account for the large range of estimates. Overall, this suggests that our basic findings are not a result of the assumed price deflators. However, if the price deflators systematically underestimate the value of intangible product change over time or between firms, our measure of output will be understated, implying that the actual return for computer capital is higher than our estimates.

To assess the third source of error, possible understatement or misclassification of computer capital, we consider three cases: (1) hidden computer spending exists, but does not show up elsewhere in the data; (2) hidden computer spending exists and shows up in the "NON-IS LABOR AND EXPENSE" category; or (3) hidden computer spending exists and shows up in the "OTHER CAPITAL" category. If the hidden IS costs do not show up elsewhere in the firm (e.g., software development or training costs from previous years), then the effect on the estimated returns is dependent on how closely correlated these costs are to our measured COMPUTER CAPITAL. If they are uncorrelated, our estimate for the elasticity and the return to COMPUTER CAPITAL is unbiased. If the missing costs are perfectly correlated with the observed costs, then, because of the logarithmic form of our specification, they will result only in a multiplicative scaling of the variables, and the estimated elasticities and the estimated standard error will be unchanged.[22] For the same reason, the sign and statistical significance of our results for the returns to COMPUTER CAPITAL and IS STAFF will also be unaffected. However, the denominator nominator used for the MP calculations will be affected by increasing computer capital so the estimated MP will be proportionately lower or higher. For instance, if the hidden costs lead to a doubling of the true costs of computer capital, then the true MP would fall from 81% to just over 40%. Finally, if the hidden costs are negatively correlated with the observed costs, then the true returns would be higher than our estimates.

A second possible case is that hidden IS capital expenses (e.g., software) show up in the NON-IS LABOR AND EXPENSE category. To estimate the potential impact of these omissions, we estimate the potential size of the omitted misclassified IS capital relative to COMPUTER CAPITAL using data from another IDG survey (IDC 1991) on aggregate IS expenditures, including software as well as hardware. To derive a reasonable lower bound on the returns to COMPUTER CAPITAL, we assume that the misclassified IS capital had an average service life of three years, and further make the worst-case assumption of perfect correlation between misclassified IS capital and COMPUTER CAPITAL (and reduce proportionally the amount of NON-IS LABOR AND EXPENSE). In this scenario, our estimates for the amount of COMPUTER CAPITAL in firms roughly doubles, yet the rates of return are little unchanged from the basic analysis that does not include misclassified IS capital (68% vs. 81%). This surprising result appears to be due to the fact

Hypothesis	Description of Test (alternative hypothesis)	Test Statistic
H1	Positive marginal product for COMPUTER CAPITAL	$t = 3.92, P < 0.01$
H1	Positive marginal product for IS STAFF	$t = 3.38, p < 0.01$
H1	Simultaneous test for positive marginal product for COMPUTER CAPITAL and IS STAFF	$\chi^2(2) = 43.9, p < 0.01$
H2	Positive net marginal product for COMPUTER CAPITAL (see TABLE 6), Cost @ 14% (7-year average life)	$t = 3.24, p < 0.01$
H2	Positve net marginal product for COMPUTER CAPITAL (see TABLE 6), Cost @ 33% (3-year average life)	$t = 2.32, p < 0.05$
H2	Positive net marginal product for IS STAFF	$\chi^2(1) = 4.4, p < 0.05$
H2	Marginal product of COMPUTER CAPITAL exceeds marginal product of OTHER CAPITAL	See Table 4
H2	Marginal product of IS STAFF exceeds marginal product of OTHER LABOR AND EXPENSE	$\chi^2(1) = 4.0, p < 0.05$
Extension of H1	Marginal Product of COMPUTER CAPITAL change across time	$\chi^2(4) = 11.2, p < 0.02$
Extension of H1	Marginal Product of COMPUTER CAPITAL change across sectors	$\chi^2(4) = 6.6, p < 0.2$
Extension of H1	Marginal Product varies by mainframes as a percentage of total COMPUTER CAPITAL	$\chi^2(2) = 6.9, p < 0.03$

Table 7 Summmary of hypothesis tests.

that the return on NON-IS LABOR AND EXPENSE is at least as high as the return on COMPUTER CAPITAL, so moving costs from one category to another does not change overall returns much.

Alternatively, a third possible case is that the hidden IS capital expenditures show up in OTHER CAPITAL. This would apply to items such as telecommunications hardware, which would normally be classified as a capital expenditure. In this case, the marginal product of COMPUTER CAPITAL will be reduced proportionally to the amount of the misclassification. Intuitively, this case is similar to the case discussed earlier in which the hidden costs are perfectly correlated with measured costs but do not appear elsewhere. Our simulation results indicate that the elasticities on computer capital vary less than 5% even between assumptions of 0% to 100% of computer capital being misclassified.

Irrespective of these sensitivity calculations, it should be noted that the definition of COMPUTER CAPITAL used in this study was fairly narrow and did not include items such as telecommunications equipment, scientific instruments, or networking equipment. The findings should be interpreted accordingly and do not necessarily apply to broader definitions of IT. However, to the extent that the assumptions of our sensitivity analysis hold, the general finding that IT contributes significantly to output is robust (H1), although the actual point estimates of marginal product may vary, possibly resulting in no statistical difference between returns to computer capital and returns to other capital.

4. Discussion

4.1. Comparison with Earlier Research

Although we found that computer capital and IS labor increase output significantly under a variety of formulations (see summary Table 7), several other studies have failed to find evidence that IT increases output. Because the models we used were similar to those used by several previous researchers, we attribute our different findings primarily to the larger and more recent data set we used. Specifically, there are at least three reasons why our results may differ from previous results.

First, we examined a later time period (1987–1991) than did Loveman (1978–1982), Barua et al. (1978–1982), or Berndt and Morrison (1968–1986). The massive build-up of computer capital is a relatively recent phenomenon. Indeed, the delivered amount of computer power in the companies in our sample is likely to be at least an order of magnitude greater than that in comparable firms from the period studied by the other authors. Brynjolfsson (1993) argues that even if the MP of IT were twice that of non-IT capital, its impact on output in the 1970s or early 1980s would not have been large enough to be detected with available data by conventional estimation procedures. Furthermore, the changes in business processes needed to realize the benefits of IT may have taken some time to implement, so it is possible that the actual returns from investments in computers were initially fairly low. In particular, computers may have initially created organizational slack which was only recently eliminated, perhaps hastened by the increased attention engendered by earlier studies that indicated a potential productivity shortfall and suggestions that "to computerize the office, you have to reinvent the office" (Thurow 1990). Apparently, an analogous period of organizational redesign was necessary to unleash the benefits of electric motors (David 1989).

A pattern of low initial returns is also consistent with the strategy for optimal investment in the presence of learning-by-using: short-term returns should initially be lower than returns for other capital, but subsequently rise to exceed the returns to other capital, compensating for the "investment" in learning (Lester and McCabe 1993). Under this interpretation, our high estimates of computer MP indicate that businesses are beginning to reap rewards from the experimentation and learning phase in the early 1980s.

Second, we were able to use different and more detailed firm-level data than had been available before. We argue that the effects of computers in increasing variety, quality, or other intangibles are more likely to be detected in firm level data than in the aggregate data. Unfortunately, all such data, including ours, are likely to include data errors. It is possible that the data errors in our sample happened to be more favorable (or less unfavorable) to computers than those in other samples. We attempted to minimize the influence of data errors by cross-checking with other data sources, eliminating outliers, and examining the robustness of the results to different subsamples and specifications. In addition, the large size of our sample should, by the law of large numbers, mitigate the influence of random disturbances. Indeed, the precision of our estimates was generally much higher than those of previous studies; the statistical significance of our estimates owes as much to the tighter confidence bounds as to higher point estimates.

Third, our sample consisted entirely of relatively large "Fortune 500" firms. It is possible that the high IS contribution we find is limited to these larger firms. However, an earlier study (Brynjolfsson et al. 1994) found evidence that smaller firms may benefit disproportionately from investments in information technology. In any event, because firms in the sample accounted for a large share of the total U.S. output, the economic relevance of our findings is not heavily dependent on extrapolation of the results to firms outside the sample.

4.2. Managerial Implications

If the spending on computers is correlated with significantly higher returns than spending on other types of capital, it does not necessarily follow that companies should increase spending on computers. The firms with high returns and high levels of computer investment may differ systematically from the low performers in ways that cannot be rectified simply by increasing spending. For instance, recent economic theory has suggested that "modern manufacturing," involving high intensity of computer usage, may require a radical change in organization (Milgrom and Roberts 1990). This possibility is emphasized in numerous management books and articles (see, e.g., Malone and Rockart 1991, Scott Morton 1991) and supported in our discussions with managers, both at their firms and during a workshop on IT and Productivity attended by approximately 30 industry representatives.[23]

Furthermore, our results showing a high gross marginal product may be indicative of the differences between computer investment and other types of investment. For instance, managers may perceive IS investment as riskier than other investments, and therefore require higher expected returns to compensate for the increased risk. Finally, IS is often cited as an enabling technology which does not just produce productivity improvements for individuals, but provides benefits by facilitating business process redesign or improving the ability of groups to work together. In this sense, our results may be indicative of the substantial payoffs to reengineering and other recent business innovations.

5. Conclusion

We examined data which included over 1000 observations on output and several inputs at the firm level for 1987–1991. The firms in our sample had aggregate sales of over $1.8 trillion in 1991 and thus account for a substantional share of the U.S. economy. We tested a broad variety of specifications, examined several different subsamples of the data, and validated the assumptions of our econometric procedures to the extent possible. Overall, we found that computers contribute significantly to firm-level output, even after accounting for depreciation, measurement error, and some data limitations.

There are a number of other directions in which this work could be extended. First, the data set could be expanded to include alternative measures of output, such as value added, and to include additional inputs, such as R&D, that have been explored in other literature (see Brynjolfsson and Hitt 1995). Second, although our approach allowed us to infer the value created by intangibles like product variety by looking at changes in the revenues at the firm level, more direct approaches might also be promising such as directly accounting for intangible outputs such as product quality or variety.

Finally, the type of extension which is likely to have the greatest impact on practice is further analysis of the factors which differentiate firms with high returns to IT from low performers. For instance, is the current "downsizing" of firms leading to higher IT productivity? Are the firms that have undertaken substantial "reengineering efforts" also the ones with the highest returns? Since this study has presented evidence that the computer "productivity paradox" is a thing of the past, it seems appropriate that the next round of work should focus on identifying the strategies which have led to large IT productivity.[24]

NOTES

1. An observation is one year of data on all variables for a specific firm. We did not have all five years of data for every firm, but the data set does include at least one year of data for 367 different firms.
2. The database contains standard financial information, IT spending data, and other economic measures such as product prices and quality for 60 business units of 20 firms over the period 1978–1984. See Loveman (1994) for a more detailed description.
3. For instance, the 95% confidence interval exceeded ±200% for the Marginal Product of IT implied by the estimates in Loveman (1994).
4. Another common way to operationalize the theory is to use the production function to derive a "cost function" which provides the minimum cost required for a given level of output. While cost functions have some attractive features, they require access to firm-level price information for each input, which are data we do not have.
5. A more complete model might include other variables describing management practices or lags of IT spending. We do not consider lags because we already use an IT stock variable, and the panel is too short to consider lags.
6. Formally, the output elasticity of computers, E_c, is defined as: $E_c = (\partial F / \partial C)(\partial C / F \partial \partial)$. For our production function, F, this reduces to:

$$E_c = \beta_1 e^{\beta_0} C^{\beta_1 - 1} K^{\beta_2} S^{\beta_3} L^{\beta_4} \frac{C}{e^{\beta_0} C^{\beta_1} K^{\beta_2} S^{\beta_3} L^{\beta_4}} = \beta_1.$$

The MP for computers is simply the output elasticity multiplied by the ratio of output to computer input:

$$MP_c = \frac{\partial F}{\partial C} = \frac{\partial F}{\partial C} \frac{CF}{FC} = E_c \frac{F}{C}.$$

7. Sometimes also called IZEF, the iterated version of Zellner's efficient estimator. By leaving the covariance matrix across years unconstrained this procedure implicitly corrects for serial correlation among the equations, even when there are missing observations for some firms in some years.
8. Specifically, the survey targets Fortune 500 manufacturing and Fortune 500 service firms that are in the top half of their industry by sales (see Table 2a).
9. Compustat II provides financial and other related information for publicly traded firms, primarily obtained through annual reports and regulatory filings.
10. This matching procedure may be sensitive to possible reverse causality between output and IS labor as is shown by our Hausman test in Table 6.
11. Specifically, we estimated the value of terminals and the value of PCs and then weighted them by the proportion of PCs versus terminals. For terminals, we estimated the

value as the average list price of an IBM 3151 terminal in 1989 which is $609 (Pelaia 1993). For PCs we used the average nominal PC cost over 1989–1991 of $4,447, as reported in Berndt and Griliches (1990). These figures were then weighted by the proportion of PCs to terminals in the 1993 IDG survey (58% terminals). The resulting estimate was $0.42^* \$609 + 0.58^* \$4{,}447 = \$2{,}835$.

12. An alternative measure of capital stock was computed by converting historical capital investment data into a capital stock using the Winfrey S-3 table. This approach was used in earlier versions of this paper (Brynjolfsson and Hitt 1993) with similar results. However, the calculation shown above is more consistent with previous research (see, e.g., Hall 1993).

13. The impact of these changes in both cases was to lower the return to COMPUTER CAPITAL slightly as compared to the results on the full sample.

14. As noted in footnote 6, *supra*, $MP_c = E_c(F/C)$, which in this case is $0.0169/0.0209 = 0.8086$, or about 81%.

15. It is worth noting that our approach provides estimates of the *marginal* product of each input: how much the last dollar of stock or flow added to output. In general, inframarginal investments have higher rates of return than marginal investments, so the return to the average dollar invested in computers is likely to be even higher than the marginal returns we reported.

16. We thank Dan Sichel for pointing this out.

17. Technically, "negative capital gains" may be a more accurate term than "depreciation," since computer equipment is more likely to be replaced because of the arrival of cheaper, faster alternatives than because it simply wears out.

18. A decline in the returns to COMPUTER CAPITAL between 1989 and 1990 is also evident in a balanced panel of 201 firms in the sample for 1989–1991.

19. Note that if we had used OLS, further assumptions would be required: that all error terms are independent and constant variance over time.

20. If an input variable is systematically understated by a constant multiplicative factor, then the coefficient estimates would be unchanged.

21. However, in the presence of individual firm effects, lagged values are not valid instruments. While we did not test for firm effects, we suspect they may be important, and so the results of our 2SLS estimates should be interpreted with caution.

22. This is because multiplicative scaling of a regressor in a logarithmic specification will not change the coefficient estimate or the standard error. All the influence of the multiplier will appear in the intercept term, which is not crucial to our analysis.

23. The MIT Center for Coordination Science and International Financial Services Research Center jointly sponsored a Workshop on IT and Productivity which was held in December, 1992.

24. This research has been generously supported by the MIT Center for Coordination Science, the MIT Industrial Performance Center, and the MIT International Financial Services Research Center. We thank Martin Neil Baily, Rajiv Banker, Ernst Berndt, Geoff Brooke, Zvi Griliches, Bronwyn Hall, Susan Humphrey, Dan Sichel, Robert Solow, Paul Strassmann, Diane Wilson, three anonymous referees, and seminar participants at Boston University, Citibank, Harvard Business School, the International Conference on Information Systems, MIT, National Technical University in Singapore, Stanford University, the University of California at Irvine, and the U.S. Federal Reserve for valuable comments, while retaining responsibility for any errors that remain. We are also grateful to International Data Group for providing essential data. An earlier, abbreviated version of this paper was published in the *Proceedings of the International Conference on Information Systems*, 1993, under the title "Is Information Systems Spending Productive? New Evidence and New Results."

REFERENCES

Baily, M. N. and R. J. Gordon, "The Productivity Slowdown, Measurement Issues, and the Explosion of Computer Power," in W. C. Brainard and G. L. Perry (Eds.), *Brookings Papers on Economic Activity*, The Brookings Institution, Washington, DC, 1988.

Barua, A., C. Kriebel and T. Mukhopadhyay, "Information Technology and Business Value: An Analytic and Empirical Investigation," University of Texas at Austin Working Paper, Austin, TX, May, 1991.

Berndt, E., *The Practice of Econometrics: Classic and Contemporary*, Addison-Wesley, Reading, MA, 1991.

Brynjolfsson, E., "The Productivity Paradox of Information Technology," *Comm. ACM*, 35 (1993), 66–77.

———, "Technology's True Payoff," *Information week* (October 10, 1994), 34–6.

——— and L. Hitt, "Is Information Systems Spending Productive? New Evidence and New Results," *Proc. 14th International Conf. on Information Systems*, Orlando, FL, 1993.

Brynjolfsson, E. and L. Hitt, "Information Technology as a Factor of Production: The Role of Differences Among Firms," *Economics of Innovation and New Technology*, 3, 4 (1995), 183–200.

———, T. Malone, V. Gurbaxani and A. Kambil, "Does Information Technology Lead to Smaller Firms?", *Management Sci.*, 40, 12 (1994).

Bureau of Economic Analysis, U. S. D. o. C., *Fixed Reproducible Tangible Wealth in the United States, 1925–85*, U.S. Government Printing Office, Washington, DC, 1987.

Christensen, L. R. and D. W. Jorgenson, "The Measurement of U.S. Real Capital Input, 1929–1967," *Review of Income and Wealth*, 15, 4 (1969), 293–320.

Dos Santos, B. L., K. G. Peffers and D. C. Mauer, "The Impact of Information Technology Investment Announcements on the Market Value of the Firm," *Information Systems Res.*, 4, 1 (1993), 1–23.

Harris, S. E. and J. L. Katz, "Profitability and Information Technology Capital Intensity in the Insurance Industry," in *Proc. Twenty-First Hawaii International Conf. on System Science*, 1988.

International Data Corporation (IDC), "U.S. Information Technology Spending Patterns, 1969–1994," IDC Special Report 5368, 1991.

Kmenta, J., *Elements of Econometrics*, Second edition, Macmillan, New York, 1986.

Loveman, G. W., "An Assessment of the Productivity Impact on Information Technologies," in T. J. Allen and M. S. Scott Morton (Eds.), *Information Technology and the Corporation of the 1990s: Research Studies*, MIT Press, Cambridge, MA, 1994.

Malone, T. and J. Rockart, "Computers, Networks and the Corporation," *Scientific American*, 265, 3 (1991), 128–36.

Milgrom, P. and J. Roberts, "The Economics of Modern Manufacturing: Technology, Strategy and Organization," *American Economic Rev.*, 80, 3 (1990), 511–28.

Morrison, C. J. and E. R. Berndt, "Assessing the Productivity of Information Technology Equipment in the U.S. Manufacturing Industries," National Bureau of Economic Research Working Paper 3582, January, 1990.

Pelaia, E., "IBM Terminal Prices," personal communication with IBM Representative, 1993.

Popkin, J. and Company, "The Impact of Measurement and Analytical Issues in Assessing Industry Productivity and its Relation to Computer Investment," Washington, DC, Mimeo October, 1992.

Quinn, M. A. et al., "Critical Issues of Information Systems Management for 1993," CSC Index, The Sixth Annual Survey of Information Systems Management Issues, Cambridge, MA, 1993.

Scott Morton, M. (Ed.), *The Corporation of the 1990s: Information Technology and Organizational Transformation*, Oxford University Press, New York, 1991.

Strassmann, P. A., *The Business Value of Computers*, Information Economics Press, New Canaan, CT, 1990.

Thurow, L., "Are Investments in Information Systems Paying Off?," *MIT Management* Spring (1990).

Varian, H., *Microeconomic Analysis*, Third Edition, W. W. Norton and Company, Inc., New York, 1992.

Wilson, D., "Assessing the Impact of Information Technology on Organizational Performance," in R. Barker, R. Kauffman and M. A. Mahmood (Eds.), *Strategic Information Technology Management*, Idea Group, Harrisburg, PA, 1993.

Information Technology and the Nature of Managerial Work: From the Productivity Paradox to the Icarus Paradox?[1,2]

Alain Pinsonneault and Suzanne Rivard

École des Hautes Études Commerciales, IT Department, 3000 Chemin de la Côte-Sainte-Catherine, Montreal, Québec H3T 2A7, Canada, alain.pinsonneault@hec.ca
École des Hautes Études Commerciales, IT Department, 3000 Chemin de la Côte-Sainte-Catherine, Montreal, Québec H3T 2A7, Canada, suzanne.rivard@hec.ca

Abstract

Modern organizations are investing heavily in information technology (IT) with the objective of increasing overall profitability and the productivity of their knowledge workers. Yet, it is often claimed that the actual benefits of IT are disappointing at best, and that IT spending has failed to yield significant productivity gains—hence the productivity paradox. Evidence is fragmented and somewhat mitigated. This paper argues that the current state of empirical research results from a failure to understand the interplay between IT and managerial work. It addresses this issue by analyzing patterns of association between IT usage and the nature of managerial work in different organizational contexts. Fifty-nine semi-structured interviews were conducted with middle line managers in three large companies: a Bank, a Telecommunications company, and a Utility. In addition, daily activities and IT usage were logged. The data indicate that the relationship between the level of IT usage and the nature of managerial work was stronger in the two organizations that were reorienting their strategies (Bank, Telecommunications) than in the one pursuing its existing strategy (Utility). It was also found that the pattern of the relationship between IT usage and the nature of managerial work depended on the kind of strategic reorientation implemented by the firm. For instance, in the Bank, the level of IT usage was associated with the amount of time spent by managers on information-related activities (e.g., reading reports, gathering information) and on disturbance handling activities (e.g., resolving conflicts, managing crises). In the Telecommunications company, IT usage was associated with more time spent on information-related activities and less on negotiation-related activities (e.g., discussions with colleagues on resource sharing, discussions with subordinates on performance standards). This finding suggests that heavy IT users paid greater attention to and spent more time on the roles they performed best with the technology (information-related activities) and may in fact have been embarking on an over-specialization trajectory.

Keywords: IS evaluation, IS impacts, management roles, organizational strategies, IS usage

ISRL Categories: AF09, DA08, EI02, EI0211, G

Introduction

In recent years, firms have been investing substantial amounts of money in information technology (IT). In 1991, for instance, American service sector companies spent over $100 billion on hardware—more than $12,000 per information worker (Roach 1991)—and almost 40% of United States capital spending was being used to acquire IT. In 1996, American banks alone spent almost $18 billion on IT, while American and European financial institutions together invested over $75 billion (*The Economist* 1996). Much of this investment was directed at modifying office work and improving productivity (Applegate 1998; Brandt 1994; Drucker 1988, 1993). Yet, many authors claim that the benefits of IT are disappointing at best, and that IT spending has failed to yield significant productivity gains—hence the productivity paradox (Barua et al. 1991; Franke 1987; Loveman 1988; Parson et al. 1990; Roach 1985, 1991; Strassman 1990; Weill 1992). On the other hand, IT spending has also been linked to significant productivity improvements (Brynjolfsson and Hitt 1993, 1995; Osterman, 1986).

This paper addresses the productivity paradox by trying to understand the interplay between IT and the work process. More precisely, it reports the results of a study which examined the relationships between the level of IT usage and the nature of middle management work in different organizational contexts. The need for a change of focus in research emerges from two main observations that can be made regarding the empirical evidence provided for the productivity paradox. First, there may have been important *measurement problems of either inputs or outputs* in industry and economy-level studies. For example, IT stock may have been overestimated because many firms overstate the decline in the computer price deflator, thus artificially increasing the number of computers purchased in recent years compared with past figures (Brynjolfsson 1993). The purchase of complementary software or training might also artificially inflate short-term investments in computerization. Output measurement has also been problematic, in particular due to the absence of a natural unit of output allowing meaningful comparisons between different products and services (Panko 1991). Several authors argue that, rather than examining the relationship between the amount of IT investment and productivity, it might be more appropriate to consider the relationship between IT usage and the work process (Barua et al. 1991; Davis 1991; Kelley 1994; Panko 1991; Strassman 1990).

Second, the lack of convergent findings regarding the IT-productivity relationship might simply reflect the fact that *IT has often been studied without taking into account the organizational context in which it was deployed and used*. Several authors argue that the organizational context is in fact a fundamental determinant of outcomes associated with IT implementation (Barley 1986; Markus and Robey 1988; Pinsonneault and Kraemer 1993, 1997; Robey and Sahay 1996). Productivity improvement is just one of many IT investment objectives. For example, some organizations might use IT to help enhance the quality of services and increase their flexibility and responsiveness to changing environmental conditions, while others might use it to increase efficiency (Quinn and Baily 1994). Lumping together firms with different objectives may blur the analysis.

These two observations suggest that a complementary understanding of the productivity paradox might be obtained through studying IT usage in light of the organizational context, focusing on work process rather than on work output. The study reported in this paper follows this line of thought and takes a first step in that direction by analyzing the relationship between the level of IT usage and the nature of middle management work in different organizational contexts. It first assesses the literature on IT and managerial work, drawing on this body of work and on the strategic reorientation/convergence literature (Freeman and Cameron 1993) to present three propositions. It then reports the results of a field study conducted with 59 middle managers in three firms of similar size with different strategic contexts. The paper ends by discussing the findings and presenting the limitations of the study.

Information Technology and Managerial Work

Interestingly, the empirical evidence concerning the relationship between IT and managerial work parallels that concerning the productivity paradox: it is mixed and inconclusive. In fact, IT has been found to be associated with both upgraded and deskilled work. For instance, Hoos (1960) suggested that Leavitt and Whisler's (1958) prediction that much middle management work would be deskilled was rapidly being realized. Several of the managers interviewed in that study felt that IT took away numerous decision-making opportunities and limited their exercise of initiative and judgment, as well as their span of control (Hoos 1960). IT was also found to be associated with greater centralization of decision authority at top management levels, standardization and regulation of middle management work, and an increased number of rules and procedures (Mann and Williams 1960). Similarly, the study of a radio and television company (Bjørn-Andersen and Pedersen 1980) found that IT was associated with loss of managers' discretion over their jobs (deciding whether or not to perform a particular task and when and how to perform it). Finally, professional and clerical workers in 38 work groups reported that desktop computerization was transforming their work and had generated higher expectations from supervisors, more work for the groups, increased time pressures, harder work, and longer hours. Work was thus becoming more "sweat intensive" (Zmuidzinas et al. 1990).

On the other hand, some studies found that IT usage was associated with upgraded work. To explain this finding, various authors have argued that by absorbing the information-intensive tasks, IT leaves middle managers with more time to concentrate on the conceptual and decision-making aspects of their jobs and allows them to perform more unstructured activities. For instance, IT was found to be associated with greater decentralization of decision-making authority, a broadening of middle managers' work, and greater flexibility and decision orientation (George 1986; Klatzky 1970; Whisler 1970). IT was also found to be associated with greater autonomy for middle managers in performing their jobs and with less standardized, predetermined decision procedures (Pfeffer and Leblebici 1977). Another study found that middle managers perceived office automation as enriching their jobs, making them more important to the organization, and increasing personal and departmental effectiveness (Millman and Hartwick 1987). In other instances, middle managers felt that IT improved their confidence in making decisions, helped them remove uncertainty from decisions, and, overall, increased their role in the organization (Buchanan and McCalman 1988). One study found that IT was perceived by middle managers as allowing them to become more involved in strategy implementation and in the development of the organizational structure and reward systems (Wooldridge and Floyd 1990). Finally, IT was also found to be associated with middle managers spending more time on people management and on the strategic aspects of their work (Dopson and Stewart 1993). In addition, these middle managers reported that IT provided better information faster, allowing them to deal with routine tasks more efficiently.

While the final outcome associated with IT usage—upgraded or deskilled work—may differ, both sets of findings are explained by a similar argument. It is suggested that, because of its comparative advantage in handling and processing information, IT takes over most information-related activities (Dopson and Stewart 1993; George 1986; Hoos 1960; Klatzky 1970; Leavitt and Whisler 1958). Hence, the first proposition:

Proposition 1: The level of IT usage will be associated with spending less time on information-related activities.

A key question arises at this point: why is IT sometimes associated with deskilled work and sometimes with upgraded work? Several authors suggest that the answer to this question resides in the context in which IT was deployed and used (Markus and Robey 1988). Studies indicate that IT itself is not a determinant of organizational or individual outcomes, but rather an enabler whose effects are dependent on how it is implemented and used (Pinsonneault and Kraemer 1993, 1997; Robey and Sahay 1996). The nature of managerial work associated with IT usage is thus likely to be different depending on the organizational context.

Two generic models of organizational change, which have been developed in the literature, are useful in understanding the organizational context in which IT usage occurs (Freeman and Cameron 1993; Miller and Friesen 1980; Pettigrew 1985; Tushman and Romanelli 1985). These models emphasize the differences between evolutionary, incremental, or convergent change on the one hand, and revolutionary, metamorphic, or discontinuous change on the other. *In a convergence context*, incremental and adaptive change focuses on improving the efficiency of existing operations and achieving greater consistency among the organization's internal activities. The desire to refine and adjust the organization's systems and improve consistency actually hinders radical and discontinuous change (Freeman and Cameron 1993). On the other hand, a *reorientation context* is characterized by simultaneous and abrupt shifts in strategy, power distribution, structure, and control systems. In such a context, top managers mediate between internal and external institutional forces for inertia and the competitive or technological forces for fundamental changes (Freeman and Cameron 1993). Reorientation contexts are thus characterized by the desire to do different things and may include substantially modifying structures and work design.

IT is a tool that organizations use to facilitate change and generate opportunities for organizational change (Robey and Sahay 1996). As such, it is likely to be more closely associated with the nature of managerial work when an organization is in the process of reorienting its strategy than when it is in a convergence mode. In a reorientation context, IT can be used to facilitate major strategy and structural shifts and to produce managerial activities that are better aligned with the firm's new strategy. In a convergence context, IT usage is likely to reinforce the current structure and strategy, as well as the current nature of managerial work. Hence, the second proposition:

Proposition 2: There will be a stronger relationship between the level of IT usage and the nature of managerial work in a reorientation context than in a convergence context.

One study found that a given technology (a geographical information system) had fundamentally different consequences depending on the organizational context in which it was implemented and used (Robey and Sahay 1996). The study concluded, along with others (Barley 1986; Orlikowski 1993), that different contextual elements interact with technical initiatives to produce different consequences. This suggests that technological initiatives can be expected to interact differently with managerial work, depending on the organizational context in which the interplay takes place. Hence, a given IT initiative would be associated differently with the nature of managerial work depending on the kind of strategic reorientation of the firm. For example, in an organization reorienting itself to create closer relationships with its clients, managers are likely to use IT to understand their customers better (e.g., segmentation, client profiles) and to spend a greater proportion of their work time on developing the organization's customer base and tightening its relationship with clients, and less time on other aspects of their work. On the other hand, in an organization refocusing its strategy on a low cost approach, IT might be used by managers to control and streamline operations and increase efficiency, rather than to enhance the quality of the firm's services. Managers will spend more time controlling operations and less on developing new business opportunities. In a convergence context, IT is not likely to be associated with the time allocation pattern of managers because the organization is not modifying its strategy or its structure. Hence, the third proposition:

Proposition 3: The pattern of the relationship between IT usage and the nature of managerial work depends on the kind of reorientation the firm has undertaken.

Research Method

Concepts and Measures

The present study revolves around two main concepts: the nature of managerial work and IT usage. More precisely, it examines the relationship between the level of IT usage and the time spent by middle managers on various managerial activities. Following previous work, middle managers are defined as managers above first-level supervisors but below department heads (Pinsonneault and Kraemer 1997).

While several studies have analyzed the relationship between IT and *managerial work* (Dopson and Stewart 1993; Hoos 1960; Millman and Hartwick 1987; Wooldridge and Floyd 1990), none has proposed a formal measure of this concept. However, in the management field, numerous conceptualizations of managerial work have been proposed. Among them, Mintzberg's (1973) framework is the one most frequently used to study managerial work in general (Kotter 1982; Kurke and Aldrich 1983; McCall and Segrist 1980), and middle managerial work in particular (Tsui 1984) (see Appendix A for a summary of Mintzberg's roles and examples of the activities within each role). In the present study, the nature of managerial work was assessed by measuring the number of minutes middle managers spent on different roles, as defined by Mintzberg, during a typical working day. *IT usage* refers to interactions with the computer and was measured by the number of minutes of hands-on usage during a typical working day.

Data Collection

The relationship between IT usage and the nature of managerial work was explored through semistructured interviews with 59 middle managers in three comparable large firms (between 22,000 and 25,000 employees) in the service sector.

Selection of Participating Firms

The firms were selected on the basis of an interview with the vice-president of operations and the vice-president of information technology in each organization. A second interview was conducted with these respondents to understand the structure, environment, strategy, IT implementation process and orientation, and overall IT usage of each firm. Three large organizations were chosen for this study: one in a reorientation context (Bank), the second in a prereorientation context (Telecommunications), and the third in a convergence context (Utility).

When the study took place, the Bank had suffered major losses, mainly in some of its largest accounts, and a new president with a radically different vision from his predecessor had been appointed. The new president decided to reorient the Bank's strategy from a traditional banking mode, with emphasis on large clients, accounts, and equity, to a new mode, segmenting its customer base and establishing a privileged relationship with all profitable customers. A vice-president was in charge of the reorientation project, although the president was actively and directly involved. The new customer orientation required substantial changes throughout the Bank. Tellers became "relationship bankers," or customer sales representatives (CSRs), cross-selling products and services and offering advice to customers, using new computerized customer information files. Middle managers, essentially branch managers, became responsible on the one hand for reinforcing relationships with existing clients and establishing new ones, and on the other for monitoring and controlling the different accounts. From a mainly internal focus, the Bank shifted to a customer orientation. Intensive computerization, perceived by top management as leverage facilitating the Bank's reorientation, was also carried out at all levels of the organization.

The Telecommunications company was in a "prereorientation" context. Significant changes were taking place in the telecommunications industry (increased competition, deregulation), but the company had not yet defined its new strategic orientation, although the need to do so was recognized throughout the firm. The company was actually trying to understand how the industry was changing and had focused essentially on boundary-spanning activities and gathering external information. The middle managers were the organization's gatekeepers in their respective domains, gathering data and endowing them with relevance for the firm.

The Utility company was in a convergence context. It had been in a monopolistic situation for over 30 years and enjoyed a very stable customer base and environment. Its thrust was essentially to increase operational efficiency, for example by decreasing the number of shutdowns and improving internal consistency among the different units. The organization was overwhelmed by guidelines, standards, and procedures for managing budgets, human resources, and operations, which were written and distributed throughout the organization.

Selection of Respondents

In each firm, the middle managers who would participate in the study were identified during the interviews with the vice-president of operations and the vice-president of IT. To maximize sample homogeneity regarding the nature of work and to facilitate comparisons, middle line managers in the operations department of each firm were selected. Line managers were favored over staff professionals because the scope of their work is broader, and differences in the time allocation pattern for the various activities were thus likely to be easier to observe. Staff professionals typically focus on a few specific roles. Operations managers were selected because their work was highly comparable in the three firms. In all three cases, middle managers in the operations departments were responsible for units of comparable size (called divisions, sections, or branches, depending on the firm) providing direct services to clients. All middle managers had one supervisory level under them, managing the employees in direct contact with clients. Studies have suggested that the 10 roles are present in all managerial jobs and that their relative importance varies only with different functions and hierarchical levels (Mintzberg 1973; Kurke and Aldrich 1983). Hence, the middle managers who participated in this study should have had similar roles. The identified middle managers were called by a member of the research team, who briefly presented the research project, verified their eligibility, and identified the software applications used. All the managers contacted agreed to participate in the study.

Data Collection

The data on IT usage and on the time spent on different managerial activities were collected by means of two logs and were validated in follow-up interviews. Once respondents had agreed to participate in the study, they were given instructions on how to complete the IT logs. The logs were then sent to them along with written instructions. Respondents were asked to register each computer usage (number of minutes actively using the computer, the software used, and the purpose of each usage; see Appendix B) during a day that was representative of a typical working day. They were also asked to keep a detailed log of activities during the same day. Finally, they were asked to return the two logs to the researchers at the end of the day. The measure of IT usage was obtained by adding the number of minutes managers spent on each application.

Follow-up interviews were conducted by a hypothesis and theory-blind research professional the day following receipt of the logs.[3] The interviews were divided into three parts. The first part discussed the respondent's jobs and general responsibilities (e.g., the role and importance of the unit within the organization, how the unit had evolved in recent years, how many people worked in it). During the second part, the interviewer validated the data on IT usage. The third part was concerned with the time devoted to the managerial activities. Respondents first described the activities they had registered in the logs. They then read a summary of Mintzberg's definitions of managerial roles and the examples of activities of each role proposed by Mintzberg (1973), McCall and Segrist (1980), and Tsui (1984) (Appendix A). The interviewer further described and clarified each role. Based on the questionnaires developed by McCall and Segrist and by Tsui, the interviewer then grouped the different activities into the role categories in the presence of the manager, who was invited to express an opinion as the classification progressed. This procedure ensured that the activities were grouped consistently into the different roles across interviews. The interview ended with a freer discussion aimed at gauging the respondent's opinions about computer use. After each interview, the number of minutes spent on each activity was added by role to obtain the time managers spent on each role. The activities performed using IT were also carefully coded by the interviewer into the different roles based on their nature and purpose as reported by the respondents (described in the column "Purpose" of the IT log; see Appendix B).

The procedure was pretested by the research professional on a sample of five managers. One researcher then met with the five respondents to validate the logs and the classification of activities into roles. No erroneous classifications were found.

For analysis purposes, the time spent on different roles and using IT was divided by the number of minutes worked in a particular day. This ratio measure facilitates comparisons of time allocation patterns between managers and across

	Bank	Telecommunications	Utility	Overall
Number of Respondents (N)	16	13	30	59
Experience (Years)				
Working	22.19	22.23	22.93	22.58
In the current job	1.48	5.48	2.03	2.64
Managerial	8.01	13.46	11.70	11.10
Daily IT Usage				
Average (minutes)	57.87	80.39	36.33	51.78
Standard deviation	72.27	48.41	56.16	60.86
Minimum/maximum	0 to 270	0 to 160	0 to 160	0 to 270
Average (proportional*)	.16	.18	.07	0.11
Standard deviation	.21	.11	.11	0.14
Minimum/maximum	0 to .69	0 to .36	0 to .34	0 to .69
Experience With IT (Year)				
Average	9.67	14.19	8.54	9.92
Standard deviation	6.79	9.69	8.17	8.18
Minimum/maximum	1 to 28	4 to 30.50	0 to 32	0 to 32

Table 1 Descriptive statistics.

Note
* Minutes of IT usage divided by minutes worked per day.

organizations. For example, comparing a manager who spends one hour per day on information gathering with one who spends 30 minutes on the same activity might be misleading if the two managers do not work the same number of hours per day (for example, one hour in a 12-hour work day is less important than 30 minutes in a five-hour working day).

Table 1 presents the descriptive statistics for each of the three participating firms. Respondents had an average work experience of 22.58 years, including 11 years as managers. They had been in their current positions for over 2.5 years. T-tests indicate that the Bank managers had slightly less managerial experience than the managers in the Telecommunications company (t = −2.08, p < .048) and that they were newer to their current jobs (t = −2.47, p < .029). Active IT usage averaged 51 minutes per day, or about 10% of working hours, and ranged from zero minutes to a maximum of 4.5 hours per day. T-tests indicated that the Utility company managers used IT less than the Telecommunications firm managers (t = 2.46, p < 0.018), but that there was no significant difference otherwise. Overall, managers had slightly less than 10 years of experience using IT, with no significant difference across organizations.

Results

Figure 1 presents the approach used in the statistical analyses. The aggregate informational roles, rather than the component roles, were used here because the interest was in studying how IT usage was related overall to the informational roles, and how in turn the informational roles were related to the other roles, rather than how IT usage was related to each individual informational role.[4]

To test the relationships illustrated in Figure 1, the path analysis approach was used (Pedhazur 1982). To obtain the path coefficients, each endogenous (dependent) variable was regressed on its independent variables (e.g., regressing

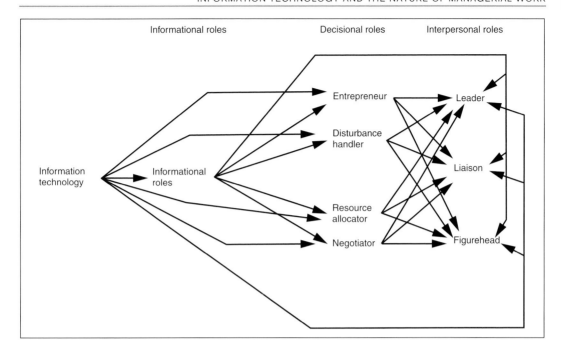

Figure 1   Statistical approach : The generic model.

leader on entrepreneur, disturbance handler, resource allocator, negotiator, informational roles, and IT usage). The path coefficients are the regression coefficients. The error term associated with an endogenous variable is equal to the square root of $(1-R^2)$ for a given regression. Using path analysis enables the relationship between the variables of a model to be broken down into different components:[5] direct effects (e.g., the relationship between the level of IT usage and the time devoted to the leader role, in Figure 1) and indirect effects (e.g., the relationship between the level of IT usage and the time devoted to the leader role as mediated by the time spent on the informational, entrepreneur, disturbance handler, resource allocator, and negotiator roles). The sum of the direct and indirect effects of an independent variable constitutes its total effect, or effect coefficient, and represents its overall relationship with the dependent variable.

The statistical analysis followed a two-step procedure. First, the path coefficients for the generic model (Figure 1) were calculated. Second, as suggested by prior research (Duncan 1975; Heise 1969; Pedhazur 1982), path coefficients that did not meet statistical significance criteria or did not significantly increase the chi square were deleted where theoretically justified. The path coefficients for the new model were then recalculated. The following statistical test was used to determine whether the nested model fit the data better than its "parent model" (the model from which the nested model was derived) (Specht 1975):

$W = -(N-d)\log_e((1-M_1)/(1-M_2))$

W   has a chi square distribution
N   sample size
d   difference between the numbers of over-identifying restrictions of the two models
M_1 $1- (1-R1^2)(1-R_2^2)(1-R_3^2)(1-R_4^2)$ ... (or the parent model)
M_2 $1- (1-R_1^2)(1-R_2^2)(1-R_3^2)(1-R_4^2)$ ... (for the nested model)

Because regression coefficients change when a variable is deleted from an equation, it was sometimes necessary to repeat the second step in order to delete new coefficients that had become non-significant. Following Pedhazur, this second step was performed until an optimal model was obtained—in other words, a model in which no path coefficient could be deleted without

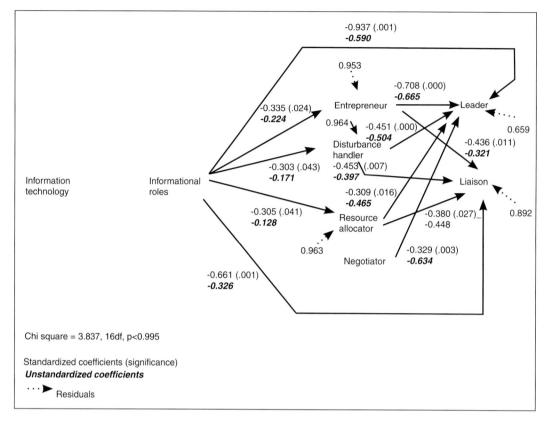

Figure 2 Overall sample.

significantly losing fit to the data. The procedure was applied first to the whole sample, then to each of the three organizational subsamples. The final models, along with the direct, indirect and total effect coefficients of IT usage and of the informational roles, are presented in Figures 2, 3, 4, and 5 for the sample as a whole, and for the Telecommunications, Bank, and Utility companies separately. Tables 2, 3, 4, and 5 present the means, standard deviations, and correlation coefficients for

	Mean[1]	Standard deviation	1	2	3	4	5	6	7	8
1. Information Technology	.112	.137								
2. Informational Roles	.477	.180	.257							
3. Entrepreneur	.103	.119	.009	−.332*						
4. Resource Allocator	.081	.075	.054	−.298*	−.142					
5. Disturbance Handler	.085	.101	−.164	−.297*	−.096	−.207				
6. Negotiator	.044	.057	−.084	−.122	−.073	.093	.109			
7. Leader	.091	.112	−.123	−.429**	−.214	.197	−.068	−.163		
8. Liaison	.076	.088	.064	−.262	−.034	.024	−.147	−.048	−.009	
9. Figurehead	.007	.037	−.160	−.184	.212	−.024	−.152	−.129	.044	−.129

Table 2 Means, correlation, standard deviation: overall sample.

Note
[1] Means represent the number of minute spent on a managerial role (or using IT) in a day divided by the total number of minutes worked in that day.
* $p < .05$; ** $p < .01$; *** $p < .001$.

the sample as a whole, and for the Telecommunications, Bank, and Utility companies separately.

The data do not support Proposition 1. The data presented in the four figures do not support the first proposition, which stated that as the level of IT usage increases, the time spent by middle managers on informational roles decreases. When the data are analyzed using the whole sample, no significant relationship is found between the level of IT usage and the time devoted to information-related activities (see Figure 2). From this first result, one would be tempted to conclude that, for the sample studied, there was no relationship between the level of IT usage and the time spent on managerial activities. However, a more detailed analysis produced some additional and interesting results. Indeed, when each firm is analyzed independently, relationships between the level of IT usage and the time spent on informational roles are found in two of the three companies (Figures 3, 4, and 5). Interestingly, and contrary to Proposition 1, IT was found to be associated with spending more time on the informational roles in the Telecommunications company (1.037, $p < .026$) and in the Bank (0.884, $p < .003$). No effect was found in the Utility (Figure 5).

The data provide preliminary support for Proposition 2, which stated that the relationship between the level of IT usage and the nature of managerial work is stronger in a reorientation context than in a convergence context. As discussed above, while no significant relationship was found between IT usage and the time spent on information-related activities in the Utility (which was in a convergence context), significant positive relationships were found in the Telecommunications firm and the Bank (which were in a prereorientation and a reorientation context, respectively). The level of IT usage was significantly related to the time allocated to four roles in the Telecommunications company (informational, negotiator, leader, liaison) and to six roles in the Bank (informational, negotiator, leader, liaison, entrepreneur, disturbance handler), but it had no significant relationship with any of the managerial roles in the Utility company. These results provide support for Proposition 2.

Moreover, in the Bank and in the Telecommunications firm, the level of IT usage is not only related to the informational roles, but is also, and significantly, related to the time spent on the decisional and interpersonal roles. More precisely, the level of

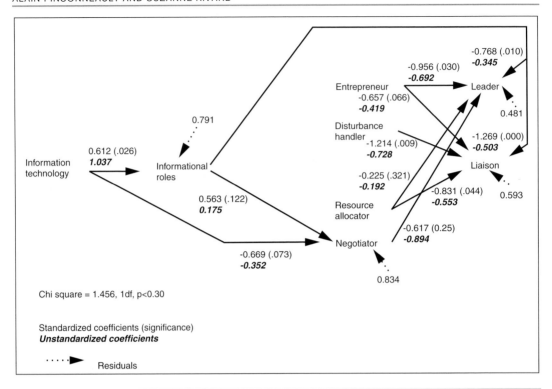

Total effect of information technology on:	Total effect	=	Direct effect	+	Indirect effect
Informational Roles	0.612	=	0.612	+	0.000
	1.037	=	*1.037*	+	*0.000*
Decisional Roles					
Negotiator	−0.324	=	−0.669	+	0.612* 0.563
	−0.171	=	*−0.352*	+	*1.037* 0.175*
Interpersonal Roles					
Leader	−0.270	=	0.000	+	(0.612* −0.768) + (0.612* 0.563* −0.617) + (−.669* −.617)
	−0.205	=	*0.000*	+	*(1.037* −0.345) + (1.037* 0.175* −0.894) + (−.352* .894)*
Liaison	−0.777	=	0.000	+	0.612* −1.269
	−0.522	=	*0.000*	+	*1.037* −0.503*
Total effect of the informational roles on:	Total effect	=	Direct effect	+	Indirect effects
Decisional Roles					
Negotiator	0.563	=	0.563	+	0.000
	0.175	=	*0.175*	+	*0.000*
Interpersonal Roles					
Leader	−1.115	=	−0.768	+	0.563* −0.617
	−0.501	=	*−0.345*	+	*0.175* −0.894*
Liaison	−1.269	=	−1.269	+	0.000
	−0.503	=	*−0.503*	+	*0.000*

Figure 3 Telecommunications company.

INFORMATION TECHNOLOGY AND THE NATURE OF MANAGERIAL WORK

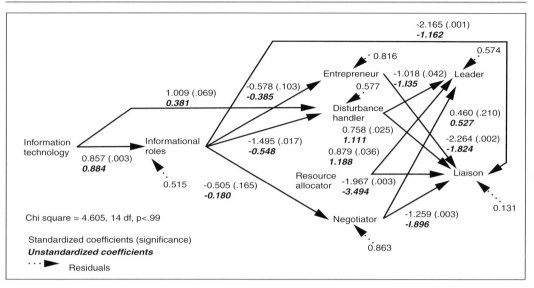

Total effect of information technology on:	Total effect	=	Direct effect	+	Indirect effect
Informational Roles	0.857	=	0.857	+	0.000
	0.884	=	*0.884*	+	*0.000*
Decisional Roles					
Entrepreneur	−0.495	=	0.000	+	0.857* −0.578
	−0.340	=	*0.000*	+	*0.884* −0.385*
Negotiator	−0.433	=	0.000	+	0.857* −0.505
	−0.159	=	*0.000*	+	*0.884* −0.180*
Disturbance Handler	−0.272	=	1.009	+	0.857* −1.495
	−0.103	=	*0.381*	+	*0.884* −0.548*
Interpersonal Roles					
Leader	0.078	=	0.000	+	(0.857* −1.495* −1.018) + (0.857* −0.505* 0.460) + (1.009* −1.018)
Liaison	*0.034*	=	*0.000*	+	*(0.884* −0.548* −1.135) + (0.884* −0.180* 0.527) + (.381* −1.135)*
	−1.160	=	0.000	+	(0.857* −0.578* −2.264) + (0.857* −1.495* 0.758) + (0.857* −0.505* −1.259) + (0.857* −2.165)
	−0.643	=	*0.000*	+	*(0.884* −0.385* −1.824) + (0.884* −0.548* 1.111) + (0.884* −0.180* −1.896) + (0.884* −1.162)*

Total effect of the informational roles on:	Total effect	=	Direct effect	+	Indirect effect
Decisional Roles					
Entrepreneur	−0.578	=	−0.578	+	0.000
	−0.385	=	*−0.385*	+	*0.000*
Disturbance Handler	−1.495	=	−1.495	+	0.000
	−0.548	=	*−0.548*	+	*0.000*
Negotiator	−0.505	=	−0.505	+	0.000
	−0.385	=	*−0.385*	+	*0.000*
Interpersonal Roles					
Leader	1.290	=	0.000	+	(−1.495* −1.018) + (−0.505* 0.460)
	0.527	=	*0.000*	+	*(−0.548* −1.135) + (−0.180* 0.527)*
Liaison	−1.354	=	−2.165	+	(−0.578* −2.264) + (−1.495* 0.758) + (−0.505* −1.259)
	−0.727	=	*−1.162*	+	*(−0.385* −1.824) + (−0.548* 1.111) + −0.180* −1.896)*

Figure 4 Bank.

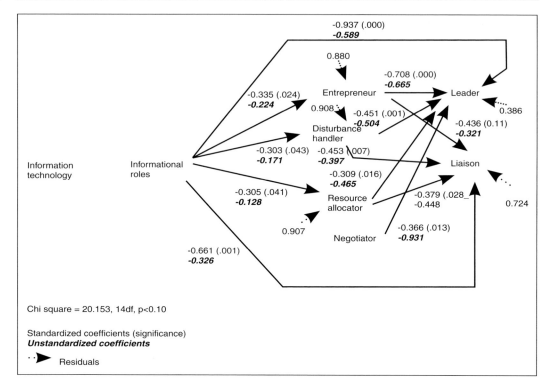

Figure 5 Utility company.

IT usage is negatively related to the time spent on these roles, overall, in both companies (see Figures 3 and 4: −0.171 with negotiator; −0.205 with leader; −0.522 with liaison in the Telecommunications company; −0.159 with negotiator; 0.034 with leader; −0.643 with liaison; −0.340 with entrepreneur; -0.103 with disturbance handler in the Bank). Therefore, in these two firms, IT usage was associated with spending more time on the informational roles and less time on the decisional and interpersonal roles—except for the leader role in the Bank. In the Utility company, no relationship was found between the level of IT usage and the time spent on any of the managerial roles (see Figure 5).

The data support Proposition 3, which stipulated that the pattern of the relationship

	Mean[1]	Standard deviation	1	2	3	4	5	6	7	8
1. Information Technology	.183	.105								
2. Informational Roles	.465	.179	.612*							
3. Entrepreneur	.130	.111	−.091	−.351						
4. Resource Allocator	.106	.106	−.031	−.366	−.126					
5. Disturbance Handler	.092	.118	−.308	−.322	−.201	−.372				
6. Negotiator	.045	.055	−.325	−.153	−.275	−.390	.041			
7. Leader	.059	.080	−.172	−.556*	−.134	.343	.319	−.469		
8. Liaison	.098	.071	−.363	−.343	.138	.166	−.364	.321	−.207	
9. Figurehead	.002	.008	−.236	−.227	.633*	−.123	−.076	−.082	.057	−.097

Table 3 Means, correlation, standard deviation: telecommunications company.

Note
[1]Means represent the number of minutes spent on a managerial role (or using IT) in a day divided by the total number of minutes worked in that day.

	Mean[1]	Standard deviation	1	2	3	4	5	6	7	8
1. Information Technology	.132	.194								
2. Informational Roles	.492.	.211	857**							
3. Entrepreneur	.103	.137	.418	−.578						
4. Resource Allocator	.067	.065	−.258	−.290	−.442					
5. Disturbance Handler	.051	.077	−.272	−.630	.297	.540				
6. Negotiator	.059	.073	−.364	−.505	.179	.302	.594			
7. Leader	.099	.089	−.367	−.057	−.241	.427	−.270	.161		
8. Liaison	.067	.109	−.164	−.128	.008	−.198	−.361	−.282	−.033	
9. Figurehead	.008	.026	−.204	−.179	.623	−.437	−.038	−.148	−.482	.041

Table 4 Means, correlation, standard deviation: bank.

Note
[1]Means represent the number of minutes spent on a role (or using IT) in a day divided by the total number of minutes worked in that day.
* $p < .05$; ** $p < .01$; *** $p < .001$.

	Mean[1]	Standard Deviation	1	2	3	4	5	6	7	8
1. Information Technology	.070	.111								
2. Informational Roles	.477	.175	−.333							
3. Entrepreneur	.090	.118	.245	−.203						
4. Resource Allocator	.074	.059	.235	−.279	−.107					
5. Disturbance Handler	.094	.099	−.046	−.190	−.164	−.289				
6. Negotiator	.037	.052	−.158	−.080	−.144	.540**	.062			
7. Leader	.102	.133	.053	−.548**	−.206	.203	−.177	.190		
8. Liaison	.067	.087	.336	−.297	−.168	−.036	.041	−.070	.087	
9. Figurehead	.009	.048	−.145	−.216	.129	.107	−.209	−.161	.102	−.183

Table 5 Means, correlation, standard deviation: utility company.

Note
[1]Means represent the number of minutes spent on a managerial role (or using IT) in a day divided by the total number of minutes worked in that day.
* p< .05; ** p< .01; *** p< .001.

between IT usage and the nature of managerial work depends on the kind of reorientation upon which the firm has embarked. It should be remembered that the Telecommunications company was in a "preorientation" phase, where top management had recognized that the environment was changing rapidly and drastically, but had not yet defined a new strategic thrust. The main issue was to understand the changing environment; the results obtained reflect this. IT usage was positively associated with the time devoted to the informational roles by middle managers (1.037, p < .026) and negatively related to one decisional role (negotiator, −0.171), and two interpersonal roles (leader, −0.205; liaison, −0.522). The interviews suggested that the larger amount of time spent on the informational roles associated with IT usage was a result of top management's efforts to reorient middle management work. Top managers made it very clear through memos and direct conversations with middle managers that obtaining, interpreting, and distributing information on competitors and on the environment was crucial, in view of the Telecommunications company's environment. During the interviews, all middle managers emphasized the fact that, due to the firm's changing environment, it was their responsibility to obtain such information in their respective domains and that they used IT to do so. For example, a manager who was responsible for the installation, maintenance, and repair of telecommunications cables used IT (i.e., direct access to different stock exchanges, access specialized journals and databases) to analyze the behavior of the firm's competitors and to benchmark his unit's performance with that of other firms. This manager indicated that he did not perform these tasks at all before the introduction of IT because the information sources were not readily available and because he was not expected to do so.

The Bank was already well into a strategic reorientation. Its new customer focus, with emphasis on tighter relationships with clients and closer monitoring of accounts, and its accompanying structural and operational changes had just been implemented. Branch managers (the respondents in this study) were much more involved in the development of new markets, the consolidation of existing client groups, and the monitoring of different accounts than before the reorientation. Here again, the relationship between IT usage and the nature of managerial work reflects the new strategic thrust of the organization, different from the Telecommunications firm. While the level of IT usage was also positively associated with the time spent on informational roles (0.884, p < .003), it was negatively related to the time spent on three of the decision-making roles (negotiator, –0.159; entrepreneur, -0.340; and disturbance handler, –0.103), and on one interpersonal role (liaison, –0.643). On the other hand, the level of IT usage was positively related to the time spent on one of the interpersonal roles (leader, total effect = 0.034).

Interviews indicated that IT was used by branch managers to support and reinforce the informational roles (e.g., gathering market information, monitoring customer accounts) and, in turn, to support leader roles (e.g., customer segmentation, new market development, leading employees to reinforce customer relationships). One manager described this new strategy, together with the effect it had on his IT usage and on his job. He emphasized the fact that, as branch manager, he had two main responsibilities: first, to ensure that the customer sales representatives (CSRs, formally referred to as bank tellers) established and maintained strong customer relationships and that they sold the bank's products and services efficiently; and second, to monitor closely the accounts opened by the CSRs. As he said: "I will be promoted on the basis of my branch's market share growth and profitability, and my job is on the line if my branch gets a few bad accounts." The data also indicate that the time spent on the entrepreneur role by branch managers was negatively associated with the level of IT usage. This may seem contradictory to the bank's overall strategy, but in fact it is not. It simply reflects the fact that IT gave managers direct access to data that facilitated the monitoring and control of CSR sales efforts. The more managers used IT, the more they felt comfortable delegating business development to CSRs (the entrepreneur role).

The Utility company was in a very stable, long-lasting convergence context in which attention was given to improving the efficiency of existing procedures and operations and improving consistency across units. In this company, the level of IT usage was not significantly related to the time spent on any particular role.

A close look at the results provides more support for Proposition 3. As shown in Figures 2, 3, 4, and 5, the data was also used to analyze the relationships between the time spent on information-related activities and the time spent on other roles (total effects of informational roles). These results help in understanding the "dynamics of time allocation among roles." In the case of the Utility, it was found that a larger proportion of time spent on the informational roles was negatively related to the proportion of time spent on five other roles (three decisional roles: entrepreneur, disturbance handler, and resource allocator; two interpersonal roles: leader and liaison). In other words, in that organization, the more time spent on information handling, the less time spent on the other roles. The results are somewhat different in the two other firms.

In the Telecommunications company, for instance, the informational roles have a significant, positive relationship with one of the decisional roles (negotiator), but are not related to the other roles in that category. Also, the time spent on informational roles has a significant negative relationship with the time spent on two of the three interpersonal roles (leader and liaison). In other words, in that firm, an increase in the time spent on informational roles is related to a reduction in the time spent on the interpersonal roles, but to an increase in the time spent on one decisional role (negotiator). These results indicate that information will be handled in relation to decision making—in other words, as discussed earlier, as managers obtain more information, they are in a better position to negotiate, probably obtain, and commit resources within the organization (negotiator role).

In the Bank, the increased amount of time spent on informational roles is related to the time spent on other roles in a different fashion. On the one hand, it is negatively related to the time spent on the decisional roles and on one interpersonal role (liaison), and positively related to the time spent on one interpersonal role (leader). As previously indicated, this is consistent with the strong customer orientation introduced by the Bank's management.

These results strongly suggest that when strategic changes are underway at the organization level (reorientation contexts), IT is used mainly to reinforce and support the specific roles perceived as critical by managers.

Discussion

Overall, the study provides support for the reorientation/convergence framework and for the fact that relationships between the level of IT usage and the nature of managerial work are fundamentally dependent on the context in which IT is implemented and used. Indeed, while significant relationships were found between IT usage and managerial roles in the two organizations operating within a strategic reorientation (Telecommunications and Bank), no relationship was found in the firm operating within a convergence context (the Utility). It may be argued that the lack of significant relationships in the Utility company was due to insufficient IT usage (36.33, 57.87, and 80.39 minutes of daily IT usage in the Utility, the Bank, and the Telecommunications company, respectively). However, path analyses using the model shown in Figure 1 were conducted with the 15 highest users from the Utility company (average IT usage of this subsample: 77.86 minutes per day, standard deviation: 59.63 minutes, no statistical difference between IT usage in the three firms). The results indicated that, even then, the level of IT usage was not significantly related to any managerial role, suggesting that the lack of a relationship in that company was not due to insufficient usage.

The results in this study support the claim made by several authors to the effect that research on IT usage must take into account the organizational context in which it is deployed and used. Lumping together firms with different strategic orientations in an aggregate-level study would have blurred the analysis. The results also show the importance of focusing on work process rather than on work output. For instance, it might be beneficial in the Telecommunications company for middle managers to focus on informational activities at the expense of other roles, and in the Bank for middle managers to spend more time on the informational and leader roles. However, this appears to be difficult to include in a study assessing productivity at the industry or economy level. This paper argues that a better understanding of how IT usage is related to the nature of managerial work and the context in which it is deployed will help untangle the productivity paradox.

In the above discussion, the limitations of the study have been borne in mind. First, although the study provides some indications of the importance of the organizational context in determining the strength of the relationship between the level of IT usage and the nature of managerial work, the small sample size (both in terms of number of respondents and number of firms) precludes generalization of the findings to other industries and to managers other than middle managers. More research is needed to test this result further. Ideally, a quasi-experimental study comparing IT usage in a 2 X 2 factorial approach (high and low IT usage, reorientation and convergence), or a survey with a large enough sample to allow for such analysis, should be conducted before drawing further conclusions. Second, data on IT usage and time allocation between roles were gathered during a single working day. It should be noted, though, that great care was taken to ensure that this day was typical for the respondents. Third, the study was cross-sectional rather than longitudinal—that is, it did not compare how managers allocated their work time before and after IT implementation, but examined the relationships between the level of IT usage and the time spent on various roles by managers. Finally, alternative hypotheses for explaining the results obtained cannot be completely ruled out. One such hypothesis may be that the Bank and the Telecommunications firm are in information-intensive industries, while the Utility is not, and that this environmental characteristic plays a more important role than strategic orientation. However, it appears that this is not the case here. A t-test was performed in order to determine whether there was a significant difference in the average time spent on information roles by managers from the three firms, and the results clearly indicate that no such difference exists ($t = .011$, $p < .992$)

From the Productivity Paradox to the Icarus Paradox?

In the two organizations in a reorientation context, IT usage was associated with a concentration of work time by middle managers on the few roles they perceived as critical to their success. In the Bank, IT usage was positively associated with the time spent on informational and leader roles; in the Telecommunications firm, IT was positively associated with the time spent on informational roles. The data indicate that this concentration resulted from both the strategic reorientation of the firm and IT usage, the latter being an enabler facilitating the concentration.

Contrary to expectations, higher levels of IT usage were associated with spending more time on informational roles in the Bank and in the Telecommunications company. The interviews indicated that managers focused their attention and spent their time on the informational roles because they felt their performance in these roles was enhanced by using IT. For instance, most managers in the Telecommunications firm emphasized the fact that they were using IT because they felt it provided better information, enhancing their decision making and, in turn, leading them to use IT even more. However, they claimed that, after a while, they realized they were spending too much time searching for available information and not enough time on the other activities of their jobs. Several managers from the Bank described a similar experience of IT usage. Given the importance of managing and monitoring the accounts held in their branches (managers stated in the interviews that "their jobs were on the line" as soon as a few accounts became delinquent), they were constantly looking for new information on these accounts. For example, a manager told us that he had to refrain consciously from using IT because he felt he was neglecting the rest of his job, and this had begun to affect his branch's performance. He told us that frontline employees (CSRs) had said they felt he was not supporting and coaching them adequately in learning their new jobs.

A greater concentration on a small number of roles may seem beneficial at first, as managers focus on one or two areas crucial to the firm's operations and strategy. However, it may also contain the seeds for an over-specialization beyond strategic necessity, in which the manager's decision making and flexibility are impaired. Although there is no empirical evidence to explain what appears to be an over-concentration on a few roles, a preliminary explanation based on the Icarus paradox is offered:

> The fabled Icarus of Greek mythology is said to have flown so high, so close to the sun, that his artificial wings melted and he plunged to his death in the Aegean Sea. The power of Icarus' wings gave rise to the abandon that so doomed him. The paradox, of course, is that his greatest asset led to his demise. And that same paradox applies to many outstanding companies today: their victories and their strengths often seduce them into the excesses that cause their downfall (Miller 1990, p. 3).

The Icarus fable used to explain the rise and fall of several modern organizations may well be relevant in explaining the relationship found between the level of IT usage and the nature of middle managerial work. Miller posits that organizations achieve outstanding performances by focusing on a few distinctive competencies and winning strategies. They react to their success by putting still more emphasis on the same competencies and by using the same strategies—in other words, by embarking on a trajectory that will ultimately lead to their own demise. Similarly, it was found that managers tend to exploit more intensively the one or two activities at which they are best, and this induces them to neglect their other activities (Miller and Chen 1996).

The mechanism observed in this study resembles the dynamics of the Icarus paradox. As is the case for successful organizations in Miller's analysis, the middle managers studied here may well have been launching themselves into a specialization trajectory. They focused on the activities and roles that were identified as important by senior management—mostly the informational roles—hence limiting their involvement in the other dimensions of their work. Miller argues that, in such a situation, managers develop the specialized knowledge of *how* things are done rather than *why* they are done the way they are. Organizations reinforce this pattern by hiring specialized workers and failing to recruit and retain people with different talents and skills. Managerial work might thus become ever more specialized and eventually fall into a specialization spiral, just as organizations launch themselves into a similar specialization spiral illustrated by the Icarus paradox.

This does not imply that specialization is dysfunctional by nature. Rather, it suggests that greater specialization of middle line managers' jobs contradicts the very essence of their mission, which lies not in specialization but in the integration of a wide range of managerial functions (Kotter 1982; Kurke and Aldrich 1983; Mintzberg 1973; Stewart 1989; Tsui 1984).

The data gathered in this study do not allow for a formal testing of the hypothesis that managers' job may have been launched into a specialization trajectory similar to that described by Miller.

However, the results are intriguing enough to lead to pursuit of investigations in that direction.

NOTES

1. Lynda Applegate was the accepting senior editor for this paper.
2. An earlier version of this paper was presented at the Americas Conference on Information Systems in 1996.
3. No logs were taken on a Friday to prevent interviewing respondents on a Monday and asking them to recall events that occurred three days earlier. Using a hypothesis and theory-blind research professional to conduct the interviews assured that the classification of activities into roles was not biased toward or against the propositions.
4. In addition to the theoretical explanation provided above, the inclusion of the three informational roles would have made the path analysis very complex, adding 16 new paths as well as three independent variables in every regression of the model, which, given the limited sample size, was not recommended.
5. A model might also include spurious and unanalyzed components, but these are of no interest to the present study and will not be included in the analysis.

ACKNOWLEDGMENTS

This study was supported financially by SSHRC of Canada and FCAR of Quebec. We would like to thank Anne Beaudry for her helpful assistance. We also thank the senior editor, the associate editor, and the reviewers for their helpful comments.

REFERENCES

Applegate, L. M. "In Search of a New Organizational Model: Lessons from the Field," in *Communication Technology and Organizational Forms*, G. DeSanctis and J. Fulk (eds.), Sage Publications, Berkeley, CA, 1998.

Barley, S. R. "Technology as an Occasion for Structuring: Evidence from Observations of CT Scanners and the Social Order of Radiology Departments," *Administrative Science Quarterly* (31), 1986, pp. 78–108.

Barua, A., Kriebel, C. and Mukhopadhyay, T. "Information Technology and Business Value: An Analytic and Empirical Investigation," working paper, University of Texas, Austin, May 1991.

Bjørn-Andersen, N. and Pedersen, P. H. "Computer Facilitated Changes in the Managerial Power Structure," *Accounting, Organizations and Society* (5:2), 1980, pp. 203–16.

Brandt, J. R. "Middle Management: Where the Action Will Be," *Industry Week*, May 2, 1994, pp. 30–6.

Brynjolfsson, E. "The Productivity Paradox of Information Technology," *Communications of the ACM* (26:12), 1993, pp. 67–77.

Brynjolfsson, E. and Hitt, L. "Is Information Systems Spending Productive? New Evidence and New Results," in *Proceedings of the Fourteenth International Conference on Information Systems*, J. I. DeGross, R. P. Bostrom and D. Robey (eds.), Orlando, FL, 1993, pp. 47–64.

Brynjolfsson, E. and Hitt, L. "The Productive Keep Producing," *Information Week*, September 18, 1995.

Buchanan, D. and McCalman, J. "Confidence, Visibility and Pressure: The Effects of Shared Information in Computer Aided Hotel Management," *New Technology, Work, and Employment* (3:1), 1988, pp. 38–46.

Davis, R. V. "Information Technology and White-Collar Productivity," *Academy of Management Executive* (5:1), 1991, pp. 55–67.

Dopson, S. and Stewart, R. "Information Technology, Organizational Restructuring and the Future of Middle Management," *New Technology, Work and Employment* (8:1), 1993, pp. 10–20.

Drucker, P. F. "The Coming of the New Organization," *Harvard Business Review*, January-February, 1988, pp. 45–53.

Drucker, P. F. "Restructuring Middle Management," *Modern Office Technology*, January 1993, pp. 8–10.

Duncan, O. D. *Introduction to Structural Equation Models*, Academic Press, New York, 1975.

Franke, R. H. "Technological Revolution and Productive Decline: Computer Introduction in the Financial Industry," *Technological Forecasting and Social Change* (31), 1987, pp. 143–54.

Freeman, S. J. and Cameron, K. S. "Organizational Downsizing: A Convergence and Reorientation Framework," *Organization Science* (4:1), 1993, pp. 10–29.

George, J. F. *Computer and the Centralization of Decision-Making in U.S. City Governments*, unpublished Ph.D. dissertation, University of California, Irvine, 1986.

Heise, D. R. "Problems in Path Analysis and Causal Inference," in *Sociological Methodology*, E. F. Borgatta and G. W. Bohrnstedt (eds.), Josey-Bass, San Francisco, 1969.

Hoos, I. R. "When the Computer Takes Over the Office," *Harvard Business Review* (38:4), 1960, pp. 102–12.

Kelley, M. R. "Productivity and Information Technology: The Elusive Connection," *Management Science* (40:11), 1994, pp. 1406–25.

Klatzky, S. R. "Automation, Size, and the Locus of Decision Making: The Cascade Effect," *Journal of Business* (43:2), 1970, pp. 141–51.

Kotter, J. P. "What Effective General Managers Really Do," *Harvard Business Review* (60:6), November-December, 1982, pp. 156–67.

Kurke, L. B. and Aldrich, H. E. "Mintzberg Was Right! A Replication and Extension of the Nature of Managerial Work," *Management Science* (29:8), 1983, pp. 975–84.

Leavitt, H. J. and Whisler, T. L. "Management in the 1980s," *Harvard Business Review*, November-December, 1958, pp. 41–8.

Loveman, G. W. "An Assessment of the Productivity Impact of Information Technologies," MIT Management in the 1990s, Working Paper, 88–054, July 1988.

Mann, F. C. and Williams, L. K. "Observations on the Dynamics of a Change to Electronic Data Processing Equipment," *Administrative Science Quarterly*, September, 1960, pp. 217–56.

Markus, M. L. and Robey, D. "Information Technology and Organizational Change: Causal Structure in Theory and Research," *Management Science* (34:5), 1988, pp. 583–98.

McCall, M. W. Jr and Segrist, C. A. "In Pursuit of the Manager's Job: Building on Mintzberg," Technical Report 14, Center for Creative Leadership, Greensboro, NC, 1980.

Miller, D. *The Icarus Paradox: How Exceptional Companies Bring About Their Own Downfall*, Harper Collins, New York, 1990.

Miller, D. and Chen, M. J. "The Simplicity of Strategic Repertoires: An Empirical Analysis," *Strategic Management Journal* (17:16), 1996, pp. 419–39.

Miller, D. and Friesen, P. H. "Momentum and Revolution in Organizational Adaptation," *Academy of Management Journal* (23), 1980, pp. 591–614.

Millman, Z. and Hartwick, J. "The Impact of Automated Office Systems on Middle Managers and Their Work," *MIS Quarterly* (11:4), 1987, pp. 479–91.

Mintzberg, H. *The Nature of Managerial Work*, Harper and Row, New York, 1973.

Orlikowski, W. J. "CASE Tools as Organizational Change: Investigating Incremental and Radical Changes in Systems Development," *MIS Quarterly* (17), 1993, pp. 309–40.

Osterman, P. "The Impact of Computers on the Employment of Clerks and Managers," *Industrial and Labor Relations Review* (39), 1986, pp. 175–86.

Panko, R. R. "Is Office Productivity Stagnant?" *MIS Quarterly* (15:2), June 1991, pp. 191–203.

Parson, D. J., Gottlieb, C. C. and Denny, M. "Productivity and Computers in Canadian Banking," University of Toronto, Department of Economics Working Paper 9012, June 1990.

Pedhazur, E. J. *Multiple Regression in Behavioral Research*, Holt, Rinehart and Winston, New York, 1982.

Pettigrew, A. M. *The Awakening Giant: Continuity and Change at ICI*, Basil Blackwell, Oxford, England, 1985.

Pfeffer, J. and Leblebici, H. "Information Technology and Organizational Structure," *Pacific Sociological Review* (20:2), 1977, pp. 241–61.

Pinsonneault, A. and Kraemer, K. L. "The Impact of Information Technology on the Middle Management Workforce," *MIS Quarterly* (17:3), September 1993, pp. 271–92.

Pinsonneault, A. and Kraemer, K. L. "Middle Management Downsizing: An Empirical Investigation of the Impact of Information Technology," *Management Science* (43:5), 1997, pp. 659–79.

Quinn, J. B. and Baily, N. "Information Technology: Increasing Productivity in Services," *Academy of Management Executive* (8:3), 1994, pp. 28–47.

Roach, S. S. "The New Technology Cycle," *Economic Perspectives*, Morgan Stanley and Co., New York, September 11, 1985.

Roach, S. S. "Services Under Siege: The Restructuring Imperative," *Harvard Business Review*, September-October, 1991, pp. 82–92.

Robey, D. and Sahay, S. "Transforming Work Through Information Technology: A Comparative Case Study of Geographic Information Systems in County Government," *Information Systems Research* (7:1), 1996, pp. 93–110.

Specht, D.A. "On the Evaluation of Causal Models," *Social Science Research* (4), 1975, pp. 113–33.

Stewart, R. "Studies of Managerial Jobs and Behavior: The Ways Forward," *Journal of Management Studies* (26:1), 1989, pp. 1–10.

Strassman, P.A. *The Business Value of Computers*, Information Economics Press, New Canaan, CT, 1990.

The Economist. "Turning Digits into Dollars," October 26, 1996, pp. 3–22.

Tsui, A. "A Role Set Analysis of Managerial Reputation," *Organizational Behavior and Human Performance* (34), 1984, pp. 64–96.

Tushman, M. L. and Romanelli, E. "Organizational Evolution: A Metamorphosis Model of Convergence and Reorientation," in *Research in Organizational Behavior* (7), L. L. Cummings and B. M. Staw (eds.), JAI Press, Greenwich, CT, 1985.

Weill, P. "The Relationship Between Investment and Information Technology and Firm Performance: A Study of the Valve Manufacturing Sector," *Information Systems Research* (3:4), 1992, pp. 307–33.

Whisler, T. I. *The Impact of Computers on Organizations*, Praeger Publishers, New York, 1970.

Wooldridge, B. and Floyd, S. W. "The Strategy Process, Middle Management Involvement, and Organizational Performance," *Strategic Management Journal* (11), 1990, pp. 231–41.

Zmuidzinas, M., Kling, R. and George, J. "Desktop Computerization as a Continuing Process," in *Proceedings of the Eleventh International Conference on Information Systems*, J. I. DeGross, M. Alaui and H. J. Oppelland (eds.), Copenhagen, 1990, pp. 125–37.

ABOUT THE AUTHORS

Alain Pinsonneault is an associate professor at the École des Hautes Études Commerciales (HEC), Montreal, and director of the Ph.D. program. He holds a Ph.D. in administration from the University of California at Irvine and an M.Sc. in management information systems from HEC. His current research interests include the organizational and individual impacts of information technology, the strategic alignment of information technology, group support systems, and IT department management. He has published articles in *Management Science, MIS Quarterly, Journal of Management Information Systems, Decision Support Systems*, and *European Journal of Operational Research*.

Suzanne Rivard is professor of information technology at the École des Hautes Études Commerciales (HEC), Montréal. She holds an MBA from HEC and a Ph.D from the Richard Ivey School of Business, the University of Western Ontario. Her research interests focus on software project risk management, outsourcing of information systems services, and IT impacts. She has published in *Communications of the ACM, Decision Support Systems, Journal of Management Information Systems, MIS Quarterly*, and *Omega*.

APPENDIX A

Mintzberg's Role Typology

Roles	Description and examples of activities
Informational	
Monitor	Managers scanning the environment for information, interrogating liaison contacts and subordinates, and receiving unsolicited information. **Examples of Activities*:** Keeping up with information on the progress of operations in the firm Gathering information about trends outside the organization Gathering information about customers, competitors, associates, etc. Touring facilities for observational purposes Reading reports on activities in this and other companies Scanning the environment for opportunities
Disseminator	Managers sharing and distributing information to subordinates or to other managers. **Examples of Activities:** Briefing subordinates of the organization on the results of activities (e.g., trips, conversations, meetings) Distributing information on a specialized events (i.e., conference, meeting) Diffusing information to subordinates on standards or procedures for decision making Forwarding mail into the organization for informational purposes
Spokesman	Managers sending information to people outside the unit, informing and satisfying the influential people who control the organizational unit. **Examples of Activities:** Serve as an expert to people outside the unit Answer inquiries about the unit Presiding at meetings as a representative of the organization Informing others of the organization's future plans Answering letters or inquiries on behalf of the organization Keeping other people informed about the organization's activities
Decisional	
Entrepreneur	Managers initiating a project and trying to improve the work unit by adapting it to the work environment. **Examples of Activities:** Planning and implementing change Initiating controlled change in the unit Solving problems by instituting needed changes in the organization
Disturbance handler	Managers responding to external pressures and disturbances. **Examples of Activities:** Resolving conflicts between subordinates (e.g., resource demand, personality clashes) Managing an unexpected problem in operations (e.g., breakdown) Managing potential resources losses (e.g., important customer leaving)

Resource allocator	Managers deciding who will get what in the organizational unit and authorizing important decisions.
	Examples of Activities:
	  Distributing budget resources
	  Making decisions about time parameters for upcoming programs
	  Allocating monies within the unit
	  Deciding which programs to provide resources (manpower, dollars, etc.)
	  Allocating equipment or materials	

Negotiator	Managers committing organizational resources in "real time" and exchanging resources within and outside the unit.
	Examples of Activities:
	  Negotiating the price and services offered with a consultant team
	  Negotiating with subordinates for the setting of performance standards
	  Discussing with colleagues for sharing common resources (e.g., personnel, funds, material, offices, computers)	

Interpersonal

Figurehead	Managers performing duties of a ceremonial nature.
	Examples of Activities:
	  Representing the unit in clients' activities
	  Writing and signing letters of recommendations for employees
	  Entertaining important clients (e.g., touring of the facilities)	

Leader	Managers being responsible for the work of people in the unit, including hiring, training, motivating, and encouraging employees.
	Examples of Activities:
	  Evaluating the quality of subordinate job performance
	  Resolving conflicts between subordinates
	  Allocating manpower to specific jobs or tasks
	  Seeing to it that subordinates are alert to problems that need attention
	  Maintaining supervision over changes in the organization
	  Directing the work of subordinates	

Liaison	Managers making contacts outside the vertical chain of command.
	Examples of Activities:
	  Attend meetings in other units
	  Attend social functions as a representative
	  Maintaining a personal network of contacts (visits or phone calls)
	  Developing new contacts by answering requests for information
	  Developing contacts with important people outside the organization	

[*]The list of activities used in the interviews was based on McCall and Segrist's (1980) and Tsui's (1984) questionnaires of managerial roles and on Mintzberg (1973).

Appendix B

Example of an IT Usage Log

IT usage Log

Every time you use a computerized application, please indicate what application you used, for what purpose, and for how many minutes you **actively used it (that is, you interacted with the application).**

Respondent: _____

Date: _____

		Application	Purpose
Example:	Number of minutes: _____	_____	
1st usage	Number of minutes: _____	_____	
2nd usage	Number of minutes: _____	_____	
3rd usage	Number of minutes: _____	_____	
4th usage	Number of minutes: _____	_____	
5th usage	Number of minutes: _____	_____	
6th usage	Number of minutes: _____	_____	
7th usage	Number of minutes: _____	_____	
8th usage	Number of minutes: _____	_____	
9th usage	Number of minutes: _____	_____	
10th usage	Number of minutes: _____	_____	

-
-
-
-
-

| 19th usage | Number of minutes: _____ | _____ | |
| 20th usage | Number of minutes: _____ | _____ | |

Please fax the log at the end of the day to the following number: 340–5634.

Please fax the log at the end of the day to the followign number: 340-5634.

IT Value: The Great Divide between Qualitative and Quantitative and Individual and Organizational Measures

Yolande E. Chan

Abstract:

A comprehensive review was conducted of IT value articles in the *Communications of the ACM, Information Systems Research, Journal of Management Information Systems*, and *MIS Quarterly* from 1993 to 1998. IT-value measures published during this period were documented, classified, analyzed, and reported. The review of these journal articles revealed a schism between the use of organization-level measures and other measures. *Communications of the ACM* and *Information Systems Research* also provided strong evidence of a schism between the use of quantitative and qualitative measures in IT-value research. The *Journal of Management Information Systems* and *MIS Quarterly* data provided more limited evidence of this schism as well. These schisms have become more pronounced over time. This may be due partly to an increasing reliance on secondary data set analyses that use only quantitative measures and organization-level analyses. The current research confirmed what many researchers suspect—schisms exist, and may be deepening, in IT-value research.

Keywords and phrases: information technology productivity, information technology investment value.

THERE HAS BEEN MUCH RECENT DISCUSSION OF THE "PRODUCTIVITY PARADOX" in the information technology (IT) literature [14, 38]. A great deal of energy has been focused on describing the paradox, denying the paradox, solving the paradox, and burying the paradox [15, 34, 38, 52]. The debate may have, paradoxically, legitimized the very measures that have not served the IT community particularly well—measures that paint a bleak picture of the value of IT investments.

How so? With so much MIS researcher and practitioner attention focused on the IT productivity paradox, a great deal of energy has been poured into studies that seek to demonstrate positive relationships between IT investment and organizational performance [7, 77, 78, 101]. In an attempt to provide evidence that is credible to an executive audience, many of these studies have focused exclusively on quantitative measures of performance. Several have underemphasized the role of individual-level IT benefits and focused almost exclusively on benefits of IT investments that may be observed at organizational and industrial levels. The IT researcher's lens has grown bigger, if not better, over time. With the IT productivity paradox hype, the focus has been on "hard" numbers, not qualitative judgments, and "big IT wins," not incremental process and product-service improvements that may occur one employee at a time.

This study examines IT value articles published in the *Communications of the ACM, Information Systems Research, Journal of Management Information Systems*, and *MIS Quarterly*—four leading North American MIS journals[1]—in recent years (1993–98). IT value measures published in these journals during this period are documented, classified, analyzed, and reported. Based on this analysis, it is argued that more balanced perspectives of IT value [61] are required.

Discussion of Related Literature

The IT Productivity Paradox

The relationship between information technology (IT) and productivity is widely discussed but little understood. Delivered computing power in the U.S. economy has increased by more than two orders of magnitude since 1970 yet productivity, especially in the service sector, seems to have stagnated. Given the enormous promise of IT to usher in "the biggest technological revolution men have known," disillusionment and even frustration with the technology is increasingly evident in statements like "No, computers do not boost productivity, at least not most of the time."

SO BEGINS BRYNJOLFSSON'S [14] WIDELY CITED ARTICLE DISCUSSING "The Productivity Paradox of Information Technology." Brynjolfsson highlights earlier studies [75, 103, 104, 105, 115] that suggest an *apparent* IT investment paradox with respect to economy-wide productivity (e.g., total IT investment in relation to gross national product), the productivity of IT capital in manufacturing, and the productivity of IT capital in services. Brynjolfsson states:

Productivity is the fundamental economic measure of a technology's contribution. With this in mind, CEOs and line managers have increasingly begun to question their huge investments in computers and related technologies. [14, p. 67]

Although the IT productivity paradox was originally defined at the economy level and some studies have been carried out at national and industrial levels, most MIS researchers have addressed the productivity question at the organization level. Several MIS researchers have tried to produce hard evidence of productivity gains afforded to firms as a result of IT investments. Mahmood [77] writes:

Strategic managers clearly need a better understanding of the impact of IT investment on organizational strategic and economic performance. Clearer understanding of the factors that drive such performance could help a firm better utilise resources dedicated to the relevant delivery process, and increase the firm's position vis-à-vis its competitors....Pressures have, therefore, been mounting on information systems researchers to validate empirically the relationship between IT investment and organizational strategic and economic benefits. Kauffman et al. (1988) and Banker and Kauffman (1988) have urged that "hard" evidence be provided that relates IT investment to organizational economic outputs. [pp. 185–186]

The IT Productivity Paradox—Past Measures and Current Results

In his review of research studies investigating the IT productivity paradox, Mahmood [77] suggests that there have been three main categories of studies: those using a "key ratios" approach, others using a "competitive interaction approach," and finally others relying on a "microeconomic" approach. Mahmood does not consider "soft" approaches, although this may be because of his attempt to respond specifically to Kauffman's calls for "hard" evidence. Mahmood focuses on organization-level studies.

Examples of the "key ratios" approach include calculations of the ratio of IT expense to total operating expense and annual IT budget as a percentage of revenue. Mahmood illustrates the "competitive interaction approach" by describing the Banker and Kauffman [6] study that found, while ATM network membership could increase a bank's local deposit market share, at the same time the presence of an ATM contributed little to a

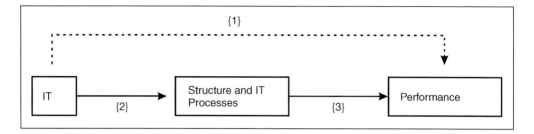

Figure 1 Areas for IT impact research (adapted from [2]. Reprinted with permission.

bank's economic performance. In the "microeconomic theory-based approach," researchers use microeconomic theory to formulate models to investigate IT's organizational impacts. Variables such as product/service demand, capital costs, labor costs, and the total costs of doing business are examined.

Studies examining these kinds of "hard" organization-level evidence have at times lent support to (i.e., not refuted) the IT productivity paradox. Brynjolfsson [14] provides four possible explanations for this:

- Mismeasurement of inputs and outputs
- Lags due to learning and adjustment
- Redistribution and dissipation of profits
- Mismanagement of information and technology.

Other researchers [38, 52, 121] provide additional reasons why hard evidence may not explain away the paradox (e.g., inadequate traditional accounting systems, IT capital spent primarily to take market share away from competing firms and not to increase the size of the market, and IT investments that merely fuel the need for further IT investments and do not increase productivity outside the computer manufacturing industry). Overall, Brynjolfsson [14] concludes:

> After reviewing and assessing the research to date, it appears that the shortfall of IT productivity is as much due to deficiencies in our measurement and methodological toolkit as to mismanagement by developers and users of IT. [p. 67]

The closer one examines the data behind the studies of IT performance, the more it looks like mismeasurement is at the core of the "productivity paradox." Rapid innovation has made IT-intensive industries particularly susceptible to the problems associated with measuring quality changes and valuing new products.... Increased variety, improved timeliness of delivery and personalized customer service are additional benefits that are poorly represented in productivity statistics. These are all qualities that are particularly likely to be enhanced by IT. [p. 74]

Researchers must not overlook [the] fact that our tools are still "blunt." ...The business transformation literature highlights how difficult and perhaps inappropriate it would be to try to translate the benefits of IT usage into quantifiable productivity measures of output ... Researchers [must] be prepared to look beyond conventional productivity measurement techniques. [p. 76]

The IT Productivity Paradox—Other Lessons from the MIS Literature

Bakos [2] also issues a cautionary warning to MIS researchers:

> In the context of organizational impacts of information technology, alternative perspectives[2] lead to different dependent variables and suggest the use of different theoretical tools for the study of these impacts. Studies based on different perspectives have used different vocabularies and, as a result, have often *talked past each other*. A simple model for the impact of information technology is shown in Figure 1.

The technology has an impact on organizational structure and process, thereby affecting organizational performance.... The majority of impacts research will belong to one of the first two areas: impact of information technology on (1) organizational performance and on (2) organizational structure and processes. The difference between the two areas can be visualized as whether the structure and process box in Figure 1 is seen as a system that can be modeled and probed, or as a "black box" whose inputs and outputs are the only observable variables. [pp. 12–13, emphasis added]

It is possible that much of the IT value research (i.e., studies that examine the benefits of IT investments) using soft measures "talks past" research emphasizing objective numeric assessments, and vice versa. Although some researchers do use both qualitative and quantitative measures (even in the same studies), others do not and appear to participate in what may best be described as "camps" that are unreceptive to certain research methods and measures.

Despite the call for hard measures of economic impact, the value of IT may not be fully understood without incorporating, at some point, qualitative, individual, and group-level measures. If this were not the case, we would be subscribing to "black box" approaches where only macro-level inputs and outputs are observed.

Some of the research conducted specifically as part of the IT productivity paradox debate has, in fact, emphasized individual and group-level outcome measures and process measures. For example, Barua et al. [7] examined the effect of IT on "intermediate-level variables" such as capacity utilization, inventory turnover, relative quality, relative price, and new products. They have related these intermediate variables to final performance variables such as market share and ROA. Barua et al. [7] document that other researchers [29, 63, 87, 88] have also found that the effects of IT on organization performance can be best identified through a "web of intermediate level contributions." They argue that these "lower-level impacts" should, in turn, affect organizational/higher-level performance measures [67]. Barua et al. [7] write:

> Our basic thesis is that primary economic impacts or contributions (to performance) of information technologies (if any) can be measured at lower operational levels in an enterprise, at or near the site where the technology is implemented. To capture these impacts, measurements should be taken in the organization where the potential for first-order effects exists. These effects may then be traced through a chain of relationships within the organizational hierarchy to reveal higher order impacts (if any) on enterprise performance.... We suspect that as the distance between a first-order effect and higher levels increases, the ability to detect and measure an impact decreases (perhaps rapidly). For this reason, we believe prior research based on conventional microeconomic production theory (attempting to relate variables such as MIS budgets and market share directly) does not have the power to reveal an association with high statistical significance. [pp. 6–7]

Given the numerous recommendations and cautions regarding the study of IT value that have appeared in the MIS literature, one might expect to find an increasing number of articles that examine first-order and intermediate IT effects. One might expect to see researchers developing less conventional and less "blunt" investigative tools. This study's review of recent IT value articles documents the extent to which this has, in fact, been the case.

The IT Productivity Paradox—Other Issues Raised in the Organization Development Literature

It can be argued that much of the IT productivity paradox debate has been couched in a rational-economic paradigm. However, task interdependence in organizations makes collaboration a necessary prerequisite for ongoing organizational effectiveness [110, p. 172], suggesting that, in evaluations of long-term organization performance, human relations and task issues need to be reviewed along with short-term economic outcomes.

Organizations accomplish their work through motivated people [122]. Generally, information systems are used by people (e.g., customers, suppliers, employees). IT investments can be used to alter tasks, customer interactions, employee psychological contracts, expectations, motivation, and productivity. IT value measures may then usefully assess organizational processes and tasks, and organizational health and renewal [73].

Because an organization is a complex system, when one factor is changed, meaningful assessment may need to go beyond immediate, isolated outcomes, to encompass long-term system changes as well. Longitudinal IT evaluation studies may be required. Schein [110] writes:

> One rarely, if ever, finds a real-life situation in which there is only *one* goal operating. It is a characteristic of all human systems to have multiple goals, all of which are generally operating simultaneously, and among which the priorities are shifting constantly. Progress toward any goal can be measured, and that measure has usually been defined as the *efficiency* of an organization. But choosing the right priorities among goals, ensuring that the ultimate [purposes] of the organization are met, is a more complex process, one that approximates the concept of effectiveness.
>
> ... Organizations do have multiple functions and multiple goals, ... some of these are actually in conflict with each other.... The dilemma of effectiveness, then, is clear. Is effectiveness the ability to maximize profit in the short run (which would require a definition of "short run"), or does effectiveness have something to do with the ability to *maintain* profits over some longer period of time to which the concepts of survival and growth are more applicable?
>
> ... One attempted resolution ... has been to define effectiveness in terms of systems-level criteria ... A system's effectiveness can be defined as its *capacity to survive, adapt, maintain itself, and grow* ... [a] more general concept of "health." [pp. 230–231]

Schein's remarks point out the limitations of assessing IT impact with only an organization-level approach to analysis, or with any single number (e.g., ROI or NPV). A more complete assessment of technology innovations might involve several levels of analysis (e.g., individual and group) and several sets of "numbers." Unfortunately, the difficulties encountered in responsibly integrating findings at various analytic levels are not insignificant. For instance, if individuals are highly satisfied with a system but there is no visible short- or long-term economic benefit, can the system be described as successful? Or, conversely, if the "bottom line" is vastly improved through radical reengineering using technology but employee morale is at an all-time low, is the organization more effective? To some extent, these questions involve difficult value judgments. Perhaps part of the challenge associated with technology evaluations is the need to let go of narrow, one-dimensional, win/lose pronouncements, and to accept instead mixed, multidimensional, multistakeholder, explicitly value-based assessments. In doing so, it may be necessary to examine researcher and practitioner assumptions and biases [50].

Schein's comments also lead us to question the appropriate boundaries for IT investments. Perhaps investments do not originate when funds are formally approved for new systems, but earlier, for example, when proposed systems are seriously being considered and employees are reacting, possibly negatively. Researchers conducting IT value studies may consider explicitly identifying appropriate boundaries or limits of the impacts to be investigated. Also, because the organization is a dynamic system with feedback loops, secondary, tertiary, and other indirect impacts may be measured if this is deemed appropriate. In order to do this, however, the relevant environments need to be identified. If IT evaluation approaches are designed with static, closed systems in mind, they may be inadequate.

Technology investments generally are initiated by one or more individuals who seek to make system changes in order to accomplish certain objectives. Much of the recent discussion in the literature on alignment focuses on the context of the IT investment [19]. The technology is often expected to leverage business strategic orientation [124], streamline tasks, and leverage human capital. Thus, similar technology investments (e.g., similar hardware-software installations using the same systems development methodology) frequently have quite different outcomes. This raises the issue of whether IT investments can be characterized adequately outside their organizational and industrial settings. In order to make accurate evaluations, strategic contexts and human contexts may need to be documented also.

It is difficult, however, for any single study to investigate and measure a complete sociotechnical system and its environments. Social science research can be conducted carefully, though, with the recognition of ever-present research limitations. At times, apparent paradoxes may simply be the result of these limitations.

Research Objectives

A KEY PURPOSE OF THE CURRENT STUDY HAS BEEN TO INVESTIGATE a possible trend in "IT value measurement" (i.e., the documentation of benefits provided by IT investments) to examine only hard, organization-level measures of value. Such a trend, as we have seen above, can be shortsighted but may be a direct result of the amount of press that has been given to the apparent IT productivity paradox (see, e.g., [121]). However, much of the organization development literature stresses the importance of the human resource function (e.g., individuals, teams, and networks), which uses business processes, in combination with technology, to achieve organizational goals. The MIS literature also underscores the value of technology in the management of human/intellectual capital (e.g., individual and group knowledge). It would seem that hard and soft measures, and organizational, group, and individual-level measures, all have the potential to inform the discussion of IT value.

For this reason, this article focuses not on the many strengths of "hard" IT value research streams, but on their weaknesses. Certainly, there are many limitations of soft or subjective measures (see [20] and [84] for criticisms of the user satisfaction construct, for example, and questions raised in [16] regarding weak relationships between job satisfaction and job performance). The article does *not* call for an exclusive return to the use of soft, individual, and group-level measures or process-focused measures but instead reminds us of the importance of these measures and examines their usage in recent studies of IT value.

Research Design

IN ORDER TO SYSTEMATICALLY REVIEW MEASURES USED IN RECENT IT VALUE research, the author, with the assistance of two MIS graduate students, examined all studies discussing IT impacts published in four top North American MIS journals—*Communications of the ACM (CACM), Information Systems Research (ISR), Journal of Management Information Systems (JMIS),* and *MIS Quarterly (MISQ)*. These journals were chosen because they are regarded primarily as *MIS* (as opposed to management) journals and are consistently highly ranked (e.g., [125]). Time and resources did not permit a review of a wider selection of journals. In order to determine current trends in IT value research, all studies published in these journals between 1993 and 1998 (inclusive) were examined. Initially articles were selected for consideration, and their measures—if any—examined, only if they involved research in business settings, *and* their titles, abstracts, or key words emphasized computers, systems, technology,[3] *and* also evaluation, efficiency, investment, payoffs, productivity, performance, usefulness, or value. Because some articles appeared to be IT value articles but did not have any of the latter key words, the following key words were also eventually added: benefits, competitiveness, competitive advantage, effectiveness, and innovation.[4] Because many *CACM* articles had no abstracts or key words, title information often had to be supplemented with a scan of the body of the article. Appendices A–D document the *CACM, ISR, JMIS,* and *MISQ* articles that were classified as IT value articles.

Articles were classified as "related empirical" articles if their titles, abstracts, and key words emphasized other effects, impacts, or improvements (e.g., decision-making quality) due to the use of systems or technology, but the articles, although empirical, were not concerned *primarily* with demonstrating the value of IT investments. Measures used in "related empirical" studies were not analyzed. (A number of software-development articles were excluded because they addressed the issue of IT value indirectly or not at all. A number of group support systems studies were classified as "empirical related" articles because there was some discussion of IT value, but this was still not their primary goal—see appendices A–D.) A number of IT value articles focused on the derivation of theoretical proofs. These articles were classified as "related theoretical" articles. Generally, there were no measures in these articles to document or analyze.

If any uncertainty existed about the correct classification of an article based on the information contained in the title, abstract, and key words, the researchers read the full article. In order to be particularly careful in the identification and classification of articles related to IT value, the procedure carried out was as follows:

1. Initial meetings were held to discuss the classification process and the handling of articles that did not clearly fit main categories.
2. The author and graduate students examined the journals independently and identified all articles on the subject of IT value/impacts. The author reviewed all articles in all four journals. The graduate students each reviewed articles in two journals. To ensure that there would be no bias in the selection of articles, initially the graduate students were not told how the data gathered from the IT value articles would be used.
3. The author and graduate students independently classified journal articles as articles to be analyzed, related empirical articles, related theoretical articles, and unrelated articles.
4. Later, the author and graduate students reviewed each others' article classifications.
5. Where there was disagreement among two researchers about the correct classification of an article, the article was also reviewed by the third researcher (a graduate student) who was not told how the article had previously been classified. This researcher then presented to the other two researchers his final classification decision.
6. Graduate students documented and analyzed measures used in the IT value articles. The full text of each IT value article was examined during this analysis.
7. The author reviewed step 6.
8. Final project debriefing sessions were held.

This process, although time-consuming, reduced error in the identification and classification of IT value articles (see the appendices) and increased the validity of the research findings. The author and the graduate student reviewing *CACM* and *JMIS* disagreed on the classifications of six (out of 1,060) articles—in other words, they were in agreement almost 100 percent of the time. The third researcher reviewed these six articles independently and classified them in a manner similar to the author's classification. This graduate student reviewed *ISR* and *MISQ* articles. There was 100 percent agreement between his classification of these articles and the author's classification.

Research Findings

AS TABLES 1 AND 2 SHOW, ONLY 2 PERCENT OF THE ARTICLES PUBLISHED in *CACM* since 1993 addressed the topic of IT value. However, significantly more *ISR, JMIS,* and *MISQ* articles—19

Journal	Period	No. journal issues examined	No. articles examined	No. articles on the topic of IT value or addressing "related" topics
CACM	January 1993–December 1998	72	843	14
ISR	March 1993–December 1998	24	118	23
JMIS	March 1993–December 1998	24	217	30
MISQ	March 1993–December 1998	24	126	31

Table 1 Journal issues and articles reviewed.

Journal	No. articles analyzed in detail	No. related empirical articles	No. related theoretical articles	Total no. articles
CACM	7	7	0	14
ISR	5	9	9	23
JMIS	11	10	9	30
MISQ	15	13	1	31

Table 2 Classification of IT value and related articles.

percent, 14 percent, and 25 percent, respectively—published during the same period addressed this topic. The relatively scant attention paid by *CACM* to IT value may reflect its broad readership base, as described in the *CACM* information provided to prospective authors.[5]

In contrast, the significant attention paid to IT value studies by *MISQ* no doubt reflects the journal's explicit emphasis on publishing research of managerial relevance. It follows that *MISQ* would devote relatively more pages to the benefits of IT. *ISR* and *JMIS* fall closer in their IT value publication profiles to *MISQ* than to *CACM*. Interestingly, although *ISR* published significantly fewer IT value articles than *JMIS* in the 1993–98 period (23 versus 30), because *JMIS* publishes more articles per issue, a greater proportion of *ISR* articles focused on IT value.

ISR, although somewhat concerned with managerial relevance, has historically sought to publish particularly rigorous research. It is described as "a leading international journal of theory, research, and intellectual development focused on information systems in organizations, institutions, the economy, and society" (summary statement on the editorial page, September 1996 issue). Perhaps not surprisingly, given its theoretical bent, 9 of the 23 IT value articles published in this journal (i.e., 39 percent) could not be analyzed in terms of measures because they focused on the development of proofs and were entirely theoretical. Similar figures for *CACM, JMIS,* and *MISQ,* respectively, were 0 percent, 30 percent, and 3 percent.

The *JMIS* editorial statement describes the journal as "a widely recognized forum for the presentation of research that advances the practice and understanding of organizational information systems. It serves those investigating new modes of information delivery and the changing landscape of information policy making, as well as practitioners and executives managing the information resource. A vital aim of the quarterly is to bridge the gap between theory and practice of management information systems" (editorial statement, Fall 1998 issue). With respect to the publication of IT value articles, *JMIS* appears to be slightly less receptive to theoretical proofs than *ISR*, but significantly more receptive than *CACM* and *MISQ*.

The Use of Quantitative Versus Qualitative Measures

Table 3 shows that all five of the *ISR* IT value articles published during the 1993–98 period used secondary analyses (e.g., of Compustat data) and drew conclusions based largely, if not only, on an examination of quantitative measures. This is despite the fact that:

> IT is said to enhance organizational capabilities, resulting in improved product variety, quality, and customer satisfaction, while enabling the streamlining of administrative processes and facilitating improved labor and management productivity. However, such improvements are often not reflected in improved financial performance, as benefits may be redistributed within or across organizations or passed on to consumers.
>
> ... Hitt and Brynjolfsson (1994) argue that IT has the capacity to lower and increase entry barriers and to intensify and reduce competitive rivalry. They also cite this equivocal effect of IT on competitive strategy and industry structure as an important reason for the lack of relationships between IT investment and measures of profitability, such as ROA and ROE. Our results also suggest that while various measures of IT investment can increase firm output and lower firm costs, their effect on financial measures of business performance is less consistent. [101, pp. 90, 91, 95]

The data in Table 3 describing IT value articles in the other three journals paint a somewhat more balanced picture of the use of hard and soft

Journal	Research methods used in IT value articles*	Quantitative and/or qualitative measures used	Finacial and/or nonfinancial measures used
CACM	4 secondary data analyses;	5 studies used quantitative measures only;	2 studies used financial measures only;
	2 case studies;	2 studies used quantitative and qualitative measures	1 study used nonfinancial measures only;
	1 survey		4 studies used financial and nonfinancial measures
ISR	5 secondary data analyses	5 studies used quantitative measures only	2 studies used financial measures only;
			3 studies used financial and nonfinancial measures
JMIS	4 secondary data and market data analyses;	4 studies used quantitative measures only;	5 studies used financial measures only;
	5 case studies;	5 studies used qualitative measures only;	5 studies used nonfinancial measures only;
	4 surveys;		
	1 historical analysis	2 studies used quantitative and qualitative measures	1 study used financial and nonfinancial measures
MISQ	3 secondary date analyses;	5 studies used quantitiative measures only;	6 studies used nonfinancial measures only;
	8 case studies	6 studies used qualitative measures only;	9 studies used financial and nonfinancial measures
	4 surveys	4 studies used quantitative and qualitative measures	

Table 3 Research methods and measures used in IT value articles.

Note
* Several studies used more than one research method, so column totals are unequal.

measures. To some extent, *CACM* favored the use of quantitative measures. Five of the seven studies relied on quantitative measures only. In *JMIS* and *MISQ*, however, roughly equal numbers of articles used only quantitative measures or only qualitative measures. Several articles used both quantitative and qualitative measures.

It is interesting to reflect on differences in the prevalence of hard measures and the reliance on secondary data analyses in *ISR* and *CACM* relative to *JMIS* and *MISQ*. IT value articles in the former two journals relied primarily on secondary data analyses and quantitative measures. However, the IT value articles in *JMIS* and *MISQ*, on average, tended to be balanced in their use of a variety of research methods and their reliance on quantitative and qualitative measures. No doubt this difference may be related to the editorial statements and policies published by these journals during the period examined:

> *CACM* general interest articles ... cover material of substance and emphasize concepts and principles. An article sets the background, defines fundamental concepts, compares alternate approaches, and explains the significance or application of a particular technology or result by means of well-reasoned text and pertinent graphical material. ... All submissions in this category are reviewed for technical accuracy. [*CACM* Information for Authors][6]

Information Systems Research (ISR) is dedicated to advancing the understanding and practice of information systems in organizations through theoretical and empirical research.... Submitted articles should make a contribution to knowledge in the field. Either or both quantitative and qualitative research methods may be employed.... Acceptable research articles will most frequently join theoretical analysis with empirical investigation.... Rigorous argument and presentation are expected throughout; however, the use of more complex mathematics and statistics than is necessary is discouraged. [*ISR*, March 1993]

ISR's interests are wide ranging, seeking contributions that build on established lines of work as well as well as break new ground. High-quality work from any analytical or research tradition is welcome, including theoretical, analytical, and empirical studies. [*ISR*, September 1998]

[*JMIS*] accepts empirical and interpretive submissions that make a significant contribution to the field of management information systems. Such contributions may present:

- experimental, survey-based, or theoretical research relevant to the progress of the field
- paradigmatic designs and applications
- analyses of informational policy making in an organizational, national, or international setting
- investigations of social and economic issues of organizational computing. [*JMIS*, Fall 1998]

> On the empirical side, we [at *MISQ*] welcome research based on positivist, interpretive, or integrated approaches. Traditionally, *MIS Quarterly* has emphasized positivist research methods. Though we remain strong in our commitment to hypothesis testing and quantitative data analysis, we would like to stress our interest in research that applies interpretive techniques, such as case studies, textual analysis, ethnography, and participant observation. [*MISQ*, March 1993]

The above statements suggest greater explicit receptiveness, on the part of *JMIS* and *MISQ*, to interpretive and other nonpositivist approaches. It would appear that, while recent IT value articles in *ISR* and *CACM* (especially the former) suggest a "divide" between quantitative and qualitative measures, with the use of quantitative measures being viewed particularly favorably, this pattern is only partially supported by the data gathered from *JMIS* and *MISQ*. It is supported in these latter journals to the extent that only a minority of recent articles use both quantitative and qualitative measures within the same study.

The greater receptivity, on the part of *JMIS* and *MISQ*, to nonpositivist approaches is also seen in the use of financial and nonfinancial measures in IT value articles. In *JMIS* and in *MISQ*, a large number of studies relied solely on nonfinancial measures (see Table 3). In fact, in *MISQ*, no studies used only financial measures. However, in *CACM* and in *ISR*, the reverse was true—almost no studies relied solely on nonfinancial measures.

Investigating Links Between Research Methods and the Use of Quantitative and Qualitative Measures

As Table 4 demonstrates, in IT value studies, the choice of research methods and measures was interdependent. All 16 studies using secondary data analyses relied entirely on quantitative measures only. Interestingly, a number of the surveys used soft measures (e.g., user-satisfaction measures) and a number of case studies incorporated hard measures. Almost half of the surveys and case studies used both quantitative and qualitative measures. The single historical analysis used qualitative measures. The "divide" then may be most apparent with respect to studies using secondary data analyses.

The Use of Individual, Organizational, and Other Levels of Analysis

Let us now examine the frequency of individual-level, group-level, organization-level, and industry-level analyses in IT value studies. In all four journals, IT value articles used organization-level analyses in the main, either solely or in conjunction with other analytic approaches (see Table 5). Six of the seven *CACM* articles, all 5 *ISR* articles, 7 of the 11 *JMIS* articles, and 10 of the 15 *MISQ* articles used organization-level measures. This is not in itself problematic. However, it suggests that the IT productivity paradox discussion may indeed have helped shift researcher attention to organization-level outputs. As the organization development literature cited above indicates, however,

Research methods used in IT value articles	Quantitative and/or qualitative measures used	Levels of analysis used
16 secondary data and market analyses	All 16 studies used quantitative measures only	1 study examined international-level analyses; 11 studies used organization-level analyses only; 1 study used organization and national-level analyses; 2 studies used organization and industry-level analyses; 1 study used organization and group-level analyses
9 surveys	2 studies used quantitative measures only; 3 studies used qualitative measures only; 4 studies used quantitative and qualitative measures	3 studies used organization-level analyses only; 1 study used organization and industry-level analyses; 5 studies used individual-level analyses only
15 case studies	2 studies used quantitative measures only; 7 studies used qualitative measures only; 6 studies used quantitative and qualitative measures	1 study used nation-level analyses only; 1 study used national- and individual-level analyses; 1 study used industry-level analyses; 1 study used industry- and organization-level analyses; 8 studies used organization-level analyses only; 1 study used organization-group-, and individual-level analyses; 2 studies used organization- and individual-level analyses
1 historical analysis	The study used qualitative measures	The study used national- and individual-level analyses

Table 4 Research methods, measures and levels of analysis.

organization effectiveness is achieved, and IT contributions are made, at many different levels (e.g., the individual and group).

Rai et al. [101], in their commentary on IT value research, write:

> In various studies, there is no uniform conceptualization of IT investment or identification of appropriate performance measures. For instance, if IT investments are conceptualized at the firm level, the value of IT needs to be measured at the firm level as well. On the other hand, if IT investments are conceptualized at the activity or department level, performance should be measured at these lower levels. [p. 90]

Barua et al. [7] also argue that the effects of IT on organization performance can best be identified through a "web of intermediate level contributions." However, the data indicate that this intermediate (e.g., process, individual, and group) approach to analysis has *not* been the norm. Instead, a "black box," input–output approach currently appears to dominate the IT value literature. Although it can be difficult to combine multiple levels of analysis (e.g., group and organizational) within the same study, a small number of the articles examined [7, 10, 31, 123] demonstrate that it can be done.

In all four journals, organization-level analyses were carried out significantly more often on their own than in conjunction with other (e.g., individual, group, industry, or national) approaches. Relatively few studies combined multiple approaches (e.g., analyses at the individual, group, and organization levels). This suggests a divide between the use of organization-level variables and other variables in recent IT value research.

One might think that, given the macroeconomic origins of the IT productivity paradox debate (see, e.g., [75, 103, 104, 105, 115]), in the past, quantitative, organization-level measures have not served researchers particularly well in their search for IT productivity gains. Interestingly enough, instead of reevaluating our reliance on these measures and promoting new concepts and measures of IT value, several researchers appear to have redoubled their efforts to uncover quantitative, organization-level evidence of IT value. Certainly, IT value studies using organization-level analyses appear to be the ones primarily being published in North American journals today.

Investigating Links Between Research Methods and Levels of Analysis Used

Table 5 reveals that IT value studies using secondary data analyses relied primarily on organization-level analyses only. A small number of these studies conducted analyses at other levels also. Surveys appeared to be split roughly equally between the use of organization-level analyses and individual-level analyses. No surveys incorporated analyses at both levels. Case studies focused on organization-level analyses. A very small number of these studies addressed both organization- and individual-level variables. The single historical analysis that was reviewed addressed both national-level and individual-level phenomena. These findings suggest strong ties between levels of analysis and research methods. In some ways, this is not surprising. Certain research methods may be better suited to investigate individual-level or organization-level issues. What may be surprising, however, is the depth of the divide between specific research methods and levels of analysis. For instance, one might have expected to find more surveys and case studies that used both organization- and individual-level analyses.

Interestingly, journals had a significant impact on the findings here. For instance, in studies using the survey research method, when the use of levels of analysis is examined (see the appendices also), we find that all four surveys reported in *MISQ* on IT value, during 1993–98, used individual levels of analysis only. The other five surveys reported in *CACM* and *JMIS* (*ISR* published no surveys on the subject during this period) used organization-level analyses primarily. When we examine case studies on IT value during 1993–98, we see that 8 (just over

Journal	Level(s) of analysis used in IT Value articles
CACM	1 study used international-level analyses
	1 study used national- and organization-level analyses
	4 studies used organization-level analyses
	1 study used organization- and individual-level analyses
ISR	4 studies used organization-level analyses
	1 study used organization- and group-level analyses
JMIS	2 studies used national- and individual-level analyses
	1 study used industry-level analyses
	3 studies used industry- and organization level analyses
	4 studies used organization -level analyses
	1 study used individual-level analyses
MISQ	1 study used national-level analyses
	8 studies used organization-level analysis
	1 study used organization-, group-, and individual-level analyses
	1 study used organization- and individual-level analyses
	4 studies used individual-level analyses

Table 5 Levels of analysis used in IT value articles.

half) of the 15 studies were published by *MISQ* alone. Of these case studies, most relied only on organization-level analyses. However, of the five case studies published by *JMIS* (*ISR* published no IT value case studies, and *CACM* published two during 1993–98), several relied on industry- and national-level analyses. This once again underscores the strong links seen between journals examined and the kinds of analyses published.

In the case of IT value research, there appear to be complex interactions among journals, research methods, the use of quantitative and qualitative measures, and levels of analysis. The gatekeepers of IT value research (i.e., the journals) may themselves be divided in terms of the research that is published. Journal editors may find it useful to review their journal's positioning in the MIS "research industry" periodically, and their journal's explicit or implicit role in promoting or eliminating research "divides."

Examining Trends over Time

Table 6 examines the emergence of trends over time in the kinds of IT value articles that have been published by North American journals. First, it is clear that there has been no noticeable surge or tapering off of interest in the subject. With the exception of 1997, approximately seven articles have been published each year between 1993 and 1998 in the four journals reviewed. Second, prior to 1996, the quantitative–qualitative pendulum swung backward and forward. In different years, different measures were seen most commonly. However, from 1996 onward, studies using quantitative measures appear to have dominated the IT value literature. Third, the data suggest that organization-level analyses have continually dominated the IT value literature throughout the six-year period examined. Between 1993 and 1996, in each year, roughly half the studies relied only on organization-level analyses. In 1997, there was an interesting anomaly where the divide between organization-level analyses and other analyses appeared to have been bridged. Several studies combined organization-level analyses with analyses at other levels. In 1998, however, the divide was once again very apparent and perhaps wider than seen previously. Five of the seven studies published used organization-level analyses only.

Summary: Hard Versus Soft? High Versus Low?

The review of recent *CACM*, *ISR*, *JMIS*, and *MISQ* articles on IT value revealed a schism between the use of organization-level measures and other measures. *CACM* and *ISR* also provided strong evidence of a schism between quantitative and qualitative measures. The *JMIS* and *MISQ* data provided more limited evidence of this schism. The data suggested that the schisms are getting more noticeable over time. This may be partly due to an increasing reliance on, and receptivity to, secondary data set analyses that tend to use only quantitative measures and organization-level analyses. The current research confirms what many researchers suspect—schisms exist, and may be deepening, in IT value research.

The *CACM*, *ISR*, *JMIS*, and *MISQ* data suggest a need for renewed recognition by MIS researchers of the importance of using a *variety* of measures and levels of analysis when conducting IT value studies. In order to promote rich understanding and meaningful analyses of the benefits of IT investments, more balanced perspectives of IT value (e.g., combinations of organization and nonorganization level analyses, and hard and soft measures) are required.

Research Limitations

BEFORE CLOSING, A NUMBER OF LIMITATIONS OF THIS RESEARCH must be acknowledged. First, this article draws its conclusions from studies published in only four North American journals since 1993. Admittedly, these publications are leading MIS publications. Possible additional extensions to this research, however, could include analyses covering longer time periods (say, ten years), and/or examining additional journals, such as research published in European journals on the subject of IT value.

Another limitation of the current study is one of "small numbers." Thirty-eight articles were examined in detail, which precludes broad generalizations about the subject of IT value research. The findings discussed above are intended primarily to raise the awareness, and heighten the sensitivity, of MIS researchers to trends in the methods and measures used to investigate IT value. The findings provide some evidence of a deepening

Year*	Quantitative and/or qualitative measures used	Levels of analysis used
1993 (6 IT value articles)	3 studies used quantitative measures only; 1 study used qualitative measures only; 2 studies used quantitative and qualitative measures	1 study used national- and organization-level analyses; 1 study used national- and individual-level analyses; 3 studies used organization-level analyses only; 1 study used individual-level analyses only
1994 (7 IT value articles)	1 study used quantitative measures only; 4 studies used qualitative measures only; 2 studies used quantitative and qualitative measures	1 study used national-level analyses; 1 study used national- and individual-level analyses; 4 studies used organization-level analyses only; 1 study used organization-, group-, and individual-level analyses
1995 (7 IT value articles)	3 studies used quantitative measures only; 3 studies used qualitative measures only; 1 study used quantitative and qualitative measures	1 study used industry- and organization-level analyses; 3 studies used organization-level analyses only; 1 study used organization and group-level analyses; 2 studies used individual-level analyses only
1996 (7 IT value articles)	4 studies used quantitative measures only; 2 studies used qualitative measures only; 1 study used quantitative and qualitative measures	1 study used industry-level analyses; 1 study used industry- and organization-level analyses; 4 studies used organization-level analyses only; 1 study used individual-level analyses only
1997 (4 IT value articles)	2 studies used quantitative measures only; 1 study used qualitative measures only; 1 study used quantitative and qualitative measures	1 study used industry- and organization-level analyses; 1 study used organization-level analyses only; 2 studies used organization- and individual-level analyses
1998 (7 IT value articles)	6 studies used quantitative measures only; 1 study used quantitative and qualitative measures	1 study used international-level analyses; 5 studies used organization-level analyses only; 1 study used individual-level analyses only

Table 6 Longitudinal view of measures and levels of analysis utilized.

Note
* 1993–94 data were included in 1993. 1994–95 data were included in 1994. 1995–96 data were included in 1995. 1996–97 data were included in 1996.

analytic divide, despite repeated calls in the literature for the use of multiple methods and measures.

An additional limitation of this study involves the subjective judgments made by the author and two graduate students (e.g., about which articles qualified as "IT value" articles and which articles were "related'). However, the process followed in selecting, classifying, and analyzing articles was designed to be as rigorous as time and resources would allow. Several independent checks were carefully built into the article selection, classification, and analysis process.

Yet another limitation is that this study focused on *published research*. It did not examine all IT value research submitted to journals for their review. So it may tell us more about powerful editors' and reviewers' views of valid IT value measures than about those of IT value researchers. Similarly, the study has not examined IT value research that is currently under way (i.e., still to be submitted to journals). It may therefore tell us more about research undertaken several years ago than about current research on IT value, because of the significant publishing time lags.

Finally, the study tells us little about the use of IT value measures in MIS *practice*. Questions such as the following can usefully be addressed in future studies: To what extent do business managers look to published research as sources of information on IT value measures? How strong are the links between IT value research and practice? And do business managers experience similar schisms in their corporations?

Research Implications

SEVERAL IMPLICATIONS FOR IT VALUE RESEARCH ARISE from this study. The data suggest that researchers, in the future, may be better served by:

- *Emphasizing theory generation*, and reducing the reliance on isolated, input–output "black box" approaches. It may be that more concepts in IT value research can usefully be identified at individual and group (i.e., intermediate) levels. Innovative models (e.g., dynamic, process-focused, open system models of IT investments) may be quite helpful. As Kauffman and Weill [65, p. 385] argue, "IT value research is still in its adolescence." There are many promising reference disciplines (e.g., organization development, psychology, sociology, and industrial relations) that researchers can draw on also as they carry out future IT value studies.
- *Explicitly recognizing the limitations of current methods and measures* in IT value research, and focusing on creating additional, unconventional methods and measures. It is expected that new measures would complement (not replace) existing conventional (e.g., microeconomic) measures. For example, IT value studies could explicitly monitor messy phenomena such as culture—the set of shared, taken-for-granted implicit assumptions that determine how a group perceives and reacts to its environments [109] and its investments. As Schein [108, p. 229] writes: "I believe our failure to take [phenomena like] culture seriously enough stems from our methods of inquiry, which put a greater premium on abstractions that can be measured than on careful ethnographic or clinical observation of organizational phenomena.... I also hope that we as researchers will come to recognize how much our own methods and concepts are a product of our own culture."
- *Becoming more aware as researchers of our own assumptions and biases*, periodically challenging these views, and examining our receptivity to change. One might expect that the current study would paint a very different picture—one with a great deal of innovation in IT value research, as researchers heeded recommendations made in earlier studies. Instead, the study has served to highlight recommendations that have been made previously, but that have not been acted on, in the main. Unless we are willing to change, our research camps may remain divided, our methods fossilized, and our tools blunt.

Management Implications

THIS STUDY ALSO HAS SEVERAL IMPLICATIONS FOR MANAGERS, ARISING both from the literature that has been reviewed and from the data analyses that have been conducted. They are as follows:

- *IT value is discussed meaningfully in the context of the organization's goals, strategies, culture, structure, and environment. IT investments can usefully be viewed as organization change initiatives* [74]. The management task related to obtaining benefits from IT investments involves facilitating ongoing system adaptation and continuous learning. System boundary identification is a challenging, but necessary task, if IT paybacks are to be correctly assessed. A variety of internal and external stakeholder (e.g., employee and customer) impacts should be monitored.
- *Because systems are dynamic, an assessment of IT value that relies heavily on a few key numbers at a single point in time will be incomplete and possibly misleading*. Managers evaluating IT investments may wish to identify and report on a number of performance dimensions (e.g., customer impacts, profitability, stock prices, and employee satisfaction), at different points in time [61].
- *In order to fully harvest economic benefits of IT investments, ongoing management processes must be established*. IT investments unfold, and must be managed, over time. This requires open systems planning [110]. Unfortunately, while many organizations are prepared to spend large sums on technology, at the same time they may resist spending even modest sums on ongoing

management systems required to ensure that expected IT paybacks are realized. What we often have are short-term "transaction" (single event) approaches to obtaining IT value, when what we often need are long-term "relationship" (multiple event) approaches. Perhaps, in the final analysis, IT valuation is less concerned with producing a single number and more concerned with promoting informed, thought-provoking, and ongoing discussion about IT investments.

- *IT evaluation approaches are also systems. They should evolve with the organization, and be adapted to specific information systems under consideration.* Evaluation approaches themselves need to be periodically reviewed and redesigned [74].

Closing Remarks

In summary, whereas most current IT value research appears to address the question "*what* value do IT investments provide?" this research may not yet be adequately addressing the related set of questions, "*why, where, when, how*, and *to whom* do these investments provide value?" These questions in turn may require an examination of a variety of qualitative and quantitative measures, and the use of individual, group, process, and organization-level measures. Meaningful and rich documentation of the value of IT investments may ultimately require us to unite the "hard" and "soft" camps, and the "high" and "low" camps, and to bridge the great divide.

NOTES

The author gratefully acknowledges funding provided by the Social Sciences and Humanities Research Council of Canada; research assistance provided by Ph.D. candidates, Peter Gray and Yann Malara; and administrative assistance provided by Linda Freeman. A subset of this article was previously published in the *Proceedings of the Fifth (1998) European Conference on the Evaluation of IT.*

1 For MIS journal rankings, see ISWorld Net <http://is.lse.ac.uk/iswnet/profact/journal.htm>.
2 The rational, goal-oriented perspective is just one of three organizational perspectives outlined by Bakos.
3 The technology set of key words screened out non-IT value articles such as those focused on the performance of meeting facilitators or the usefulness of a particular methodology.
4 Innovation has multiple meanings. Here it was used strictly to refer to the adoption of new technology.
5 See <http://catt.bus.okstate.edu/isworld/journal2.htm>.
6 See <http://catt.bus.okstate.edu/isworld/journal2.htm>.

REFERENCES

1 Abdul-Gader, A. H. and Kozar, K. A. The impact of computer alienation on information technology investment decisions: an exploratory cross-national analysis. *MIS Quarterly, 19*, 4 (December 1995), 535–59.
2 Bakos, J. Y. Dependent variables for the study of firm and industry-level impacts of information technology. *Proceedings of the Eighth International Conference on Information Systems.* Pittsburgh, December 1987, pp. 10–23.
3 Bakos, J. Y. The emerging role of electronic marketplaces on the internet. *Communications of the ACM, 41*, 8 (August 1998), 35–42.
4 Bakos, J. Y. and Brynjolfsson, E. Information technology, incentives and the optimal numbers of suppliers. *Journal of Management Information Systems, 10*, 2 (1993), 37–53.
5 Bakos, J. Y. and Nault, B. R. Ownership and investment in electronic networks. *Information Systems Research, 8*, 4 (December 1997), 321–41.
6 Banker, R. D. and Kauffman, R. J. Strategic contributions of information technology: an empirical study of ATM networks. *Proceedings of the Ninth International Conference on Information Systems.* Minneapolis, December 1988.
7 Barua, A., Kriebel, C. H. and Mukhopadhyay, T. Information technologies and business value: an analytic and empirical investigation. *Information Systems Research, 6*, 1 (March 1995), 3–23.
8 Barua, A. and Lee, B. An economic analysis of the introduction of an electronic data interchange system. *Information Systems Research, 8*, 4 (December 1997), 398–22.
9 Barua, A.; Lee, C. H. S. and Whinston, A.B. The calculus of reengineering. *Information System Research, 7*, 4 (December 1996), 409–28.
10 Belcher, L. W. and Watson, H. J. Assessing the value of Conoco's EIS. *MIS Quarterly, 17*, 3 (September 1993), 239–53.
11 Bensaou, M. Interorganizational cooperation: the role of information technology–an

empirical comparison of U.S. and Japanese supplier relations. *Information Systems Research, 8*, 2 (June 1997), 107–24.
12. Brown, R.M.; Gatian, A. W.; and Hicks, J. O. Jr. Strategic information systems and financial performance. *Journal of Management Information Systems, 11*, 4 (1995), 215–48.
13. Brynjolfsson, E. The contribution of information technology to consumer welfare. *Information Systems Research, 7*, 3 (September 1996), 281–300.
14. Brynjolfsson, E. The productivity paradox of information technology. *Communications of the ACM, 36*, 12 (December 1993), 67–77.
15. Brynjolfsson, E. and Hitt, L. M. Beyond the productivity paradox. *Communications of the ACM, 41*, 8 (August 1998), 49–55.
16. Campbell, J. P.; Dunnette, M. D.; Lawler, E. E.; and Weick, K.E. *Managerial Behavior, Performance, and Effectiveness.* New York: McGraw-Hill, 1970.
17. Caron, J. R.; Jarvenpaa, S. L.; and Stoddard, D.B. Business reengineering at CIGNA Corporation: experiences and lessons learned from the first five years. *MIS Quarterly, 18*, 3 (September 1994), 233–50.
18. Cats-Baril, W. L. and Jelassi, T. The French Videotex System Minitel: a successful implementation of a national information technology infrastructure. *MIS Quarterly, 18*, 1 (March 1994), 1–20.
19. Chan, Y. E.; Huff, S. L.; Barclay, D. W. and Copeland, D.G. Business strategic orientation, information systems strategic orientation, and strategic alignment. *Information Systems Research, 8*, 2 (June 1997), 125–50.
20. Chismar, W. G. and Kriebel, C. H. A method for assessing the economic impact of information systems technology on organizations. *Proceedings of the Sixth International Conference on Information Systems.* Indianapolis, December 1985, pp. 45–56.
21. Choe, J. M. The relationship among performance of accounting information systems, influence factors, and evolution level of information systems. *Journal of Management Information Systems, 12*, 4 (1996), 215–39.
22. Clark, T. H. and Stoddard, D. B. Interorganizational business process redesign: merging technological and process innovation, *Journal of Management Information Systems, 13*, 2 (1996), 9–28.
23. Clemons, E. K.; Croson, D. C.; and Weber, B. W. Market dominance as a precursor of a firm's failure. *Journal of Management Information Systems, 13*, 2 (1996), 59–75.
24. Clemons, E. K.; Reddi, S. P.; and Row, M. C. Information technology and the organization of economic activity: the "move to the middle" hypothesis. *Journal of Management Information Systems, 10*, 2 (1993), 9–35.
25. Clemons, E. K. and Weber, B. W. Alternative securities trading systems: tests and regulatory implications of the adoption of technology. *Information Systems Research, 7*, 2 (June 1996), 163–88.
26. Clemons, E. K. and Weber, B. W. Segmentation, differentiation, and flexible pricing: experiences with information technology and segment-tailored strategies. *Journal of Management Information Systems, 11*, 2 (1994), 9–36.
27. Clemons, E. K. and Weber, B. W. Restructuring institutional block trading: an overview of the OptiMark system. *Journal of Management Information Systems, 15*, 2 (1998), 41–60.
28. Coopersmith, J. Texas politics and the fax revolution. *Information Systems Research, 7*, 1 (March 1996), 37–51.
29. Crowston, K. and Treacy, M. E. Assessing the impact of information technology on enterprise level performance. *Proceedings of the Seventh International Conference on Information Systems.* San Diego, 1986, pp. 299–310.
30. De, P. and Ferrat, T. W. An information system involving competing organizations. *Communications of the ACM, 41*, 12 (December 1998), 90–98.
31. Desmaris, M. C.; Leclair, R.; Fiset, J.-Y. and Talbi, H. Cost-justifying electronic performance support systems. *Communications of the ACM, 40*, 7 (July 1997), 39–48.
32. Dewan, R. M.; Freimer, M. L.; and Seidmann, A. Internet service providers, proprietary content, and the battle for users' dollars. *Communications of the ACM, 41*, 8 (August 1998), 56–62.
33. Dewan, S. Pricing computer services under alternative control structures: tradeoffs and trends. *Information Systems Research, 7*, 3 (September 1996), 301–307.
34. Dewan, S. and Kraemer, K. L. International dimensions of the productivity paradox. *Communications of the ACM, 41*, 8 (August 1998), 56–62.

35 Dewan, S.; Michael, S. C. and Min, C-K. Firm characteristics and investments in information technology: scale and scope effects. *Information Systems Research, 9*, 3 (September 1998), 219–32.

36 Diebold, J. How computers and communications are boosting productivity: an analysis. *International Journal of Technology Management, 5*, 2 (1990), 141–52.

37 Dos Santos, B. L.; Peffers, K.; and Mauer, D. C. The impact of information technology investment announcements on the market value of the firm. *Information Systems Research, 4*, 1 (March 1993), 1–23.

38 Due, R. T. The productivity paradox revisited. *Information Systems Management, 4*, 1 (Winter 1994), 74–6.

39 Duchessi, P. and Chengalur-Smith, I. Client/server benefits, problems, best practices. *Communications of the ACM, 41*, 5 (May 1998), 87–94.

40 Edberg, D. T. and Bowman, B. J. User-developed applications: an empirical study of application quality and developer productivity. *Journal of Management Information Systems, 13*, 1 (1996), 167–85.

41 El Sawy, O. A. and Bowles, G. Redesigning the customer support process for the electronic economy: insights from storage dimensions. *MIS Quarterly, 21*, 4 (December 1997), 457–83.

42 Finlay, P. N. and Mitchell, A. C. Perceptions of the benefits from the introduction of CASE: an empirical study. *MIS Quarterly, 18*, 4 (December 1994), 353–70.

43 Francalanci, C. and Galal, H. Information technology and worker composition: determinants of productivity in the life insurance industry. *MIS Quarterly, 22*, 2 (June 1998), 227–41.

44 Gill, T. G. Early expert systems: where are they now? *MIS Quarterly, 19*, 1 (March 1995), 51–81.

45 Gill, T. G. Expert systems usage: task change and intrinsic motivation. *MIS Quarterly, 20*, 3 (September 1996) 301–29.

46 Goodhue, D. L. and Thompson, R. L. Task-technology fit and individual performance. *MIS Quarterly, 19*, 2 (June 1995), 213–36.

47 Grover, V.; Teng, J. T. C.; and Fiedler, K. D. IS investment priorities in contemporary organizations. *Communications of the ACM, 41*, 2 (February 1998), 40–48.

48 Gurbaxani, V. and Mendelson, H. Modeling vs. forecasting: the case of information systems spending (Research Report). *Information Systems Research, 5*, 2 (June 1994), 180–90.

49 Henderson, J. C. and Lentz, C. M. A. Learning, working, and innovation: a case study in the insurance industry. *Journal of Management Information Systems, 12*, 3 (1995–96), 43–64.

50 Henderson, J. C. and Sifonis, J. G. The value of strategic IS planning: understanding consistency, validity, and IS markets. *MIS Quarterly, 12*, 2 (June 1988), 186–200.

51 Hess, C. M. and Kemerer, C. F. Computerized loan origination systems: an industry case study of the electronic markets hypothesis. *MIS Quarterly, 18*, 3 (September 1994), 251–75.

52 Hildebrand, C. Resounding maybe. *CIO* (February 1, 1994), 35–37.

53 Hitt, L. and Brynjolfsson, E. Information technology and internal firm organization: an exploratory analysis. *Journal of Management Information Systems, 14*, 2 (1997), 81–101.

54 Hitt, L. and Brynjolfsson, E. Productivity, business profitability, and consumer surplus: three *different* measures of information technology value. *MIS Quarterly, 20*, 2 (June 1996), 121–42.

55 Hitt, L. and Brynjolfsson, E. The three faces of IT value: theory and evidence. *Proceedings of the Fifteenth International Conference in Information Systems*. Vancouver, BC, December 1994, pp. 263–77.

56 Holden, T. and Wilhemij, P. Improved decision making through better integration of human resource and business process factors in a hospital situation. *Journal of Management Information Systems, 12*, 3 (1995–96), 21–41.

57 Iacovou, C. L.; Benbasat, I.; and Dexter, A. S. Electronic data interchange and small organizations: adoption and impact of technology. *MIS Quarterly, 19*, 4 (December 1995), 465–86.

58 Jarvenpaa, S. L. and Leidner, D. E. An information company in Mexico: extending the resource-based view of the firm to a developing country context. *Information Systems Research, 9*, 4 (December 1998), 342–61.

59 Jelassi, T. and Figon, O. Competing through EDI at Brun Passot: achievements in France and ambitions for the single European market. *MIS Quarterly, 18*, 4 (December 1994), 337–52.

60 Kambil, A. and van Heck, E. Reengineering the Dutch flower auctions: a framework for

analyzing exchange organizations. *Information Systems Research, 9*, 1 (March 1998), 1–19.

61 Kaplan, R. S. and Norton, D. P. The balanced scorecard: measures that drive performance. *Harvard Business Review* (January–February 1992), 71–9.

62 Karami, J.; Gupta, Y. Y. and Somers, T. M. Impact of competitive strategy and information technology maturity on firm's strategic response to globalization. *Journal of Management Information Systems, 13*, 1 (1996), 63–88.

63 Kauffman, R. J. and Kriebel, C. H. Modeling and measuring the business value of information technology. In ICIT Research Study Team no. 2 (eds.), *Measuring the Business Value of Information Technologies*. Washington, DC: ICIT Press, 1988.

64 Kauffman, R. J.; Kriebel, C. H.; and Zajonc, P. C. Measuring business value for investments in point of sale technology. Working paper no. 193. Center for Research on Information Systems, Stern School of Business, New York University, December 1988.

65 Kauffman, R. J. and Weill, P. An evaluative framework for research on the performance effects of information technology investment. *Proceedings of the Tenth International Conference on Information Systems*. Boston, December 1989, pp. 377–88.

66 Kettinger, W. J.; Grover, V.; Guha, S. and Segars, A.H. Strategic information systems revisited: a study in sustainability and performance. *MIS Quarterly, 18*, 1 (March 1994), 31–58.

67 King, J. L. and Kraemer, K. L. Implementation of strategic information systems. In K. C. Laudon and J. A. Turner (eds.), *Information Technology and Management Strategy*. Englewood Cliffs, NJ: Prentice-Hall, pp. 78–91.

68 King, W. R. and Teo, T. S. H. Key dimensions of facilitators and inhibitors for the strategic use of information technology. *Journal of Management Information Systems, 12*, 3 (1996), 35–53.

69 Kraemer, K. L.; Danziger, J. N.; Dunkle, D. E.; and King, J. L. The usefulness of computer-based information to public managers. *MIS Quarterly, 17*, 2 (June 1993), 129–48.

70 Kraemer, K. L. and Dedrick, J. Globalization and increasing returns: implications for the U.S. computer industry. *Information Systems Research, 9*, 4 (December 1998), 303–22.

71 Kumar, R. L. A note on project risk and option values of investments in information technologies. *Journal of Management Information Systems, 13*, 1 (1996), 187–93.

72 Lee, H. G. and Clark, T. H. Market process reengineering through electronic market systems: opportunities and challenges. *Journal of Management Information Systems, 13*, 3 (1996–97), 113–36.

73 Lippitt, G. L. *Organizational Renewal: A Holistic Approach to Organization Development*. Englewood Cliffs, NJ: Prentice-Hall, 1982.

74 Lippitt, G. L.; Langseth, P.; and Mossop, J., eds. *Implementing Organizational Change: A Practical Guide to Managing Change Effort*. San Francisco: Jossey-Bass, 1985.

75 Loveman, G. W. An assessment of the productivity impact of information technologies. MIT Management in the 1990s Working paper no. 88–054, July 1988.

76 Lucas, H. C. Jr.; Berndt, D. J.; and Truman, G. A reengineering framework for evaluating a financial imaging system. *Communications of the ACM, 39*, 5 (May 1996), 86–96.

77 Mahmood, M. A. Associating organizational strategic performance with information technology investment: an exploratory research. *European Journal of Information Systems, 2*, 3 (1993), 185–200.

78 Mahmood, M. A. and Mann, G. J. Measuring the organizational impact of information technology investment: an exploratory study. *Journal of Management Information Systems, 10*, 1 (Summer 1993), 97–122.

79 Maier, J. L.; Rainer, R. K. Jr.; and Snyder, C. A. Environmental scanning for information technology. *Journal of Management Information Systems, 14*, 2 (Fall 1997), 177–200.

80 Manning, P. K. Information technology in the police context: the "sailor" phone. *Information Systems Research, 7*, 1 (March 1996), 52–62.

81 Massetti, B. An empirical examination of the value of creativity support systems on idea generation. *MIS Quarterly, 20*, 1 (March 1996), 83–98.

82 Massetti, B. and Zmud, R. W. Measuring the extent of EDI usage in complex organizations: strategies and illustrative examples. *MIS Quarterly, 20*, 3 (September 1996), 331–45.

83 Mata, F. J.; Fuerst, W. L.; and Barney, J. B. Information technology and sustained competitive advantage: a resource-based analysis. *MIS Quarterly, 19*, 5 (December 1995), 487–506.

84 Melone, N. P. A theoretical assessment of the user-satisfaction construct in information systems research. *Management Science, 36*, 1 (January 1990), 76–91.

85 Mitra, S. and Chaya, A. K. Analyzing cost-effectiveness of organizations: the impact of information technology spending. *Journal of Management Information Systems, 13*, 2 (1996), 29–57.

86 Mookerjee, V. S. and Dos Santos, B. L. Inductive expert system design: maximizing system value. *Information Systems Research, 4*, 2 (June 1993), 111–40.

87 Mukhopadhyay, T. and Cooper, R. B. A microeconomic production assessment of the business value of management information systems. *Journal of Management Information Systems, 10*, 1 (Summer 1993), 33–55.

88 Mukhopadhyay, T. and Cooper, R. B. Impact of management information systems on decisions. *Omega, 20*, 1 (1992), 37–49.

89 Mukhopadhyay, T.; Kekre, S.; and Kalathur, S. Business value of information technology: a study of electronic data interchange. *MIS Quarterly, 19*, 2 (June 1995), 137–56.

90 Nam, K.; Rajagopalan, H.; Raghav, R.; and Chaudhury A. A two-level investigation of information systems outsourcing. *Communications of the ACM, 39*, 7 (July 1996), 36–44.

91 Nault, B. R. Research report: information technology and investment incentives in distributed operations. *Information Systems Research, 8*, 2 (June 1997), 196–202.

92 Nault, B. R. and Dexter, A. S. Added value and pricing with information technology. *MIS Quarterly, 19*, 4 (December 1995), 449–64.

93 Nelson, P.; Richmond, W.; and Seidmann, A. Two dimensions of software acquisition. *Communications of the ACM, 39*, 7 (July 1996), 29–35.

94 Newman, J. K. and Kozar, K. A. A multimedia solution to productivity gridlock: a reengineered jewelry appraisal system at Zale corporation. *MIS Quarterly, 18*, 1 (March 1994), 21–30.

95 Nidumolu, S. R. and Knotts, G. W. The effects of customizability and reusability on perceived process and competitive performance of software firms. *MIS Quarterly, 22*, 2 (June 1998), 105–37.

96 Pinsonneault, A. and Kraemer, K. L. The impact of information technology on middle managers. *MIS Quarterly, 17*, 3 (September 1993), 271–92.

97 Pinsonneault, A. and Rivard, S. Information technology and the nature of managerial work: from the productivity paradox to the Icarus paradox? *MIS Quarterly, 22*, 3 (September 1998), 287–311.

98 Pitt, L. F.; Watson, R. T.; and Kavan, C. B. Service quality: A measure of information systems effectiveness. *MIS Quarterly, 19*, 2 (June 1995), 173–88.

99 Post, G. V.; Kagan, A.; and Lau, K-N. A modeling approach to evaluating strategic uses of information technology. *Journal of Management Information Systems, 2*, 1 (1995), 161–87.

100 Premkumar, G. and King, W. R. Organizational characteristics and information systems planning: an empirical study. *Information Systems Research, 5*, 2 (June 1994), 75–109.

101 Rai, A.; Patnayakuni, R.; and Patnayakuni, N. Technology investment and business performance. *Communications of the ACM, 40*, 7 (July 1997), 89–97.

102 Rice, D.E. Relating electronic mail use and network structure and R&D work networks and performance. *Journal of Management Information Systems, 11*, 1 (1994), 9–29.

103 Roach, S. S. America's technology dilemma: a profile of the information economy. Morgan Stanley's economics newsletter series, April 22, 1987.

104 Roach, S. S. America's white-collar productivity dilemma. *Manufacturing Engineering* (August 1989), 104.

105 Roach, S. S. Services under siege—the restructuring imperative. *Harvard Business Review* (September–October 1991), 82–92.

106 Robey, D. and Sahay, S. Transforming work through information technology: a comparative case study of geographic information systems in county government. *Information Systems Research, 7*, 1 (March 1996), 93–110.

107 Sampler, J. L. and Short, J. E. An examination of information technology's impact on the value of information and expertise: implications for organizational change. *Journal of Management Information Systems, 11*, 2 (1994), 59–73.

108 Schein, E. H. Culture: The missing concept in organization studies. *Administrative Science Quarterly, 4*, 2 (1996), 229–40.

109 Schein, E. H. *Organizational Culture and Leadership*. San Francisco: Jossey-Bass, 1992.
110 Schein, E. H. *Organizational Psychology*. Englewood Cliffs, NJ: Prentice-Hall, 1980.
111 Seddon, P. B. A respecification and extension of the DeLone and McLean model of IS success. *Information Systems Research, 8*, 3 (September 1997), 240–53.
112 Seidmann, A. and Sundararajan, A. Competing in information-intensive services: analyzing the impact of task consolidation and employee empowerment. *Journal of Management Information Systems, 14*, 2 (1997), 33–56.
113 Sheffield, J. and Gallupe, B. R. Using electronic meeting technology to support economic policy development in New-Zealand: short-term results. *Journal of Management Information Systems, 10*, 3 (1993–94), 97–116.
114 Sheffield, J. and Gallupe, B. R. Using group support systems to improve the New Zealand economy. *Journal of Management Information Systems, 11*, 3 (1994–95), 135–53.
115 Strassmann, P. A. *The Business Value of Computers*. New Canaan, CT: Information Economics Press, 1990.
116 Subramanian, G. H. and Zarnich, G. E. An examination of some software development effort and productivity determinants in ICASE tool projects. *Journal of Management Information Systems, 12*, 4 (1996), 143–60.
117 Tam, K. Y. Dynamic price elasticity and the diffusion of mainframe computing. *Journal of Management Information Systems, 13*, 2 (1996), 163–83.
118 Tam, K. Y. The impact of information technology investments on firm performance and evaluation: evidence from newly industrialized economies. *Information Systems Research, 9*, 1 (March 1998), 85–98.
119 Teng, J. T. C.; Jeong, S. R.; and Grover, V. Profiling successful reengineering projects. *Communications of the ACM, 41*, 6 (June 1998), 96–102.
120 Teo, H-K.; Tan, B. C. Y.; and Wei, K-K. Organizational transformation using electronic data interchange: the case of TradeNet in Singapore. *Journal of Management Information Systems, 13*, 4 (1997), 139–165.
121 *The Economist*, How real is the new economy? and the new economy, work in progress. July 24, 1999, pp. 17–18 and 21–24.
122 This, L. and Lippitt, G. L. Managerial guidelines to sensitivity training. In G. L. Lippitt, L. E. This and R. G. Bidwell, Jr. (eds), *Optimizing Human Resources*. Reading, MA: Addison-Wesley, 1971.
123 Vandenbosch, B. and Huff, S. L. Searching and scanning: how executives obtain information from executive information systems. *MIS Quarterly, 21*, 1 (March 1997), 81–107.
124 Venkatraman, N. Strategic orientation of business enterprises: the construct, dimensionality, and measurement. *Management Science, 35*, 8 (August 1989), 942–962.
125 Walstrom, K. A.; Hardgrave, B. C.; and Wilson, R. L. Forums for management information systems scholars. *Communications of the ACM, 38*, 3 (1995), 93–107.
126 West, L. A. Jr. Researching the cost of information systems. *Journal of Management Information Systems, 11*, 2 (1994), 75–107.
127 Wong, P.-K. Leveraging the global information revolution for economic development: Singapore's evolving information industry strategy. *Information Systems Research, 9*, 4 (December 1998), 323–341.
128 Yoon, Y.; Guimaraes, T.; and O'Neal, Q. Exploring the factors associated with expert systems success. *MIS Quarterly, 19*, 1 (March 1995), 83–106.
129 Ytterstad, P.; Akselsen, S.; Svendsen, G.; and Watson, R. T. Teledemocracy: using information technology to enhance political work. *MIS Quarterly, 20*, 3 (September 1996), 347–348.

ABOUT THE AUTHOR

Yolande E. Chan is an Associate Professor of Management Information Systems at Queen's University. She holds a Ph.D. from the University of Western Ontario, an M.Phil. in management studies from Oxford University, and S.M. and S.B. degrees in electrical engineering and computer science from the Massachusetts Institute of Technology. Prior to joining Queen's, Dr Chan worked for several years with Andersen Consulting. She has published in *Information Systems Research*, *Journal of Strategic Information Systems*, *Academy of Management Executive*, and *Information and Management*. Her research interests include knowledge management and information technology strategy, alignment, and performance.

APPENDIX 1

Communications of the ACM

1. ARTICLES ANALYZED

STUDY	RESEARCH METHOD(S)	MEASURES USED TO ASSESS DEPENDENT VARIABLES	QUANTITATIVE and/or QUALITATIVE MEASURES	FINANCIAL and/or NON-FINANCIAL	LEVEL(S) OF ANALYSIS—individual/group/organizational/industry/national/international	RESULTS
[14]: Brynjolfsson, E. (1993)	Secondary Data Analysis (Literature Review)	Labour productivity Output	Quantitative	Financial and Non-financial	Organizational and National	Apparent lack of productivity is due to mismeasurement of outputs and inputs, lags due to learning and adjustment, redistribution and dissipation of profits, and mismanagement of information and technology.
[76]: Lucas, H.C. Jr.;Berndt, D.J.; Truman, G. (1993)	Case Study	Changes in organizational structure Changes in workflows and functions Changes in interface operations Changes in technology Numerical measures of stability, obsolescence, change, extent of automation, system-wide change (based on data flow diagram analysis)	Qualitative and Quantitative	Financial and Non-financial	Organizational	Introduction of financial imaging system resulted in improvements to customer service, control of certificates, higher quality images, improved search speed, cost reduction, research time reduction, staff reduction.
[31]: Desmaris, M.C.; Leclair, R.; Fiset, J-Y.; Taïbi, H. (1997)	Case Study	Cost-benefit analysis Software development costs Operating costs Reduction in training time Annual monetary benefits	Quantitative	Financial	Individual and Organizational	Introduction of an electronic performance support system is expected to reduce employee training time, resulting in a financial break-even point between 1 and 3 years.
[101]: Rai, A.; Patnayakuni, R.; Patnayakuni, N (1997)	Seconday Data Analysis (Information week and Compustat)	Labor and related expenses Total property, plant, and equipment Total number of employees Company sector Sales Return on assets Return on equity Labor productivity Administrative productivity	Quantitative	Finacial	Organizational	All measures of IT investment are positively associated with firm output. IT capital and client/server expenditures are positively associated with return on asets. Most expenditures except software and telecom are associated with increased labor productivity. IS staff, hardware, software, and telecom expenditures are negatively related with administrative productivity.

Study	Method	Measure	Level	Findings
[47]: Grover, V.; Teng, J.T.C.; Fiedler, K.D. (1998)	Survey	Ranking of importance among investments in strategic systems, traditional development, decision support systems, infrastructure, business process redesign, and maintenance.	Qualitative and Quantitative	Non-financial	Organizational	When IS is in a support role and when there is a lack of broad managerial attention, companies tend to develop transaction procesing systems and information reporting systems An IS planning culture among top management is associated with strategic systems investments. Diversity of types of IT is associated with BPR and infrastructure investment and does not favor traditional systems investment. Managing IT requires change management skills. Both IS and business inputs need to be used in prioritizing investments.
[15]: Brynjolfsson, E.; Hitt, L.M. (1998)	Secondary Data Analysis (Literature Review)	Productivity Decentralization IT spending	Quantitative	Financial and Non-financial	Organizational	Investment in computers does not automatically increase productivity, but is part of a broader system of organizational changes that does increase productivity.
[34]: Dewan, S.; Kraemer, K.L. (1998)	Secondary Data Analysis (Labor productivity data)	Gross domestic product IT stock Non-IT stock Number of workers GDP per worker IT capital per worker Non-IT capital per worker	Quantitative	Financial and Non-financial	International	Increases in IT capital spending per worker are associated with an increase in GDP per worker, on average. Developed countries are receiving a positive and significant return on their IT investments.

II. Related Empirical Studies

STUDY
[93]: Nelson, P.; Richmond, W.; Seidmann, A. (1996)
[90]: Nam, K.; Rajagopalan, H., Raghav. R.; Chaudhury, A. (1996)
[3]: Bakos,Y. (1998)
[32]: Dewan, R.M.; Freimer, M.L.; Seidmann, A. (1998)
[119]: Teng, J.T.C.; Jeong, S.R.; Grover, V. (1998)
[39]: Duchessi, P.; Chengalur-Smith, I. (1998)
[30]: De, P.; Ferrat, T.W. (1998)

Appendix II

Information Systems Research

1. Articles Analyzed

Study	RESEARCH METHOD(S)	MEASURES USED TO ASSESS DEPENDENT VARIABLES	QUANTITATIVE and/or QUALITATIVE MEASURES	FINANCIAL and/or NON-FINANCIAL	LEVEL(S) OF ANALYSIS— individual/group/ organizational/ industry/national/ in ternational	RESULTS
[37]: Dos Santos, B.L; Peffers, K.; Mauer, D.C. (1993)	Secondary Data Analysis (PR Newswire, PTS Prompt)	Stock price reactions around announcements of IT investments (abnormal daily stock returns)	Quantitative	Financial	Organizational	On average, IT investments are zero net present value investments; they are worth as much as they cost. Innovative IT investments increase the value of the firm.
[7]: Barua, A; Kriebel, C.H.; Mukhopadhyay, T. (1995)	Secondary Data Analysis (Strategic Planning Institute MPIT database)	Five intermediate variables: Capacity utilization, inventory turnover, relative price, relative inferior quality and new products. Final performance variables: market shares, return on assets	Quantitative	Financial and Non-financial	Group and Organizational	Partial support was received for the positive impacts of the economic input variables on five intermediate variables. The five intermediate variables had significant positive impacts on the final performace variables of the strategic business units.
[13]: Brynjolfsson, E. (1996)	Secondary Data Analysis (U.S. Bureau of Economic Analysis; government GDP data)	Consumer welfare: Marshallian surplus, non-parametric estimates, value based on the index number	Quantitative	Financial and Non-financial	Organizational	IT investments generate approximately three times their cost in value for consumers.
[35]: Dewan, S.; Michael, S.C.; Min, C-K. (1998)	Secondary Data Analysis (Computerworld and Commpustat data)	Demand for IT investment (total stock of IT capital, net of depreciation)	Quantitative	Financial and Non-financial	Organizational	The level of IT investment is positively related to the degree of firm diversification. Furthermore, related diversification demands greater IT than unrelated diversification. Firms that are less vertically integrated have a higher level of IT investment. Finally, firms with fewer growth options in their investment opportunity set tend to have a higher IT investment.
[118]: Tam, K.Y. (1998)	Secondary Data Analysis (Asia Computer Directory; PACAV and GV financial databases)	Total Shareholder return Return on equity Return on assets Return on sales Book value of assets Market value	Quantitative	Financial	Organizational	IT investment is not correlated with shareholder return. Level of computerization is not valued by the stock market in developed and newly developed countries. There is no consistent measurement of IT investment.

II. Related Empirical Studies

STUDY

[100]: Premkumar, G.; King, W.R. (1994)
[28]: Coopersmith, J. (1996)
[80]: Manning, P.K. (1996)
[106]: Robey, D.; Sahay, S. (1996)
[11]: Bensaou, M. (1997)
[60]: Kambil, A.; Van Heck, E. (1998)
[70]: Kraemer, K.L.; Dedrick, J. (1998)
[127]: Wong, P-K. (1998)
[58]: Jarvenpaa, S.L.; Leidner, D.E. (1998)

III. Related Theoretical Studies

STUDY

[86]: Mookerjee, V.S.; Dos Santos, B.L. (1993)
[48]: Gurbaxani, V.; Mendelson, H. (1994)
[25]: Clemons, E.K.; Weber, B.W. (1996)
[33]: Dewan, S. (1996)
[9]: Barua, A; Lee, C.H.S; Whinston, A.B. (1996)
[91]: Nault, B.R. (1997)
[5]: Bakos, Y.J; Nault, B.R. (1997)
[8]: Barua, A.: Lee, B. (1997)
[111]: Seddon, P.B. (1997)

Appendix III

Journal of Management Information Systems

I. Articles Analyzed

STUDY	RESEARCH METHOD(S)	MEASURES USED TO ASSESS DEPENDENT VARIABLES	QUANTITATIVE and/or QUALITATIVE MEASURES	FINANCIAL and/or NON-FINANCIAL	LEVEL(S) OF ANALYSIS—individual/ group/ organizational/ industry/ national	RESULTS
[78]: Mahmood, M.A.; Mann, G.J. (1993)	Field Survey & Secondary Data Analysis (Computerworld "Premier' 100")	Return on investment, return on sales, growth in revenue, sales by total assets, sales by employee, market value to book value.	Quantitative	Financial	Organization	Individual IT investment variables were found to be weakly related to orgnizational strategic and economic performance. However, they were significantly related to performance when grouped and analyzed by canonical correlation.
[113]: Sheffield, J.; Gallupe, R.B. (1993-94)	Multiple Case Study	Meeting effectiveness; Overall effectiveness, effectiveness of facilitation, effectiveness of technology, reducing barriers, participation, information exchange, meeting outcomes, and average effectiveness.	Qualitative	Non-Financial	Individual and National	Study participants thought that the use of group support technology was effective and efficient in supporting economic development processes. GSS was helpful in meetings where participants came from a variety of backgrounds (e.g., business competitors, different ethnic groups) and where meeting urgency and efficiency were of prime importance.
[114]: Sheffield, J.; Gallupe, R.B. (1994-95)	Modified Historical Analysis	Link between action plans and competitive advantage. Implementation activities and outcomes: projects that became inactive within 1 month, projects that became inactive after 1-18 months, continuing joint projects, continuing stand alone projects. Meeting as "unfreezing" events: absence of perceived conflict, participation, information exchange, consensus for cooperative action. Change: additional electronic meetings held to involve related groups. Refreezing: recommended organizational form adopted. Perceived success in plan implementation.	Qualitative	Non-financial	Individual and National	The electronically assisted meetings promoted interorganizational learning and were effective catalysts of industry-wide change in situations previously characterized by dysfunctional conflict.

Reference	Method	Variables	Research Approach	Dependent Variable Type	Level of Analysis	Key Findings
[12]: Brown, R.M; Gatian, A.W.; Hicks, J.O. (1995)	Event Study, Market Data (Compustat)	Announcements that firms are using information systems (investment)	Quantitative	Financial	Organization and Industry	The stock market reacted favorably to announcements that firms were using successful strategic information systems (SIS). In subsequent years these firms tended to be more productive and more profitable than other firms in their respective industries.
[49]: Henderson, J.C.; Lentz, C.M.A. (1995-96)	Case Study	Organizational learning New products and services Improved operating effectiveness	Qualitative	Non-financial	Organization	The benefits anticipated from IT investments (e.g. innovation) are marginal unless integrated, dynamic processes exist to actively manage and adapt these investments.
[21]: Choe, J.-M. (1996)	Survey	User accounting information system (AIS) satisfaction: information system satisfaction resulting from the correspondence between the job requirements and system functionality. User AIS use: frequency and willingness of use	Qualitative	Non-financial	Individual	There are significant positive correlations between the performance of an AIS and influence factors such as user involvement, capability of IS personnel and organization size.
[22]: Clark, T.H.; Stoddard, D.B. (1996)	Case Studies, Survey	Interorganizational redesign, use of electronic data interchange (EDI) and continuous replenishment (CRP)	Qualitative and Quantitative	Financial	Organization	It is important to merge technological and process innovations. Interorganizational business process design, in the form of CRP using EDI, represented a dramatic performance improvement for the channel overall, benefiting both retailers and manufacturers.
[85]: Mitra, S.; Chaya, A.K. (1996)	Secondary Data Analysis (Computerworld)	Level of IT investments made by the firm IT budget as a percentage of sales (ITB/S), averaged over a period of time.	Quantitative	Financial	Organization and Industry	Higher IT investments are associated with lower average production costs, lower average total costs, and higher average overhead costs. Larger companies spend more on information technology as a percentage of their revenues than smaller companies. There was no evidence that IT reduces labor costs in organizations.
[117]: Tam, K.Y. (1996)	Secondary Data Analysis. (Bureau of Economic Analysis (BEA); Computerworld)	Organizational adoption of IT Mainframe purchases Price elasticity of mainframe computing	Quantitative	Financial	Organization	Price is an important factor in the innovation diffusion process. Organizations' reactions to price changes (i.e., price elasticity) are not constant. Elasticity dynamics can serve as an innovation attribute that provides a continuous characterization of adoption behavior over the life cycle of an innovation.
[72]: Lee, H.G.; Clark, T.H. (1996-97)	Case Study	Innovation in traditional market transaction processes via the use of electronic markets. Three transaction process dimensions: information gathering, contract formation, and trade settlement.	Qualitative	Non-financial	Industry (electronic markets)	Successful deployment of electronic markets requires consideration of barriers resulting from market process reengineering along with projected economic benefits. Most risks and barriers stem from social and economic factors, rather than IT-related obstacles. Success is as dependent on the management of barriers as it is on the economic benefits enabled by IT.

[20]: Teo, H.-K.; Tan, B.C.Y.; Wei, K.-K.(1997)	Case Study, Survey, Change Point Analysis	Changes in organizational structure Business process changes Business network changes Business scope changes Efficiency Effectiveness	Qualitative and Quantitative	Financial and Non-financial	Organization and Industry	The use EDI in conjunction with organizational transformation can lead to phenomenal gains in organization efficiency and effectiveness.

II. Related Empirical Studies

STUDY

[102]: Rice, D.E. (1994)
[56]: Holden, T.; Wilhemij, O. (1995-6)
[62]: Karami, J.; Gupta, Y.Y.; Somers, T.M. (1996)
[116]: Subramanian, G.H.; Zarnich, G.E. (1996)
[68]: King, W.R.; Teo, T.S.H. (1996)
[40]: Edberg, D.T.; Bowman, B.J (1996)
[23]: Clemons, E.K.; Croson, D.C.; Weber, B.W. (1996)
[53]: Hitt, L.M.; Brynjolfsson. E. (1997)
[79]: Maier, J.L.; Rainer, K.Jr; Snyder, C.A. (1997)
[27]: Clemons, E.K.; Weber, B.W. (1998)

III. Related Theoretical Studies

STUDY

[87]: Mukhopadhyay, T.; Cooper, R.B. (1993)
[24]: Clemons, E.K.; Reddi, S.P. Row, M.C. (1993)
[4]: Bakos, J.Y.; Brynjolfsson, E. (1993)
[26]: Clemons, E.K.; Weber, B.W. (1994)
[107]: Sampler, J.L.; Short, J.E. (1994)
[126]: West, L.A. Jr. (1994)
[99]: Post, G.V.; Kagan, A.; Lau, K.-N. (1995)
[71]: Kumar, R.L. (1996)
[112]: Seidmann, A.; Sundararajan, A. (1997)

Appendix IV

Management Information Systems Quarterly

I. Articles Analyzed

STUDY	RESEARCH METHOD(S)	MEASURES USED TO ASSESS DEPENDENT VARIABLES	QUANTITATIVE and/or QUALITATIVE MEASURES	FINANCIAL and/or NON-FINANCIAL	LEVEL(S) OF ANALYSIS— individual/group/ organizational/ industry/national	RESULTS
[69]: Kraemer, K.L.; Danziger, J.N.; Dunkle, D.E.; King, J.L. (1993)	Survey	Perceived usefulness of computer based information (CBI) for financial management Perceived usefulness of CBI for operations management	Qualitative and Quantitative	Non-financial	Individual	Computer based information is important for most managers, and many report they are extremely dependent on it. The managers surveyed found CBI more valuable for the control of financial resources than the management of operations. Quality and accessibility of CBI and manager's style of computer use affected the manager's perception of usefulness. Managers most satisfied with CBI used support staff to mediate the CBI environment rather than using the computer to access information directly.
[10]: Belcher, L.W.; Watson, H.J. (1993)	Case Study	Productivity improvements Decision making improvements Information distribution cost savings Services replacement cost savings Software replacement cost savings Other intangible benefits Out of pocket direct costs Indirect personnel costs	Qualitative and Quantitative	Financial and Non-financial	Individual, group, and organizational	Benefits included improved productivity, improved decision making, information distribution cost savings, services replacement cost savings, and software replacement cost savings. Costs included the direct costs of maintaining the EIS and the indirect costs absorbed by operating groups who provided personnel to perform ESI-related tasks. Benefits were found to exceed the system's costs.
[18]: Cats-Baril, W.L.; Jelassi, T. (1994)	Case Study	Existence of: Subsidies to end users State-of-the-art telephone and data transmission network Easy-to-use interface Inexpensive terminals Transparent billing system	Qualitative	Financial and Non-financial	National	Building an advanced national information technology infrastructure can provide a competitive advantage for the countries that develop it as well as for the companies that operate in those countries. The French national videotex system was profitable and successful.

Reference	Method	Measures	Qual/Quant	Financial/Non-financial	Level	Results
[94]: Newman, J.; Kozar, K.A. (1994)	Case Study	Positive identification of jewelry; Time required for item evaluation; Availability of decision support for gemologist throughout evaluation process	Qualitative	Financial and Non-financial	Organizational	System resulted in: Better asset management and financial control; Increased productivity; Reduced costs and increased revenue; Better quality merchandise
[66]: Kettinger, W.J.; Grover, V.; Guha, S.; Segars, A.H. (1994)	Content Analysis and Secondary Data Analysis (COMPUSTAT II)	Relative profitability; Relative market share	Quantitative	Financial and Non-financial	Organizational	Establishment of technological base and capital availability are both needed for sustainability of competitive advantage.
[59]: Jelassi, T.; Figon, O. (1994)	Case Study	Number of customers using EDI; Return on investment; Cost comparisons; Marked share	Qualitative and Quantitative	Financial and Non-financial	Organizational	Implementation of EDI system improved relationship with customers, lowered costs, improved speed of internal order processing, reduced errors, increased productivity, and provided competitive advantage.
[51]: Hess, C.M.; Kemerer, C.F. (1994)	Case Study	Development of electronic markets for home mortgages; Changes in market structure; Customer driven movement toward electronic markets; Evolution of electronic markets	Qualitative	Non-financial	Organizational	CLOs provided limited support for the establishment and evolution of electronic markets.
[128]: Yoon, Y.; Guimaraes, T.; O'Neal, Q. (1995)	Survey	Expert system success measured by user satisfaction	Qualitative	Non-financial	Individual	Expert system success was found to be positively related to developer skill, end-user characteristics, desirability, shell characteristics, user involvement, problem difficulty, domain expert quality, and management support.
[89]: Mukhopadhyay, T.; Kekre, S.; Kalathur, S. (1995)	Case Study	Inventory turnover; Obsolete inventory; Premium freight; % of material dollars under EDI program; Annual production volume; Parts variety; New parts introduction	Quantitative	Financial and Non-financial	Organizational	EDI resulted in cost reductions ($100 savings per vehicle, annual savings of $220 million)
[46]: Goodhue, D.L.; Thompson, R.L. (1995)	Survey	Perceived effectiveness, productivity and performance; Utilization or perceived system dependence	Qualitative	Non-financial	Individual	For IT to have a positive impact on individual performance: Technology must be utilized; Technology must fit task

[92]: Nault, B.R.; Dexter. A.S. (1995)	Case Study	Price of fuel Convenience, credit and control provided to customers	Qualitative and Quantitative	Financial and Non-financial	Organizational	Application of IT yielded price premiums between 5% and 12% of the retail fuel price.
[54]: Hitt,L.M.; Brynjolfsson, E. (1996)	Secondary Data Analysis (IDG Annual IT Spending Survey)	Production function: Productivity Business profitability Consumer surplus	Quantitative	Financial and Non-financial	Organizational	IT increased productivity and consumer value, but did not result in supranormal business profitability. There is no inherent contradiction between increased productivity, increased consumer value, and unchanged business profitability.
[123]: Vandenbosch, B.; Huff, S.L. (1998)	Case Study	Perceived improvements in organizational performance: efficiency and effectiveness	Quantitative	Non-financial	Individual and Organizational	EISs contributed to gains in efficiency more frequently than to gains in effectiveness. However EISs could also be used to help formulate problems and foster creativity.
[43]: Francalanci, C; Galal, H. (1998)	Secondary Data Analysis (LOMA)	Productivity: Premium income per employee Total operating expense to premium income	Quantitative	Financial and Non-financial	Organizational	Increases in IT investment were associated with productivity benefits when accompanied by changes in worker composition.
[97]: Pinsonneault, A., Rivard,S. (1998)	Survey	Logs of time spent in various managerial activities Logs of time spent online	Quantitative	Non-financial	Individual	Managerial IT usage is sometimes, but not always, associated with spending more time in information roles and less time in decisional and interpersonal roles. Companies that are experiencing discontinuous change in strategy are likely to exhibit this pattern, while those that are focused on incremental change are not.

II. Related Empirical Studies

STUDY

[96]: Pinsonneault, A; Kraemer, K.L. (1993)
[17]: Caron, J.R.; Jarvenpaa, S.L.; Stoddard, D.B. (1994)
[42]: Finlay, P.N.; Mitchell, A.C. (1994)
[44]: Gill, T.G. (1995)
[98]: Pitt, L.F.; Watson, R.T.; Kavan, C.B. (1995)
[57]: Iacovou,C.L.: Benbasat, I.; Dexter, A.S. (1995)
[1]: Abdul-Gader, A.H.; Kozar, K.A. (1995)
[81]: Massetti, B. (1996)
[129]: Ytterstad, P.; Akselsen, S.; Svendsen, G.; Watson, R.T. (1996)
[45]: Gill, T.G. (1996)
[82]: Massetti, B.; Zmud, R.W. (1996)
[41]:El Sawy, O.A.; Bowles, G. (1997)
[95]: Nidumolu, S.R.; Knotts, G.W. (1998)

III. Related Theoretical Studies

STUDY

[83]: Mata, F.J.; Fuerst, W.L.; Barney, J.B (1995)

The Impact of Information Technology Investment on Enterprise Performance: A Case Study

Melinda K. Cline and C. Steve Guynes

Abstract

Investment in information technology is steadily increasing, but many organizations find it difficult to formally assess the value of IT investments because the latter are often incorporated into broad management initiatives. The authors believe that the results of the research study reported on here can help firms to develop a better understanding of the dynamic relationship between IT investment and performance at both the firm and industry levels of analysis. This study clearly demonstrates the importance of adopting an organizational change perspective when assessing the impact of IT investment on firm performance.

THIS STUDY USES AN IN-DEPTH approach to measure the effects of IT investment on related performance returns at a major eastern railroad. The railroad provides rail transportation and distribution services over a 23,000 route-mile network, and serves 23 states, the District of Columbia, and Canada.[1] Ten years of software development effort is analyzed and related to the railroad's performance. The study examines approximately 5.6 million hours of applications development effort on 8,700 projects over a 10-year period and relates this investment in IT to organizational performance at the firm and industry levels of analysis.

Investment in information technology is steadily increasing even though many organizations report that it is difficult, if not impossible, to formally assess the value of IT investments. The determination of benefits from IT investments depends largely on accurately associating and measuring the investments and related returns, and this has proven to be a difficult undertaking since IT investments are often incorporated into broad management initiatives. These initiatives may include responding to competitive threats, increasing operational control, decreasing expenses, and improving customer satisfaction. With IT investment motivated by differing, and sometimes conflicting, organizational objectives, changes in performance resulting from IT investment must be evaluated based on expected results. It is the recommendation of this study that the role of the IT investment be considered when evaluating returns. By evaluating results based on managerial objectives, companies will understand how and why IT investments contribute to improved firm performance. The authors believe that the results of this study can be applied to firms other than those in the railroad industry.

The central issue

The central issue addressed in this article is: has information technology investment contributed to the productivity improvements in the rail industry

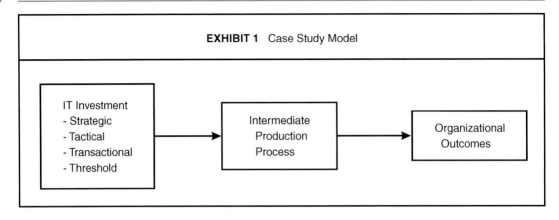

during the last decade, and, if so, what types of software investments have proven most beneficial? The crux of the problem is whether IT investment makes a difference in firm performance. To determine if IT has a significant and unique effect on firm performance, this study examines the effect of software development investment on changes in productivity. Methodologically, archival data is used to develop a relationship between software development intensity and multiple intermediate process measures. Software development intensity is defined as a measure of the effort committed over time to develop and maintain applications software. To explore the nature of this relationship, software development initiatives are divided into categories, based on prior literature, and the relationship of each type of software investment is compared to the productivity measures.

American railroads are a particularly good industry in which to analyze IT-related productivity gains because railroads have invested heavily in IT since deregulation. It is generally known that forces such as deregulation and intense competition create a strong impetus for firms to gain efficiency. In their effort to improve efficiency, railroads have invested heavily in software systems and the hardware and telecommunications equipment to support the related processing and data transmission requirements. In this study, IT investment is broadly defined as the identification, justification, and leveraging of information technology expenditures to create value for a firm. IT investment is considered an element of organizational change where the change sought is increased value for the firm through management's ongoing improvements to the firm's organizational processes. By conceptualizing IT investment as managerial action, this study recognizes the variety of roles IT may be intended to play. The study then explores the dynamics of the relationship between IT investment and organizational outcome. IT investment is categorized as strategic, tactical, transactional, or threshold.[2] The relationship between each category of IT investment is related to firm and industry performance measures. The case study model is illustrated in Exhibit 1 and the constructs are defined in Exhibit 2.

Strategic Investment

Strategic investments support efforts to develop new, improved, and customized products. These IT investments are expected to positively influence market share and sales growth. Strategic IT investment is used for expansion rather than efficiency. Investments of this type enable organizations to offer new and extended services to customers and may be measured by analyzing changes in the revenue stream of the organization.

Tactical Investment

Tactical IT investments are directed toward incremental product or service adjustments. They include IT processes that are analytical in nature and support improved managerial decision making, such as decision support systems. Performance, especially poor performance, is thought to drive tactical adjustments. This type of IT investment helps a firm better understand itself so that it can perform its transformation processes faster and at less cost in the future.

EXHIBIT 2	Case Study Constructs
IT Investment Categories	**Definition**
IT investment	Management's allocation of IT resources over time
Strategic IT investment	IT investment that extends services to new customers, develops customer prices, or supports the negotiation of customer contracts
Tactical IT investment	IT investment devoted to analytical systems used to support managerial decision making for the purpose of improving the operations of the railroad
Transactional IT investment	IT investment to support the day-to-day operations of the firm
Threshold IT investment	IT investment to support processes that are ancillary to rail operations yet required for the firm to be in business. These business processes are primarily related to accounting, risk management, and human relations

Transactional Investment

Transactional investments support day-to-day operations of the firm. These IT investments support operational management, and the motivation for investment is usually cost cutting by substituting capital for personnel or equipment. Transactional IT investment is the type most associated with productivity gains and has been found to provide the firm with both short-term and compounded investment returns.

Threshold Investment

Threshold IT investment supports business processes mandated by external authorities and is directed toward processes ancillary to the value transformations of the firm. Examples of such external authorities include the Internal Revenue Service, regulatory agencies, and state and local governments. Such requirements have increased as IT capabilities have made them more feasible. Threshold investments, by definition, are not expected to contribute to productivity gains. Instead, they are considered a cost of doing business, and IT is an element in lowering associated costs.

Intermediate Production Processes and Enterprise Performance

Intermediate production processes and enterprise performance include measures standard to the railroad industry. In this industry, a detailed list of standardized industry indicators are reported annually to the Surface Transportation Board (STB), formerly known as the Interstate Commerce Commission (ICC). Reports provided to the STB are collected, archived, aggregated, and verified by representatives of the Association of American Railroads (AAR), which is headquartered in Washington, D.C. The AAR publishes reports for government and industry distribution, and sells them for private use.[3] The data reported to the STB is sworn to be accurate by the chief accounting officer and the chief executive officer of each reporting railroad.

The case study process

This study examines the relationship between IT investment and firm performance by comparing software development effort aggregated by category in annual increments over a 10-year period to related intermediate and enterprise measures of firm performance. Methodologically, archival data is used to develop a statistical relationship between software development intensity, and firm- and industry-level productivity and financial measures. Software development intensity is defined as a measure of the effort committed over time to develop and maintain application software. The study's methodology includes a longitudinal, in-depth examination of a single major railroad's complete software development applications portfolio. This railroad is among the largest within the industry and alone represents approximately 15 percent of the industry's total revenue.[4]

This study uses detail rather than aggregated data to measure IT investment and related performance results. This method avoids many of the measurement problems that have made it difficult to draw consistent conclusions from past empirical studies that have used only aggregated data. The firm's entire portfolio of IT projects for the 10-year study period are included in the analysis. This method supports a micro-level analysis of how IT investment affects firm performance, while still focusing on impacts at the organizational level. In this study, IT investment is measured at the business process application level and then aggregated to the investment type. It is derived by assigning actual applications development hours to each of the four investment categories (strategic, tactical, transactional, and threshold).

The study includes the following three research propositions:

1 *Proposition One*: The relationship between aggregated software development intensity and firm-level performance will be positive but weak.
2 Proposition Two: IT investment, categorized by management objective, will have the following relationship with associated intermediate production process measures:
A. *Strategic IT investment* will exhibit a positive effect on revenue-related process returns.
B. *Tactical IT investment* will exhibit a postive effect on yield management process measures.
C. *Transactional IT investment* will exhibit a positive effect on expense reduction process returns.
D. *Threshold IT investment* will not exhibit a positive or negative effect on revenue, yield, or expense-related process returns.
3 *Proposition Three*: IT Investment, categorized by management objective, will have the following temporal relationship with associated intermediate production process measures:
A. Strategic IT investments will have the highest positive correlation to associated intermediate returns when modeled with a two-year lag.
B. Tactical IT investment will have the highest positive correlation to associated intermediate returns when modeled with a one-year lag.
C. Transactional IT investment will have the highest negative correlation to associated intermediate returns (expense measures) when modeled with no lag.
D. Threshold IT investment will have no higher correlation to associated intermediate returns when modeled with no lag, a one-year lag, or a two-year lag.

IT Investment

This study uses hours to measure IT investment. Using hours to estimate investment and examine returns is a common approach in economic analysis. For instance, the U.S. Bureau of Economic Analysis (BEA) calculates productivity statistics using hours. Output per gross national product (GNP) is calculated as gross national product divided by the total number of hours that wage and salary workers were paid for and that self-employed persons and unpaid family members worked.

There are several benefits to measuring investment in hours, as opposed to dollars. First, a 60-minute hour remains constant over time, so there is no need for present or future value adjustments. Second, hours are the same unit of measure across firms. This avoids the confounding effect of differing cost calculation methods when observing IT investment in multiple organizations. Third, by law, most companies are required to record the hours worked by employees, so this type of data should be available across firms for future study replication. In total, approximately 5.6 million hours of applications development effort on 8,700 projects over a 10-year period is included in the analysis.

Railroad Performance Data

The performance measures used in the assessment of IT investment impact are chosen to reflect the business objective. In this study, the IT investment measures, discussed in the previous section, are related to measures of revenues, expenses, and resource utilization and are adjusted to remove inflationary and externally prescribed cost factors. The dependent measures are

IT investment categories	Associated performance measures
Strategic performance	Revenue per revenue ton mile
	Revenue per employee
	Difference between firm and industry
	Revenue per revenue ton mile
	Difference between firm and industry
	Revenue per employee
Tactical performance	Revenue ton miles per mile of road
	Revenue ton miles per employee
	Difference between firm and industry
	Revenue ton miles per mile of road
	Difference between firm and industry
	Revenue ton miles per employee
Threshold performance	Adjusted general and administrative
	Expense per revenue ton mile
	Revenue ton miles per administrative employee
	Difference between firm and industry
	General and administrative expense per revenue ton mile
	Difference between firm and industry
	Revenue ton miles per administrative employee
Transactional performance	Adjusted total transportation expense
	Per revenue ton mile
	Adjusted total transportation expense
	Per employee
	Difference between firm and industry
	Adjusted total transportation expense per employee

Exhibit 3 Railroad Performance Measures.

summarized in Exhibit 3 and are discussed in detail in the remaining paragraphs of this section.

The first dependent measures are related to the strategic IT investment objective and are adjusted revenue per revenue ton mile and adjusted revenue per employee. Each is measured first at the strategic business unit (SBU) level and then by determining the difference between the SBU performance and the average Class I railroad performance, with the subject firm removed from the industry analysis. A strategic business unit (SBU) is defined as a unit of a firm that sells a distinct product or service to an identifiable set of customers within a well-defined set of competitors.[5] Together these measures support the investigation of IT investment returns in both an absolute fashion (firm performance) and relative fashion (how the firm performs relative to other Class I railroads).

Adjustment to the revenue measures is required to eliminate inflationary effects. The adjustment restates revenue per year in constant 1986 dollars using the chain-based GDP index listed in the 1996 Railroad Fact Book.[4] Revenue per ton mile is a standard industry measure used as a surrogate for rail rates. These similar measures capture whether or not the amount billed per revenue ton is increasing or decreasing. Reasons for changes in revenue per ton mile and per employee are arguable. Perhaps changes are due to deregulation driving down rates, perhaps changes are due to competitive forces among railroads and from other modes affecting rate structures, or perhaps changes reflect the return of cost savings to customers.

Tactical IT investment returns are investigated using multiple measures of resource utilization efficiency, and each is calculated as an absolute measure and by determining the difference between the SBU performance and the other Class I railroads, with the subject firm removed from the industry analysis. Revenue ton miles per mile of road measures the utilization of the rail network. Revenue ton miles per employee is an aggregate measure of employee productivity. Revenue ton miles per administrative employee measures backoffice employee productivity. Revenue tons per train hour is a measure of train operations efficiency. Revenue ton miles per train and engine employee measures train crew utilization. Revenue ton miles per operations support employee measures the scheduling efficiency of dispatchers, clerks, and operations supervisory personnel. Percent of loaded freight car miles (as opposed to empty car miles) measures equipment utilization. Revenue ton miles per yard switch hour measures yard operations efficiency. Revenue ton miles per in-service locomotive measures locomotive utilization. Together, these measures provide a multidimensional view of the tactical operations of a railroad and capture resource utilization increases and decreases.

The threshold category of railroad performance is investigated with the financial measure adjusted general and administration expense per revenue ton mile. Included in this measure are general administrative labor and materials expenses. In this study, excluded from general and administration expenses are advertising and public relations charges, restructuring charges, charges due to changes in accounting methods, taxes, and fringe benefits. Included salaries and expenses are adjusted to constant 1986 dollars.

The transactional category of performance includes two measures: adjusted total transportation expense per revenue ton mile and adjusted total transportation expense per employee. Both are designed to measure changes in the level of productivity associated with the day-to-day operations of the SBU. This measure is related to the total applications development effort categorized as Transactional investment.

Results of the study

At least one finding resulted from this research effort for each category of IT investment by using the research propositions to guide the inquiry. The examination of total IT investment in proposition one indicates that total IT investment is related to firm-level performance, but that a two-year lag is required for the effect. This finding is supported by Quinn and Baily[6] who, after interviewing more than 100 executives representing numerous industries, concluded that returns from IT investment may be delayed for years.

Proposition two, which had four parts, examined each category of IT investment. Part A indicates that strategic IT investment is related to volume-related measures, but not revenue-related measures. This may be the most important of the findings because it implies that measures using revenue (e.g., sales revenue) or a revenue-related measure (e.g., return on assets or net income) may not find IT investment effects, even when they are present, should the firm be operating in an industry characterized by declining price structures. Part B indicates that tactical IT investment is related to measures of yield management. Barua[7] also found a positive relationship between IT investment and intermediate variables, such as capacity utilization and inventory turnover. The findings from Part C confirm that transactional IT investment is correlated with firm-level expense reductions. An interesting finding is that when transactional IT investment is significantly reduced, expense reductions continue to occur, but at a slower pace than the industry average. This finding indicates that firms that are leaders in implementing IT are more profitable and productive than other firms within their industry. Part D, which examined Threshold IT investment, was partially supported. Threshold IT investment was not related to revenue, yield management, or expense-related measures of firm performance; however, there did appear to be a relationship with expense-related measures at the industry level of analysis.

Proposition Three, which examines the temporal effects of IT investment, indicates that there are multiple lag effects to consider. The lag effects tend to support the idea that the more the performance data is aggregated, the longer the lag effect is. The detail measures tended to have lag effects of zero or one year. Total IT investment demonstrates its clearest relationship to firm performance when modeled with a two-year lag; tactical IT lag is best at zero years; transactional exhibits mixed lags of zero and one year; and threshold is clearest using a zero-year lag. The mix of lag effects is likely influenced by the specific initiatives in each of the categories.

Conclusion

This research provides important insights into the relationship between IT investments and firm performance. It investigates this relationship by categorizing IT investment objectives as strategic, tactical, transactional, and threshold. Findings indicate each type has unique implications for directing IT investment toward performance improvements.

Strategic IT investment is found to be related to an increase in sales volume, but not to an increase in revenue. Drawing from this finding, it may be postulated that firms that extend themselves into new markets through the Internet or other technologies may see an increase in sales volume, but not necessarily an increase in revenue and profit. This is particularly true if the firm operates in an industry sector characterized by declining rates.

Tactical investment is the type found to be most correlated with improved firm profitability. As managers gain improved understanding and control of business operations, they are able to make decisions that lead to a decline in expenses and improved profitability. This finding confirms the importance of IT investments directed toward decision support, knowledge management, and enterprise resource planning.

Transactional investment is confirmed to be expense reducing. With this type of IT investment, firms gain process improvements in day-to-day operations. This research demonstrates that if a firm reduces investment in this area, it may reduce its competitive position within its industry sector, assuming its competitors continue to use IT to improve their processes.

Threshold IT investment is found to be a required cost of doing business, and investment in this area is unlikely to affect firm performance since the savings are largely an issue of cost avoidance.

This research also has implications for total IT investment with a firm. Total IT spending is subject to changes in strategy by firm management. Technology investment may be drastically cut. In the short term, core systems may be somewhat insulated from the cuts, but in time, IT investment in these processes may also decline. Changes in firm performance based on total IT investment are best evaluated over time, and this research indicates a two-year lag is required.

The study of IT investment should consider the differing roles of IT investment within an organization when evaluating IT investment effectiveness. IT investment may help a firm increase sales volume, improve productivity and profit, reduce expenses, and avoid increased costs. How an investment will affect firm performance is determined by the organization's ability to effectively plan and implement new technology, and by the managerial objective of the investment.

The purpose of this article was to develop a better understanding of the dynamics of the relationship between IT investment and performance at both the firm and industry levels of analysis. This article demonstrates, in detail, the complexity of the relationship between IT investment and firm performance. Forces exogenous to the firm, intra-organizational capabilities and constraints, management commitment to IT investment, and technological capabilities and costs are all factors that influence the relationship. This article clearly demonstrates the importance of adopting an organizational change perspective when assessing the impact of IT investment on firm performance. The authors believe that while some of their findings may be generalized to other industries, IT investment remains an important subject for future study.

REFERENCES

1   CSX Midweek Report (1998) CSX Corporate Communications and Public Affairs, January 14

2 Weill, Peter (1992) "The Relationship Between Investment in Information Technology and Firm Performance: A Study of the Valve Manufacturing Sector," *Information Systems Research*, 3(4), pp. 307–33.
3 AAR Railroad Annual Report (1986–1995) CSX R-1 Schedule 512.
4 Railroad Facts (1996) Washington, D.C.: AAR Publications.
5 Loveman, Gary W. (1994) "Information Technology and the Corporation of the 1990s," in Allen, Thomas J. and Morton, Michael S. Scott Eds, *Research Studies*, New York, NY: Oxford University Press.
6 Quinn, James Brian and Bailey, Martin Neil (1994) "Information Technology: Increasing Productivity in Services," *Academy of Management Executive*, 8(3), pp. 28–51.
7 Barua, Anitesh, Kriebel, Charles H. and Mukhopadhyay, Tridas (1995) "Information Technologies and Business Value: An Analytic and Empirical Investigation," *Information Systems Research*, March, 6(1), pp. 3–23.

ABOUT THE AUTHORS

Dr Melinda K. Cline is an assistant professor of information and management science at the University of North Texas, Denton, Texas.

Carl Stephen Guynes is a Regents Professor of business computer information systems at the University of North Texas.

PART THREE

Inter-organizational systems and process improvements

INTRODUCTION TO PART THREE

Part Three of the book widens our focus to the role and impact of information systems that connect organizations together in a purposeful manner. Electronic data interchange (EDI), inter-organizational systems (IOS), and IT for competitive advantage became topics of interest in the IS field in the 1980s, stimulated by the publication of now-classic cases such as American Airlines' Sabre system and American Hospital Supply's ASAP system. The topics remain interesting today, although theoretical perspectives and framing may vary. (Today, these topics are more likely to be framed in terms of electronic commerce, standards, architecture and infrastructure than in terms of competitive advantage.)

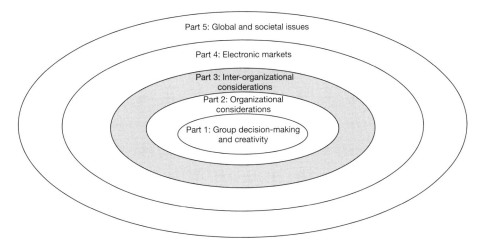

Figure 3.1 Inter-organizational considerations within the broadening focus of Information Systems research.

The four papers included in this section illustrate the range of perspectives and methods present in the literature. The first two papers by Jelassi and Figon, and Mukhopadhyay, et al. are single-company case studies featuring EDI-based procurement systems, but they could not be more different from each other in terms of methodology. The Jelassi and Figon study is qualitative and descriptive; the study by Mukhopadhyay, et al. employs economic modeling and econometric analysis. The second pair of studies, by Palmer and Markus, and by Kraut, et al. looks at patterns of technology use across companies but employs very different perspectives. Palmer and Markus adopt a strategic alignment perspective to analyze the adoption of Quick Response in the specialty retailing industry; Kraut, et al. examine the use of electronic networking between buyers and suppliers in four industries through the lenses of transaction cost and coordination theories.

The Jelassi and Figon study was published at a time when *MISQ* solicited two genres of manuscript: "theory and research" papers, which required a strong theoretical foundation and rigorous

	Jelassi and Figon (1994)[1]	Mukhopadhyay, Kekre and Kalathur (1995)[2]	Palmer and Markus (2000)[3]	Kraut, Steinfield, Chan, Butler and Hoag (1999)[4]
Phenomenon of interest	How a small company (Brun Passot, a French distributor of office supplies) can use IT for competitive advantage; adoption and diffusion of EDI.	The business value of investing in EDI at Chrysler; economic impacts on inventory carrying costs and administrative paperwork.	The effects on firm performance of investing in the Quick Response program for the specialty retailing industry.	The effects of use of electronic networks for transactions with suppliers on firms' degree of virtualization (e.g. increased outsourcing) and the outcomes of transactions with suppliers (such as coordination success and quality of personal relationships).
Perspective or theory	Implicit theoretical framework: organizational and business impacts of inter-organizational systems, including the customers' perspective.	Business value literature; hierarchical inter-organizational systems.	Strategic alignment theory versus practitioner claims about the value of investing in the Quick Response program.	Transaction cost theory and electronic coordination theory.
Research method	Holistic descriptive case study; single company study, retrospective longitundinal.	Single company study; economic modeling and econometric analysis of secondary data; longitudinal analysis over ten years.	Cross-sectional survey in a single industry; self-reported strategy and IT use; subjective and objective performance measures including operational measures (sales per sq. foot) and accounting measures (profitability).	Cross-sectional telephone survey of 250 firms in four industries; 102 item survey instrument with multi-item scales for key concepts.
Technology	Videotext, PC-based, and EDI-based catalog and application for customer use in purchasing.	EDI used for ordering from suppliers.	Quick Response – a suite of technologies, applications and business practices (e.g. point-of-sale technology, automatic inventory replenishment, pre-season planning, and vendor-managed inventory).	Electronic networks – any type of computer communication used to exchange information with suppliers.

	Jelassi and Figon (1994)[1]	Mukhopadhyay, Kekre and Kalathur (1995)[2]	Palmer and Markus (2000)[3]	Kraut, Steinfield, Chan, Butler and Hoag (1999)[4]
Findings and contribution	Company tripled gross revenues in five years holding staffing constant; in-depth description of competitive advantages of IT in European context; an early work that foreshadows current work on strategic IT architecture.	Inventory savings of $60 per vehicle from improved information exchange; additional document preparation savings of $40 per vehicle; Chrysler should force suppliers to participate; quantifies the effects of a particular kind of IT investment controlling for a number of factors likely to affect outcomes; shows that IT provides. opportunities for process streamlining that may be important in achieving the benefits of "IT".	Implementation of QR program at minimal level yielded higher performance, but higher levels yielded no significant improvements, and evidence of alignment with strategy was weak; few companies adopt the full QR program as prescribed; shows the interdependence between IT investments, applications, and business practices in the QR approach; industry-specific conditions may influence pattern of investment, deployment and outcomes.	Use of electronic networks was *negatively* associated with order quality and efficiency and satisfaction with customers, whereas more reliance on personal linkages was associated with better outcomes and mitigated the negative effects of using electronic networks; suggests that use of personal relationships and electronic networks are complementary; findings are inconsistent with hypothesis that increased use of networking leads to outsourcing and consistent with notion that networks are more associated with hierarchical relations than with market relations; virtual organization is a matter of degree rather than kind.

Table 3.1 Comparison of chapters in Part 3.

Notes
1. Jelassi, T. and Figon, O. (1994) "Competing through EDI at Brun Passot: Achievements in France and ambitions for the Single European Market," *MIS Quarterly*, 18(4), November–December, 337–52
2. Mukhopadhyay, T., Kekre, S. and Kalathur, S. (1995) "Business value of Information Technology: A study of electronic data interchange," *MIS Quarterly*, 19(2), June, 137–56.
3. Palmer, J. W. and Markus, M. L. (2000) "The perforamnce impacts of quick response and strategic alignment in specialty retailing," *Information Systems Research*, 11(3), September 241–59.
4. Kraut, R., Steinfield, C., Chan, A. P., Butler, B. and Hoag, A. (1999) "Coordination and virtualization: The role of electronic networks and personal relationships, *Organization Science*, 10(6), 722–40.

research methods, and "application" papers, which carefully documented an innovation in practice, such as a new use of IT or a better IT management practice. Over time, *MISQ* editors increased the theoretical and methodological requirements for application papers to the point where these papers became indistinguishable from theory and research papers. Confusion on the part of authors and reviewers eventually led *MISQ* editors to abolish the applications category. Therefore, the Jelassi and Figon paper would likely not be published today in its original form, because the authors do not explicitly present a theoretical framework for the research nor document their research approach. At the same time, the paper remains worthy of study as an unusually detailed positivist qualitative study that richly describes a company, its business context, the systems it developed over a ten-year period, and their impacts at various levels and on several dimensions. The authors' discussion of Brun Passot's strategic initiatives toward the European economic union anticipates by several years today's IT management research theme relating strategic IT architectures to business flexibility (Ross, 2003).

By the mid-1990s, excessive hype around the competitive benefits of IT investments had given to way to disillusionment, owing in large part to research on what came to be known as the "productivity paradox". Mukhopadhyay, *et al.* motivated their study of the impacts of EDI at Chrysler with reference to the controversy over the business value of IT, and particularly of EDI. They provide a good qualitative description of Chrysler's EDI program before developing a model of EDI's effects on four categories of cost and then testing the model with ten years of company-collected data from nine automobile assembly plants. The results show that Chrysler obtained considerable benefits from their EDI investments. The authors enumerate the ways in which their study overcomes limitations of prior research, but they do not highlight one important issue that the Jelassi and Figon study does – the impacts of EDI from the business partner's point of view. In contrast to the Mukhopadhyay, *et al.* study, which quantitatively examines benefits only from the focal firm's point of view, recent studies highlight the costs and benefits experienced by suppliers (cf., Subramani, 2004).

Comparing and contrasting these two papers raises a number of challenging questions:

- Mukhopadhyay, *et al.* argue that their study is stronger than many studies in the IT impacts tradition because of their theory base in business value research and hierarchical interorganizational systems. How much stronger do you believe the Jelassi and Figon paper would be if it had an explicit theoretical foundation?
- Do you believe that there is a place in IS research for richly descriptive studies without an explicit theoretical foundation such as Jelassi and Figon?
- Yin (1999) argues that a key characteristic of good descriptive case studies is a carefully crafted descriptive framework (set of topics addressed by the description). What is the descriptive framework of Jelassi and Figon? Is it complete in your view? If not, what's missing?
- Do you believe the Mukhopadhyay, *et al.* study is more credible than Jelassi and Figon because of the formers' use of quantitative data and quantitative analysis? Why or why not?
- How would you need to modify the research design of Mukhopadhyay, *et al.* to incorporate the suppliers' costs and benefits?
- How did Mukhopadhyay, *et al.* control for changes that occurred over the ten years of their study?
- The Mukhopadhyay, *et al.* study employs secondary data; in this case, data collected by Chrysler. What are the pros and cons of using secondary data?
- How credible are single case studies, even high-quality ones such as Jelassi and Figon and Mukhopadhyay, *et al.*? Are multi-unit studies inherently better than single unit studies?

The study by Palmer and Markus is what some researchers refer to as a "phenomenon-centered" study as opposed to a "concept- or theory-centered" study. That is, the purpose of the Palmer and Markus study is to learn about a particular phenomenon, in this case Quick Response (a package of technologies, applications and management techniques developed for the retailing industry) more than it is to learn about a particular theory, e.g., strategic alignment theory. Nevertheless, the paper develops a theoretical argument and derives hypotheses for testing, just as it would if it were a theory-centered article. The theoretical argument involves surfacing assumptions implicit in the practitioner literature on Quick Response and contrasting them with hypotheses drawn from the academic literature on strategic alignment. The study surveyed companies in the specialty retailing industry, asking questions about business strategy, IT use and firm performance (operationalized with measures that are commonly used in the industry). Study findings provided some support for both the academic and practitioners' perspectives — but overwhelming support for neither.

Kraut, et al. set out to study virtual organizations — companies that outsource all non-core functions — but were unable to find many in the industries they had selected for study: apparel manufacturing, pharmaceuticals, magazine publishing, and advertising. Consequently, they shifted their focus to mechanisms for coordinating relationships with suppliers; these mechanisms were suggested by prior theory and research as related to firms' use of outsourcing and coordination success. Specifically, Kraut, et al. examined the interactions between personal relationships and the use of electronic networks of all kinds, including email and fax, as well as EDI using a research design involving telephone interviews with managers in 250 firms in four industries. Their findings were contrary to prior theory on a number of key points. A particularly interesting finding was that, although electronic networks and personal relationships were generally used together, use of electronic networks had negative consequences that were mitigated by better relationships.

Among the questions posed by this pair of articles are the following:

- The Palmer and Markus study did not explore personal relationships between retailers and their suppliers. How might personal relationships have affected the results obtained by Palmer and Markus?
- Similarly, the Kraut, et al. study did not examine company strategy. Do you think their findings would have been different if they had?
- What are the pros and cons of using mailed (or online) surveys versus structured telephone interviews? What are the pros and cons of structured versus unstructured telephone interviews?
- Is the Kraut, et al. study more credible than the Palmer and Markus article, because the former involved four industries versus one industry in the latter? What are the pros of single-industry studies?
- How did Kraut, et al. control for industry differences?

Taken as a group, the four studies in this section also raise issues about the phenomenon of information technology and systems. In the Kraut, et al. study, electronic networking was conceptualized at a very abstract level as "any type of computer communication that allows you to exchange information with this supplier" (p. 730). Their operationalization of electronic networks would not have allowed the researchers to analyze their results according to type of electronic networking used, e.g. email versus EDI. This issue is interesting, because some researchers have argued that the commitment to using IT in relationships between buyers and suppliers can offset certain governance problems (Grover, Teng and Fiedler, 2002). Presumably this argument would apply in greater force for IT that is challenging to implement jointly, such as EDI, than to IT that can be adopted casually, such as email.

Of the four studies, only Mukhopadhyay, et al. focuses on a single, fairly well-bounded, information technology — EDI. Even here, however, the technology was used in different ways at different points in time and used with complementary changes in business practices, such as pickup schedules. The other two studies explicitly involve diverse collections of information technologies, applications, and business practices. The Jelassi and Figon study describes three IT-supported purchasing processes, one based on videotext favored by customers with no computer equipment, one based on PCs in either standalone or networked configurations, and an EDI version with systems integration for the largest customers. The Palmer and Markus study involves technologies such as point-of-sale scanning and EDI, IT applications such as order entry and inventory systems, and business philosophies and practices such as vendor-managed inventory, cross-docking, and seasonless retailing. These differing conceptualizations of the phenomenon of interest make it challenging for researchers to generalize across the studies, and they explain continuing calls for the IS community to develop a theory of the "IT artifact" (Orlikowski and Iacono, 2001; Weber, 1987). They also raise concerns about the appropriate level of aggregation in studies of IT use (Fichman, 2001).

REFERENCES

Fichman, R. G. (2001) "The role of aggregation in the measurement of IT-related organizational innovation", *MIS Quarterly*, 25(4), 401–29.

Grover, V., Teng, J. T. C. and Fiedler, K. D. (2002) "Investigating the role of information technology in building buyer–supplier relationships", *Journal of the Association for Information Systems*, 3(7).

Jelassi, T. and Figon, O. (1994) "Competing through EDI at Brun Passot: Achievements in France and ambitions for the Single European Market", *MIS Quarterly*, 18(4), November–December, 337–52.

Kraut, R., Steinfield, C., Chan, A. P., Butler, B. and Hoag, A. (1999) "Coordination and virtualization: The role of electronic networks and personal relationships", *Organization Science*, 10(6), 722–40.

Mukhopadhyay, T., Kekre, S. and Kalathur, S. (1995) "Business value of information technology: A study of electronic data interchange", *MIS Quarterly*, 19(2), June, 137–56.

Orlikowski, W. J. and Iacono, C. S. (2001) "Research commentary: Desperately seeking the 'IT' in IT research – A call to theorizing the IT artifeact", *Information Systems Research*, 12(2), 121–34.

Palmer, J. W. and Markus, M. L. (2000) "The performance impacts of quick response and strategic alignment in specialty retailing", *Information Systems Research*, 11(3), September, 241–59.

Ross, J. (2003) "Creating a strategic IT architecture competency: Learning in stages", *MIS Quarterly Executive*, 2(1).

Subramani, M. (2004) "How do suppliers benefit from information technology use in supply chain relationships?", *MIS Quarterly*, 28(1), 45–73.

Weber, R. (1987) "Toward a theory of artifacts: A paradigmatic basis for information systems research", *Journal of Information Systems*, 1(2), 3–19.

Yin, R. K. (1999) *Case Study Research: Design and Method*, third edition, Thousand Oaks, CA: Sage.

Competing through EDI at Brun Passot: Achievements in France and Ambitions for the Single European Market[1]

Tawfik Jelassi and Olivier Figon

INSEAD, Technology Management Area, Boulevard de Constance, 77305 Fontainebleau, France, jelassi@insead.fr
CISI Group, 3 Rue Maryse Bastier. BP 23, 69675 Bron Cédex, France

Abstract

To differentiate its customer service, Brun Passot, a small French company specializing in the distribution of office supplies, developed a set of telepurchasing applications. In 1982, it launched Bureautel, a videotex-based service that allows customers to electronically place their orders. In 1986, at the request of its large customers, it developed a PC-based service, then in 1989 an advanced electronic data interchange (EDI) application linking customers to its supply information system. These services allow data on product availability, price lists, orders, acknowledgement receipts, delivery notices, invoices, and related bank payments to be electronically transmitted. Using ISDN, they also make it possible to look up the photos of the 12,000 products that Brun Passot markets. This article illustrates how a small-sized company has used IT to improve the quality of its customer service, shorten lead time and reduce management costs, as well as create new business opportunities in France. It also raises some issues related to the adoption and diffusion of EDI and presents Brun Passot's ambitions to use this technology as an essential enabler to expand its geographical coverage. The 1993 fall of mobility barriers within the European Community, leading to the formation of the single European market, presented for Brun Passot a unique business opportunity to further leverage its IT infrastructure and gain new markets.

Keywords: strategic information systems; IT-enabled competitive advantage, electronic data interchange, single European market

ISRL Categories: CP110, GA0101, GA0102, GC01, GC06, HA07

Background

The office supplies industry in France is highly fragmented; the principal players are the manufacturers, the distributors, and the final customers. Many highly specialized manufacturers are often dedicated to a single product line. The distributors are of different sizes and degrees of specialization. The total number of distributors in France in 1992 was about 5,000; this figure sharply contrasts with the one in Great Britain, where approximately 100 distributors shared a slightly larger market.

Approximately 25 percent of the French office supplies distribution market is held by the four main companies: Guilbert, Gaspard, Saci, and Brun Passot; the remaining 75 percent of this FF11 billion[2] market is divided among numerous small players. The size of the European Community (EC) office supplies market is FF175 billion. The two main players are Germany and Great Britain, who have a share of FF35 billion and FF15 billion respectively. The 1993 fall of mobility barriers within the EC nations leading to the formation of the single European market has allowed some large American, German, and British firms to enter the French market. The high volume and global operating capabilities of these firms are likely to have a strong effect on the French industry.

Brun Passot

Brun Passot is a family business founded in 1949 near Lyon, France. The company initially specialized exclusively in paper processing. In 1970, it started diversifying into the distribution of office supplies and products related to computer and office equipment. In 1992, Brun Passot employed 160 people including a salesforce of 22 people. It had recently significantly enlarged its direct customer base to include major industrial and service organizations (e.g., Renault, Alcatel, Dassault, Péchiney, Crédit Lyonnais, Shell, Philips, and DEC France) as well as several government agencies (such as Electricité de France, France Télécom, the French Armed Forces, and the national railroad company SNCF). Through its network of 11 branches and one warehouse center, Brun Passot offered 12,000 products to 6,000 customers at 15,000 delivery locations throughout France. From a mere FF15 million in 1970, the company's turnover reached FF254 million in 1991.

The growth of Brun Passot over the years, coupled with higher diversification and more products and partners (customers and wholesalers), had increased the business complexity for the company. In 1978, the company networked the corporate headquarters with the central warehouse. Aware of the potential of this IT platform and at the stimulus of several large customers, Brun Passot established an electronic link between their purchasing departments and its supply information system. Several routine tasks, including order generation, inventory inquiries, and statistics, could now be handled in a more efficient, paper-free manner resulting in a number of benefits for both parties.

The Business Strategy

According to a national study,[3] an employee of the service or manufacturing sector uses on average FF2,200 of stationery (i.e., writing materials) per year; this figure excludes the purchase of paper, preprinted forms, and computer-related equipment. Purchasing this stationery requires, on average, 16 purchase orders, each containing 70 product lines. Brun Passot has estimated the costs for companies, to process these orders and manage the subsequent inventory, to range from 38 to 145 percent of the purchase value[4] (see Table 1).

In 1980, aiming at enhancing its competitiveness, Brun Passot's top management sought to distinguish the company from its rivals by offering a distinctive customer service based on the concept of telepurchasing.[5] The vision came from

Activity	Percent %
Purchasing	61
Storing	16
Administration	14
Management	6
Negotiation	3

Table 1 Costs of office supplies for customers (for a purchase of FF2,200/office worker/year).

Source: Brun Passot.

Jean Philippe Passot, deputy general manager and then head of IT, who thought that the company should offer a service and not just a product. Because of telepurchasing's potential for reducing the costs of acquiring, storing, and managing office products, Brun Passot saw it as a means to win the loyalty of existing customers (Christopher, 1993). Other customers, it was hoped, could be stolen away from its rivals. Such benefits are especially important in the highly competitive office supplies market where profit margins are small (approximately 3–4 percent) and price sensitivity very high.

Telepurchasing Applications

From 1983 through 1990, Brun Passot developed three telepurchasing applications: Bureautel, SICLAD, and Advanced EDI.

Bureautel

In 1983, Brun Passot adopted Minitel, the publicly available videotex platform, to develop its first telepurchasing application. There were three reasons that motivated management: First, France Télécom provided the terminal free of charge; second, Minitel was increasingly used throughout France;[6] and third, the Minitel terminal allowed for connection to a computer network. Bureautel was the first Minitel-based telepurchasing service offered in France.

Bureautel was developed in one year by four members of Brun Passot's nine-person information technology group. The system allowed the sending of electronic orders in a validated and secure way (each customer has an identification number and a password). The system also let Brun Passot managers make routine inquiries of Brun Passot's inventory and provided them with reports on the status of purchases to date as well as cash flow.

An enhanced version of Bureautel, developed in 1989, served as a management reinforcement tool; it supported customers in following up on their supplies. Based on the LECAM[7] technology, it gave users direct access to Brun Passot's order entry application. Brun Passot issued its own credit card that had a predefined maximum purchase limit per customer department for a certain time period. As orders were placed, the value of the items was subtracted from the department budget. Using reports provided by Bureautel, users/departments were able to trace their expenses. This card was not used for actual payment; instead, orders resulted in the issuance of a regular invoice. The benefits of the system included: (1) it substituted for a purchase order and hence, reduced paper work, (2) users no longer needed to request management approval or go through a centralized purchasing department to order office supplies, and (3) users were able to carefully monitor the use of their office supplies budget since they could not exceed it without getting their supervisor's approval.

SICLAD

Customers with no computer equipment were attracted to Bureautel, but others found it less appealing due to the limited functionality offered. Some larger customers pushed Brun Passot into developing a PC-based telepurchasing system. This system, called SICLAD (Systéme Informatisé de Commande Locale pour Approvisionnement Décentralisé), was developed in-house in 1985 by a five-member team; it had several advantages over Bureautel: (1) it was cheaper for customers since, with the PC, data input is free;[8] (2) compared to Minitel, the PC is faster and more user-friendly; and (3) it offers memory storage and networking capabilities (allowing file management operations and LAN configuration set up).

The LAN version of SICLAD supported up to 32 customer PCs, with anyone permitted to access the external network. This provided centralized control over placing orders, while still giving customers the convenience of generating orders from multiple offices. SICLAD allowed customers three ways to access the Brun Passot server by way of the customer's private automatic branch exchange: first, over the telephone network through the use of a modem; second, over the TRANSPAC[9] network; and third, over the French ISDN network Numéris (see Figure 1). The choice of the path depended mainly on the volume of transactions that a customer had with Brun Passot. Customers could use SICLAD to send purchase orders electronically and to receive receipt acknowledgements. Invoices and catalogues were not available over the network.

Three technical limitations restricted electronic distribution of the catalogue and invoices. First,

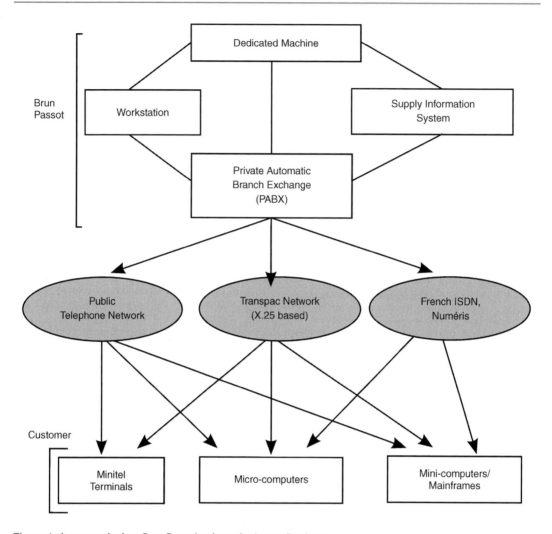

Figure 1 Access methods to Brun Passot's telepurchasing applications.
Source: Adapted from *Laidet*, (1990).

the typical PC did not have sufficient memory space to store a huge volume of data.[10] Second, more sophisticated software would have been required. Third, incompatibility of data formats would have required customers to rekey invoice data.

EDI system

An enhanced version of SICLAD, developed in 1989, used Numéris, the French ISDN service. It provided color photos of each product using an image database. Customers accessed this database either by locally looking up the images of the 200 products[11] stored on the hard disk of their PC, or by remotely connecting to Brun Passot's workstation. In the latter case, the entire image database for 12,000 products was accessible. In spite of the added functionality, ease-of-use, and convenience that the various versions of SICLAD brought over Bureautel, some of Brun Passot's large customers still did not adopt it because they wanted multi-supplier telepurchasing services rather than proprietary, bilateral services such as Bureautel and SICLAD. In order to further enhance its telepurchasing service, Brun Passot developed, in late 1989, an advanced EDI application through which it electronically sent product files, delivery status reports, purchase quotes, shipping notices, invoices, payments and related bank details, as well as e-mail messages. However, a hard copy of each invoice was generated for archival purposes since the French judicial system did not yet recognize electronic invoices.

The French subsidiary of Digital Equipment Corporation (DEC) was the first Brun Passot customer to use the advanced EDI application.[12] Shortly after this pioneering implementation, other large customers connected to the system including Electricité de France, Elf Aquitaine, Péchiney, Matra, and Spie Batignolles.

The EDI linkage between Brun Passot and its customers was made via a value-added network (VAN), France Télécom's ATLAS 400. VANs are best suited for a company that deals electronically with hundreds of business partners since they provide good security measures.[13]

Establishing an EDI link between a customer purchasing department and Brun Passot's order entry system required commitment and trust from both sides as well as a good understanding of the customer operating procedures. As Brun Passot's deputy managing director put it: "In a business as banal as that of office supplies, you tend to get a lot of what I call flirtation between big companies and their suppliers. With EDI, you need the commitment of true love. Before we set up an EDI link with one of our customers, we study their logistics for as long as a year. This requires trust and openness from both parties. In the end, we know their supply patterns better than they do. In order for the system to really take root in major companies, we need to set up a real partnership with the Computing Department as well as the Purchasing and Finance Divisions of our customers. This means that the system is integrated into the client company so it can evolve while taking into account the future needs of the users."[14]

Organizational changes induced by EDI at Brun Passot

Three actions by Brun Passot top management helped diffuse customers' adoption of the telepurchasing applications while building internal commitment. These actions, which also led to some organizational changes inside the company, were:

- **Creating a new Marketing unit (1989)** exclusively in charge of promoting the diffusion of SICLAD and, in particular, its Numéris version. This unit, which had 3 full-time members, had been participating in a variety of fairs and industry shows throughout France, hence, helping the company salesforce.
- **Establishing a new financial bonus (1990)** to reward each salesperson who would convince a customer to adopt the basic EDI system (SICLAD) or the advanced one. The bonus was paid in addition to the already existing financial reward for winning new customers.
- **Offering SICLAD free of charge.** Brun Passot top management believed that their business was to sell office supplies, not computer software, and that by giving the software and its related services (training, update, maintenance) for free, the company could attract new customers.

Over a two-year period (from September 1990 through September 1992), the number of corporate SICLAD users drastically increased, from 15 to almost 100. Moreover, all the new large customers[15] have adopted either SICLAD (80 implementations) or the advanced EDI service (seven implementations). The larger number of SICLAD implementations was due to one of three factors: First, companies who adopted SICLAD did not have the required computer equipment for the advanced EDI service. Second, they were in the process of restructuring their information systems and did not want to add a new major operation. Or third, they found the investment too heavy.

The advanced EDI service had also affected the organizational relationship of Brun Passot with some of its customers, especially the large ones. The company became in 1990 the single supplier of office products to DEC France, and shortly after that, a similar change in the relationship[16] happened with other large customers (such as Péchiney in Grenoble and Matra Espace in Toulouse). Customer benefits from having Brun Passot as a single supplier result from eliminating warehousing and inventory management, getting lower prices due to larger order quantities, and reducing negotiation time and effort since they have only one supplier to deal with.

Investment in telepurchasing and resulting benefits for Brun Passot

For Brun Passot, the initial investment made for Bureautel and SICLAD amounted to a total of FF250,000 (FF150,000 for acquiring additional hardware to the existing large computers[17] and FF100,000 for developing the software). Subsequent investment to purchase microcomputers as well as to use EDIFACT and Numéris amounted to FF300,000. Maintenance costs reach approximately FF100,000 per year, an expense covered by the FF280 monthly subscription fee to the system that only Bureautel users pay.

The return on this investment became visible rapidly. In 1984, Bureautel contributed 2 percent (or FF4.5 million) to total turnover, with 18,000 electronic orders processed, a figure that reached 22 percent (or FF27 million) in 1988, corresponding to a volume of 180,000 electronic orders. In 1992, the contribution of all three telepurchasing applications reached 40 percent of total turnover[18] or a value of FF120 million, with Bureautel contributing FF41 million, SICLAD FF28 million, and the advanced EDI application FF44 million. Brun Passot's management thought that, although the contribution of Bureautel reached a ceiling, SICLAD and especially the advanced EDI application would continue to increase. For the 1993–1996 period, they expect a gradual slight decrease of the contribution of Bureautel because some of its users will switch from the Minitel-based application to the PC-based SICLAD, and also some SICLAD users will migrate to the advanced EDI application. Moreover the company management expects the number of SICLAD and EDI users to almost double by 1996.

The introduction of the telepurchasing applications at Brun Passot simplified the supply procedure and the related administrative work. This freed up 25 people to do more sales and customer visits. Telepurchasing also enabled the company to predict more accurately customer needs and, consequently, to have a better idea of what goods to order from the wholesalers and when it should be done. This improvement led to faster stock rotations (from nine times in 1977 to 11 times in 1983 to 16 times in 1989) and to reduced inventory management costs (7 percent).

Qualitative benefits were also achieved. The telepurchasing applications helped Brun Passot improve its personnel productivity; for example, salespeople no longer make field visits to take orders but to promote products and sell more. Telepurchasing also enabled the company to differentiate itself from its competition (Porter and Millar, 1985) by first establishing Brun Passot as an innovative user of new technologies and then by sustaining this advantage over time through the continuous enhancement of these applications (Feeny and Ives, 1990). The telepurchasing service has also helped Brun Passot attract new customers, as well as create and maintain client loyalty.

Customers' use of telepurchasing and resulting benefits

Today, Brun Passot's telepurchasing applications are used by 1,120 customers who connect to the system on an average of 400 times every day (approximately 10,000 times each month) for a duration of about seven minutes per connection. The applications are mainly used for placing orders (78 percent of the traffic), but also for generating control reports (8 percent), sending e-mail messages (8 percent), and getting cash flow statements (6 percent).

Brun Passot claims that, based on a survey of 50 of its customers,[19] its telepurchasing services can save companies 20 to 60 percent of their present office supplies budget.[20] Compared to the traditional paper-based procedure, these services decrease the lead time by two-four days and reduce the rate of errors (due to rekeying the information contained in the paper documents) by a factor of five.[21]

Figure 2 shows costs incurred by Brun Passot customers through the four different ways of acquiring office supplies: Economat (which refers to the traditional paper-based method), Bureautel, SICLAD, and the advanced EDI application. These costs are related to a purchase value of FF2,200 and are given for each associated function, i.e., purchasing, storing, distribution, and management accounting. The costs of the EDI acquisition method are only a small fraction of the corresponding Economat costs.

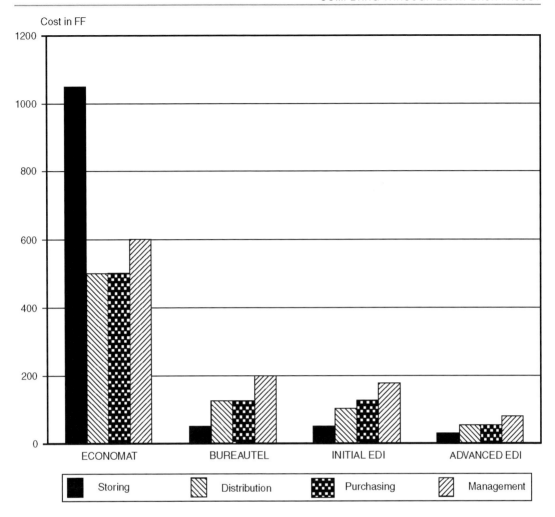

Figure 2 Costs of office supplies for customers based on four purchasing methods (for a purchase value of FF2,200).
Source: Brun Passot.

Users' Perspective on SICLAD

Customers are convinced of the benefits of adopting the telepurchasing applications. For example, COGEMA (Compagnie Générale des Matiéres Atomiques) has for the last two years been using the simplest version of SICLAD, which operates on a stand-alone PC station. According to COGEMA's Purchasing Department manager, users in his company like the Brun Passot system because it is a good and fast service, based on a simple procedure. The investment was very minimal (mainly the cost of a modem and telephone charges), but annual savings were quite significant: 30 to 40 percent of the cost of the paper-based procedure. The savings are due to the reduction of inventory and the elimination of one staff position as well as mailing costs. Aware of the additional benefits to be gained from eliminating the paper-based documents filled by the internal users, COGEMA implemented the network version of SICLAD in late 1993.

The Toulouse division of Matra Espace, an aeronautics company employing 2,000 people, installed the network version of SICLAD. The company, which purchases office supplies worth FF2 million from Brun Passot annually, has been using SICLAD as part of its new purchasing procedure. Throughout the week, secretaries key in their office supplies orders on the company computer network. On Friday evening, the purchasing manager reviews these orders and then transmits approved orders to the Brun Passot server. The following Tuesday, Brun Passot delivers the ordered products to the company offices in Toulouse. Since the adoption of SICLAD, the benefits to Matra Espace have reportedly been impressive: an annual savings of FF700,000 to 800,000, which corresponds to 35–40 percent of the office supplies budget. Benefits are attributed to the reduction of personnel needed to prepare the paper-based documents and to the elimination of duplicating and mailing costs. The system has also eliminated the need to follow up, by phone or fax, on the purchased orders. Moreover, Matra management find the statistics they get from SICLAD very helpful in knowing what has been expended. In the Summer of 1993, in order to benefit from the enhanced telepurchasing capabilities, the Toulouse division of Matra Espace started using the advanced EDI application of Brun Passot, now its single supplier of office products.

The successful experience of the Toulouse division of Matra Espace with Brun Passot has attracted other divisions of the company. For example, Matra Vélizy has recently adopted SICLAD, and other companies of the Matra Group are considering switching from their current supplier (who uses the traditional, paper-based approach) to Brun Passot.

Users' Perspective on the Advanced EDI Application

The Research Center of Péchiney, a major chemicals company, employs 400 people in its Grenoble offices. A pilot installation of the Brun Passot advanced EDI application was set up over an 18-month period; then, three months later, the use of the system to the entire Center was generalized. The company employees are predominantly knowledge workers, mainly engineers and technicians, who, because of the nature of their work, are big users of office products. A lot of coordination and time was needed to acquire and manage these low-priced products. Now, with telepurchasing, users are responsible for their own purchases.

Today, purchasing of office supplies is decentralized at Péchiney with each department managing its own budget. Once a week, each department secretary looks up the Brun Passot catalogue on his or her computer screen, keys in the products to order, and transmits them via the EDI system to Brun Passot. The latter delivers the ordered products directly to each requesting department. According to a corporate manager, the EDI investment was small, but the reduction of the overall office supplies budget was significant; for example, there are no misuses or abuses (such as "the start of schools" phenomenon) anymore. There was also significant time savings since everything is now done directly between Brun Passot and the final user without going through the Purchasing Department.

Péchiney stopped acquiring office products from the small suppliers it used to deal with and now does all its business with Brun Passot. However, the company does not think that the EDI system caused a "lock-in" effect vis-à-vis Brun Passot. It believes that it can easily switch to other players in the market such as Guilbert or Gaspard.

The issue of customer independence/lock-in has been central to the ongoing debate at Brun Passot. Some managers prefer to further "push" SICLAD because they think the proprietary nature of this software would lock in customers. Other managers favor diffusing the advanced EDI application because of the additional capabilities and enhanced customer service it provides.

DEC France, another user of EDI with Brun Passot, has an annual volume of 8,000 orders, averaging a value of FF700 per order. These orders total about 60,000 item lines generated from over 1,000 internal departments within DEC France. In the past, four paper-based documents were generated per order: the purchase order, the receipt acknowledgement, the shipping notice, and the invoice. The associated procedure was error-prone (due to rekeying the data), costly, and time consuming. Since October 1989, about 1,100 terminals located in 24 sites within DEC France have been connected through the company network to the Brun Passot server. Through these terminals, users place their office supplies orders in an autonomous, yet controlled manner, without having to go through a centralized purchasing department. According to a manager at DEC France headquarters in Evry, the achieved benefits to the company consist of a time saving of eight-12 days for processing an order, which corresponds to a gain of FF400,000 to FF700,000 per year.

Brun Passot guarantees delivery of the ordered products to the customer premises within 48 hours of receipt of the electronic purchase order. This factor allowed Spie-Batignolles, a major construction company employing 3,500 people, to go one step further than DEC France and the other customers. It decided to abolish its FF2 million stock of office supplies, which required 10 full-time employees (the freed individuals were given other job assignments). Since then, Brun Passot delivers three to four tons of products daily to Spie-Batignolles.

Facilitators and Barriers to the Use and Diffusion of the EDI Applications

The analysis of the telepurchasing applications at Brun Passot, from the idea creation stage through system implementation, reveals that several facilitators and barriers have helped or hindered the development, use, and diffusion of the company's various EDI-based services.

Facilitators

Some of the facilitators were due to a clear business strategy and sound management decisions; others were just the result of good timing and luck. These were:

- **A strong business pull** (as opposed to a technology push) at the very start and throughout the development of all the telepurchasing applications. This pull came from some large customers who recognized a need for shifting to just-in-time procurement and believed in the benefits of establishing an electronic link with Brun Passot. For the top management at Brun Passot, the strong business pull constituted the first and most important facilitator for developing the EDI-based services.
- **The availablity of new technologies** (such as TRANSPAC, Numéris and ATLAS 400) developed by a public third-party (France Télécom). This factor, thought of by Brun Passot top management as the second most important facilitator, has made the development of SICLAD and the advanced EDI application easy, fast, as well as quite inexpensive, and therefore feasible for the company.

The two factors mentioned above formed the critical pillars on which the Brun Passot telepurchasing strategy rested. In this context, the company's vision consisted of anticipating the increasing importance of these factors and taking the leadership in France (in spite of Brun Passot's small size, its financial weakness, and non-strategic business sector) to develop telepurchasing applications. Had the company failed to first recognize and then capitalize on these factors, not only would it not have been able to be a first-mover in the industry and therefore seize new business opportunities, but it would have been put in a harder survival position than the one it was in during the early 1980s. Other facilitators were:

- **The perception of telepurchasing and EDI as the core of a business strategy** and not just an IT project. This perception represented a major attitude change within the company since top management moved from considering IT as just a support function to thinking of it strategically (Cash, et al., 1992). (For a discussion and specific examples of the strategic/competitive use of IT in Europe, see Jelassi (1994) as well as Ciborra and Jelassi (1994).) Moreover, some education and training programs were set up in order to enforce this attitude change especially in the Marketing and Sales unit.
- **The long-term commitment and involvement of Brun Passot top management.** Jean-Philippe Passot, the 39-year-old deputy general manager with a background in law and management, has been a fervent champion of the telepurchasing projects since he joined the company in 1980. For example, he was the key sponsor of these projects at Executive Committee meetings, defending them and winning approval for their development and funding.[22]
- **A "motivated" organizational environment** for developing the EDI applications, due to the already available Bureautel service. Moreover, SICLAD helped launch the advanced EDI application. This motivation also stemmed from the strong message sent throughout the company by the top executives who were able to convince Brun Passot employees of the crucial importance of telepurchasing.

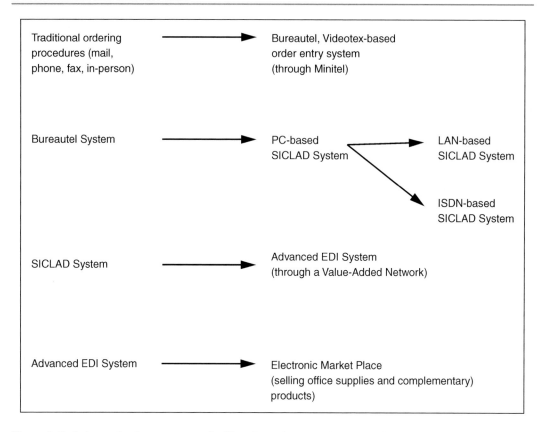

Figure 3 Evolutionary development approach of Brun Passot interorganizational applications.

- **The adoption by Brun Passot of an evolutionary approach** to allow for future enhancement and growth of its inter-organizational relationship (see Figure 3). This approach was coupled, on one hand, with strategic technology alliances with France Télécom (for adopting the Minitel platform to develop Bureautel and for using Numéris as the basis for the ISDN-based SICLAD) as well as DEC (for setting up the Advanced EDI service); and, on the other hand, with a continuous enhancement of the skills of the IT Group to keep it abreast of the latest advances in the field. The IT Group was divided into three teams, each having a specific field of expertise (i.e., videotex, PC, or EDI systems) and focusing on one telepurchasing application (Bureautel, SICLAD, or Advanced EDI respectively).
- **The close Interaction with customers** to define the "what, where, when and how" of the product supply chain so that both customer and supplier can benefit from the added value; also customers' reactions to a promising tool that simplifies procedures and reduces time and cost were important in developing the systems and their enhancements.
- **Competitors' late development of telepurchasing.** The other major players in the French office supplies market already have their own Bureautel-like system but not yet an EDI-based service. Guilbert, the market leader, has just developed (with the help of a software development company) GUILTEX400, a SICLAD-like system; however, it has not really attracted many customers so far because of its high cost and the 486-chip PC required.

Barriers

There were only a few barriers to the use and diffusion of Brun Passot's EDI applications. Although important, these barriers did not critically affect the company's successful implementation of its telepurchasing strategy. They were:

- **The rapid success of Bureautel** (i.e., its wide adoption by customers and its rapid contribution to Brun Passot total turnover) as well as the decision to keep offering it (even after introducing SICLAD and Advanced EDI) constituted a barrier for the diffusion of the EDI applications. The rationale for this decision was keeping customers, who had a small transactions volume and who were especially sensitive to costs, for whom Bureautel was best suited. Although Brun Passot did not discontinue its Bureautel service, it has heavily promoted the SICLAD and Advanced EDI applications by demonstrating to customers their enhanced capabilities (as compared to those of Bureautel) and explaining the differential benefits they can get from them.
- **The relatively weak bargaining power of Brun Passot** vis-à-vis its customers because of the non-strategic nature of the products it markets. This barrier was especially strong in the early 1980s; however, the company was able to alleviate it starting in 1989 when it became the exclusive provider of office supplies to several large firms.
- **The relatively heavy investment needed on the customer side** to use the Advanced EDI application (Scala and McGrath, 1993). This investment has been decreasing mainly due to the ever-improving IT cost-performance ratio, hence reducing the importance of this barrier.

Strategic Business Initiatives

Brun Passot has already started leveraging its EDI infrastructure through several ongoing projects. As reflected in an internal document, the company intends to use EDI as an essential enabler for further growth. "At Brun Passot, EDI spells the future. It is 'paperless trading' relying on 'peopleless administration'... The beauty of these [EDI] applications is that they need not be confined to the procurement of office supplies, but can be developed to encompass all purchasing undertaken by the company." Brun Passot has actually started expanding into other business lines by offering complementary products such as office furniture and cleaning materials as well as computer accessories. This larger product portfolio will allow Brun Passot to extend its telepurchasing applications into a broader electronic market place (Malone, et al., 1987; 1989), hence implementing its "one-stop shopping" concept.

Expand product diversity and penetration through tight linkage to suppliers

Brun Passot has started extending its information system backwards to suppliers (the wholesalers) in order to get access to a more diversified product offering (from the current 12,000 to 120,000 products). Both parties would benefit from this electronic linkage since Brun Passot could increase the products penetration rate with its customers.

Reduce inventories through tight linkage to suppliers

A just-in-time (JIT) purchasing system can generate savings (due to reduced inventories) for both Brun Passot and its suppliers. Operationally, this JIT purchasing system is used as follows. Because of its strong knowledge of the nature and quantity of products its customers order, Brun Passot needs to send, for replenishment purposes, electronic orders to its suppliers only once a week. In some rare cases where a customer requests an exceptional quantity of products, Brun Passot places an urgent order immediately with its supplier(s) without waiting for the regular weekend consolidation.

Brun Passot aims by the end of 1994 to conduct 80 percent of its sales turnover with suppliers through EDI. It considers setting up an electronic link with a supplier as an opportunity to assess its business performance in terms of logistics costs, quickness of delivery, and quality of service. In some cases, this assessment led the company to stop doing business with some traditional suppliers.

Answering customer requests through rerouting of product information

In order to provide customers with accurate, up-to-date product information (extracted directly from its source), Brun Passot has opted for "rerouting." The idea consists of setting up "electronic bridges" using TRANSPAC or Minitel, which would allow customers to access, through a single connection to the Brun Passot network, different servers related to a given market. For example, a user connected to one of Brun Passot's telepurchasing applications and requesting some information on product lines (e.g., those of 3M France) that Brun Passot markets gets automatically rerouted to the server of that company. Rerouting takes place while the customer is still logged on to the Brun Passot system; once he/she completes all inquiries about those product lines, he/she is disconnected from the host server and taken back to the original telepurchasing application.[23] Moreover, this new capability alleviates Brun Passot from having to include such data on its server.

New business opportunity: Productizing EDI expertise

The availability in France of a large, diversified telecommunications network[24] allows many companies to install or to enhance intercompany electronic communication through EDI. However, many suppliers are small or medium-sized enterprises that often lack the financial basis and the technical expertise necessary for implementing EDI systems. Having been a pioneer and an innovative user of telepurchasing over the last decade, Brun Passot has decided to leverage its expertise in this area through SATELITE, a new subsidiary set up to offer services in the development and implementation of EDI systems.

Eliminating the costs of banking transactions

Brun Passot aims at eliminating the costs of banking transactions for both customers and suppliers by offering "Financial EDI" applications. The company believes that it makes no sense to separate the commercial and physical exchange of documents from the financial payment. Once business partners communicate with each other using the same mode, e.g., EDIFACT, they can process all their transactions operations.

Brun Passot Ambitions for the Single European Market

In order to seize the new business opportunities that the single European market offers outside France, Brun Passot has initiated a set of actions aimed at exploiting the elimination of trade and customs barriers within the EC member states. Moreover, these actions are also aimed at strengthening the company's position in the new competitive landscape.

Formation of the single European market

The single European market, established by the 12 EC nations on January 1st, 1993, consists of 344 million consumers, which is 50 percent more than in the U.S., and has the potential to grow even larger.[25] Although the formation of this $4-trillion market seems inevitable and beneficial to the European economy (Cecchini, 1988; MAC Group, 1988), full implementation is being delayed because of many remaining fiscal (taxation policy), legal (antitrust law), monetary (possible single EC currency) and operational problems (e.g., passport controls). Nevertheless, the elimination of customs and all other barriers that prevent the free flow of goods and capital has already started, and many companies have prepared themselves for increasing competition as new players (both European and non-European) enter or expand their operations in the EC market (Gogel and Larréché, 1991). The formation of the single European market has resulted in major investments in Europe and in a wave of corporate restructuring and mergers within those industries most directly affected, such as banking, insurance, and airlines (Héau, 1991).

Brun Passot action plan

- **Raise Entry Barriers by Forming Strategic Alliances for Sharing Telepurchasing.** In the office supplies market and taking advantage of the 1993 event, some American companies that have already established themselves in England (such as BasicNet) as well as some British and German firms (e.g., Spicers and Herlitz, respectively) have expanded their operations in Europe. In response to the threat such a move represents to the market share of French companies, Brun Passot merged in 1992 with SACI, another distributor of office supplies with similar market share. Fiducial, the new, larger group, aims at increasing profitability margins by benefiting from economies of scale, strengthening bargaining power vis-à-vis wholesalers and customers, as well as further leveraging Brun Passot telepurchasing applications.

- **Expand Business Scale Beyond France Through Multilingual Telepurchasing.** After having strengthened its competitive position in its home market and in order to become a European service provider, Brun Passot is working on expanding its geographical coverage to other European markets. As part of its action plan, it developed a multilingual (English and Spanish, in addition to French) version of its telepurchasing applications that uses the X.25 packet-switched networks already available in several EC member states. This new application is needed in order to provide integrated service to national as well as pan-European corporate customers.
- **Increase Market Share Through the European Subsidiaries of Multinational Customers.** Brun Passot approached some of its multinational customers who have expressed their interest in reducing the number of suppliers they are dealing with across Europe. It plans to start its European operations with DEC, who has decided to centralize on a single computer (located in Geneva, Switzerland) all the purchase requests generated at its different European subsidiaries. The information system residing on this computer would then select, based on the geographic location of the requesting party, the best-suited supplier to provide the goods. Brun Passot considers "winning" the European subsidiaries of its present multinational customers as a good business opportunity for quick penetration of the single European market.
- **Enhance Logistics Operations to Meet European Market Needs.** In preparation for its European expansion and the subsequent increase in customer orders, Brun Passot has made significant investments to upgrade the performance of its logistics center located in Heyrieux (France).[26] It doubled the size of the facility and installed a fully automated picking system managed by a computer through barcode readers located alongside the conveyor belt. It also installed an automated packing system using a thermo-fusion procedure and a set of robots that automatically wrap the packages.
- **Acquire Some National Companies and Integrate Them Through IT.** Among management plans to implement the geographical coverage expansion are acquiring, or joint venturing with some national companies, as well as setting up some distribution centers near potential new European customers. The challenge for Brun Passot is to be able to move products around the continent as efficiently as it is done at present within the French borders and to offer bottom-line savings to the new European customers. Meeting this challenge requires an intense physical and informational interdependence among geographical units (Doz, 1991), including a sound organizational design for IT (Jarvenpaa and Ives, 1993). For Brun Passot, as well as for any company that wants to be pan-European, these are strategic ingredients for any future gain from the development of the single EC market.

Summary and Concluding Remarks

In the early 1980s, Brun Passot, a French family business with a small market share in the distribution of office supplies, sought a differentiation strategy that would allow it to offer a superior customer service. Central to this strategy was the development of a set of telepurchasing applications through which customers can electronically view catalogue and product images, send purchase orders, as well as receive acknowledgement receipts, delivery notices, and invoices. These applications are based on videotex and EDI technology; the most advanced one was completed late 1989.

Telepurchasing allowed customers to eliminate paper work, improve data accuracy and timeliness, as well as reduce (sometimes even abolish) inventory, resulting in significant savings. It has also helped Brun Passot gain market share by winning new customers or becoming the single provider of office supplies to several large companies. The results achieved by Brun Passot (tripling gross revenues in five years, while maintaining constant manpower) are rather remarkable given the non-strategic nature of the products it markets. Critical success factors of the telepurchasing project at Brun Passot include: business pull (as opposed to technology push) as the main EDI driver, top management commitment and involvement, an evolutionary approach to adopting and diffusing new technologies, and the perception of EDI as a business/marketing project.

The formation in 1993 of the Single European Market and the consequent fall of entry barriers within the European Community member states represent a unique business opportunity. Aware of the potential this opportunity presents, Brun Passot top management has taken several steps to foster the firm's competitiveness: they merged with another distributor, broadened the company's product portfolio, and developed a multilingual version of the telepurchasing applications. Brun Passot wants to use EDI as an essential enabler to expand its geographical coverage to other European markets. By the end of 1994, it aims at achieving a turnover of FF800 million, with 80 percent of its transactions electronically made and processed, and with only 15 percent of personnel increase!

NOTES

1 A previous version of this paper was an award winner in the Society for Information Management's Annual Paper Award Contest.
2 Though the exchange rate of the French franc (FF) has fluctuated in the last decade between 4 and 10 FF per American dollar (US$), its trading "band" is typically between 5 and 6 FF/US$. The average value over the last decade has been close to FF5.15 per US$.
3 A study made in France in 1989 by the Institut National des Statistiques et des Etudes Economiques (INSEE), Paris.
4 These figures are based on a representative sample of 80 customers, with a total number of employees ranging from 300 to 5,000.
5 For a discussion of how to gain a competitive advantage with inter-organizational information systems, see Johnston and Vitale (1988).
6 There were 120,000 Minitel terminals distributed in 1983. This number increased to 531,000 by December 1984, to over 2 million in 1986, and has reached 7 million today. In addition to the electronic telephone directory, Minitel terminals offer information services, professional databases, banking services, electronic mail, order processing, cash management, portfolio management, and accounting. (For more information on the development and diffusion of Minitel, see Cats-Baril and Jelassi (1994); for examples of business applications of Minitel, see Jelassi and Loebbecke (1994) and Jelassi and Murthy (1994).)
7 LECAM (Lecteur de Carte à Mémoire) is a device that can be attached to a Minitel terminal to read magnetic-stripe cards.
8 With Minitel, customers pay for the phone connection while keying in their purchase orders. With the PC, these orders are keyed in a file and then electronically transmitted over the network.
9 TRANSPAC (Transmission par Pacquets) is based on the X.25 packet-switching standard.
10 Storing just the 12,000 products catalogue would have required a minimum of 10 megabytes.
11 This figure represents the average number of office supplies frequently purchased by large customers and that correspond to products of ongoing consumption. These products slightly differ by customer (by a factor of 10 percent).
12 It was also the first EDI experience of this nature for the French subsidiary of DEC.
13 Exchanged messages between the sender and the receiver were on the EDIFACT format.
14 "Electronic Documentation Offers Greater Efficiency," *The International Herald Tribune*, March 14, 1991.
15 Large customers account for about 90 percent of Brun Passot's client base.
16 For a discussion of the change of buyer-seller relationships affected by EDI, see Cunningham and Tynan (1993).
17 The telepurchasing applications run on a PRIME 6350 computer (with a processing power of 10 MIPS), connected locally to a VAX 3400 (having 4.5 MIPS) and remotely to five other PRIME computers. There are 150 terminals, local and distant, connected to the network, as well as over 1,000 videotex terminals.
18 The remaining contribution comes from sales made through the traditional modes (i.e., mail, telephone, and fax).
19 In 1989 Brun Passot commissioned a French business school, the "Ecole Supérieure de Commerce de Lyon," to conduct this survey. The latter was based on a mail questionnaire that, in some cases, was followed up by telephone interviews.
20 Source: *Une Entreprise, Une Application Télétel, France Télécom*, No. 19, February 1990.
21 ibid.
22 The corporate IT budget at Brun Passot has been since 1982 approximately 4–5 percent of turnover, a figure that is double the average IT budget in the industry.
23 Rerouting can be thought of as a multiwindowing facility through which, for example, a software package gets called upon or executed from an already activitated application.
24 Industry analysts consider the French telecommunications system better than that in other western countries (see, for example, Nguyen (1988)). This is due to the availability of a fully digitized telephone network as well as a nation-wide videotex, ISDN, and packet-switched networks.

25. Jacques Delors, European commission president, is already envisioning a European Community that will eventually include Western European countries and Eastern Europe, as well as the former Soviet republics.
26. This Brun Passot logistics center is geographically close to the European headquarters and purchasing center of DEC, located in Geneva.

ACKNOWLEDGMENTS

The authors would like to thank the senior editor for many valuable comments and suggestions, as well as three anonymous referees for their substantive feedback on an earlier version of this paper. Helpful discussions with members of the SISNet Group, particularly Jon Turner (New York University) and Claudia Loebbecke (University of Cologne, Germany), are acknowledged.

This research work has been conducted under the auspices of the 1992 Trade EDI Systems (TEDIS) program of the Commission of the European Communities. One output of this research program is the book edited by Krcmar, et al. (1995) in which an earlier version of this paper will appear.

REFERENCES

Cash, J. I., McFarlan, F. W., McKenney, J. L. and Applegate, L.M. *Corporate Information Systems Management: Text and Cases* (3rd edition), Irwin, Homewood, IL, 1992.

Cats-Baril, W. and Jelassi, T. "The French Videotex System Minitel: A Successful Implementation of a National Information Technology Infrastructure," *MIS Quarterly* (18:1), March 1994, pp. 1–20.

Cecchini, P. "The European Challenge: 1992," *The Benefits of a Single Market*, Wildwood House, Aldershot, England, 1988.

Christopher, M. "Logistics and Competitive Strategy," *European Management Journal* (11:2), 1993, pp. 258–61.

Ciborra, C. and Jelassi, T. (eds). *Strategic Information Systems: A European Perspective*, John Wiley & Sons, London, 1994.

Cunningham, C. and Tynan, C. "Electronic Trading, Inter-Organizational Systems and the Nature of Buyer-Seller Relationships: The Need for a Network Perspective," *International Journal of Information Management* (13:1), 1993, pp. 3–28.

Doz, Y. "Aligning Strategic Demands and Corporate Capabilities," in *Single Market Europe: Opportunities and Challenges for Business*, S. Makridakis and Associates, Jossey-Bass, San Francisco, CA, 1991, pp. 141–66.

Feeny, D. and Ives, B. "In Search of Sustainability—Reaping Long Term Advantage from Investments in Information Technology," *Journal of Management Information Systems* (7:1), 1990, pp. 27–46.

Gogel, R. and Larréché, J. C. "Pan-European Marketing: Combining Product Strength and Geographical Coverage," in *Single Market Europe: Opportunities and Challenges for Business*, S. Makridakis and Associates, Jossey-Bass, San Francisco, CA, 1991, pp. 99–118.

Héau, D. "Seizing Opportunities—The Changing Role of Strategy in European Companies," in *Single Market Europe: Opportunities and Challenges for Business*, S. Makridakis and Associates, Jossey-Bass, San Francisco, CA, 1991, pp. 167–91.

Jarvenpaa, S. L. and Ives, B. "Organizing for Global Competition: The Fit of Information Technology," *Decision Sciences* (24:3), 1993, pp. 547–80.

Jelassi, T. *Competing through Information Technology: Strategy and Implementation*, Prentice Hall, London, 1994.

Jelassi, T. and Loebbecke, C. "Home Banking: An IT-Based Business Strategy or A Complementary Distribution Channel – CORTAL versus Crédit Commercial de France," in *Competing through Information Technology: Strategy and Implementation*, Prentice Hall, London, 1994, pp. 244–66.

Jelassi, T. and Murthy, G. "Minitel, A Home Retailing," in *Competing through Information Technology: Strategy and Implementation*, Prentice Hall, London, 1994, pp. 199–215.

Johnston, R. H. and Vitale, M. R. "Creating Competitive Advantage with Inter-Organizational Information Systems," *MIS Quarterly* (12:2), June 1988, pp. 153–65.

Krcmar, H., Bjørn-Andersen, N., and O'Callaghan, R. (eds.). *EDI in Europe: Lessons from the Field*, John Wiley & Sons, London, forthcoming, 1995.

Laidet, A. "Une PME dans la caverne d'EDI Babas," *Télécoms Magazine* (No. 33), April 1990, p. 64.

MAC Group, *Commission of the European Communities. The Cost of Non-Europe*, Vol. 12b, Official Publication of the European Community, Luxembourg, 1988.

Makridakis, S. and Associates. *Single Market Europe: Opportunities and Challenges for Business*, Jossey-Bass Publishers, San Fransisco, 1991.

Malone, T. W., Yates, J. and Benjamin, R.I. "Electronic Markets and Electronic Hierarchies," *Communications of the ACM* (30:6), 1987, pp. 484–97.

Malone, T. W., Yates, J. and Benjamin, R. I. "The Logic of Electronic Markets," *Harvard Business Review*, May–June 1989, pp. 166–70.

Nguyen, G. D. "Telecommunications in France," in *European Telecommunications Organization*, J. Foreman-Peck and J. Muller (eds), Nomos Verlagsgesellschaft, Baden-Baden, 1988, pp. 132–54.a.

Porter, M. E. and Millar, V. E. "How Information Gives You Competitive Advantage," *Harvard Business Review* (63:4), July–August 1985, pp. 149–61.

Scala, S. and McGrath, R., Jr. "Advantages and Disadvantages of Electronic Data Interchange: An Industry Perspective," *Information and Management* (25:2), 1993, pp. 85–91.

ABOUT THE AUTHORS

Tawfik Jelassi is associate professor of information systems at INSEAD (Fontainebleau), where he also served as coordinator of the Technology Management Area. He received a Ph.D. in MIS from New York University and degrees from the Université de Paris-Dauphine and the Université de Tunis. His recent research focuses on the competitive use of information technology in Europe. His most recent books are: *Competing through Information Technology: Strategy and Implementation* (Prentice Hall, 1994) and *Strategic Information Systems: A European Perspective* (co-edited with Claudio Ciborra, Wiley, 1994). Dr. Jelassi contributed articles to several academic journals including *MIS Quarterly, Journal of MIS, Information and Management, Journal of Strategic Information Systems, Decision Sciences, Decision Support Systems, European Journal of Operational Research*, and *OMEGA*. He is editor/associate editor of several academic journals, chairman of the EURO Working Group on DSS, and vice president of the TIMS College on Group Decision and Negotiation.

Olivier Figon is EDI expert at CISI, a leading European consultancy and software engineering group, and coordinator at the Université de Lyon of the "Networks and EDI" course in the postgraduate European diploma on computer-aided management. Prior positions include: manager of the Information Systems Department at Brun Passot (office supplies) and Schneider Group (electromechanical engineering branch). He was also responsible for the EDI relationship with customers at EUROTEC (a manufacturer of plastic components for the automotive industry). Mr. Figon holds a degree in information systems management from the Université de Lyon and is a member of the network GITASS (Group International de recherche en Théorie et Analyse Scientifique de Systémes).

Business Value of Information Technology: A Study of Electronic Data Interchange

Tridas Mukhopadhyay, Sunder Kekre and Suresh Kalathur

Graduate School of Industrial Administration, Carnegie Mellon University, Pittsburgh, PA 15213, U.S.A., tm25@andrew.cmu.edu
Graduate School of Industrial Administration, Carnegie Mellon University, Pittsburgh, PA 15213, U.S.A., sk0a@andrew.cmu.edu
Cimnet Systems, Inc., 601 Oakmont Lane, Westmont, IL 60559, U.S.A.

Abstract

A great deal of controversy exists about the impact of information technology on firm performance. While some authors have reported positive impacts, others have found negative or no impacts. This study focuses on Electronic Data Interchange (EDI) technology. Many of the problems in this line of research are overcome in this study by conducting a careful analysis of the performance data of the past decade gathered from the assembly centers of Chrysler Corporation. This study estimates the dollar benefits of improved information exchanges between Chrysler and its suppliers that result from using EDI. After controlling for variations in operational complexity arising from mix, volume, parts complexity, model, and engineering changes, the savings per vehicle that result from improved information exchanges are estimated to be about $60. Including the additional savings from electronic document preparation and transmission, the total benefits of EDI per vehicle amount to over $100. System wide, this translates to annual savings of $220 million for the company.

Keywords: business value, electronic data interchange, information technology, inventory costs, transportation costs, information handling costs

ISRL Categories: AD0507, AD0511, AD0517, AI0108, HA07, HB18, HB24

Introduction

Information technology (IT) has become a matter of serious concern for management today. The spectacular growth of IT has enormous potential for improving the performance of organizations. However, the huge investment made in IT puts increasing pressure on management to justify the outlay by quantifying the business value of IT.

Unfortunately, the results of recent studies of IT business value are at best inconclusive. While some authors have reported positive impacts, others have found negative or no impacts. At least two limitations are posed in those recent works. First, IT is often treated as a single factor. Given the complexity of the technology and the difficulty of implementing it in organizations, some systems may be effective, while others may bring negative returns. Therefore, by aggregating over all systems, the favorable impact of effective systems may be nullified by poorly designed systems. Second, many of the earlier studies use cross-sectional or short-time series data. If there is a lag in achieving IT business value, data covering a limited time period may not reveal the impact.

To overcome such problems, a longitudinal study of information technology was undertaken. The study focused on Electronic Data Interchange (EDI) technology and examined its use at the assembly centers of Chrysler Corporation for two primary reasons. First, EDI is a key technology of the 1990s and merits an investigation on its own. It is estimated that more than 40,000 organizations had adopted EDI by 1994, and many more are expected to adopt it soon (Verity, 1994). Additionally, EDI technology has been in use long enough at Chrysler to allow us to assess its impact.

Researchers have claimed that interorganizational systems using EDI, e-mail, etc., lead to "vertical information integration" between trading partners along the value chain.[1] By improving the accuracy and timeliness of information exchanged over manual methods, EDI is believed to significantly change how organizations conduct business with their suppliers and customers. Although the effect of EDI on business is undeniable, no rigorous studies have yet been attempted to quantify the financial returns from improved information exchanges due to EDI.

Data were collected at Chrysler assembly centers and key measures were obtained on the implementation process and use of EDI with Chrysler's suppliers over the past decade. Our field study is an ex post analysis of the EDI program at Chrysler and assesses management's goal of reducing manufacturing and logistics costs and streamlining operations for JIT (just-in-time) implementation. The impact of EDI systems on inventory, obsolescence, and transportation costs was analyzed for the model years 1981 to 1990. After controlling for variations in operations complexity arising from mix, volume, parts complexity, model, and engineering changes, the savings from improved information exchanges are currently estimated to be over $60 per vehicle for a typical assembly plant. Adding the additional savings from electronic document preparation and transmission, the total benefits of EDI per vehicle amounts to over $100. Across the entire Chrysler assembly system, annual savings amount to $220 million at the current production level.

Our results underscore the potential benefit and value of EDI and to some extent, IT in general. Although no attempt should be made to directly generalize the gains to other sites or other technologies, the lessons from this study are significant. When an IT is used for a specific goal, it can achieve substantial business value. However, caution should be used against making any hasty claim about the nature of the impact of IT without a careful assessment of the technology implementation, operational complexity, and usage over a sufficient timeframe.

This paper is organized as follows. First, prior work on the business value of IT and EDI is briefly reviewed. Next, the research site and the research framework are described. Then the data collection procedure and the results and management implications of our analysis are discussed. Finally, the limitations of the study and the concluding remarks are presented.

Business Value Research

One common management practice of evaluating an investment project is to assess its impact on firm performance in terms of financial criteria. In recent years, IT investments have come under close scrutiny because several studies have reported contradictory results on the business value of IT. Four excellent reviews of this literature are now available.[2] Our goal is not to review the prior work in detail, but rather to highlight key findings and discuss major hurdles researchers and practitioners face in this area. The relatively limited literature on the business value of EDI is reviewed.

Business value of IT

The majority of studies in this area report little or no impact of IT on firm performance. Two such studies are discussed below. The first examines the productivity impact of IT in the manufacturing sector (Loveman, 1994). The results indicate that "the marginal dollar would best have been spent on non-IT inputs into production, such as non-IT capital" (p. 27). The second study compares the productivity of information and production workers (Roach, 1991). An important finding of this study is that information worker productivity has either declined in some sectors or not kept the pace of production worker productivity in the manufacturing sector.

In contrast, two more-recent studies show evidence of positive IT impacts. One study classifies IT into three categories based on the management goals supported by the system (Weill, 1992). This study of the valve manufacturing sector identifies significant productivity gains that result from transactional IT, but did not find any positive impacts for strategic or informational (IT infrastructure) systems. A second study proposes a two-stage analysis of intermediate and final output variables for measuring IT contributions in the manufacturing sector (Barua, et al., 1995). This study found significant impact of IT on intermediate variables such as capacity utilization, inventory turnover, and product quality, but found little impact on return on assets or market share.

The contradictory findings on the value of IT are symptomatic of the difficulties of conducting such an analysis. The four reviews mentioned above delineate the problems and difficulties associated with this line of work. Together the reviews examine 43 studies. The earliest critique deplores the lack of theoretical analysis and quality data from industry applications (Crowston and Treacy, 1986), but a recent review is more optimistic about the progress in this area (Bakos and Kemerer, 1992). However, the overall findings of the four meta-studies are very similar—a majority *did not* find strong evidence of IT impact. The concerns expressed by the authors are also in broad agreement. Overall, they question the validity of prior findings and call for more rigorous research to pinpoint the true costs and benefits.

Why is there so little consensus or cumulative knowledge about the impact of IT despite substantial research efforts in this area? First, it is a difficult problem. One has to contend with two complex issues: measurement and data. Since the IT impact cannot be assessed in isolation, confounding effects of other inputs (e.g., labor and non-IT capital) must be accounted for. The problem is compounded by the lack of quality data on IT investment. Often, firms are reluctant to divulge this data for competitive reasons. Second, the complexity of teasing out the benefits and value necessitates the use of rigorous scientific methods that are often missing in prior work.

Business value of EDI

Only a handful of studies are available on this topic. One early survey found that the majority of EDI users were asked by their trading partners (initiators) to adopt this technology, and they believed their partners were gaining more from the system (Stern and Kaufmann, 1985). Research from another survey found that the involvement of trading partners is crucial for implementation success (Hwang, et al., 1991).

Three field studies on EDI conducted recently have reported some encouraging results. A study at LTV Steel on its EDI linkages with outside processors establishes that compared to the old manual mode of information exchange, EDI transactions have more favorable effects on quality and inventory measures (Kekre and Mukhopadhyay, 1992). Another field study of a U.S. automobile manufacturer and its first-tier suppliers reports that the use of integrated EDI to share schedule information lowers the level of shipping discrepancies (Srinivasan, et al., 1994). Finally, a third study of order processing systems found that combining EDI with business process reengineering can lead to faster recovery of payments from customers as well as fewer errors in order processing (Mukhopadhyay, et al., 1995).

Unfortunately, none of these studies directly addresses the business value of EDI. The first two survey studies are based on perceptual data, not on financial data. The three field studies measure EDI impact using intermediate variables, but do not examine the *dollar impact* of EDI. For example, the automobile industry study measures the impact of EDI on shipment discrepancy, but does not quantify the impact on the financial bottom line. In addition, all of the studies surveyed use cross-sectional data and thus fail to capture the gains over time. Our study was motivated to overcome the deficiencies of earlier work on EDI by attempting to gauge the dollar impact arising from improved information exchanges as a result of using EDI over a decade. We pooled longitudinal data obtained from nine Chrysler assembly plants. Our detailed discussions with plant personnel helped us systematically assess the drivers of performance and quantify financial gains.

EDI at Chrysler

Chrysler launched electronic communication with a few suppliers in 1969. At that time, this system lacked adequate communication abilities at both ends, which resulted in excessive human interventions. In 1976, Chrysler implemented an online receiving system to track shipping information and uncover discrepancies. The EDI program, as it exists today, was launched in 1984 with the introduction of the Supplier Delivery Schedule transaction, which gives the precise shipping quantities to suppliers electronically. The impetus for the program came from management directives to implement JIT practices in assembly centers using EDI. Finally, in 1990, the ANSI X12 standard was mandated by Chrysler and adopted almost universally by assembly plant suppliers.

As part of the corporate goal to reduce the cost per vehicle by 30 percent, the procurement, manufacturing, and logistics areas were asked to coordinate and improve the efficacy of the materials management process. It was evident that coordination with suppliers was unsatisfactory; suppliers neither had current information about Chrysler's requirements, nor did the assembly plants know about the exact content and status of shipments. To cope with the uncertainty, large safety buffers had to be maintained at assembly centers. When shortages occurred, emergency deliveries resulted to avert costly line stoppages. Finally, with 70 percent of components and subassemblies outsourced, Chrysler took a bold initiative to correct this problem. EDI technology was harnessed to change material flows and implement JIT practices. The movement of materials had to be supported by rapid, reliable, and frequent communication with suppliers. Chrysler management believed that with EDI and JIT practices, the plants could reduce inventory levels and the need for rush deliveries.

Chrysler developed a well-planned strategy in implementing EDI systems at the assembly centers. All suppliers were notified in detail about Chrysler's targets for fully electronic communication. Although the implementation was to be phased over time, all suppliers were asked to streamline their information handling procedures. Future business was conditional on a supplier's ability to send accurate, appropriate documents (hard copy or electronic copy). Chrysler management felt these changes would enforce "discipline" before the EDI program was launched.

The suppliers also saw benefits from the EDI program. The Supplier Delivery Schedule, for example, provided precise production schedule information for a 10-day period. Suppliers were also given long-term forecasts (weekly schedules for six to eight months) through the Material Release transaction set. Thus, the vendors were expected to handle short-term variations in the production schedule as well as effectively plan their internal operations by taking into account long-term demand trends.

From the supplier end, the EDI system permitted the transmission of Advance Shipping Notices (ASN) to Chrysler assembly plants. These notices, sent on a real-time basis, used the Chrysler electronic network. The supplier received a verification of the content of each ASN from Chrysler through the Application Advice transaction set. Any discrepancy detected in the shipment was subsequently communicated to the supplier using the Receiving Advice transaction set. In many cases, fast response was mandatory because it took less than 15 minutes for the material to arrive at Chrysler assembly docks (Sherwood and Marocco, 1989). Without the electronic link, it was not possible to implement frequent and accurate delivery schedules nor to dramatically reduce buffer inventories.

Chrysler changed the mode of logistics operations to fully exploit the improved quality and quantity of information available. For instance, under the old mode, the inbound material flow was inefficient and resulted in less than truckloads (LTL). Carriers hauled materials, mixing them as many as four times while moving from suppliers' docks to destination. Under the new system, using EDI capability with trading partners, Chrysler has changed the operating mode to the "Scheduled Delivery Program." The new operating premise, according to E. Krajca of Chrysler, is a scheduled

BUSINESS VALUE OF INFORMATION TECHNOLOGY

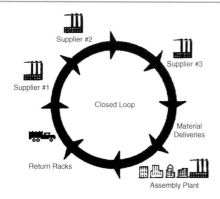

Figure 1 Scheduled Delivery Program.

pickup loop, with the carrier picking from several suppliers and returning the reusable containers (Figure 1). This mode substantially improved and simplified the planning and control function of route managers.

The inventory management function also changed considerably with changes in IT. Before 1984, scores of stockchasers examined the daily schedule and physically counted necessary parts in order to flag potential shortages and, in extreme cases, to pull incomplete vehicles out of assembly sequence. But the main strategy to avoid stockout in the old mode was to maintain large incoming buffer inventories. The new practice with improved information attempts to stabilize assembly line production rate and provide precise requirements to suppliers electronically. In addition, once a car enters the assembly line, it cannot be removed until it is completed. This enforced in-sequence production further reduces uncertainty for inventory management. Finally, visual inspection has been heightened to ensure that stacked incoming materials and work in process inventory remains below eye level, allowing an unobstructed view of the shop floor. Overall, the information-rich environment of the supply chain is expected to make material flow reliably and efficiently.

To analyze the impact of EDI at Chrysler, data were collected at the plant level from nine assembly facilities during the period 1981–1990 (Chrysler acquired three other plants from AMC in the late 1980s). The data were obtained from computerized historical records at Chrysler. Site visits were also conducted at selected supplier facilities to gain a deeper understanding of the information flows. At Chrysler, input was sought from managers and supervisors from the procurement, logistics, and production control departments.

The analysis of the impact on costs was complicated since it covered a period of 10 years. During this period, the system complexity and factors related to the Chrysler logistic system changed significantly. Thus, in order to pinpoint the impact of EDI, there had to be control for several factors. For instance, there had been gradual though significant changes in mix, volume, parts complexity, model and engineering changes, and the use of various transportation modes at the assembly plants. Therefore, the controls in our study are for variations in these factors. The study highlights the savings accrued in major categories of costs as a result of the support provided by interorganizational information exchange between Chrysler and its assembly plant suppliers.

Modeling the EDI Impact

The major benefits of the EDI system at Chrysler can be divided into two categories. First, improved information exchanges between Chrysler and its suppliers because of EDI may lead to cost savings. This section develops models to quantify the dollar benefits that result from the improved information quality. Second, preparing and processing documents electronically should also lead to savings over the manual mode. These savings accrue primarily from reduced personnel cost and lower transmission charges. A subsequent section estimates the savings from electronic data handling.

Our goal in this section is to estimate the benefits of improved information exchanges between Chrysler and its suppliers that result from using EDI. We build on the literature on interorganizational systems (IOS) shared by separate business entities. Of particular interest are hierarchical IOSs involving trading partners with a pre-existing, ongoing relationship prior to electronic integration. The literature on hierarchical IOSs asserts that electronic communication would improve coordination between trading partners because of superior speed and accuracy of information exchange.[3] In particular, Mukhopadhyay (1993) has proposed a framework for assessing the business value of EDI. His framework is used in order to assess the impacts at Chrysler. This framework consists of four steps and is summarized in Table 1 below.

As discussed earlier, Chrysler assembly centers utilize EDI in two applications supporting the materials management function. In materials planning, Chrysler uses Material Release and Supplier Delivery Schedule transaction sets to communicate material requirements to suppliers. In the shipping and receiving application, suppliers send an Advance Shipping Notice for each shipment and receive a verification notice (Application Advice) and an error message (Receiving Advice) in case a discrepancy is detected in the shipment. In summary, the locus of the EDI impact at Chrysler should be found in the materials management function.

Before the EDI program was launched, the poor quality of information exchange at both Chrysler and supplier ends led to inventory buffer buildup at the assembly plants. In essence, the risk of costly line stoppages was reduced using inventory buffers. However, the inability to effectively control inventory also led to large write-offs at the end of the model year. In addition, expenses in premium freight accrued as dispatchers frequently rushed materials from suppliers on short notice using air transportation or non-scheduled carriers.

Our theoretical premise is that the EDI program should make the management of inventory more effective by improving the quality of information exchanged at both suppliers' and Chrysler's ends. Inventory turns should improve with the usage of EDI, and the write-offs of inventory at the assembly centers should decrease due to effective information exchanges resulting in closer monitoring of stock. In addition, a reduction in premium freight is anticipated. However, a key question is whether this inventory reduction is achieved at the expense of higher transportation costs. Did the EDI program affect total transportation costs? To answer this question, both regular transportation cost and premium freight were investigated in this study. Our analysis thus focuses on the impact of EDI on the following four cost categories at the assembly centers:

- Inventory Holding Cost
- Obsolete Inventory Cost
- Transportation Cost
- Premium Freight

Step 1.	Examine the transaction sets and the application systems interfacing with EDI to determine the locus of impact.
Step 2.	Find out the extent of the EDI network (e.g., number of trading partners) to assess the magnitude of the impact.
Step 3.	Study the production/operation process to identify control variables.
Step 4.	Determine the exogenous industry or economy-wide trends that may mitigate the results.

Table 1 A framework for assessing the business value of EDI.

The second step of our framework assesses the extent of the EDI network (see Table 1). Recall that at Chrysler a phased implementation of EDI was undertaken. The first step was to improve information accuracy at both ends before bringing suppliers on board. Next, Chrysler selected high-value items as candidates for leading the EDI program. Subsequently, each year, more items were added to the list of EDI items. This phased implementation strategy suggested that the EDI program should be quantified in two ways. First, launching the program itself implied system-wide efficiency gains. We expect this brought substantial benefits as a result of the enforced discipline in information exchanges between suppliers and the assembly plants. The suppliers (manual or EDI) had to streamline their operations in order to meet program launching capabilities and requirements. Second, as EDI penetrated more into the operations and changed the nature of material and information flows, additional favorable impacts were felt. This two-step impact of the EDI program was modeled in this study.

The specification of the models is discussed next. For each model, the relevant production characteristics are identified as control variables (Step 3, Table 1). The auto industry trends during the last decade are also captured in terms of reduced parts variety and model changes. A list of all variables is presented in Table 2, and a schematic diagram of our research model is shown in Figure 2.

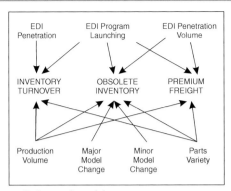

Figure 2 Research model.

past decade for each plant, the summary measure could be derived from past records.

Thus, inventory turnover (INV) is specified as a linear function of PROGRAM, EDIP, PARTS, and VOLUME. We expect the effects of the technology variables (PROGRAM and EDIP) to be positive. Note, as discussed earlier, that EDI required a system-wide discipline to be enforced before any implementation could begin. This in itself led to an improvement and is captured by PROGRAM. In addition, the effects of PARTS and VOLUME should be negative on INV.

Obsolete inventory cost

Inventory holding cost

First, the effect of EDI implementation is modeled on inventory turnover at assembly centers. Subsequently, the dollar savings resulting from the improvement in the inventory turnover (INV) are assessed.

Our hypothesis states that INV is a function of the following:

- Launching of the EDI program (PROGRAM);
- Level of EDI penetration (EDI), measured by the percentage of material dollars procurement under the EDI program.[4]

To isolate the effects of EDI, we must control for the effects of plant characteristics that impact the level of inventory turnover. For example, it may be difficult to attain high inventory turnover for large plants because of operational complexity. Similarly, if an assembly plant handles a huge variety of parts, components, and subassemblies, the level of inventory may increase. To capture these factors, two more variables are introduced:

- Production volume (VOLUME), measured in thousands of vehicles/year;
- Unique production parts (PARTS), to capture parts variety.

Note that while VOLUME is a continuous variable, PARTS is a binary variable. PARTS assumes a value of one if parts variety exceeds the average value for the Chrysler system. While the absolute number of parts could not be reconstructed for the

The EDI technology is not only expected to lead to more effective use of inventory, but it should also have an impact on the obsolete inventory at the end of the model year. As a result of the close tracking of the flow of material from the supplier relative to production requirements, the regulated flow of material should result in lower inventory write-offs when it comes time for the yearly model change.

As before, the model controls for the effect of relevant plant characteristics on the level of obsolete materials. In particular, the effects of year-to-year model changes are considered. For example, if the automobiles assembled at a plant undergo major design changes, the level of obsolete material is likely to increase despite EDI adoption. To capture these factors, two new variables are introduced:

- Minor model changes (MINOR) between 5 percent and 15 percent new parts.
- Major model changes (MAJOR) with more than 15 percent new parts.

Note that these two new variables are binary. MINOR and MAJOR are assigned a value of one when model changes amount to the respective levels for each variable. These measures of model changes reflect both the way management attempts to control this inevitable but costly phenomenon and its effect on performance. The dependent variable is OBS, the annual material dollars (in thousands) written-off at an assembly plant.

In summary, a linear model is used for obsolete inventory cost (OBS) with six drivers that include two EDI-related variables (PROGRAM and

Symbol	Definition
INV	Inventory Turnover = Annual Production ($)/Inventory ($)
OBS	Annual material dollars (000) written off at a plant
PFRGHT	Premium freight ('000) $
PROGRAM	1 after launching of the EDI program, 0 otherwise
EDIP	Percentage of material dollars under the EDI program
VOLUME	Annual Production volume (thousands of vehicles/year)
PARTS	1 if parts variety exceeds the average level, 0 otherwise
MINOR	1 if 5%–15% new parts introduced, 0 otherwise
MAJOR	1 if more than 15% new parts introduced, 0 otherwise

Table 2 Variable definitions.

EDIP*VOLUME) and four control variables (VOLUME, PARTS, MINOR, and MAJOR). The impacts of EDI variables are likely to reduce the level of obsolete inventory, while the other factors are likely to increase it. Note that VOLUME captures the direct size-impact of a plant, while EDIP*VOLUME tracks the savings from EDI depending on the plant size. We hypothesize that higher levels of EDI penetration bring in more benefits for a larger plant.

Transportation cost

The normal transportation cost is likely to be affected by the EDI program. However, other factors such as production volume, average supplier distance from the plant, mix and prices of different transportation modes (rail and road), and changes in the management control of transportation costs also determine the transportation cost at the plant level.

Unfortunately, transportation costs data were not available at the plant level for the period under study. Thus, a detailed statistical analysis could not be conducted at the plant level. Instead, the change in the unit transportation cost was decomposed to tease out the effects of the changes in price, mix, and management control of different modes at a system-wide level. The results of this analysis are presented in a subsequent section.

Premium freight

Premium freight occurs when the normal EDI mode of operations is disrupted. Common causes include loss of material, receiving discrepancy, records problems, damage in transit, supplier-side problem, etc. To overcome these disturbances, materials are procured using special delivery methods including air transport. Obviously, these conditions lead to inefficient transport and escalate inbound freight costs.

The launch of the EDI program allowed Chrysler to control premium freight largely because of speedy information sharing with suppliers. EDI suppliers receive schedule information and have the electronic capability to receive material release and send shipment data. We hypothesize that the effect of EDI is to reduce premium freight (PFRGHT) at the plant level. The increased size and complexity of the plant, however, tend to drive up the cost. Hence, a (linear) model is proposed for premium freight dollars (PFRGHT) with two EDI variables (PROGRAM and EDI*VOLUME) and two control variables (VOLUME and PARTS). The EDI variables should reduce premium freight, while the control variables should increase it.

Data Collection and Model Estimation

Cross-sectional and time-series data were combined to assess the impacts of EDI systems. Our data covers nine assembly plants over the period 1981 through 1990 model years. At least two advantages of this data set were evident. First, it allowed us to track the same assembly plants over a decade. Second, it described several assembly sites of similar products capturing sufficient diversity without introducing unacceptable heterogeneity into the data set.

A relatively long time period is necessary to determine IT impacts. Our data set therefore tracked the plants both before and after the start of the EDI program. We used annual data for this time period to capture costs associated with a model year and to avoid irregularity in the month-to-month production levels. Moreover, certain performance variables, by definition, were measured on an annual basis (e.g., obsolete inventory) at the end of model year. A summary of the key data is given in Table 3. All dollar figures are adjusted to 1991 dollars.

Our data set was developed primarily from computerized production, accounting, and engineering data. Care was taken to assure the quality of the data; for example, the annual production volume data was directly available from past production records. This data was expected to be of good quality because Chrysler assumes the legal title of each vehicle as it comes off the assembly line. Also, the inventory turnover was used rather than inventory cost data because the

Time Period	1981–1990
Number of Assembly Plants	9
Total Transportation Costs	Over $300 million
Number of Supply Lines	7000
Number of Daily Truck Deliveries (June 1991)	1483
Obsolete Inventory Cost (1990)	$5.9 million
Gross Productive Inventory (September 1990) (Assembly Plants)	$328.7 million

Table 3 Key data.

latter is more difficult to measure.[5] The value of obsolete materials and premium freight were available at the plant level and were based on actual material and transportation costs respectively. The level of EDI penetration (EDIP) was calculated as the dollars paid to EDI suppliers as a percentage of the total dollars paid to all suppliers for each plant for each year. Finally, the degree of model changes (MAJOR and MINOR) and parts variety were obtained from engineering records.[6]

Results

Our preliminary analysis of the raw data provided evidence of strong EDI impact in most cases. Figure 3 shows two scatter plots of inventory turnover for two plants. The top plot corresponds to a smaller plant with lower parts variety. It is clear from this figure that both plants had significant improvements in inventory turnover at the launch of the EDI program in 1984. However, the rate of gain in turns subsided with time just as the rate of EDI penetration slowed down for these plants. We should caution that these plots are subject to confounding effects of other variables such as production fluctuations over time, model changes, etc. A formal analysis can be performed using the estimation results (adjusted for autocorrelation and hetero-skedasticity) summarized in Table 4. See Appendix A for a discussion of model estimation and validation.

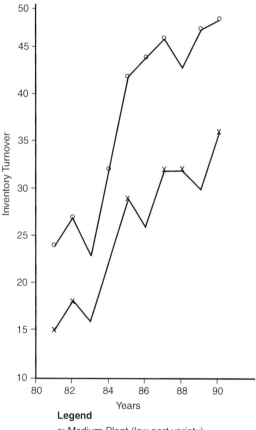

Figure 3 Scatter plot of inventory turnover.

Legend
o: Medium Plant (low part variety)
   Average Production Volume = 189,00
x: Medium Plant (high part variety)
   Average Production Volume = 211,000

Variable name	Inventory turnover (INV)	Obsolete Inventory (OBS)	Premium Freight (PFRGHT)
Constant	42.81***	253*	1849*
PROGRAM	6.80**	−168***	−2064**
EDIP	0.219***		
VOLUME	−.074**	5.633*	27.14***
EDIP*VOLUME		−0.038**	−.180***
PARTS	−9.94*	94.7	768*
MINOR		48.384**	
MAJOR		237***	
N	90	90	90
F-Stat	28.9	22.4	24.8
Adj-R^2	0.87	0.74	0.77

Table 4 Estimation results.[7]

Note
*, **, *** indicate α = .1, .05, and .01 respectively.

Inventory carrying cost

The results of this model indicate that both the launch of the EDI program and its penetration over time significantly helped improve inventory turnover. The benefits of the enforced discipline mandated before the EDI program was launched are clearly established. The significant improvement in inventory turns, almost on the initiation of the program, underscores the importance of getting all parties of the system involved and committed to the process. As expected, the marginal impact of EDI penetration was positive. The coefficients of VOLUME and PARTS were negative and statistically significant. These results concur with our expectations and capture the adverse effects of large plant size and high parts variety.

Based on our estimation results, Figure 4 depicts the impact of EDI on the inventory carrying cost for three sizes of assembly plants. The savings from the reduction in inventory were calculated based on an annual carrying cost of 20 percent.[8] The savings depended on the inventory turnover at the initiation of the EDI program. As the figure indicates, the start of the EDI program in itself resulted in substantial savings, depending on

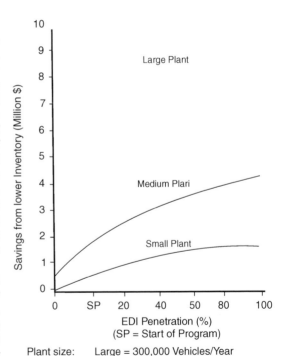

Plant size: Large = 300,000 Vehicles/Year
Medium = 200,000 Vehicles/Year
Small = 100,000 Vehicles/Year

Figure 4 Savings from lower inventory.

the size of the plant. It is clear that the savings are greater for plants with lower initial turns (larger production volumes). In other words, plants with higher turns have less to gain compared to plants that have lower inventory turns. The rate of savings tends to diminish with increased EDI penetration because inventory costs are inversely proportional to inventory turnover.

Obsolete inventory cost

The estimation results of this model largely support our expectations. The effect of EDI was to lower the obsolescence cost as is evidenced by the negative sign of the coefficients of PROGRAM and EDI*VOLUME. The effect of model change was, as expected, to escalate write offs. In fact, major model changes (more than 15 percent new parts) were found to have a much higher impact than minor model changes (5 percent to 15 percent new parts). The estimated coefficient of VOLUME was positive, confirming that larger plants are more susceptible to obsolescence problems. However, the level of parts variety captured by PARTS did not seem to have a statistically significant effect. That is, the degree of model changes rather than parts variety seems to drive obsolescence costs.

The EDI impacts on material obsolescence cost are illustrated in Figure 5. It is clear from this figure that the initiation of the EDI program leads to a reduction of material write-off by almost 0.17 million dollars. As the level of EDI penetration increases, the extent of savings increases. Accordingly, the additional gains are larger for bigger plants.

Premium freight

All parameter estimates of this model were statistically significant. Moreover, the sign of each parameter estimate was in the expected direction. Thus, for example, the effects of EDI on this cost category were negative as evidenced by the parameter estimates of PROGRAM and EDI*VOLUME variables. The size of the plant (VOLUME) and the variety of parts (PARTS), on the other hand, tended to increase premium freight.

The joint effect of EDI penetration and the variety of parts is illustrated for a typical plant production volume of 200,000 automobiles in Figure 6. Plants with low variety of parts incur less premium freight for any level of EDI penetration compared to high variety plants.

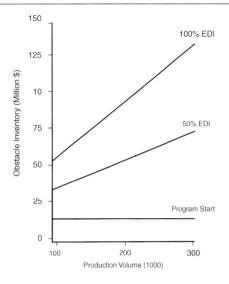

Figure 5 Obsolete inventory cost savings.

Analysis of Transportation Costs

A direct comparison of unit inbound transportation costs between 1981 and 1990 is misleading because of changes in several factors during this time period. First, the changing mix of road and rail transportation must be considered. Chrysler's use of road transportation gradually increased over these years. Thus, there is a "mix effect" in the change of unit transportation cost because of the higher proportion of road transport.

The second component is the "price effect." Road and rail transport prices have changed considerably over the years following deregulation of the trucking industry. Initially prices declined and then rose after the consolidation following the initial shake out. This price effect is determined based on the standard transportation price indices published by Transportation in America (1991).

The third component is the "systems effect" arising from better and more effective logistic capability from deploying EDI systems. This capability can be attributed to the higher quality and quantity of information available to the trading partners and to the smart use of the information available to manage material movements. EDI

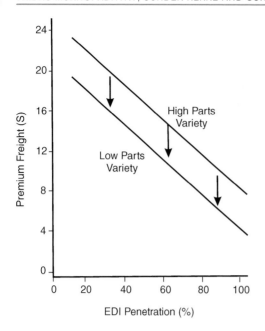

Figure 6 Impact of EDI penetration on premium freight per unit (plant production volume = 200,00).

	Combined effect of mix and price changes	System effect	Total
Rail	−20.98	2.23	−18.75
Truck	24.42	−12.86	11.56
Total	3.44	−10.63	−7.19

Table 5 Decomposition of the change in unit transportation cost.

plays a critical role in providing reliable, rapid, and frequent communication exchanges.

Our primary interest was to assess the last component—namely, the systems effect. The mix and price effects were filtered out from the change in unit transportation cost to determine the systems effect. A mathematical representation for the analysis is provided in Appendix B, and a summary of the results is shown in Table 5. The results show the decomposition of the unit cost change between 1981 and 1990 into the combined effect of mix and price changes and the systems effect.

The net change in unit transportation costs is a decrease of $7.19 per vehicle. Note in Table 5 that the Total column indicates an increase in unit road transport costs of $11.56 and a decrease in rail transport costs of $18.75. A breakup of the figure for road transport provides further insights. Out of the $11.56 increase, $24.42 can be attributed to increased shipments by road and the price differential. The noteworthy result, however, is the decrease in road transportation costs by $12.86 due to systems effect. This effect combines impacts of management practice and increased volume of EDI shipments.

This aspect of the analysis—decomposition of the cost change—needs to be emphasized. While raw statistics give the misleading conclusion that road transportation costs increased by $11.56 after EDI implementation, a closer scrutiny after decomposing the change reveals that systems effect is strong despite the increased logistics complexity. EDI and the use of information to manage the new logistics systems result in systems efficiency gains of $12.86. The comparable systems effect for rail transportation is unfavorable since EDI was not enforced for this mode of transportation. Overall, the last row indicates a net decrease in unit transportation costs that would not be possible without the improved system performance.

Savings in Information Handling Costs

An obvious source of savings that result from using EDI is the reduction in information handling costs. Typically, information-handling cost savings are realized from reduced personnel cost and lower transmission charges. The use of EDI affects a series of clerical activities normally performed in a manual system. These activities include keypunching of data, mail preparation, document filing and storage, etc.

The magnitude of savings realized from electronic document handling depends on the manual procedures followed by specific organizations. For example, RCA estimates that it saves $50 for each purchase order processed electronically, while Douglas Aircraft reports a reduction of $5 per document (Emmelhainz, 1990).

One estimate of these savings for Chrysler is $5 per document (Falvey and Wujcikowski, 1989). In 1988, Chrysler manufactured 2.2 million vehicles and had 17 million EDI transactions with its suppliers (HBS, 1991). As a result, the savings from electronic document preparation and transmission for Chrysler amounted to about $38.63 per vehicle.

Total EDI Impact

What is the cost impact of improved information exchanges due to EDI systems at Chrysler? Given the production volume and level of EDI penetration, the impact per unit vehicle cost from reduction in inventory carrying cost, obsolete inventory cost, and premium freight can be quantified. The cost reduction was calculated for the average EDI penetration in 1990, which was about 90 percent. Table 6 provides an estimate for a typical plant with a production volume of 200,000 vehicles/year and low parts variety. Also included is the reduction in transportation cost from the systems effect (see Table 5). Thus, the savings from improved information exchanges for 90 percent EDI penetration is over $60 per vehicle. Savings from inventory holding cost and premium freight are the major components of the total savings for such a situation. Including the additional savings from electronic document preparation and transmission, the total benefits of EDI per vehicle amount to over $100.

Discussion

This section discusses the strengths and weaknesses of this study in order to place the results in

	Dollar savings per vehicle
Inventory Holding Cost	20.85
Obsolete Inventory Cost	4.26
Premium Freight	26.52
Transportation Cost	10.63
Information Handling Cost	38.63
Total	100.89

Table 6 Total savings as a result of using EDI (production volume of plant = 200,000).

proper perspective. To structure this discussion, five benchmarks derived from the reviews of the business-value research discussed earlier is used to evaluate the results. Also discussed are the implications of the results given cost/benefit considerations and strategic impacts.

Benchmarks

Theory-Related Concerns

1 **Lack of theoretical analysis:** Much of the earlier work in this area is exploratory and does not have a theoretical basis. We drew on the literature on hierarchical interorganizational systems, which suggests substantive business value can be gained from EDI at our research site.
2 **Need for process orientation:** An understanding of the process by which IT is utilized is a prerequisite to assessing business value. Detailed knowledge of EDI implementation at Chrysler was obtained and the consequent changes in information flows from using EDI were identified.

Method-Related Concerns

1 **Possible confounding factors:** Most empirical studies in this area lack control, and this study is no exception. A common consideration involves management practice. In our case, a majority of the JIT principles were introduced at Chrysler prior to EDI implementation for the entire supply chain—from purchasing, component manufacturing, and assembly operations, to the delivery of finished goods. The effect of JIT was to transform the entire value chain, and its effect was not limited to the impacts analyzed in this paper.

One can argue that EDI was a vehicle for implementing JIT principles at Chrysler. However, certain JIT techniques did not depend on EDI, yet they affected the four cost categories examined in this paper. In particular, the changes in logistics and inventory management effected during the study period should have affected transportation and inventory carrying costs. However, it is not possible to tease out ex post the impact of JIT practices that could have been implemented without the EDI program. Thus, EDI should be considered a necessary but not sufficient condition for the business value quantified in this paper.

2 **Need for sophisticated modeling and analysis:** At the conceptual level, several factors have been incorporated into our models: parts variety, production volume, degree of model changes, and others. In estimating the models, possible violations of statistical assumptions were carefully analyzed, and where possible, multiple methods were used to account for such violations.

Measurement-Related Concerns

1 **Measurement of IT and outputs:** Based on our knowledge of the EDI implementation process, a two-step measurement for IT was used that captured both the launching and penetration of the EDI program. The analysis of the changes in information flows caused by the EDI program also helped pinpoint the cost categories (impacts) affected. In summary, the relevant costs were those associated with the management of the interface between Chrysler assembly centers and their suppliers.
2 **Aggregation problem:** One danger of aggregating all applications in an organization is that the impacts of effective systems are neutralized by ineffective systems (Mukhopadhyay, et al., 1995). However, the aggregation problem in this study was minimal because one specific technology was examined rather than all Chrysler applications.

Data-Related Concerns

1 **Problem of cross sectional data:** The IT impact may take some time to realize and is difficult to capture in a cross-sectional data set. Therefore, our data includes a 10-year time series for nine assembly plants. This time period adequately covers both before and after EDI implementation.
2 **Data Quality:** Obtaining quality data is a major hurdle for researchers in this area. The data set was assembled from computerized records maintained by Chrysler. The accuracy of this data was reasonably assured because only key accounting, engineering, and operational characteristics of the assembly centers were used. And the accounting procedures used were uniform across the plants.

Other Concerns

1 **Mismanagement:** One reason for the lack of IT impact is the difficulty of managing advanced technologies. While we cannot certify that EDI management at Chrysler was optimal, Chrysler has had a long experience with EDI and is acknowledged as a leader in this field (Computerworld, 1990). Our assessment is that EDI was well managed at Chrysler.
2 **Redistribution:** The IT impact at the industry level may be negligible if the benefits lead to a redistribution of economic surplus among the firms. This issue is not directly relevant to our study because IT impact was examined at the firm level. However, it does raise the question about the level of benefits obtained by Chrysler suppliers. It has been proposed that buyers and suppliers may not obtain equal benefits from EDI systems (Riggins and Mukhopadhyay, 1994).

In summary, this study overcomes many of the common deficiencies in this area of work. However, causality is not implied in our results; EDI was a major contributor. The other possible contributor was Chrysler management, in particular, through its commitment to JIT philosophy. We believe no organization with poor management can realize substantial gains from IT. However, it is difficult to quantify the effect of management quality, more so in this ex post study.

This study experienced the same limitations common to any ex post study covering a long time horizon. In particular, the available data was limited. For example, the impact of vehicle mix was not considered in this study. In addition, the effect of the acquisition of AMC by Chrysler was not addressed. AMC plants were excluded from our analysis since only recent data were available for those plants. The next section discusses the

Implications

The focus of this study was to measure the dollar impact of improved information exchanges between Chrysler and its assembly center suppliers as a result of using EDI. However, major investments in IT also demand an analysis of costs and benefits, plus strategic considerations. Each of these issues is discussed below in the context of EDI at Chrysler.

A cost/benefit analysis from Chrysler's viewpoint would require data beyond what is presented in this work. Although the stream of costs over time can possibly be identified, the incidence of benefits over time is harder to quantify (Emmelhainz, 1990). On the cost side, the major components would include additional hardware required to process EDI transactions, software development and maintenance, telecommunication, training operating personnel, and other system-related expenses. In addition, a variable cost of adding each supplier to the EDI network involving joint testing of transaction sets, negotiating a legal contract, etc. should also be considered.

As discussed earlier, the benefits of the EDI system can be divided into two categories. First, substantial cost savings result because of improved information exchanges between Chrysler and its suppliers. Our estimate of these benefits is about $62.26 per vehicle. Second, preparing and processing documents electronically also lead to savings in information handling costs. Thus, the total tangible benefits of EDI resulted in over $100 per vehicle for Chrysler Corporation (Table 6).

Note that improved information exchanges can also lead to intangible benefits over and above the benefits we have analyzed. For example, EDI and JIT systems may lead to a reduction of production cycle time and improved product quality. The gains made by Chrysler after the study period, including its successful introduction of Neon and other car lines, bear testimony to the capability of the company to launch new models. Supplier coordination, including smooth flows of information and materials, contributed toward this goal. In this sense, Chrysler and its supplier network demonstrated organizational flexibility and efficiency where information linkages played a vital role in the cost reduction and quality improvement efforts.

An important reason for the growth of EDI in this case was Chrysler management—it made EDI a condition for doing business with its assembly center suppliers. In general, a fair amount of strategic manipulations are apparent in EDI adoption by trading partners. In the past, initiators of electronic linkages had been suppliers seeking to add value to their product and to gain a competitive advantage over their rivals. Some authors, however, predict that in the long run, suppliers will be expected to do business using EDI and will have to relinquish the control over to their buyers (Benjamin, et al., 1990).

More recently, large buyers in many industries have begun initiating EDI with their suppliers in hopes of achieving greater cost savings than when the system is supplier driven (Sokol, 1989). The literature on the adoption of buyer-driven EDI by suppliers indicates that buyers may have to induce some suppliers, who may be reluctant to incur additional expenses to change over to electronic communication.[9] The adoption of the national X.12 standards may reduce the set-up cost for EDI and thus alleviate the problems facing the continuing expansion of this technology.

Conclusion

Jim Stechschulte of Chrysler Corporation characterized the frequent dilemma facing management today in the following remark (Stechschulte and Kekre, 1993). "We often question if a new IT application will be cost effective. Making that investment decision becomes even more difficult when we cannot assess the dollar impact of similar applications already implemented" (p.7). Our study shows that rigorous evaluations can be performed to help in such management decision making.

Our analysis of the last decade's performance data at Chrysler's assembly centers confirms that modern information technology such as EDI has enabled Chrysler to significantly reduce operating costs associated with carrying inventories, obsolescence, and transportation. Effective use of information to coordinate material movements by Chrysler and its suppliers has resulted in significant savings. Our estimates indicate that they are over $100 per vehicle, for a typical assembly plant, translating to an annual savings of $220 million.

The study controlled for changes in factors such as parts variety, engineering changes, and volume. The analysis shows that significant savings have been realized from reduced inventories, lower write offs each year following model change, and reduced premium freight. Furthermore, from a cost perspective (despite increased usage of trucking, a more costly, though flexible, mode of transportation than rail), Chrysler's EDI program has resulted in a net reduction of $7.19 per unit vehicle inbound transportation costs.

The study underscores the systems aspect of EDI philosophy. Bringing Chrysler and suppliers together and enforcing discipline alone brought about significant gains at the inception of the EDI program. The models reveal that in almost all the cost categories analyzed, the "discipline" imposed by EDI has its merits. With gradual penetration of the EDI program, Chrysler has been able to reduce costs through frequent, reliable, and error-free shipments. As more suppliers came on board, however, the increased penetration of EDI resulted in smaller incremental savings.

EDI systems have surely brought in the returns. The EDI technology has been the enabler—replacing inventory with information. However, determining the business value of IT in general requires much further work. Our analysis illustrates that researchers need to exercise much care in teasing out the contributions of IT to business value. Our experience also points to the high degree of cooperation needed between the industry and academia to understand the value of IT on a retrospective basis.

There are at least two lessons to be learned from the Chrysler EDI experience. First, EDI should be used as the dominant method of communication between buyers and suppliers. To achieve this goal, the buyer may have to implement multiple transaction sets. In addition, the buyer must ensure that most suppliers adopt EDI. Second, EDI provides an opportunity for process simplification and redesign. At Chrysler, the simultaneous implementation of JIT and EDI was a critical reason for the impressive benefits we found. Chrysler also benefited from integrating its EDI applications with the internal systems. Finally, management should not overlook the possibility of restructuring the supplier base in conjunction with EDI implementation.

Our analysis is based on the assembly plants of Chrysler Corporation. As the industry trend leans toward standardized transactions, our findings may apply to other automobile manufacturers as well. However, replicating the study with other firms in different industries is necessary to conclusively test the results of this paper. Given the importance of EDI for U.S. manufacturing, this line of research is certainly exigent.

NOTES

1 See, for example, Bakos and Treacy (1986).
2 Bakos and Kemerer, 1992; Brynjolfsson, 1993; Crowston and Treacy, 1986; Kauffman and Weill, 1989.
3 Bakos and Treacy (1986) show how IOSs allow firms to achieve vertical information integration. Clemons and Row (1989) explain the virtual integration between trading partners using IOS. Mukhopadhyay and Cooper (1992; 1993) examine how information accuracy and timeliness generate business value.
4 EDI penetration can be measured as the percentage of material dollars procurement under the EDI program or as the percentage of inventory items under the EDI program. The first measure is used because we are interested in the dollar impact of the EDI program.
5 Inventory Turnover = Annual Production ($)/Inventory ($) and Inventory Cost = Inventory ($) * Carrying Cost Rate. It is more difficult to assess the carrying cost rate compared to the dollar value of annual production.
6 We could not determine the absolute number of parts carried by each plant for the past decade. Chrysler management provided the summary binary measure PARTS based on available data. Note that MAJOR and MINOR model changes were measures in use at Chrysler prior to our study.
7 The correction for autocorrelation tends to inflate the value of adjusted R^2 (Maddala, 1992).
8 The estimation of inventory carrying cost is a difficult problem (Gardner and Dannenbring, 1979). It includes charges for storage and handling, property taxes, insurance, spoilage, pilferage, and capital requirements. In our case, the 20 percent rate was approved by Chrysler management. It represents the average rate for the plants included in our study. The rate depends on the nature of the plant layout, product complexity, and mix, as well as interest rate. Rates of individual plants are deemed confidential.
9 See, for example, Clemons and Row (1993); Meier and Chismar (1991); Riggins, et al., (1994) and Riggins, et al., (1995).

ACKNOWLEDGMENTS

This research was supported in part by a grant from the Automotive Industry Action Group. We acknowledge many helpful discussions with J. Stechschulte, E. Krajca, E. T. Sprock, and R. Mitzel from Chrysler and J. Phelan from Ford in the course of this study.

REFERENCES

Bakos, J. Y. and Treacy, M. E. "Information Technology and Corporate Strategy: A Research Perspective," *MIS Quarterly* (10:2), June 1986, pp. 107-19.

Bakos, J. Y. and Kemerer, C. F. "Recent Applications of Economic Theory in Information Technology Research," *Decision Support Systems* (8:5), September 1992, pp. 365-86.

Barua, A., Kriebel, C. H. and Mukhopadhyay, T. "Information Technologies and Business Value: An Analytic and Empirical Investigation," *Information Systems Research* (6:1), March 1995, pp. 1-24.

Belsley, D. A., Kuh, E. and Welsch, R. E. *Regression Diagnostics*, Wiley, New York, 1980.

Benjamin, R., DeLong, D. and Scott Morton, M. "Electronic Data Interchange: How Much Competitive Advantage?" *Long Range Planning* (23:1), February 1990, pp. 29-40.

Breusch, T. S. and Pagan, A. R. "A Simple Test for Heteroskedasticity and Random Coefficient Variation," *Econometrica* (47:9), September 1979, pp. 1287-94.

Brynjolfsson, E. "The Productivity Paradox of Information Technology," *Communications of the ACM* (36:12), December 1993, pp. 66-77.

Clemons, E. K. and Row, M. "Information Technology and Economic Reorganization," *Proceedings of the Tenth International Conference on Information Systems*, Boston, MA, 1989, pp. 341-51.

Clemons, E. K. and Row, M. "Limits to Interfirm Coordination through Information Technology: Results of a Field Study in Consumer Goods Distribution, *Journal of Management Information Systems* (10:1), Summer 1993, pp. 73-95.

Computerworld. "Among Big 3, Chrysler First in EDI Links," *Computerworld* (24:6), February 5, 1990, pp. 1.

Crowston, K. and Treacy, M. E. "Assessing the Impact of Information Technology on Enterprise Level Performance," *Proceedings of the Seventh International Conference on Information Systems*, San Diego, 1986, pp. 299-310.

Doran, H. E. and Griffiths, W. E. "On the Relative Efficiency of Estimators Which Include the Initial Observations in the Estimation of Seemingly Unrelated Regression with First-order Autoregressive Disturbances," *Journal of Econometrics* (23:2), October 1983, pp. 165-91.

Emmelhainz, M. *Electronic Data Interchange: A Total Management Guide*, Von Nostrand, New York, 1990.

Falvey, R. and Wujcikowski, R. V. "Chrysler Announces its Aggressive EDI Implementation Plans," *Actionline* (8:1), January 1989, pp. 16-19.

Gardner, E. S. and Dannenbring, D. G. "Using Optimal Policy Surfaces to Analyze Aggregate Inventory Tradeoffs," *Management Science* (25:8), August 1979, pp. 709-20.

Goldfeld, S. M. and Quandt, R. E. *Nonlinear Methods in Econometrics*, North-Holland, Amsterdam, 1972.

HBS. "Chrysler Corporation: JIT and EDI (A)," Harvard Business School Case Study 9-191-146, Cambridge, MA, 1991.

Hwang, K. T., Pegels, C. C., Rao, H. R. and Sethi, V. "Evaluating the Implementation Success and Competitive Impact of Electronic Data Interchange Systems," report, School of Management, SUNY, Buffalo, 1991.

Judge, G. G., Hill, R. C., Griffiths, W. E., Lutkepohl, H. and Lee, T. *The Theory and Practice of Econometrics*, John Wiley, New York, 1985.

Kauffman, R. J. and Weill, P. "An Evaluative Framework for Research on the Performance Effects of Information Technology Investments," *Proceedings of the Tenth International Conference on Information Systems*, Boston, 1989, pp. 377-88.

Kekre, S. and Mukhopadhyay, T. "Impact of Electronic Data Interchange Technology on Quality Improvement and Inventory Reduction Programs: A Field Study," *International Journal of Production Economics* (28:3), December 1992, pp. 265-82.

Kmenta, J. *Elements of Econometrics*, MacMillan, New York, 1986.

Loveman, G. "An Assessment of the Productivity Impact of Information Technologies," in n *Information Technology and the Corporation of the 1990s: Research Studies*, T. J. Allen and M. S. Scott Morton (eds), MIT Press, Cambridge, MA, 1994.

Maddala, G. S. *Introduction to Econometrics*, Macmillan, New York, 1992.

Meier, J. and Chismar, W. G. "A Formal Model of the Introduction of a Vertical EDI System," *Proceedings of the Twenty Fourth Hawaii International Conference on Systems Sciences* (IV), January 1991, Koloa, HI, pp. 508–23.

Mukhopadhyay, T. "Assessing the Economic Impacts of Electronic Data Interchange Technology," in *Strategic and Economic Impacts of Information Technology Investment*, R. Banker, R. J. Kauffman, and M. A. Mahmood, Idea Publishing, Middletown, PA, (1993) pp. 241–64.

Mukhopadhyay, T. and Cooper, R. B. "A Microeconomic Production Assessment of the Business Value of Management Information Systems: The Case of Inventory Control," *Journal of Management Information Systems* (10:1), Summer 1993, pp. 33–55.

Mukhopadhyay, T. and Cooper, R. B. "Impact of Management Information Systems on Decisions," *Omega* (20:1), January 1992, pp. 37–49.

Mukhopadhyay, T., Lerch, F. J. and Gadh, V. Assessing the Impact of Information Technology on Labor Productivity," *Decision Support Systems* (in press), 1995.

Mukhopadhyay, T., Kekre, S. and Pokorney, T. "Strategic and Operational Benefits of Electronic Data Interchange," GSIA Working Paper, Carnegie Mellon University, Pittsburgh, PA, 1995.

Riggins, F. J. and Mukhopadhyay, T. "Interdependent Benefits from Interorganizational Systems: Opportunities for Business Partner Reengineering," *Journal of MIS* (11:2), Fall 1994, pp. 37–57.

Riggins, F. J., Kriebel, C. H. and Mukhopadhyay, T. "The Growth of Interorganizational Systems in the Presence of Network Externalities," *Management Science* (40:8), August 1994, pp. 984–98.

Riggins, F. J., Mukhopadhyay, T. and Kriebel, C. H. "Optimal Policies for Subsidizing Supplier Interorganizational System Adoption," *Journal of Organizational Computing* (in press), 1995.

Roach, S. "Services Under Siege—the Restructuring Imperative," *Harvard Business Review* (69:5), September-October 1991, pp. 82–91.

Sherwood, B. and Marocco, L. "Chrysler: Reaching for new EDI Milestones," *Actionline* (8:9), September 1989, pp. 16–19.

Sokol, P. *EDI: The Competitive Advantage*, McGraw-Hill, New York, 1989.

Spencer, D. E. and Berk, K. N. "A Limited Information Specification Test," *Econometrica* (49:7), July 1981, pp. 1079–85.

Srinivasan, K., Kekre, S. and Mukhopadhyay, T. "Impact of Electronic Data Interchange Technology on JIT Shipments," *Management Science* (40:10), October 1994, 1291–1304.

Stechschulte, J. M. and Kekre, S. "EDI Systems: Analysis of the Last Decade at Chrysler," Symposium on Global Distribution Management, Columbia University, New York, 1993.

Stern, L. and Kaufmann, P. "Electronic Interchange in Selected Consumer Goods Industries: An Interorganizational Perspective," in *Marketing in an Electronic Age*, R. Buzzel (ed.), Harvard Business School Press, Cambridge, MA, 1985, pp. 52–73.

Verity, J. W. "Truck Lanes for the Info Highway," *Business Week*, April 18, 1994, pp. 112–114.

Weill, P. "The Relationship between Investment in Information Technology and Firm Performance: A Study of the Valve Manufacturing Sector," *Information Systems Research* (3:4), December 1992, pp. 307–33.

White, H. "A Heteroskedasticity-consistent Covariance Matrix Estimator and a Direct Test for Heteroskedasticity," *Econometrica* (48:4), April 1980, pp. 817–38.

Zellner, A. "An Efficient Method of Estimating Seemingly Unrelated Regressions and Tests of Aggregation Bias," *Journal of the American Statistical Association* (57:6), June 1962, pp. 348–68.

ABOUT THE AUTHORS

Tridas Mukhopadhyay is an associate professor of industrial administration at Carnegie Mellon University. He received his Ph. D. in computer and information systems from the University of Michigan in 1987. His research interests include business value of information technology, economic impacts of electronic data interchange, software development productivity, and cost analysis. His primary area of interest is in the economics of information technology. His research appears in *Information Systems Research, Journal of Manufacturing and*

Operations Management, MIS Quarterly, Omega, IEEE Transactions on Software Engineering, Journal of Operations Management, Accounting Review, Management Science, Journal of Management Information Systems, Decision Support Systems, Journal of Organizational Computing, and other publications. He is an associate editor for Information Systems Research.

Sunder Kekre is an associate professor of industrial administration at Carnegie Mellon University. He received his Ph. D. in production and operation management from the University of Rochester in 1984. His research interests include problems at the interface of manufacturing with information systems, accounting, and marketing. His research appears in IIE Transactions, Interfaces, Journal of Manufacturing Systems, Operations Research, Journal of Accounting and Economics, Journal of Manufacturing and Operations Management, IEEE Transactions on Software Engineering, Journal of Operations Management, Accounting Review, Management Science, European Journal of Operations Research, International Journal of Forecasting, International Journal of Production Economics, and other publications. He is an associate editor for Management Science.

Suresh Kalathur is currently employed with Cimnet Systems, Inc. He received his Ph.D. in industrial engineering from the University of Pittsburgh in 1994. His research interest is at the interface of information systems and manufacturing management.

Appendix A

Since we have a panel data set, it is important to consider violations of the ordinary least squares assumptions that can result in biased and inconsistent estimates for our models. Since the data contain a time series, one can expect some serial correlation. However, annual data is not subject to severe serial correlation because random shocks may last for a few months, but possibly not much longer. Model changes also decouple one year to the next. Thus, first-order autocorrelation is examined for each model, using the well-known Durbin-Watson statistic to test the presence of autocorrelation. For inventory turns and premium freight, the value of the statistic was smaller than the lower limit, indicating autocorrelation. For obsolete inventory, the statistic is not available; the number of independent variables in this equation is higher (six) than other equations. However, autocorrelation was corrected for all three models. We used a non-iterative procedure in the correction for serial correlation (Kmenta, 1986). Coefficients computed for each assembly plant were used to transform the data.

As discussed earlier, the models were specified to account for differences in plants with respect to production volume, variety of parts used, etc. However, all possible differences could not be accounted for. Breusch-Pagan (1979) and Goldfeld-Quandt (1972) tests were used to examine heteroskedasticity. While the first test rejected the null hypothesis of homoskedastic error for INV and PFRGHT models, the second test rejected the null hypothesis for OBS and PFRGHT models (alpha = 0.01). However, heteroskedastic consistent estimates were obtained for each equation using a common procedure (White, 1980). Belsely, et al. (1980) condition indices and eigen vector values indicated no muticollinearity problems.

Contemporaneous correlation between the residuals across the plants may occur because of the commonalties in management policy and procedures. For example, major changes at Chrysler in management labor negotiations, and supplier and carrier problems may evoke similar responses from all plants. Similarly, external events such as changes in federal regulations, new Japanese transplants, etc., may also affect plants in an analogous manner. Thus, if contemporaneous correlation is present, the estimates may be unbiased and consistent, but inefficient. However, the efficiency gain from seemingly unrelated regression (SUR) estimates is likely to be small in this case for three reasons (Doran and Griffiths, 1983; Judge, et al., 1985): (1) the three equations contain similar explanatory variables; (2) the estimation is based on trended data, and (3) the correlation coefficients among the explanatory variables, by year, are often greater than 0.50, and sometimes even higher. Moreover, in this case, the correlations between residuals were small, indicating little gain in efficiency could be achieved by estimating the equations as a set of seemingly unrelated regressions (Zellner, 1962). Nonetheless, SUR estimates were obtained for these equations. The results were similar to the estimates obtained after adjusting for autocorrelation and heteroskedasticity. A series of tests was also performed to determine if any of the dependent variable specifications should contain another dependent variable (Spencer and Berk, 1981). None of these tests violated our original specifications.

Each of the models, was examined to see whether the inclusion of the two EDI variables was statistically justified. The calculated value of the F statistic always exceeded the critical value ($F_{2,85}$ for INV and PFRGHT, $F_{2,83}$ for OBS) at alpha = .01, supporting our model specifications.

Appendix B

The change in unit transportation cost was decomposed over the period of study to filter out the mix and price effects and to capture the systems efficiency effect. Two subscripts were used—the first subscript referred to the time period (1 for pre-EDI and 2 for post-EDI), while the second subscript denoted the transportation mode (1 for rail and 2 for truck). Let

C_i = total per unit transportation cost for period i.
C_{ij} = mode j unit transportation cost for period i.
   Thus $C_i = C_{i1} + C_{i2}$.
m_{ij} = proportion of material dollars transported by mode j for period i.
p_{ij} = price index of transportation mode j for period i.
η_j = efficiency in managing mode j. Note that a decrease in η_j implied a reduction of costs and an improvement in efficiency.

Consider the ratio of unit transportation cost for mode j = C_{2j}/C_{1j}. This ratio can be greater than one or less than one. Furthermore, it is driven by three factors: (1) increased or decreased role of mode j ($m_{2j}/m_{1j} > 1$ or < 1); (2) increase or decrease in prices of mode j ($p_{2j}/p_{1j} >$ or < 1), and finally, (3) due to change in systems efficiency for mode j ($\varsigma_j > 1$ or < 1). Thus

$$\frac{C_{2j}}{C_{1j}} = \frac{m_{2j}}{m_{1j}} \frac{p_{2j}}{p_{1j}} \eta_j$$

We assumed that $\eta_j = 1$ such that the change in unit transportation cost was attributable to the combined effect of changes in mix and price. Thus

$$C_{2j} = C_{1j} \frac{m_{2j}}{m_{1j}} \frac{p_{2j}}{p_{1j}}$$ or

$$C_{2j} - C_{1j} = C_{1j} \frac{m_{2j}}{m_{1j}} \frac{p_{2j}}{p_{1j}} - C_{1j} = C_{1j} \left[\frac{m_{2j}}{m_{1j}} \frac{p_{2j}}{p_{1j}} - 1 \right]$$

This expression captured the combined effect of mix and price changes. Next, the systems efficiency change was also accounted for:

$$C_{2j} - C_{1j} = C_{1j} \frac{m_{2j}}{m_{1j}} \frac{p_{2j}}{p_{1j}} \eta_j - C_{1j}$$

$$= C_{1j} \left[\frac{m_{2j}}{m_{1j}} \frac{p_{2j}}{p_{1j}} - 1 \right] + C_{1j} \frac{m_{2j}}{m_{1j}} \frac{p_{2j}}{p_{1j}} [\eta_j - 1]$$

where the first term indicated the combined effect of price and mix changes as above, and the second term captured the residual system effect. Summing over all modes,

$$C_2 - C_1 = \sum_{j=1}^{2} \left(C_{1j} \left[\frac{m_{2j}}{m_{1j}} \frac{p_{2j}}{p_{1j}} - 1 \right] \right)$$

$$+ \sum_{j=1}^{2} \left(C_{1j} \frac{m_{2j}}{m_{1j}} \frac{p_{2j}}{p_{1j}} [\eta_j - 1] \right)$$

The Performance Impacts of Quick Response and Strategic Alignment in Specialty Retailing

Jonathan W. Palmer and M. Lynne Markus

Decision and Information Technologies, Robert H. Smith School of Business, University of Maryland, 4348 Van Munching Hall, College Park, Maryland 20742, jpalmer@rhsmith.umd.edu
Peter F. Drucker Graduate School of Management, Claremont Graduate University, 1021 North Dartmouth Street, Claremont, California 91711, m.lynne.markus@cgu.edu

Abstract

The Quick Response (QR) program is a hierarchical suite of information technologies (IT) and applications designed to improve the performance of retailers. Consultants advise retailers to adopt the program wholesale, implying that more and higher levels of technology are better than less technology and lower levels. Academicians, on the other hand, argue that good technology is "appropriate" technology. That is, firms should adopt only those technologies that suit the specific strategic directions pursued by the firm. Who is right? Which approach to investing in IT yields better performance results? Surprisingly, this cross-sectional survey of 80 specialty retailers found more support for the practitioners' claims than for the academicians'. Adoption of the QR program at a minimal level was associated with higher performance, although there was no performance impact due to higher levels of QR use. Firms did appear to match their IT usage to their business strategies, but there was no linkage between strategic alignment and firm performance, and there was surprisingly little variation in business or IT strategy. In short, the findings of our study suggest that both practitioners and academicians need to refine their theories and advice about what makes IT investments pay off.
(*Performance Impacts of IT; Quick Response; Retailing; Strategic Alignment; Strategic Use of IT*)

Introduction

Today, the word retailing evokes selling on the web. With all the excitement about shopping on the Internet, it is easy to lose sight of the fact that the majority of electronic commerce is, and likely will remain, business-to-business transactions (Emerging Digital Economy 1998). The sheer volume of electronically mediated business buying and selling makes it important to assess the impacts of information technology on the performance of retailing organizations.

Like other service industries, retailing is a relative newcomer to the use of IT (National Research Council 1994). On the other hand, the labor-intensive nature of the industry creates favorable opportunities for IT use. The retailing industry is actively seeking new ways to use IT effectively in support of distribution, inventory management, planning, and sales functions. A package of information technologies called "Quick Response" has been claimed by consultants and practitioners to have enormous potential for improving organizational performance in retailing (Retail Information Systems News 1993, 1995).

The obvious question is: has the adoption of Quick Response improved the business performance of retailing firms? But past research has shown how difficult it is to pin down the links between information technology and organizational performance. While many people believe that IT has a strong effect on business success, empirical support remains mixed (Barua and Lee 1997; Boynton et al. 1993; Brynjolfsson 1993; Clemons and Row 1989, 1991; Crowston and Treacy 1986; Hitt and Brynjolfsson 1994; Harris and Katz 1989, 1991; Markus and Soh 1993; Quinn and Baily 1994; Sethi et al. 1993; Weill 1992).

Many factors contribute to the difficulty of finding strong associations between IT use and firm performance. One is the practical definition of "IT use." What precisely constitutes use? Can an organization truly be said to be "using" an integrated enterprise resource planning package if it only employs the financial modules, for example? How much of the enterprise package does the organization need to adopt before it can be said to benefit from integration? And does the organization get more benefit from adopting more modules?

Definitional questions such as these are particularly evident with Quick Response technology in the retailing industry. As described by consultants and practitioners, the Quick Response "program" consists of four levels of successively more sophisticated technologies and applications (*Retail Information Systems News* 1993, 1995). Level 1 includes point-of-sale technology and price lookup. Level 2 includes automatic inventory replenishment and sales and inventory forecasting. Level 3 includes pre- and post-season planning and support for cross-docking. Level 4 involves seasonless retailing and the transfer of inventory management functions to suppliers. The higher level technologies are generally believed to provide more benefits, but retailers are expected to start with the lower level technologies. Thus, Quick Response technology allows us to explore different definitions of IT adoption and use.

A second factor contributing to the difficulty of finding strong associations between IT use and firm performance is theoretical. Many theories of IT impact assume that impacts are contingent upon, or moderated by, various organizational and environmental conditions. One of the most frequently postulated contingencies is the degree of fit or alignment between the specific IT adopted and the business strategy of the firm (Henderson and Venkatraman 1993a, b, Lentz and Henderson 1998, Luftman 1995, Luftman et al. 1993, Venkatraman 1989). This theory holds that firms using IT in ways consistent with their business strategies will achieve the best performance results.

Again, the Quick Response program in retailing affords a useful opportunity to examine the strategic fit hypothesis. Some of the technologies and applications in the program would support a strategy of supplier partnering or of "customer intimacy," whereas others are most heavily geared toward a strategy of transaction efficiency.

In this paper, we report findings about the relationship between QR technology and organizational performance from a cross-sectional survey of specialty retailers. We found an association between adoption of QR (at a minimal level) and firm performance and a high level of alignment between the IT adopted and strategic goals. However, there was no support for the hypothesis that higher levels of QR use lead to higher business performance, or that poor IT fit with strategy leads to poorer performance. These results have interesting implications for our theories of the links between IT and firm performance.

Theoretical Background

Well-managed organizations generally try to improve their performance when they invest in IT. Yet, the evidence on how well they succeed is mixed (Boynton et al. 1993; Brynjolfsson 1993; Clemons and Row 1989, 1991; Hitt and Brynjolfsson 1994; Harris and Katz 1989, 1991; Markus and Soh 1993; National Research Council 1994; Quinn and Baily 1994; Sethi et al. 1993; Weill 1992). This apparent "productivity paradox" has produced differing reactions from practitioners and academics.

Consultants and other practitioners often make relatively unqualified claims about the relationship between IT use and business benefits. In specialty retailing, the benefits of the Quick Response (QR) are believed to be so great that all firms in the industry will benefit from adopting them (*Retail Information Systems News* 1993, 1995, 1997; Eure 1991; Juneau 1992; *Chain Store Age Executive* 1991; Pastore 1994). Further, QR proponents claim that greater benefits flow from the more advanced technologies in the program, which are in turn believed to require prior investment in the basic QR technologies. Therefore, QR practitioners implicitly hypothesize that, regardless of factors like unique business strategies, specialty retailers who adopt more of the QR program will perform better than those who adopt less.

By contrast, academics have directly addressed the ambiguous empirical relationship between IT and firm performance by refining their theoretical models. They have specified various contingencies believed to moderate IT impacts. In particular, they have urged organizations to align technologies and applications with unique business strategies (Cash et al. 1992; Chan et al. 1997; Henderson and Venkatraman 1993a, 1993b; King 1978; Luftman 1997; McFarlan 1984; Parsons 1983; Tavakolian 1989; Venkatraman 1989). This theory explicitly assumes that different firms in the same industry segment will have different business strategies, requiring different combinations of IT for strategic alignment and good business performance. Thus, business strategy should determine whether a specialty retailer will achieve superior business performance by investing 1) only in basic technologies from the Quick Response program, 2) in both basic and advanced technologies, or 3) only in advanced technologies.

Each of these perspectives has the ability to shed light on our understanding of IT impact on firm performance since they address different issues. The practitioners focus on the extensiveness of IT use, whereas the academics emphasize issues of business strategy that may be unique to a specific firm. Below, each perspective is examined in greater detail for hypotheses about the impact of QR technologies on retailers' performance.

The QR Program in the Specialty Retailing Industry

Specialty retailers are firms adopting a focused approach to retailing (Standard and Poors 1993). The "focus" may lie in product offerings, product

Figure 1 The retail pyramid.

delivery or customer service. Specialty retailers occupy a unique position in "the retail pyramid." (See Figure 1.) They range in size and scope from small local chains of five stores in a single state to international giants with nearly 2,000 stores spread across the United States and beyond (Peterson 1992). The product lines of specialty retailers are as diverse as health foods, apparel, food, auto supplies, housewares, music and video, books, computer software and supplies, and bed and bath linens (Standard and Poors 1993, Peterson 1992).

Typically more focused, agile, and flexible than the major national department stores and yet more able to enjoy economies of scale than small "mom and pop" retail stores, specialty retailers are the most profitable retail segment. They are well positioned to use IT to level the playing field against major national department store chains (Peterson 1992, Cassidy 1994, Lynch 1992), while maintaining the strong local focus characteristic of mom-and-pop stores. The specialty retailing industry is an especially interesting one to study today, because, even though technology has moved rapidly since the time this study was conducted (1994), the industry has been slow to move forward on the adoption of new technologies. The technologies addressed in this study continue to be essentially the same ones utilized today, although today they are often accessed via the Internet.

According to trade literature and consulting firms, the QR program has a strong positive impact on specialty retailer performance (*Retail Information Systems News* 1993, 1995, 1997; *Chain Store Age Executive* 1991; *Executive Briefing* 1992). Potential improvements from adopting QR include "higher inventory turns, fewer markdowns, better in-stock position and higher profitability" (*Retail Information Systems News* 1993). The QR program is the retailing industry's equivalent of "just-in-time" manufacturing (*Executive Briefing* 1992, Juneau 1992). Similar concepts have been utilized in other sales environments. Effective Channel Response in grocery stores (Clark 1994) and Effective Customer Response (Fisher et al. 1994) applied to specific retailers are targeted at better defining inventory levels and working more closely with suppliers. The implementation of QR is seen as supporting multiple organizational 5 5 strategies, but **"[n]o matter which strategy a company chooses, it needs to excel in major elements of QR to implement that strategy successfully"** (*Retail Information Systems News* 1995, p. 8).

QR is positioned as "an array of technology options for supporting the company's mission" (*Retail Information Systems News* 1996, p. 7). This array is often visualized as a hierarchical suite of tools, with each level building on those below it. In an early 1990's study, Kurt Salmon & Associates (*Retail Information Systems News* 1993) identified four levels of QR based on a mix of technologies and applications employed by retailers. (See Table 1.) The first level involves automated point-of-sale (POS), bar coding, universal product codes (UPC), automatic price look-up, and electronic data interchange (EDI) for order entry and inventory management. This level mainly consists of technologies that form the infrastructure for many of the supply chain management (SCM) applications in higher levels. A second level involves automatic replenishment by suppliers, forecasting, and electronic invoicing. EDI at this level includes order status, invoicing, and advance shipment notices. Enhancements in level three include using the information gleaned through the earlier levels for pre- and post-season planning, and shipment container marking and arrangement. In the fourth level, suppliers actually take over inventory management functions (an activity that is today called vendor-managed inventory). This level also includes seasonless retailing (the ability to provide all products on a year-round basis) and space management.

Progression through the QR hierarchy implies a transition from a simple focus on transaction efficiencies to an awareness of strategic opportunities

QR Level	Applications and technologies
1	Automated point-of-sale applications supported with bar coding and universal product code (UPC) labeling
	In-store customer service including automatic price look-up
	Distribution and inventory management through electronic data interchange (EDI) for merchandise order entry
2	Automatic inventory replenishment using point-of-sale information for inventory pull
	Sales and inventory forecasting
	Inventory management and order processing through electronic invoicing, EDI for order status, invoicing and advance shipment notices
3	Pre- and post-season planning using the information gleaned through the elements of the earlier stages
	Improved distribution processing through automatic shipment container marking and arrangement enabling cross-docking of transportation fleet
4	Suppliers take over inventory management functions, accepting responsibility for yield and space management
	Seasonless retailing, the ability to provide all products on a year-round basis

Table 1 Levels of the QR program.

Note
Adapted from: *Retail Information Systems News*, Uncovering the keys to quick response, special supplement, January 1993.

from partnering with customers and suppliers. According to CSC partner Marie Beninati,

> Retailers recognize the strategic value of technology in positioning themselves for success ... in the past retailers saw I/T simply as a tool for high-speed processing of massive amounts of small transactions. Today, they recognize that information technology is a prerequisite for survival. (*Competing with Key Technologies* 1995, p. 8)

QR technologies create a type of interorganizational system (IOS) involving the retailer, manufacturers, transporters, downstream suppliers, and banks (*Executive Briefing* 1992; *Chain Store Age Executive* 1991; *Retail Information Systems News* 1993, 1995). These technologies play an important role in supply chain management. The decision to collaborate or compete in connections with customers, suppliers and (increasingly) with competitors is generally viewed as a strategic decision (Scott Morton 1991). Electronic markets made possible through EDI can be a source of competitive advantage (Bakos 1991, Barua and Lee 1997, Malone et al. 1987). Much research suggests a positive impact of EDI on firm performance (Benjamin et al. 1990, Bouchard 1992, Clemons and McFarlan 1986, Hammer and Mangorian 1987, Holland et al. 1992, Mukhopadhyay and Kekre 1995, O'Connor 1991, Rochester 1989).

These arguments suggest that information technology may be a major factor in improving firm performance by altering competitive dynamics in the retail supply chain. Given the strategic and interorganizational nature of the QR program, there is reason to credit practitioners' claims of a tight link between QR and specialty retailer performance. Overall, the practitioner perspective suggests the following hypothesis:

Hypothesis 1. Specialty retailers that adopt QR will outperform those that do not adopt QR.

The QR program is usually described as a hierarchy of technologies and applications ranging from more basic to more advanced. Each of the identified levels can have a differential impact on the adopting organization. This suggests:

Hypothesis 1a. Higher levels of QR adoption will maker greater contributions to performance than lower levels.

It is generally assumed that retailers will acquire more basic QR technologies before moving to more advanced ones (*Retail Information Systems News* 1993, 1995; *Chain Store Age Executive* 1991), possibly because of technological dependencies. This assumption may, however, be incorrect. Thus, it is useful to inquire whether cumulative adoption of the technologies in the QR hierarchy enhances specialty retailers' performance. This suggests:

Hypothesis 1b. Higher cumulative levels of QR adoption will make greater contributions to performance than lower levels.

The Strategic Alignment Perspective and Specialty Retailing

It has become almost axiomatic that a firm should align its IT strategy with its corporate strategy to achieve greatest effectiveness. "Indeed, the key strategic IT management challenge lies in the identification of those strategic dimensions that require modification under different contingencies for enhancing organizational performance" (Henderson and Venkatraman 1993a).

The strategic alignment perspective suggests that the effect of IT on performance will depend on the fit between information technology strategy and corporate strategy. This perspective has a long pedigree in diverse literatures. These include strategic management (Andrews 1980; Ansoff 1988; Chandler 1962; Buzzell and Gale 1987; Prahalad and Hamel 1990; Mintzberg 1979; Quinn 1980, 1992; Wrapp 1967; Porter 1980, 1985; Selznick 1957; Wiseman 1988), structural contingencies (Pfeffer 1982; Scott 1981; Thompson 1967; Galbraith 1973; Schoonhoven 1981; Gutek 1989), and strategic fit (Van de Ven and Drazin 1985; Venkatraman 1989).

A number of researchers have studied the alignment between organizational and IT variables as a predictor of effective IT use. There are a number of terms for alignment, including strategic alignment (Broadbent and Weill 1993; Chan et al. 1997; Henderson and Venkatraman 1992, 1993a, b; Luftman et al. 1993), linkage (Reich and Benbasat 1996), and harmony (Luftman 1997). Among the

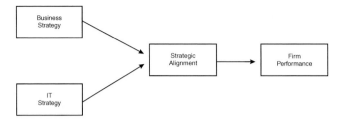

Figure 2 Basic strategic alignment model.

alignments studied are the fit between: IS structure and organizational infrastructure (Ein-Dor and Segev 1982, Olson 1980), organizational and IT strategy (McFarlan and McKenney 1983), competitive strategy and IT structure (Tavakolian 1989), business strategy, IS strategy, and strategic orientation (Chan et al. 1997), competitive strategy and strategic IT planning (King 1978), and strategy and infrastructure (Papp and Luftman 1995, Luftman 1997). The researchers in these traditions attribute variation in IT structures to differences in organizational context variables. They argue that IT systems should conform with context variables such as organizational decision-making structure, managerial philosophy, organizational form, and organizational competitive strategy (Cash et al. 1992, McFarlan and McKenney 1983).

Henderson and Venkatraman (1993a) suggest that companies can be successful in aligning their IT and business strategies in several ways. Their Strategic Alignment Model brings strategy and IT together by balancing internal and external factors and business and IT domains. The four elements of the model are business strategy, IT strategy, organizational infrastructure and processes, and IT infrastructure and processes. The authors suggest that alignment may begin in any of the four domains and proceed to others with equally effective results.

In earlier work, Venkatraman (1990) had focused particularly on the "coalignment between the strategic context and the IT infrastructure of a business" as a driver of successful exploitation of IT (p. 151). The current research maintains Venkatraman's earlier focus on the strategic coalignment of business strategy and IT, minimizing the importance of organizational infrastructures and processes. The reason is that, while the firms studied populated the same industry segment and shared similar organizational infrastructures and processes, they adopted different strategic approaches to the marketplace (Peterson 1992). Thus, strategy is more likely than infrastructure and processes to account for any observed variation in IT adopted.

For the purposes of this paper, strategic alignment is defined as the correlation between the business strategy of a firm and the firm's IT strategy. In particular, we use Venkatraman's (1989) definition of alignment (or "fit") as moderation, because we are concerned with the effect of the two coaligned variables on a third (performance). The basic research model for strategic alignment is presented in Figure 2.

Alignment is operationalized in this study using strategy descriptors appropriate to the specialty retailing context, derived from Holland, Lockett, and Blackman (1992) and Treacy and Wiersema (1993). Holland, Lockett, and Blackman's (1992) model emphasizes the three basic elements of the "retail pipeline": supplier, retailer's internal operations, and customer. Another model (Treacy and Wiersema 1993) identifies three essential customer-oriented "value disciplines" as defining a firm's strategy: operational excellence, customer intimacy, and product leadership. The Treacy and Wiersema typology is particularly appropriate for the specialty retailing area, because of its strong focus on customer-oriented issues. Consistent with the strategic alignment perspective, Treacy and Wiersema (1993 and various public presentations) also explain that different technologies, applications, and infrastructures are required to support each of their three different strategies.

Our study categorizes the business and IT strategies of specialty retailers using concepts derived from the Holland et al. and the Treacy and Wiersema models. (See Table 2. The descriptions used in Table 2 are the precise terms used in the survey instrument.) Specialty retailers can adopt a strategic focus on supplier relationships, internal

Business pipeline strategy	Referring to the Retailer's Strategic Value System derived from Holland, Lockett, and Blackman (1992)	Corresponding IT Strategy	Referring to use of IT derived from Treacy and Wiersema (1993)
Supplier Focus	Focus on interactions with suppliers and manufacturers emphasizing speed in bringing new products to market, reducing cycle times, and optimizing business across organizational boundaries through cooperative efforts.	Supplier Partnering	IT focuses on relationships with key suppliers and vendors, sharing information, environmental scanning, and building the architecture for Quick Response.
Internal Focus	Focus on internal operations, in minimizing costs, offering competitive prices, and increased efficiency and reliability of service.	Transaction Efficiency	IT focuses on achieving greater levels of internal efficiency, reducing costs, improving inventory management, and logistics.
Customer Focus	Focus on flexibility and responsiveness to customers, empowering employees working closely with customers, tailoring and shaping products to fit customer needs.	Customer Detail	IT focuses on the retail customer, providing detailed description of customers and their transactions, enhancing the buying relationship, and suppporting employees working directly with the customer.

Table 2 Business strategy (retail pipeline) and IT strategy descriptions.

activities, or on the customer. They can likewise choose to focus their IT efforts on transaction efficiency, supplier partnering, or customer detail.

According to the strategic alignment perspective, business strategies and IT strategies must be matched for optimal firm performance. Table 3 presents the conditions of match, hypothesized as yielding high performance, and mismatch, hypothesized as yielding lower performance.

In strong form, the strategic alignment variant of contingency theory posits a link between strategic match and firm performance (MacDonald 1991, Van de Ven and Drazin 1985, Venkatraman 1989). This suggests the following hypothesis.

Hypothesis 2. Specialty retailers with a match between their corporate (pipeline) strategy and their IT strategy will outperform those with a mismatch.

Many contingency theorists are satisfied with less stringent empirical support for their thesis than the match-performance link. Evidence that firms have achieved strategic alignment is often viewed as support for contingency theory, even if no performance link can be shown, since many other factors than match can affect firm performance. In the event that no link is found between match and performance, it becomes very important to know whether few firms have sought strategic alignment (suggesting that they do not see this factor as important or that match is difficult to achieve) or whether many firms

IT strategy (below)	Corporate (or pipeline) strategy		
	Supplier focus	Internal focus	Customer focus
Supplier partnering	**Match**	Mismatch	Mismatch
Transaction efficiency	Mismatch	**Match**	Mismatch
Customer detail	Mismatch	Mismatch	**Match**

Table 3 Business strategy (retail pipeline)/IT strategy alignment matrix.

have sought strategic alignment (suggesting that they take the importance of match for granted, but that it no longer differentiates high and low performance). The foregoing discussion suggests:

Hypothesis 2a. Specialty retailers will adopt an IT strategy consistent with their corporate (pipeline) strategy (e.g., they will exhibit "match" in the alignment matrix).

The strategic alignment perspective is logically and intuitively appealing to academicians, particularly those schooled in general management, strategy, and organizational behavior. The basic premise of strategic alignment is that technology yields benefits when the technology has been carefully selected to fit the firm's goals. Thus, the strategic alignment model presents a different perspective on QR technology in specialty retailing than the statements of QR practitioners. Together, they may provide a more complete picture of QR's performance impacts than either perspective alone.

Methodology

This section describes our strategy for testing hypotheses derived from practitioner and academic perspectives on the performance impacts of QR technologies. We first describe the research design. Then we discuss measurement of firm performance and the key independent variables.

Research Design

The study involves analysis of data from a cross-sectional survey of firms in the specialty retailing industry. Access to firms in the industry was obtained through RETEX (Retail Technology Consortium), a consortium of specialty retailing firms whose mission is to address common technology problems and opportunities. RETEX members are diverse, including firms with regional, national, and international focus, firms engaged in a variety of different product lines (fashion, health, food, specialty items, entertainment products), and firms with differing approaches to the use of IT.

Company	QR	Strategic Alignment	Performance
Off-Price*	No	No	High
TJ	No	Yes	High
Petrie	No	Yes	Low
WSRG	No	Yes	Low
The Clothes Horse*	Yes	Yes	High
Babbage's	Yes	Yes	High
Moto Photo	Yes	No	Low
County Seat	Yes	Yes	Low

Table 4 Specialty retail case examples.

Note
* Fictitious name used at company's request.

Preliminary case studies of eight members of RETEX's board of directors were conducted. The case sites included high and low performers and high and low users of Quick Response technologies. (See Table 4.) Conducting the case studies involved headquarters and store visits and meetings with operating personnel, executive teams, and key Information Systems executives. The results of the cases were important in three respects. First, the cases showed that it was appropriate to operationalize corporate and IT strategies using the model outlined in Table 2. Second, the cases gave face validity to the four-level QR model of Kurt Salmon and Associates (*Retail Information Systems News* 1993). Third, the information provided by the key IS executives was consistent with insights gained from other members of management, suggesting that it would be appropriate to use a single respondent in our larger-sample survey.

Data about performance, strategy, technology, and other organizational and IT dimensions were gathered through a self-administered, self-report questionnaire, completed by the top IT professional in each firm. Since the research explored the intersection of IT and corporate strategy, the top IT professional was well positioned to respond to the questions. In addition, many of the objective measures reported on the questionnaires were verified against an industry database maintained by RETEX. The database covered company

Authors	Sample Population	Performance Measures
Bender (1986)	132 life insurance companies	1. IT expense to total operating expense 2. Total operating expense to premium income
Cron and Sobol (1983)	138 medical wholesalers	1. Pre-tax return on assets 2. Return on net worth 3. Per-tax profits as % of sales 4. Average 5 years sales growth
Floyd and Woolridge (1990)	55 retail banks	1. ROA 2. Product and Process IT
Harris and Katz (1989, 1991)	40 life insurance companies	1. IT expense to total operating expense 2. Total operating expense to premium income 3. IT expense to premium income
Hitt and Brynjolfsson (1994)	367 firms from Fortune 500	1. Output 2. Labor 3. ROE, ROA, ROS
Markus and Soh (1993)	195 banks	1. Profitability as operating profit/operating revenue
Sethi, Hwang, and Pegels (1993)	*Computerworld* IS effective firms 1988 and 1989 (146 and 136, respectively)	1. ROE 2. ROS 3. Sales Growth 4. Earnings Growth
Strassman (1990)	35 manufacturing business units	1. Management productivity
Turner (1985)	58 mutual savings banks	1. Percentage of net income to total assets 2. IT expense as percentage of total assets
Weill (1992)	33 valve manufacturers	1. Sales Growth 2. ROA 3. % labor change

Table 5 Performance measures in the IT literature.

demographics, IT use measures, and company-specific information on each of the organizational performance measures we used (described more fully below). The database provided a cross-check on the reliability of the self-report data and allowed us to determine that the sample was indeed quite representative of the industry (as discussed below). The questionnaire was mailed to 200 RETEX members. Twenty-five members declined to participate. Eighty usable responses were received, yielding a 40% response rate. This response rate is quite high for mailed surveys, a fact we attribute to RETEX's sponsorship of the study.

The Performance Construct

The concept of performance underlies much research in strategic management and information science. Definitions of performance vary widely. Definitions include meeting specific goals, maintaining specific operating ratios, achieving profitability targets, and maintaining long term viability as an enterprise. The prevailing work in IT has concentrated on the use of industry-specific, short-term quantitative measures of performance, such as annual sales growth. Table 5 outlines the diversity of performance measures used in recent research on IT and performance

(Baets 1996; Bender 1986; Broadbent and Weill 1993; Cron and Sobol 1983; Floyd and Woolridge 1990; Harris and Katz 1989, 1991; Hitt and Brynjolfsson 1994; Markus and Soh 1993; Reich and Benbasat 1996; Sethi et al. 1993; Strassman 1990; Turner 1985; Weill 1992).

The performance measures used in the current study include key operating ratios used in the specialty retailing industry, including profitability, comparable store sales growth, sales per employee, sales per square foot, and stock turns. Experienced industry analysts actually use these measures in examining the retail industry (Standard and Poor's 1993). Together, our multiple measures capture several aspects of performance. "Profitability" is a standard measure of net income divided by sales. "Comparable store sales growth" measures the sales growth for stores that have been in operation for more than 12 months. This measure avoids overemphasizing the results of new store openings and is similar to a measure used by Cron and Sobol (1983), Sethi et al. (1993), and Weill (1992). "Sales per employee" measures the effectiveness of sales staff, however it may be influenced by the pricing of the merchandise. "Sales per square foot" measures the sales intensity of the specific stores. "Stock turns" measures the number of times inventory is sold over the course of a year, a key measure of how rapidly products are moving through the store. The sales per square foot and stock turns measures are specific to the specialty retailing industry; Harris and Katz (1989, 1991) used industry-specific measures in their study of insurance firms. The survey was conducted in the spring of 1994 and respondents reported each of these measures, except profitability, for a five-year period, 1989–1993. Profitability figures were provided by RETEX, the Retail Technology Consortium.

Performance means for respondents were computed as follows. Each of the six sets of 80 performance measures (one for each respondent) was sorted from highest to lowest and then divided into quartiles. The top twenty respondents on each measure were given a score of 4 points; the bottom twenty received a score of 1 point with members of the intermediate quartiles getting scores of 3 or 2. This procedure allowed us to compute and statistically compare mean performance scores for different categories of firms (e.g., those adopting QR versus those that did not).

Adoption of QR and Existence of Strategic Alignment

Data for testing Hypotheses 1, 1a, and 1b about the links between QR and firm performance were derived from a questionnaire item measuring how frequently respondent companies used each of 12 technologies and applications believed to comprise the QR hierarchy. (See Table 1.) Respondents were asked to indicate frequency of use on a 5-point scale with 1 anchored at "don't use" and 5 anchored at "mature use." (Intermediate levels were anchored as "investigating," "beginning stages," and "growing use.") Respondents were coded as using QR level 1 if they reported at least "beginning stages" of use of *all* the technologies or applications associated with QR level 1. The same procedure was used for the other QR levels in the hierarchy.

Data for testing Hypotheses 2 and 2a about the links between strategic alignment and firm performance were derived from two questionnaire items. The first asked respondents to indicate the percentages of effort (adding up to 100%) their organization devoted to each of three IT strategies (described in Table 2). The second asked respondents to allocate effort similarly among three business strategies (retail pipeline orientations—also described in Table 2). Firms reporting that they devote 50% or more of their effort to a strategy were coded as using that strategy.

Other Variables

In addition to the two explanations for firm performance addressed in this study (the QR program and strategic alignment), the literature has suggested a number of alternative explanations for (or moderators of) firm performance. The most common include various organizational and technological differences such as: organizational size, firm ownership, product mix, centralization, vertical integration, and IT investment. These factors were measured and assessed in our research design. Key variables and their measurement and source are presented in Table 6.

Variable	Measure	Source
Performance Measures		
Profitability	Net income/saes (1992, 1993)	RETEX database
Comparable Store Sales Growth	Specific number (1992, 1993)	RETEX database and Survey
	Range on survey (1989–1993)	
Sales per Employee	Specific number (1992, 1993)	RETEX database and Survey
	Range on survey (1989–1993)	
Sales per Square Foot	Specific number (1992, 1993)	RETEX database and Survey
	Range on survey (1989–1993)	
Stock Turn	Specific number (1992, 1993)	RETEX database and Survey
	Range on survey (1989–1993)	
Quick Response		
Quick Response	Multi-item (12) measures frequency of use	Survey
Strategic Alignment		
Strategic Alignment	3-item measures or corporate strategy and IT strategy	Survey
Other		
SIC	SIC code	Survey
Investment in IT	Specific number (1992, 1993)	RETEX database and Survey
	Range on survey (1989–1993)	

Table 6 Key variables, measures, and information sources.

Findings

We first discuss the representativeness of the surveyed firms relative to the population of specialty retailers. Then we present the results of the tests of hypotheses.

Respondent Profiles

The firms in the sample had approximately 3,000 employees on average. They averaged $345 million in sales, and spent 1.8% of sales on IT. Average IT investment as a percentage of sales did not differ significantly by firm size. The average firm's implementation of IT began in 1985. (See Table 7.)

Table 8 compares survey respondents with industry averages on key variables including ownership, size, SIC codes, and the performance measures of sales growth, sales per square foot,

Firm Size	#	Avg Annual Sales ($ mil 1993)	Avg Employees 1993	Avg Year IT Began
Small (under $25 million)	42	$20	250	1986
Medium ($25–250 million)	19	$182	2064	1984
Large (over $250 million)	19	$1329	10701	1983
Total	80	$345	2969	1985

Table 7 Profiles of responding firms (annual sales, employees, IT adoption).

Variable	Sample	Industry
Ownership	33% public	25% public
Size	57% under $100 million	65% under $100 million
SIC/Product		
Hardware	15%	10%
Specialty	21%	25%
Apparel	27%	25%
Furniture/Appliance	19%	15%
Book/Sport Drug	19%	25%
Sales Growth	just over 10%	8%
Comparable Store Sales Growth	just under 10%	7%
Sales per Square Foot	$200–250	$150–175
Stock Turn	4–7	5–6

Table 8 Comparison of sample firms with industry means.

and stock turns. These variables were selected for comparison because they are relevant to industry analysts and therefore are collected and publicly reported. The respondents generally reflect the specialty retailing industry on all these measures. One-third of respondents are publicly-owned companies, which is close to the mean of 25% reported for the industry (Standard & Poor's Register of Corporations 1993). The distribution of respondents across five basic SIC codes mirrors the industry-wide distribution. The sample mean sales growth was over 10%, and comparable store sales growth averaged just under 10%. These figures are similar to the specialty retailing industry averages of 8% sales growth and 7% comparable store sales growth for the same five-year period. Stock turns averaged between 4–7 turns annually, roughly comparable to the industry average of 5–6. But average sales per square foot for respondents was $200–250, well above the industry average of $150–175. These performance measures suggest that the sample is representative of the specialty retailing industry, except on the important measure of sales per square foot, where the sample firms were better performing.

Since, by and large, RETEX firms are somewhat more successful than average for their industry, we would expect to see somewhat higher use of QR technologies and somewhat higher strategic match in sample firms than in the industry overall. Nevertheless, we do not believe that these differences seriously limit our ability to generalize the results of our hypothesis tests to the entire specialty retailing industry.

Impact of QR Adoption on Firm Performance

Hypothesis 1 compares the performance of low or non-users of QR with those that make more intensive use of QR technology. Respondents reporting at least beginning stage use of all technologies in QR level 1 were identified as QR adopters. All others were classified as nonadopters. Adopters reporting at least beginning stage use of all technologies in QR level 2 were identified as QR level 2 users and so on. Table 9 shows the number and percentage of respondents adopting QR at level 1 and each subsequent level. It also shows the number and percentage of respondents at each level that had also adopted the technologies of the previous level.

The first surprise here is that only 50% of the responding firms were QR adopters at least at level 1. This finding is surprising in light of the advantages claimed for QR in virtually all specialty retailing trade publications. The second surprise is how few firms using the technologies of a particular level also used the technologies of the level below it. The hierarchical description of the QR program leads one to expect that a very high percentage of firms at each level would also report using the technologies of each lower level. But 9 of 17 firms had adopted level 3 without also adopting level 2; 5 of 9 firms had adopted level 4 of the QR program without also adopting level 3. Right away, these findings cast some doubt on the arguments of QR proponents.

But hypothesis testing revealed a somewhat different picture. Recall that Hypothesis 1 compares QR adopters with nonadopters:

Hypothesis 1. Specialty retailers that adopt QR will outperform those that do not adopt QR.

QR Level	Technologies and Applications	# (and %) of respondents using QR technologies	# (and %) of technology users at each level also using previous level
1	POS, bar coding, UPC labeling, EDI	40 (50)	—
2	Automated inventory replenishment, inventory and sales forecasting, invoiceless payments, EDI	36 (45)	23 (64)
3	Cross-docking, pre- and post-season planning	17 (21)	8 (47)
4	Supplier space management, seasonless retailing	9 (11)	4 (44)

Table 9 The QR program levels.

Note
Number (and percentage) of respondents using QR.

T-tests showed that QR use (of level 1 or higher) was positively associated with all but one measure of firm performance. (See Table 10.) There were significant positive associations between QR use and profitability, comparable store sales, sales per square foot, and stock turns, but not sales per employee. On that measure, nonadopters of QR outperformed adopters. A regression analysis showed that the results were not affected by other factors such as IT investment, industry segment, or firm size. Overall, Hypothesis 1 is strongly supported by our data.

An additional analysis was conducted to explore whether the QR program achieved performance improvements in the manner claimed by QR advocates.

Hypothesis 1a. Higher levels of QR adoption will make greater contributions to performance than lower levels.

The QR program is thought of by practitioners as a hierarchy in which higher levels build on lower levels. However, many of our respondents adopted technologies at a higher level without first

Performance Measures	QR (Level 1 Or higher) n = 40		Non-QR n = 40		t-value	Supported?
	Mean	S.D.	Mean	S.D.		
Profitability	2.83	1.04	2.20	1.14	−2.57*	Yes
Comparable Store Sales Growth	2.83	1.11	2.18	1.06	−2.68**	Yes
Sales per Employee	2.38	1.10	2.65	1.17	1.08	No
Sales per Square Foot	2.70	1.11	2.30	1.11	−1.61†	Yes
Stock Turn	2.80	1.14	2.23	1.05	−2.35*	Yes

Table 10 Hypothesis 1: The QR program and performance.

Note
significance †$p < 0.10$, *$p < 0.05$, **$p < 0.01$.
T-test comparing QR (level 1 or higher) and non-QR firms on performance.

having adopted the lower level. We wanted to examine the use of QR levels and classified firms at each level of adoption. The responses were analyzed using a multiple regression analysis with level as a dummy variable, which strengthens the power of the analysis (Cohen 1978) given the small number of responses in certain cells. (See Table 11 for the regression analysis of performance by QR level.)

There is no general trend apparent in these results. Level 1 QR is significant for profitability, comparable store sales growth, sales per square foot, and stock turn. Level 2 QR was not significantly associated with any performance measures. Level 3 QR had a significant negative association with sales per square foot and stock turn. Level 4 QR was significant for sales per employee and sales per square foot. Levels 3 and 4 involve applications aimed at enhancing efficiencies in the areas of stock, inventory, and space management; these applications appear to be having a performance impact. Overall, there were no significant differences among the four levels, providing only mixed results in support of Hypothesis 1a.

Hypothesis 1b. Higher cumulative levels of QR adoption will make greater contributions to performance than lower levels.

The previous analysis did not get at the hypothesized cumulative nature of the QR model. To determine whether performance improvement at higher levels of QR requires a base of underlying technologies, we again used regression analysis to compare firms that had cumulatively adopted QR at levels 2, 3, and 4 (see Table 12). The performance differences were slight and nonsignificant. We conclude that Hypothesis 1b is not supported by our data.

Taken together, these analyses show strong support for the overall QR hypothesis that adopting QR technology, of level 1 or higher, is associated with increased firm performance compared to nonadoption. However, while we found some support for the existence of a hierarchy of technologies and applications, we found little or no support for the notion of QR advocates that there is a hierarchy of QR usage in which use of higher QR levels is associated with better performance.

Dependent Variables	R^2	Beta	T	Sig of T
Profitability	0.21			
Level 1		0.29	1.98	0.05*
Level 2		−0.17	−1.19	0.24
Level 3		0.12	0.97	0.33
Level 4		−0.11	−0.91	0.36
Comparable Store Sales Growth	0.12			
Level 1		0.34	2.50	0.01**
Level 2		−0.08	−0.61	0.54
Level 3		0.15	1.27	0.21
Level 4		−0.05	−0.40	0.69
Sales per Employee	0.08			
Level 1		−0.06	−0.45	0.66
Level 2		−0.11	−0.82	0.41
Level 3		−0.10	−0.86	0.39
Level 4		0.28	2.36	0.02*
Sales per Square Foot	0.26			
Level 1		0.23	1.89	0.06†
Level 2		−0.08	−0.69	0.49
Level 3		−0.37	−3.50	0.001**
Level 4		0.20	1.85	0.07†
Stock Turn	0.12			
Level 1		0.29	2.01	0.05*
Level 2		0.10	0.72	0.47
Level 3		−0.21	−1.79	0.08†
Level 4		0.09	0.77	0.45

Table 11 Hypothesis 1a: QR levels and performance.

Note
significance †$p < 0.10$, *$p < 0.05$, **$p < 0.01$, ***$p < 0.001$

Performance by level of QR used ($N = 80$).

Dependent Variables	R^2	Beta	T	Sig of T
Profitability	0.04			
Cumulative Level 2		−0.05	−0.37	0.71
Cumulative Level 3		0.02	0.15	0.88
Cumulative Level 4		−0.19	−1.44	0.16
Comparable Store	0.05			
Sales Growth				
Cumulative Level 2		−0.14	−1.22	0.23
Cumulative Level 3		0.16	1.34	0.19
Cumulative Level 4		−0.16	−1.34	0.19
Sales per Employee	0.03			
Cumulative Level 2		−0.02	0.17	0.87
Cumulative Level 3		−0.06	−0.55	0.58
Cumulative Level 4		0.17	1.47	0.15
Sales per Square Foot	0.04			
Cumulative Level 2		0.16	1.12	0.27
Cumulative Level 3		−0.02	−0.13	0.90
Cumulative Level 4		0.09	0.67	0.51
Stock Turn	0.04			
Cumulative Level 2		0.08	0.58	0.56
Cumulative Level 3		−0.05	−0.38	0.71
Cumulative Level 4		0.17	1.25	0.22

Table 12 Hypothesis 1b: Cumulative QR level and performance.

Note
Significance $^{\dagger}p < 0.10$, $*p < 0.05$, $**p < 0.01$.
Performance of firms by Cumulative levels of QR ($N = 80$).

Impact of Strategic Alignment on Firm Performance

We now turn to tests of hypotheses derived from the strategic alignment perspective. We first consider the weaker form of the strategic alignment model, which accepts evidence of fit as support for the hypothesis even if there is no association between fit and performance.

Hypothesis 2a. Specialty retailers will adopt an IT strategy consistent with their corporate (pipeline) strategy (e.g. they will exhibit "match" in the alignment matrix).

Respondents were asked to identify the percentage of effort they focused on each of three business strategies (based on Holland, Lockett, and Blackman's "pipeline" model) and on each of three IT strategies (based on the Treacy and Wiersema "value disciplines" model). Using the highest reported percentage of effort, each respondent was assigned to a corporate strategy and to an IT strategy. Table 13 shows the distribution of firms across corporate and IT strategy combinations.

Two types of matches were observed: firms with both an internal business strategy focus and a transaction efficiency focus in their IT, and firms with both a customer strategy focus and a customer detail IT focus. No firms in our sample reported a major emphasis on the supplier-focused business strategy or a supplier partnering IT strategy. The lack of Supplier Focus and Supplier Partnering IT may reflect that this industry has not yet recognized the supplier as a key strategic partner. The preponderance of firms focusing internally suggests a strong industry bias. Seventy-nine percent of respondents fell into two matching cells of the Strategic Alignment Matrix. This can be interpreted as strong support for Hypothesis 2a, the weaker form of the strategic alignment hypothesis. We turn now to the stronger form of the hypothesis:

	Supplier Focus	Internal Focus	Customer Focus
Supplier partnering	0	0	0
Transaction efficiency	0	**46 (58%)**	9 (11%)
Customer detail	0	8 (10%)	**17 (21%)**

Table 13 Hypothesis 2a : Strategic alignment.

Note
Number (percentage) of respondents exhibiting alignment between their pipeline and IT strategies.

	Matching $n = 63$		Non-Matching $n = 17$			
Performance Measures	Mean	S.D.	Mean	S.D.	t-value	Supported?
Profitability	2.46	1.12	2.46	1.36	0.03	No
Comparable Store Sales Growth	2.52	1.17	2.44	0.98	−0.24	No
Sales per Employee	2.34	1.10	3.11	1.08	2.63*	No
Sales per Square Foot	2.47	1.13	2.61	1.15	0.47	No
Stock Turn	2.45	1.10	2.72	1.23	0.90	No

Table 14 Hypothesis 2: Match and performance.

Note
significance *$p < 0.05$.
T-test comparing matching and non-matching firms on performance.

Hypothesis 2. Specialty retailers with a match between their corporate (pipeline) strategy and their IT strategy will outperform those with a mismatch.

Table 14 shows the mean performance values for firms in matching cells versus the firms in the non-matching cells of Table 13. Hypothesis 2, the strong form of the strategic alignment hypothesis was *not* supported by our data. Matched and nonmatched firms differed little in profitability, comparable store sales growth, sales per square foot, and stock turns. The only statistically significant difference was in sales per employee, but this difference was in the direction *opposite* to that hypothesized: nonmatching firms actually performed better than matching firms on this dimension. A regression analysis showed that the results were not affected by other factors such as IT investment, industry segment, or firm size.

In short, we found only weak support for the strategic alignment hypothesis. While firms did appear to seek a match between their business strategy and their IT strategy, strategic alignment was not associated with higher firm performance.

This result could mean that other factors intervened to diminish the performance effects of match. But regression analyses using organization

	Strong Match (top quartile) $n = 20$		Weak Match (bottom quartile) $n = 20$			
Performance Measures	Mean	S.D.	Mean	S.D.	t-value	Supported?
Profitability	2.43	1.11	2.42	1.13	0.04	No
Comparable Store Sales Growth	2.41	1.16	2.33	1.21	−0.16	No
Sales per Employee	2.44	1.11	3.00	1.10	1.48	No
Sales per Square Foot	2.39	1.22	3.17	0.98	0.45	No
Stock Turn	2.44	1.18	2.67	1.03	1.56	No

Table 15 Hypothesis 2: Match and performance.

Note
T-test comparing strong matching and weak matching firms on performance.

size (measured as a five-year average of annual sales), IT investment (measured as a five-year average of IT spending as a percentage of sales), and SIC code ruled out this explanation. Alternatively, our results could mean that firms in the specialty retailing industry do not seek strategic alignment; instead they may simply *appear* to seek alignment because most of them are pursuing an internal business focus with information technologies designed to increase transaction efficiency. (This interpretation is not likely to gladden the hearts of the business strategists!)

We next performed a "strength of alignment" comparison with the same results (see Table 15). This analysis evaluated the differences between the percentages of effort spent on corporate versus IT strategies. The total differences were then used to identify the strength of the alignment between business and IT strategies; those firms with little difference between the effort on aligned business and IT strategies were stronger in alignment. For example, a firm spending 60% effort on Internal Focus business and 60% effort on Transaction Efficiency IT were considered very strongly aligned. The results split firms on strength of alignment into quartiles. Top and bottom quartiles were compared (see Table 15); strongly aligned firms did not outperform less strongly aligned firms on any of the performance measures. The only statistically significant difference was in sales per employee in the direction opposite to that hypothesized.

Summary of Findings

The results of our analyses are summarized in Table 16. The results show that QR adoption (at level 1 or higher) is the best predictor of most performance measures. Strategic alignment was prevalent across the sample, with nearly 80% of the firms in alignment (with a strong bias toward internal strategic focus and transaction efficient IT), but strategic alignment was not significantly associated with organizational performance.

Discussion

The findings reported above create a surprising pattern. First, our results provide some support and some disconfirming evidence for both the QR perspective and the strategic alignment perspective. There was an association between adoption of the QR program at level 1 or higher and several measures of performance, and an apparent match between IT and business strategy. But there was

	Hypothesis	Findings
H1.	Specialty retailers that adopt QR will outperform those that do not adopt QR	Supported for Profitability, Comparable Store Sales Growth, Sales per Square Foot, Stock Turn
H1a.	Higher levels of QR adoption will make greater contributions to performance than lower levels	Mixed support
		Level 3 negatively affected Sales per Square Foot and Stock Turn
		Level 4 positively affected Sales per Employee and Sales per Square Foot
H1b.	Higher cumulative levels of QR adoption will make greater contributions to performance than lower levels	Not supported
H2.	Specialty retailers with a match between their corporate (pipeline) strategy and their IT strategy will outperform those with a mismatch	Not supported
H2a.	Specialty retailers will adopt an IT strategy consistent with their corporate (pipeline) strategy (e.g., they will exhibit "match" in the alignment matrix)	Supported, with 79% (63 of 80) firms in alignment

Table 16 Summary of hypotheses and findings.

no association between strategic alignment and performance and no support for the notion of a QR hierarchy.

One possible explanation of the association of QR adoption with higher performance is that higher performing firms are better able to afford QR technologies. Case studies of eight of the firms in this study suggest, however, that cost factors and availability of slack resources were not the primary considerations in adopting or not adopting QR.

> Three high performing firms found that QR was important to supporting key operations. The other high performing firm did not adopt QR, because they saw other competitive advantages available to the firm, the technologies were too difficult, and there were not provable economic incentives for suppliers. Two low performing firms had problems in adopting QR while two others chose not to develop QR because of their supplier relationships. Adoption of QR is influenced by several factors, including level of current IT sophistication and supplier relationships. (Palmer 1995)

We believe, therefore, that the theoretically derived hypothesis that QR adoption leads to higher performance is still tenable.

The lack of association between strategic alignment and performance is not inconsistent with the work of Papp and Luftman (1995), who failed to identify any performance effect of alignment. The findings do differ from Papp and Luftman (1995) and Luftman (1997), however, in that they found few firms in alignment and we found many. Thus, our study represents more convincing evidence against strategic alignment theory than do prior studies.

Despite the overall strong association between QR and performance, the QR model did not emerge from our analysis totally unscathed. Some companies adopted higher levels of QR without adopting lower levels, despite consultants' recommendations. Further, adopting higher QR levels did not appear to lead to improved performance effects. These results suggest the need to redefine the QR program and the QR hierarchy. There are clear technology investments in levels 1 and 2 and more application specific investment in levels 3 and 4. Perhaps this suggests a simpler model of a technology level and an application level.

Finally, only half of the responding firms had adopted QR at any level. Is this because they did not believe that QR has performance impacts, or because they believed the technologies of QR are inconsistent with their strategies? If so, this explanation is especially hard to reconcile with the fact that most of our respondents pursued an internally focused strategy and that QR is generally viewed as more supportive of such a strategy than either a customer or a supplier-oriented focus. In short, the QR model clearly needs further work.

The findings of our study are consistent with earlier work showing that company investments in transaction-oriented IT, but not "strategic systems," improve organizational performance (Weill 1992). Our findings do not support Treacy and Wiersema's (1993, 1995) arguments about the importance of company focus on a single "value discipline."

One possible explanation for our surprising results may lie in the limitations of this particular study. The study involved the specialty retailing industry; this segment may differ from general retailing and other industries in its already high degree of strategic differentiation based on product line. Further, since our respondents were members of a technology-oriented consortium, they may differ in their greater use of IT (or in their greater current need for IT due to late adoption). Thus, some questions about our findings will only be settled when the study has been replicated in other industries.

Other possible explanations for our results include an industry tendency to focus on the internal perspective and concentrate on cost reduction. Alternatively, the IT specialist respondents may have underreported supplier and customer focuses. Another answer may be that, given the industry situation, firms adopting QR were actually in alignment and those that had not were misaligned. While the measurement of strategic alignment is not entirely precise, the understanding of the terms generated through the pilot use of the instrument and the use of similar respondents across the companies tend to validate the accuracy of the responses.

Although we cannot generalize our findings beyond the specialty retailing segment of the retail industry, the pattern of our findings is sufficiently surprising to raise questions about the received theory of strategic alignment. We think these questions need to be taken seriously. So, let's for the moment assume that future research corroborates the findings of our research. What could these findings mean?

If replicated, the findings of our study would suggest that, while the QR program is defined differently from firm to firm, those who adopt some version of the QR can expect performance benefits. Today, the specialty retailing industry is still new enough to the technology that most adopters appear to benefit. The concentration of the QR program is on transaction efficiency and cost reduction, areas where there may still be room for substantial improvement in the retailing industry. Whether these benefits would continue after most firms in the industry segment have adopted QR is unknown.

The limited adoption of QR technologies and applications may also provide some insights. Few companies use the entire range of QR technologies and applications. Firms' concentration on different elements of the retail pipeline may drive firms' choices of what parts of the QR program to adopt or not to adopt. Non-technological issues involved in sharing information with suppliers regarding inventory and sales have often kept suppliers from truly managing inventories and participating fully in the concepts of space management and seasonless retailing. It may also be the case that the larger the number of suppliers, the more difficult it becomes for a firm to develop an economic justification for QR (Bouchard 1992). Finally, many retailing firms appear to wait until technologies are ubiquitous and commoditized before they purchase and adopt them. POS is a good example of one of the few technologies that has reached this level. EDI, satellite, and advanced QR applications may not have reached the level of stability required for most specialty retailers to adopt them.

If the findings of our study are replicated, they might mean that strategic alignment has become institutionalized. Firms may have accepted and internalized academic exhortations to align their business and IT strategies. If so, alignment per se ceases to be a differentiating factor in firm performance. An alternative explanation is that what appears to be alignment is actually a case of strong industry bias toward internal strategic focus and transaction efficiency in the use of IT. If so, these findings are vexing for strategists who argue that healthy industry segments require firms that individually pursue differentiated business strategies while collectively pursuing a complete range of strategies. In other words, our findings may be ammunition for strategists like Cooper (1995), who argue that firms compete on multiple dimensions simultaneously, and against strategists like Miles and Snow (1986) and Treacy and Wiersema (1995), who argue that companies can and should attempt to define differentiated strategic playing fields. Put differently, a plausible explanation for our findings is that specialty retailing has been moving toward the lean enterprise model (Cooper 1995). As collisions with competitors increase, those who update their operations with QR gain, and those who do not fall behind, regardless of alignment.

Conclusion

This research set out to assess the effects of information technology on firm performance in the specialty retailing industry and to triangulate a theoretical model with practical advice about the QR program. Our findings provide surprisingly little support for either. Clearly our study requires replication within the general retailing context and in other labor-intensive industries. Another avenue of future research is the supplier-retailer relationship and its impact on QR adoption and supplier and retailer performance. Future work on QR should include monitoring the adoption of technologies over time.

The implications of our study for practice are more challenging. Our study supports a recommendation to implement at least some elements of the QR program. Conversely, we find no basis for the advice we would like to give: We cannot say that firms must tailor their QR investments to unique firm strategies. We hope that our uncomfortable findings will challenge both academicians and practitioners to refine their models of what makes IT investments worthwhile.

REFERENCES

Andrews, Kenneth R. 1980. *The Concept of Corporate Strategy*. Irwin.

Ansoff, H. Igor. 1988. *The New Corporate Strategy*. Assisted by Edward J. McDonnell. Wiley, New York.

Baets, W. 1996. Some empirical evidence on IS strategy alignment in banking. *Inform. Management* **30**(4) 155–77.

Bakos, J. Yannis. 1991. A strategic analysis of electronic marketplaces. *MIS Quart.* **15**(3) 295–310.

Barua, Anitesh, Lee Byungtae. 1997. An economic analysis of the introduction of an electronic data interchange system. *Inform. Systems Res.* **8**(4) 398–420.

Bender, David H., 1986. Financial impact of information processing. *J. Management Inform. Systems* **3**(2) 22–32.

Benjamin, Robert I., David W. de Long, Michael S. Scott Morton. 1990. Electronic data interchange: How much competitive advantage? *Long Range Planning* (UK) **23**(1) 29–40.

Bouchard, Lyne. 1992. EDI: status report. Anderson Graduate School of Management. UCLA.

Boynton, Andrew C., Bart Victor, B. Joseph Pine II. 1993. New competitive strategies: Challenges to organizations and information technology. *IBM Systems J.* 32(1) 40–64.

Broadbent, M., D. Samson. 1990. Business and information strategy alignment: Ensuring outcomes of value to the organization. Working Paper #10, University of Melbourne, Graduate School of Management.

_____, P. Weill. 1993. Improving business and information strategy alignment: Learning from the banking industry. *IBM Systems J.* **32**(1) 162–179.

Brown, C. V., S. L. Magill. 1994. Alignment of the IS functions with the enterprise: Toward a model of antecedents. *MIS Quart.* **18**(4) 371–403.

Brynjolfsson, Erik. 1993. The productivity paradox of information technology. *Comm. ACM* **36**(12) 67–77.

Buzzell, Robert D., Bradley T. Gale. 1987. *The PIMS Principles: Linking Strategy to Performance.* The Free Press.

Cash, James I., Jr., F. Warren McFarlan, James L. McKenney, Michael Vitale. 1992. *Corporate Information Systems Management.* 3rd ed., Irwin.

Cassidy, Peter. 1994. Keeping them in the loop. *CIO* July 36–40.

Chain Store Age Executive. 1991. Quick response: What it is; What it's not. (March) 48–58.

Chan, Yolande E., Sid L. Huff, Donald W. Barclay, Duncan G. Copeland. 1997. Business strategic orientation, information systems strategic orientation, and strategic alignment. *Inform. Systems Res.* **8**(2) 125–50.

Chandler, Alfred D. 1962. Strategy and Structure: Chapters in the History of the Industrial Enterprise. MIT Press.

Clark, Theodore. 1994. *Linking the Grocery Channel: Technological Innovation, Organizational Transformation, and Channel Performance.* DBA Thesis. Harvard University Graduate School of Business Administration, Boston, MA.

Clemons, Eric K., F. Warren McFarlan. 1986. Telecom: Hook up or lose out. *Harvard Bus. Rev.* **64**(4) 90–97.

_____, Michael C. Row. 1989. Information technology and economic reorganization. *Proc. Tenth Internat. Conf. Inform. Systems*, Boston, MA. 341–52.

_____, _____. 1991. Sustaining IT advantage: The role of structural differences. *MIS Quart.* **15**(3) 275–292.

Cohen, J. 1978. *Statistical Power Analyses for the Social and Behavioral Sciences*. Lawrence Erlbaum Associates, Hilldale, NJ.

Cooper, Robin. 1995. *When Lean Enterprises Collide: Competing Through Confrontation*. Harvard Business School Press, Boston, MA.

Cron, W. L., M. G. Sobol. 1983. The relationship between computerization and performance: A strategy for maximizing the benefits of computerization. *J. Inform. Management.* **6** 171–181.

Crowston, Kevin, Michael Treacy. 1986. Assessing the impact of information technology on enterprise level performance. *Proc. Sixth Internat. Conf. Inform. Systems.* 299–310.

Das, S. R., S. A. Zahra, M. E. Warkentin. 1991. Integrating the content and process of strategic MIS planning with competitive strategy. *Decision Sci.* **22**(5) 953–984.

Earl, Michael. 1989. IT and strategic advantage: A framework of frameworks. *Information Management: The Strategic Dimension*. Clarendon Press.

The Emerging Digital Economy. 1998. Secretariat for Electronic Commerce, U.S. Department of Commerce. (April).

Eure, Jack. 1991. Toward the new century in retailing: Survival strategies for an industry in turmoil. *Bus. Forum* **16**(4) 24–28.

Executive Briefing. 1992. Competing for the American consumer: Partnering for quick response. **72**(3) 1–3.

Fisher, Marshall L., Janice H. Hammond, Walter R. Obermeyer, Ananth Raman. 1994. Making supply meet demand in an uncertain world. *Harvard Bus. Rev.* **72**(3) 83–93.

Floyd, S. W., B. Woolridge. 1990. Path analysis of the relationship between competitive strategy, information technology, and financial performance. *J. Management Inform. Systems* **7** 47–64.

Galbraith, Jay R. 1973. *Designing Complex Organizations*. Addison-Wesley.

Gutek, Barbara A. 1989. Work group structures and information technology: A structural contingency approach. Galegher, J. and Kraut, R. E., eds., *Intellectual Teamwork: The Social and Technical Bases of Cooperative Work.* Lawrence Erlbaum.

Hammer, Michael, Glenn Mangurian. 1987. The changing value of communications technology. *Sloan Management Rev.* **28** 65–71.

Harris, Sidney E., Joseph L. Katz. 1989. Differentiating organizational performance using information technology managerial control ratios in the insurance industry. *J. Office Tech. People* **5**(4).

——, ——, ——. 1991. Organizational performance and information technology investment intensity in the insurance industry. *Organ. Sci.* **2**(3) 263–95.

Henderson, J. C., N. Venkatraman. 1992. Strategic alignment: A model for organizational transformation through information technology. T. A. Kochan and M. Useem, eds. *Transforming Organizations.* Oxford University Press, New York.

——, ——. 1993a. Strategic alignment: Leveraging information technology for transforming organizations. *IBM Systems J.* **32**(1) 4–16.

——, ——. 1993b. Strategic alignment: A model for corporate transformation through information technology. *Transforming Organizations.* T. A. Kochan and M. Useem, eds. Oxford University Press, New York.

Hitt, Lorin, Erik Brynjolfsson. 1994. The three faces of IT value: Theory and evidence. MIT Sloan School, Cambridge, MA.

Holland, C., G. Lockett, I. Blackman. 1992. Planning for electronic data interchange. *Strategic Management J.* **13** 539–50.

Juneau, Lucie. 1992. Retailing and wholesaling: Luring customers with conspicuous efficiency. *Computerworld* **26** (September 14) 37–41.

King, William R. 1978. Strategic planning for management information systems. *MIS Quart.* **2**(1).

Lentz, C., J. C. Henderson. 1998. Aligning IT investments and business strategy: A value management capability. Boston University Working Paper 97–21, Boston, MA.

Luftman, J. N. 1997. Align in the sand. *Computerworld* Leadership Series, **3**(2) 1–11.

——, P. R. Lewis, S. H. Oldach. 1993. Transforming the enterprise: The alignment of business and information technology strategies. *IBM Systems J.* **32**(1) 198–221.

Luftman, Jerry N. 1996. *Competing in the Information Age: Strategic Alignment in Practice.* Oxford University Press, New York.

Lynch, Alfred F. 1992. Training for a new ball game: Retailing in the 21st century. *Futurist* **26**(4) 36–40.

MacDonald, Hugh. 1991. The strategic alignment process. Scott Morton, S. Michael, eds. *The Corporation of the 1990s: Information Technology and Organizational Transformation.* Oxford University Press.

Malone, T. W., J. Yates, R. I. Benjamin. 1987. Electronic markets and electronic hierarchies. *Comm. ACM* **30** 484–97.

Markus, M. Lynne, Christina Soh. 1993. Banking on information technology: Converting IT spending into firm performance. Rajiv D. Banker, Robert J. Kauffman, and Mo Adam Mahmood, eds., *Perspectives on the Strategic and Economic Value of Information Technology Investment.* Idea Group Publishing, 364–92.

Mayor, Tracy. 1998. Talking strategy. An interview with John Henderson, *CIO Magazine* (January 15).

McFarlan, F. Warren. 1984. Information technology changes the way you compete. *Harvard Business Rev.* **62**(3) 98–103.

Miles, Raymond E., Charles C. Snow. 1986. Organizations: New concepts for new forms. *California Management Rev.* **28**(3) 63–73.

Mintzberg, Henry. 1979. *The Structuring of Organizations: A Synthesis of the Research.* Prentice-Hall.

Mukhopadhyay, Tridas, Sunder Kekre. 1995. Business value of information technology: A study of electronic data interchange. *MIS Quart.* **19** (2) 137–56.

National Research Council. 1994. *Information Technology in the Service Society: A Twenty-First Century Lever.* National Academy Press.

O'Connor, M. 1991. What every CEO should know about new merchandising technology. *Supermarket Bus.* **46**(5) 63–70.

Palmer, Jonathan. 1995. The performance impact of information technology in specialty retailing. *Proc. Assoc. Inform. Sci.*, Pittsburgh, PA.

Papp, Raymond. 1998. Business-IT alignment payoff: Financial factors and performance implications. *Proc. Americas Conf. Inform. Systems*, Baltimore, MD. 533–5.

_____, Jerry Luftman. 1995. Business and I/T strategic alignment: New perspectives and assessments. *Proc. First Americas Conf. Inform. Systems*, Pittsburgh, PA. 226–8.

Parsons, Gary L. 1983. Information technology: A new competitive weapon. *Sloan Management Rev.* **25**(1).

Pastore, Richard. 1994. Minding your business. *CIO* (July) 54–8.

Peterson, Robert A. ed. 1992. *The Future of U.S. Retailing: An Agenda for the 21st Century.* Quorum Books.

Pfeffer, Jeffrey. 1982. *Organizations and Organization Theory.* Pitman.

Porter, Michael E. 1980. *Competitive Strategy: Techniques for Analyzing Industries and Competitors.* The Free Press.

_____. 1985. *Competitive Advantage: Creating and Sustaining Superior Performance.* The Free Press.

Prairie, P. 1996. Benchmarking IT Strategic Alignment. J. N. Luftman, ed. *Competing in the Information Age: Strategic Alignment in Practice.* Oxford University Press, New York. 242–90.

Quinn, James Brian, Martin Baily. 1994. Information technology. *Brookings Rev.* **12**(3) 36–41.

_____. 1992. *Intelligent Enterprise: A Knowledge and Service Based Paradigm for Industry.* The Free Press.

Reich, B. Horner, I. Benbasat. 1994. A model for the investigation of linkage between business and information technology objectives. N. Venkatraman and J. C. Henderson, eds. *Research in Strategic Management and Information Technology*, Volume 1. JAI Press Inc., Greenwich, CT. 41–72.

_____, _____. 1996. Measuring the linkage between business and information technology objectives. *MIS Quart.* **20**(1) 55–81.

Retail Information Systems News. 1996. Customer driven retailing. Special Supplement. (January).

_____. 1995. Competing with key technologies. *Fifth Ann. RIS News/CSC Retail Tech. Stud.*, New York.

_____. 1996. QR and ECR: Survival necessities. Special Supplement. (Spring) New York.

_____. 1997. Exploring the digital future. *Seventh Ann. Retail Tech. Stud.* New York.

_____. 1998. Entering the digital future. *Eighth Ann. Retail Tech. Stud.* New York.

Peter Weill, Associate Editor. This paper was received on January 5, 1996, and has been with the authors 35 months for 5 revisions.

Coordination and Virtualization: The Role of Electronic Networks and Personal Relationships

Robert Kraut, Charles Steinfield, Alice P. Chan,
Brian Butler and Anne Hoag

School of Computer Science, Wean Hall, 5000 Forbes Avenue, Carnegie Mellon University, Pittsburgh, Pennsylvania 15213, robert.kraut@cmu.edu
Department of Telecommunications, Michigan State University, 409 Communication Arts and Sciences Building, East Lansing, Michigan 48824, steinfie@pilot.msu.edu
331 Kennedy Hall, Cornell University, Ithaca, New York 14853, apc18@cornell.edu
GSIA, Carnegie Mellon University, 5000 Forbes Avenue, Pittsburgh, Pennsylvania 15213, bbutler@mail.business.pitt.edu
College of Communications, Pennsylvania State University, University Park, Pennsylvania 16802, amh13@psu.edu

Abstract

One view holds that organizations are virtual to the extent that they outsource key components of their production processes, and that electronic networks make it easier to do this. The goal of the present paper is to examine explicitly the effects that use of electronic networks for transactions with suppliers has on firms' degree of virtualization. In so doing, we also highlight factors that influence the use of networks for coordination with suppliers, and the impact such use has on coordination success. Contrary to much recent speculation, the use of electronic networks for transactions was not associated with increased outsourcing, but rather with greater dependence on internal production. Moreover, the use of interpersonal relationships for coordination, which many think of as an alternative to electronic network use, was positively associated with greater network use. Surprisingly, use of electronic networks was negatively associated with such outcomes as order quality and efficiency, and satisfaction with suppliers, while more reliance on personal linkages was associated with better outcomes and mitigated the negative consequences of using electronic networks.
(*Virtual Organizations; Coordination; Interfirm Transactions; IT Networks; Outsourcing*)

Introduction

Much of the literature on virtual organizations rests on two assumptions. The first is that firms adopt virtual forms in order to gain benefits of acquiring goods and services from specialized producers, who are able to make these inputs more efficiently (Davidow and Malone 1992). The second assumption is that modern computer and telecommunications networks sufficiently reduce the costs of coordination, allowing firms to achieve these production benefits without incurring the higher transaction costs traditionally associated with buying from an external supplier (Malone et al. 1987). The goal of this article is to examine the validity of these assumptions by empirically testing some of their implications.

We had intended to start our research with case studies of virtual companies. We initially considered these companies to be ones that were successfully using computer and telecommunications networks to link themselves with other companies, where important design, management, and production for the firms' products were being conducted. To this end, during the summer and fall of 1995, we conducted interviews in 14 large companies in four industries—apparel manufacturing, pharmaceuticals, magazine publishing, and advertising—where we thought virtual organizations were common. In each firm, the informants included a senior manager in operations or manufacturing and a senior manager in charge of information systems. Eleven firms were interviewed on-site, and three were interviewed by telephone. In these interviews, informants described the production of their firm's most important product or service, their rationale for outsourcing or keeping in-house key elements of production, and the role that information technology in general and computer networks in particular played in production. Interviews in a single firm typically lasted about three hours and involved two to four informants.

However, we found few virtual companies in these industries. While many apparel firms outsourced key elements of manufacture and even design, they zealously guarded other production elements (e.g., selection of fabric). Magazine publishers used some freelance reporters and outsourced printing, but kept editorial decisions, layout, and a substantial amount of the writing in-house. These observations led us to reconsider the definition of virtual organization we had been using and to revisit the relevant transaction cost literature, which provides a rationale for the existence of the traditional firm and suggests some of the constraints on virtual organizing. We start our introduction by proposing that virtual organizing is a matter of degree.

Even in cases where the responsibility for design and production was spread across multiple firms, we were surprised at how unimportant computer-to-computer networking seemed to be in coordinating production. Telephones, fax machines, and express mail were much more prevalent than computer networks, and where networks were used, electronic mail connecting people was far more common than electronic document interchange connecting machines. When we identified interesting examples of computer networking (e.g., a newsweekly sending page layout information to typesetting machines in a printing plant, or graphic designers in New York and Los Angeles collaborating by using shared computer files and shared computer screens), the examples were as likely to occur within a single firm as between firms. These observations led us to consider the interplay between two mechanisms for coordinating production processes, personal relationships and electronic networks, and to examine the conditions under which firms use them for coordination. We focused on these two mechanisms because of their potential importance for interfirm coordination (e.g., Granovetter 1985; Keen and Cummins 1993) and because much of the literature on virtual organizations assumes that the two mechanisms substitute for each other and have different effects on important business outcomes (e.g., Kekre and Mudhopadhyay 1992). We review the literature on electronic networks and interpersonal communication for coordination in the next section of the introduction.

Our literature review, below, leads to research questions on the conditions under which firms use electronic networks, on the effects of electronic networks on virtualization, and on the effects of coordination mechanism choices and virtualization on performance outcomes. We examined these questions through an empirical investigation of how firms in four industries use personal connections and electronic networks to support transactions with suppliers of important inputs to their production processes. The data come from a national survey of 250 firms in apparel manufacturing, pharmaceuticals, magazine publishing, and advertising. These vary in their technological sophistication and involve products that range from tangible to information-intensive. We examine both the conditions under which firms use different coordination mechanisms and the relationships between the use of these mechanisms, organizational forms, and business process outcomes.

Organizational Forms

Defining Virtual Organizations

The term "virtual organization," as used in the literature, has no consistent meaning. The term has been applied to movie production, in which personnel come together for the duration of a project; just-in-time manufacturing operations, in which subcontractors simultaneously act as a manufacturing firm's supplier and warehouse; "adhocracies," in which specialized task forces and work groups arise and disband on demand; and informal regional consortia, in which material and personnel flow through the companies in a geographic area. Although these examples do not provide a tight definition, they suggest some of the features that underlie the concept of a virtual organization. First, production processes transcend the boundaries of a single firm and, as a result, are not controlled by a single organizational hierarchy. Second, and perhaps as a result, production processes are flexible, with different parties involved at different times. Third, the parties involved in the production of a single product are often geographically dispersed. And finally, given geographic dispersion, coordination is heavily dependent on telecommunications and data networks rather than physical travel, at least for the people involved.

For most firms, being virtual is a matter of degree. The production of any complex good or service requires combining various raw materials and modifying them in many stages, with each step adding value as the product wends its way towards the final consumer (Porter 1980). At one extreme, a firm is virtual to the extent that each of these steps is performed outside the core firm's boundaries, with the firm acting as coordinator. Some publishing operations approach this extreme, with no writing, editing, printing, and distribution done within the firm itself. But this non-value-adding structure will rarely arise in competitive business environments. Even book publishing houses typically perform manuscript selection and marketing in-house. The other extreme, the traditional, fully integrated organization—in which a single firm performs all aspects of management, production, sales, and distribution—is also unlikely to arise. In reality, most firms perform some steps internally, and make contractual and logistical arrangements to have other activities performed by one or more external supplier(s). In the case of physical production processes, this involves deciding whether to make or to buy each component of the product. For services, it is often the choice between using in-house staff or external consultants, freelancers, or specialty firms. Most organizations are situated between these extremes. Rather than virtual organizations, we can expect to find the "virtualization" of organizations, which is better viewed as a continuum. Firms become more virtual when a larger proportion of important production processes occur outside of traditional organizational boundaries.

When do firms become more virtual? By defining virtual organization in terms of the number and importance of cross-boundary transactions, one links this phenomenon to the substantial theoretical and empirical literature on transaction cost economics. Beginning with Coase (1937) and Williamson (1975), transaction cost theorists have focused on understanding why certain activities are kept within the boundaries of the firm while others are performed outside. Transaction cost models of organizations have proposed that firms make decisions about the location of business processes to minimize the combined cost of production and governance.

If only production costs were considered, many theorists would argue that these costs can be reduced as more production activities are performed outside the firm (i.e., greater virtualization, e.g., Malone 1987, Davidow and Malone 1992). By moving production outside of the core firm to other, specialized companies, the core firm gains access to more experience, makes better use of available production facilities, and capitalizes on economies of scale, all of which can lead to lowered production costs. If enough potential suppliers exist, a firm can shop around for the best combination of price, quality, or other desirable attributes. As a result, the procuring firm can take advantage of the capabilities of the most efficient producers available and also avoid being held hostage to the opportunistic behavior of any single supplier. For instance, a dress manufacturer could buy designs, fabric, buttons, and assembly services in the open market, and may find "bargains" in each area. As a result, unlike firms that follow an in-house production strategy, firms that are more virtual are expected to be more efficient, flexible, and effective due to their ability to re-form as the environment changes.

However, in making decisions about the degree of virtualization to adopt, firms must also take into account the costs of executing various transactions as well as the costs of production. Costs of governance are expected to be higher when firms purchase goods and services in the open market rather than producing them in-house (Williamson 1975, 1985, 1996). When using external suppliers, firms incur costs as they search for appropriate partners, specify agreements, enforce contracts, and handle financial settlements. For complex components or services, firms may have difficulty specifying what they want, and suppliers incur costs when advertising the availability of their goods and services to potential customers (Malone et al. 1987). After the most appropriate supplier has been identified, governance processes, such as arriving at and enforcing a contractual agreement, monitoring and controlling quality, processing orders, and settling payment, all create transaction costs.

Moreover, firms have bounded rationality and cannot know all contingencies that might arise during a transaction. Hence, when the goods or services that a firm needs change from order to order or when a firm doesn't know what it will need until the last minute, firms are more likely to choose traditional, integrated organizational forms. These conditions are further exacerbated in cases where only a small number of suppliers exist that can satisfy a firm's needs, increasing the potential for opportunism. The small numbers problem is more likely to arise when business processes require highly specific inputs and, as a result, firms do not have a large number of alternative suppliers with which to do business.

A traditional, integrated organization has the advantage of lowering the costs of certain forms of governance. Routine activities are formalized in standard operating procedures, which are known throughout the organization. Formalized communication paths and assigned roles reduce the need to search and negotiate agreements. Employment agreements and hierarchical management free firms from the need to specify future contingencies in contracts. Thus, Williamson (1975) argues that traditional organization forms exist, despite production inefficiencies, because they provide lower-cost mechanisms for governing the execution of complex business processes.

In summary, the transaction cost literature proposes that features of the goods or services being procured and of the potential suppliers directly influence whether a firm makes the good or service in-house or outsources it, thus virtualizing its production process. These features include: (a) features that make a firm vulnerable to opportunism, including the relative power of potential suppliers vis-à-vis the focal firm and the uniqueness of the object being acquired; and (b) the complexity of the transaction, including the difficulty of specifying the object to be acquired, uncertainty because the object itself changes from order to order, and unpredictability in the quantity and quality of objects a firm will need in a particular order.

Coordination Mechanisms

Electronic Networks and Organizational Forms

Many distinct mechanisms can be applied to the governance and coordination of business transactions between a firm and its suppliers, including formal legal contracts, information technologies, and informal personal relationships. The focus in transaction cost theory on opportunism and contract-based mechanisms to guard against opportunism has led to an emphasis on governance processes, such as bargaining, negotiation, and enforcement (Williamson 1985, 1991).

Throughout this article, we use the term "electronic network" broadly, to designate any type of computer or data network that allows companies to exchange information between computers. The information that flows over these networks varies and can include cases where one computer directly controls other machines involved in production (e.g., a page-layout computer in New York that directly controls typesetting machines in printing plants across the country); electronic document interchanges or funds transfers, delivering invoice or payment information; or electronic mail networks, where people exchange messages via computers. We use the term to apply both to data networks deployed within a single firm and to interorganizational networks, deployed between firms.

In a widely cited article, Malone et al. (1987) argued that firms can use electronic networks to reduce governance costs and thus moderate the effects of specificity and complexity that normally lead to market failures. The argument is that if electronic networks were used to expand the pool of potential suppliers inexpensively—what Malone et al. (1987) refer to as electronic brokerage effects—they would reduce firms' vulnerability to opportunism. With more potential suppliers, firms are less subject to threats of opportunism from any particular one. Electronic networks could also lead to less vulnerability to opportunism if firms used them to monitor a supplier's internal production processes inexpensively, as has happened among organizations that use electronic silicon foundries (e.g., Hart and Estrin 1991).

When firms execute complex business activities, they must do more than guard against opportunism. They must also solve significant logistical and communication problems, i.e., problems of coordination. Malone and colleagues (1987) argue that because electronic networks reduce the costs of coordinating business processes between firms, these networks will cause firms to conduct more transactions across organizational boundaries and thus become more virtual. Their argument is that anything that lowers the cost of interfirm coordination allows firms to exploit the lower production costs they can achieve through outsourcing. A number of authors have followed Malone et al. in hypothesizing that by lowering coordination costs, electronic networks enable virtual organizing, leading to smaller firms that outsource more elements of the production process (cf., Bradley et al. 1993, Clemons 1993, Miller et al. 1993). These arguments about both governance and coordination costs lead to the prediction that the use of electronic networks will allow firms to outsource more and thus cause them to become more virtual.

Except for a series of widely cited case studies (e.g., Malone and Rockart 1993), there are few empirical tests of the hypothesis that greater use of interorganizational networks leads to greater outsourcing, and most of the data speak to the issue only indirectly. Consistent with this hypothesis is evidence at the industry level that increases in investment in information technology are associated with declines in average firm size and increases in the number of firms (Brynjolfsson et al. 1994). Kambil (1991) shows that industries investing more of their capital stock in information technology also contract out more of the value of the goods and services they produce to external suppliers (i.e., a higher buy/make ratio in production), with a two-year lag. Limitations in the data, however, mean that the analyses are only suggestive. Most importantly, the prior studies' use of investment in general information technology as the independent variable obscures the unique role that interorganizational computer and communication networks are hypothesized to play.

Two assumptions underlie much of the argument that electronic networks increase virtualization. The first is that governance and coordination are more difficult between firms than within a single firm. The second is that the use of electronic networks decreases this gap, i.e., that computer networks reduce the costs of between-firm governance and coordination more than they reduce the costs within a single firm. The second of these assumptions, however, is probably wrong. We believe that much of the discussion about the effects of electronic networks underestimates benefits that can accrue to firms adopting computer and communication networks for internal use. In deploying networks internally, firms have greater knowledge of the business processes that could benefit and greater control over the implementation process. Historically, firms first adopt networks for use within the firm. When firms deploy networks internally, they are less vulnerable to intrusion, can enforce technical standards to ensure compatibility, and can control many other factors that can lower the cost of implementation. For example, they have a greater likelihood of enforcing standards so that computers deployed at different points in the production process can communicate with each other. If firms achieve greater net governance and coordination benefits when they deploy within-firm networks than when they deploy interorganizational networks, then network use should lead to greater internalization of production, not outsourcing. One of the goals of our research is to determine whether greater use of electronic networks is associated with more or less virtualization.

Personal Relationships

Although the literature on virtual organizations is dominated by discussion of organizational form and information technology, in many cases personal relationships, not ownership or technology, may be the key mechanism for coordinating complex business processes. Sociologists like Granovetter (1985) have argued that many economic exchanges are mediated by the social relationships among the trading partners, and at times these social relationships can be more efficient than market mechanisms for conducting transactions (c.f., Lawrence and Lorsch 1967, Tushman and Nadler 1978). Recent work by Uzzi (1997) provides a detailed analysis of the ways in which coordination via personal relationships contrasts with market or arm's-length mechanisms for coordination. People generally rely on personal relationships to resolve problems and deal with unusual situations (Krackhardt 1992). Personal relationships also serve as a valuable governance mechanism. For example, personal relationships can lead to trust between parties involved in an economic exchange, which in turn reduces the likelihood of opportunistic behavior (Granovetter 1985, Uzzi 1997, Zucker 1986).

The exploratory interviews we conducted in starting our research highlighted the importance of personal relationships between buyers and suppliers. Personal relationships were frequently called upon when firms were dealing with exceptions to routine, such as when an advertising firm needed to deliver updated ad copy after a magazine's publication deadline or when a supplier needed to provide substitute fabric to a suit manufacturer. Even for more routine tasks, such as trying to identify a new supplier, purchasers typically considered contacts with people in other firms to be more useful than structured information resources like directories and databases. For example, when a manager in a consumer pharmaceuticals firm needed to find a supplier for an unusual plastic container, he spent most of the search time on the telephone getting referrals from current suppliers and professional colleagues.

Although this discussion implies that personal relationships will be an important mechanism for interfirm coordination, it is unclear how coordination based on personal relationships and that based on information technology will interact. Studies of electronic networks often imply that information technology and interpersonal relationships compete and that firms derive more benefit from using information technology to coordinate when it replaces person-to-person contact. Yet, most research that finds substitution of electronic coordination for personal coordination focuses on the coordination of routine activities (e.g., Kekre and Mukhopadhyay 1992). It is likely that personal relationships are most valuable when dealing with nonroutine transactions. For example, while electronic data interchange systems may be very effective for arranging routine orders for standardized products, they may be less suited for supporting the negotiation surrounding the acquisition of new goods or services, or when dealing with unusual situations that fall outside of standard procedure. Chan (1997), for example, found that electronic networks are perceived to be more important for routine activities (such as ordering or getting price information) than nonroutine interactions with suppliers, such as contract negotiations and after-sale problem solving.

Because coordination using electronic networks may be more fragile than coordination based on personal contacts, it may require interpersonal backup to work successfully (e.g., Suchman and Wynn 1984). Kling and Scacchi's (1982) description of a social web of computing was an early recognition of the importance of social relationships in supporting successful computer use. More recent research (e.g., Hart and Estrin 1991, Saunders and Hart 1993) has also stressed how the introduction and effective use of electronic networks for electronic integration requires prior personal acquaintanceship and trust among the parties that will be sharing data.

Effects of Coordination Mechanisms on Coordination Success.

Empirical research on organizational use of interorganizational electronic networks finds that they positively influence business process outcomes (Kekre and Mukhopadhyay 1992, Streeter et al. 1996). It is not clear, however, whether these networks are more useful for supporting outsourcing or in-house production. Similarly, it is not clear how the use of electronic networks compares to the use of personal relationships or in-house production. Coordination through personal relationships may be less efficient, because it relies on costly and error-prone human behavior. On the other hand, if personal relationships lead to trust, thus allowing firms to avoid the need for costly monitoring efforts, or if they lead to the exchange of favors, then coordination through personal relationships may be associated with successful coordination outcomes. For example, Steinfield and Caby (1997) found that an electronic network created to improve the efficiency of selling advertising time in the French television industry failed because it was perceived as decreasing the quality of outcomes. The network reduced sales representatives' flexibility, preventing them from using their existing personal relationships to offer better times and rates to preferred and high-volume customers. They thus began to bypass the system, leading to eventual failure of the electronic market.

Research Questions

We can summarize this discussion with several research questions. Our major goals in this research are to understand how the use of electronic networks influences virtualization and how the use of electronic networks and interpersonal relationships for coordination changes production outcomes. However, to examine the influence of electronic networks, it is necessary to control for other factors that could make electronic networks useful.

RESEARCH QUESTION 1. *Under what conditions will firms use electronic networks?*

Transaction Attributes. First, we expected that when products or services (such as software or information) themselves could be transportable over electronic networks, the cost advantages of using a network would be greater. Second, we expected that the speed advantages of networks would be most beneficial when firms were under greater time pressures, or when they could not predict the timing of their needs. Third, we expected that when objects were easier to describe, it would be easier to order over an electronic network. Finally, reductions in labor costs through automation of transactions would motivate firms to employ electronic networks, but only if there were a sufficient volume of transactions for this payoff to be meaningful given the costs associated with network implementation (Keen and Cummins 1993). Thus, we expected that firms would use electronic networks more when they:

a) were under greater time pressure,
b) had greater unpredictability about their inputs,
c) placed large numbers of orders per year,
d) were acquiring intangible inputs, and
e) were acquiring inputs that were easy to describe.

Personal Relationships. While predicting the influence of some production variables, like time pressure, on use of electronic networks is straightforward, predicting the influence of personal relationships is more difficult. Our review of theory suggests that electronic networks can be used either to replace personal relationships as a mechanism for coordination between a firm and its suppliers or to supplement them.

RESEARCH QUESTION 2. *What effects do electronic networks have on firms' degree of virtualization (i.e., outsourcing of key components of production)?*

Improved Coordination. The conventional view among information systems researchers is that interorganizational computer networks enable efficient outsourcing. While use of electronic networks can facilitate coordination both within firms and between them, many information systems researchers believe that in the long run these networks will lead to greater outsourcing (e.g., Malone et al. 1987). Thus, one might expect a positive association between use of electronic networks and virtualization.

Vulnerability to Opportunism. Malone and his colleagues reason that interorganizational computer networks will lead to greater virtualization in part because they can reduce potential opportunism when outsourcing. If this is correct, one would expect statistical interactions between the degree of electronic network use and the effects of the opportunism variables (object specificity, variability, unpredictability, and complexity of product description) on outsourcing. That is, firms will outsource less when procuring highly specific products (or other products subject to opportunism), but this relationship will be reduced among firms that use electronic networks.

However, if the benefits of using electronic networks inside the firm exceed the benefits of using networks to connect to external suppliers, or if the use of electronic networks with external suppliers follows use within the firm, one might expect to see a negative association of electronic network use with virtualization and no interactions between electronic network use and the opportunism variables.

RESEARCH QUESTION 3. *What effects does the mode of coordination have on the outcomes of transactions with suppliers?*

Electronic Network Use and Coordination Success. Firms are motivated to use electronic networks for transactions because of putative gains in efficiency. Empirical evidence on this topic is limited, but, based on the information systems and transaction cost literature, we expected to find that greater use of electronic networks would be associated with fewer errors in orders, greater efficiency in orders, and more satisfaction with the suppliers with which they are used.

Personal Relationships Use and Coordination Success. We expected that use of personal relationships would influence the success firms have in coordination with suppliers (errors in orders, efficiency in orders, and satisfaction with the suppliers), but made no prediction on the direction of the effect. The direction should depend on whether the flexibility associated with personal relationships is worth the cost of providing it.

Interactions Between Use of Electronic Networks and Personal Relationships on Coordination Success. A negative interaction between the use of electronic networks and the use of personal contacts on the outcome measures would suggest that electronic networks are more effective when they substitute for and replace human contact between the supplier and purchasing firm. On the other hand, a positive interaction would suggest that electronic networks are most effective when they are supplemented by personal contacts.

Virtualization and Coordination Success. To the extent that use of in-house production improves coordination, it should have positive effects on coordination success.

Interactions Between Virtualization and Use of Electronic Networks on Coordination Success. If electronic networks are more beneficial when coordinating with external suppliers than when improving in-house production, as the literature on virtualization of organizations suggests, then we would expect to see an interaction between electronic network use and degree of outsourcing on coordination success.

Methods

Sample and Interview

This research is based on a telephone survey of managers in a sample of 250 firms from the advertising, magazine publishing, women's apparel, and pharmaceuticals industries. Firms with at least 20 employees in these industries listed in the April 1996 Dunn and Bradstreet database were ranked by number of employees, and a stratified sample of large, medium, and small firms (relative to each industry) was selected from each third of the size distribution. The goal was to sample several industries with differing product and production approaches in order to look for effects that generalize across industries. Advertising and publishing were chosen as producers of information products. Apparel and pharmaceuticals were selected as examples of manufacturing industries. In the analyses below, we control for industry by including it as a dummy variable. However, because the industries differ on many unmeasured variables, we do not delve further in trying to explain industry differences. Table 1 shows the average number of employees in each industry-size category and the number of firms sampled in each category.

The telephone survey focused on the way in which the producing firm acquired key inputs from suppliers. Respondents were the senior managers most responsible for acquiring the input under consideration. They were asked to identify their company's principal product or service and then were asked about a specific input needed to produce that product. Depending on the question, respondents were asked about the way they acquired the input in general or about the way they dealt with the "the most important supplier you've worked with within the past 12 months." We stressed that the supplier could be an employee or a department in the respondent's firm or an outside individual or company.

We identified the set of key inputs for each industry through the exploratory, semistructured interviews, which we have previously described. Inputs were chosen to vary on whether they were tangible, requiring physical transport, or intangible and suitable for transport over an electronic network. Table 2 shows the four inputs for each industry, and their classification as tangible or intangible.

Industry size	Pharmaceuticals	Advertising	Women's Apparel	Magazine Publishing
Small—number of employees	44	24	56	27
(N)	(18)	(22)	(24)	(20)
Medium—number of employees	84	38	146	35
(N)	(21)	(21)	(14)	(23)
Large—number of employees	13,262	508	261	759
(N)	(20)	(21)	(22)	(24)
Total N = 250	(59)	(64)	(60)	(67)

Table 1 Sample characteristics.

Note
Entries represent mean number of employees in the firms and the number of firms presented in industry-size category.

Respondents were questioned about one input randomly chosen from a list of four for the appropriate industry. If a particular company did not use this input or if the quota for the input had already been filled in the industry-size category, then the interviewer randomly selected another input.

The telephone survey was conducted by a professional survey firm. Phone interviews, typically lasting 30 minutes, took place in July and August 1996. In reaching the respondent, interviewers used a series of screening questions to identify "the person who is most responsible for arranging and acquiring" the specified input. Interviewees included both managers specializing in procurement (e.g., a vice president of purchasing) as well as those in operational areas (e.g., the senior editor responsible for assigning stories to staffers and freelancers). The survey process produced 250 completed interviews on a survey instrument with 102 items. Excluding firms that no longer existed, that could not be contacted after seven tries, or that failed to meet size or industry definitions, the response rate was slightly over 50%.

Measurement

To address the research questions described previously, we created multi-item scales to measure concepts related to uncertainty, opportunism, object specificity, coordination mechanisms, extent of outsourcing, and coordination outcomes. Individual items were first examined using principal components factor analyses. Items that did not load as expected, or cross-loaded, were removed and factor analyses rerun. Multi-item scales were only created using items that exhibited both discriminant and convergent validity. Scale items were included if they were: 1) included in the survey to measure a single concept; 2) loaded at 0.5 or higher on the same factor; and 3) did not cross-load (at 0.4 or higher) on any other factor. When our a priori scale items did not load together, but when there was sufficient theoretical justification for including a construct, single-item measures were used in our data analyses. All multi-item scales were subjected to reliability analyses. (Appendix A lists all the multi-item scales and their reliabilities.) Reliabilities as measured by Cronbach's alpha were between 0.65 and 0.90, sufficient for exploratory analyses.

The six sets of variables included in our analyses are described below:

Control Variables. In all analyses we included variables for *industry, size of focal firm*, and *size of supplier*. We wished to hold these variables constant, because they are likely to have wide-ranging effects on how firms organize production and the degree to which firms and their suppliers use technology. Dummy variables were used to represent each of the four *industries*. The *size of focal firm* is the average of the number of employees in the firm and the firm's annual sales,

Industry	Input	
	Tangible	Intangible
Women's Apparel	Fabric Trim (buttons, zippers, etc.) Cutting	Garment Design
Advertising		Artwork and Graphics Television Time Slots Television Ad Distribution Market Research
Pharmaceuticals	Chemical Ingredients Packaging Materials Packaging Assembly	Clinical Trial Management
Magazine Publishing	Paper Stock Printing	Stories Color Separation

Table 2 Key inputs studied in each of four industries.

Note
Women's apparel manufacturers came from the following SIC codes: 2331, 2335, 2337, 2339, 2342, 2384 and 2389. Magazine publishers' SIC code was 2721. Pharmaceuticals manufacturers' SIC codes were 2833 and 2834. Advertising agencies were from SIC code 1591.

first standardized and then logged (alpha = 0.91). The data were taken from Dunn and Bradstreet's 1996 listing for the producing firm. *Supplier size* is the average of the respondent's estimates of the number of employees the supplier has and its market share, rated on a five-point scale. Both were first standardized before averaging (alpha = 0.73). In the analyses reported below, industry effects are considered as controls, since there was no a priori reason why they should differ in terms of the processes underlying the development of organizational forms, network use, and coordination outcomes.

Product and Production Attributes Influencing the Potential Utility of Electronic Networks. In order to test the effects of electronic network use on virtualization, we must control for important features of products and production that enable or encourage firms to use networks. We reasoned that firms would get the most coordination benefit from using networks when the goods they were acquiring could be transported over the network, when they were under time pressure, and when they ordered frequently.

An input's *tangibility* refers to the ability to execute the transaction entirely through an electronic network with currently existing technology. If the completion of the transaction required the transfer of a physical object (e.g., a bolt of fabric), then the transaction was categorized as tangible. If, on the other hand, it was technologically feasible to completely eliminate the exchange of material objects and still complete the transaction (e.g., a garment design, which could be exchanged as a computer file or faxed message), then the input was considered intangible, regardless of whether a particular firm actually used electronic networks. Table 2 shows the classification of inputs.

Time pressure is the respondent's judgment of the extent to which production in the firm was rushed and subject to rigid deadlines (e.g., "To meet schedules related to this product or service, we need to use every available minute efficiently.") (two items; alpha = 0.68).

Order frequency is a single item measuring the ordering cycle for the key input, from yearly to monthly.

Product and Production Attributes Influencing Vulnerability to Opportunism[1] Transaction cost theory proposes that firms will be less likely to outsource production if they are vulnerable to exploitation by external suppliers. At a conceptual level, the key variables are specificity of the required assets and complexity of the inputs. We operationalized asset specificity as the degree to which the input was generic to the industry or unique to the acquiring firm (generic versus specific object). We operationalized complexity with three variables: change versus stability in the input acquired (object certainty), unpredictability versus predictability in the quality and quantity of input needed (object predictability), and the ease of identifying the input (ease of description).

Generic versus firm-specific object is the degree to which the key input itself is usable by any producing firm, rather than being custom-made for the respondent's firm. At one extreme are inputs like newsprint or shirt buttons, which are used by many manufacturers, while at the other are news stories or specialized fabrics, which are specific to a particular manufacturer. This dimension was measured by the extent to which respondents agreed on a five-point Likert scale with four statements, (e.g., "My firm is the only one that uses the [input]" (reversed), and "The [input] we get from this supplier is fairly standard for the industry") (alpha = 0.64).

Object certainty is the extent to which the item being ordered, its availability, and its price are stable over time, (e.g., "How much does the [input] your company needs change from order to order?" (reversed) (three items; alpha = 0.68).

Object predictability is the extent to which respondents knew in advance the specific input they would need and its quantity (e.g., "For a typical order, how far in advance can you predict what [input] your company will need?") (two items; alpha = 0.79).

For *ease of description*, respondents were asked to rate how easily the key input could be described so that it could be ordered (e.g., "It is difficult to describe the [input] we routinely acquire" (reversed).

Coordination Mechanisms. The survey asked about the importance of using *electronic networks* to acquire the input from a major supplier. This was stated as the importance of using "any type of computer communication that allows you to exchange information with this supplier" in the acquisition process. Measured on a five-point scale, we asked about the importance of using electronic links for each of six separable stages in the acquisition process (Kambil 1991), including: 1) searching for and selecting a supplier; 2) developing the specifications of the key input; 3) negotiating the terms of the acquisition such as price, delivery date, and so on; 4) ordering the input; 5) monitoring the quality of the good or service; and 6) fixing problems after the order. Respondents made their judgments on five-point Likert scales, from not at all important to very important. The composite scale has high reliability (alpha = 0.94).

We measured the importance of *personal relationships* in acquiring the key input for each of the six stages of acquisition. Personal relations were defined as including "any type of personal connection or personal knowledge between people in your firm and people in the supplier's organization." Respondents made their judgments on a five-point Likert scale of importance (alpha = 0.85).

Organizational Form. Outsourced production is a composite index that estimates the extent to which a firm outsourced the key input or produced it in-house. It is based on three items: (1) whether the major supplier for the input was external or in-house; (2) the nature of the relation between the major supplier and the producer firm, ranging from "departments within your firm" to "owned subsidiary" to "a joint venture firm" to an "outsider" firm; and (3) the percentage of the key input which was produced in-house, ranging from 0 to 100. If the index is high, then the firm outsources more of its production and is more of a virtual organization (alpha = 0.86).

Outcome Variables. Overall order quality is a six-item composite index consisting of two subscales—the extent to which orders were free from error and the extent to which they were executed efficiently. The error items included the percentage of orders arriving after the agreed-upon delivery date, the percentage of orders containing any sort of error, and the percentage of orders that failed to meet a firm's quality standards. The efficiency items included the average elapsed time between ordering and receiving the input, the number of different people who are involved in handling the order, and the number of communications with the supplier during a typical order cycle. Because these items were nonnormally distributed, estimates were logged and then standardized before averaging (six items; alpha = 0.65).

Satisfaction with a supplier is an index in which respondents indicated their overall satisfaction with the supplier, their satisfaction with six aspects of the ordering process, and their satisfaction with the supplier's ability to handle exceptional orders (nine items; alpha = 0.89).

Missing Data

Most variables had only a few missing data values. In the regression models that follow, we imputed missing values by replacing them with the industry mean. When imputing firm size we replaced the missing value with the average number of employees in the other firms in the appropriate industry-size category. Since attributes of the suppliers varied with the input being acquired, we imputed supplier size from the average of the nonmissing values from firms providing the same input (e.g., from other fabric suppliers).

Analysis

Our research questions were investigated through a series of regression analyses. First, in order to examine Research Question 1, which asked about the conditions under which firms acquire inputs from suppliers through electronic networks, the variables that indicated a need or an opportunity for coordination efficiencies (such as order frequency, time pressures, and tangibility) were entered into a regression equation predicting the importance of electronic network use (Model 1: Predicting Network Use). We entered personal relationships into this equation to examine whether these are complementary or contradictory modes of coordination. Control variables as well as other independent variables reflecting vulnerability to opportunism were also entered into this equation. We next examined Research Question 2.1, which investigates the effects of electronic linkages on organizational form, by predicting the extent of outsourcing from the importance of electronic network use, controlling for the full set of control and independent variables (Model 2: Predicting Outsourcing (all suppliers)). A more conservative test of Research Question 2.1 is also provided in the results section, by examining these relationships only for that subset of the sample of firms that used external suppliers as their primary source of supply (Model 3: Predicting Outsourcing (external suppliers)). As part of our analysis of Research Question 2, we explored the argument that electronic networks permitted greater outsourcing by moderating the factors that cause vulnerability to opportunism (Research Question 2.2). We did this by testing the interaction of network use with the opportunism variables (Model 4: Networks Moderating the Effects of Opportunism). Our final set of analyses, based on Research Question 3, asked about the effects of coordination mode on transaction outcomes. To examine this, we explored whether electronic networks and personal relationships predicted perceived order quality (Model 5: Predicting Order Quality) and satisfaction with supplier (Model 6: Predicting Satisfaction), controlling for all other predictor variables and the extent of outsourced production. We included the interaction of electronic networks with personal relationships to determine if the use of these two coordination mechanisms substitute for or complement each other. We included the interaction of electronic networks with outsourced production, to determine if outsourcing was more beneficial when used for in-house coordination or for coordination with outside suppliers.

Results

Research Question 1: Predicting Use of Electronic Networks

Model 1 in Table 3 shows standardized beta coefficients from a regression analysis predicting use of electronic networks from control variables, production efficiency variables, and opportunism variables. As expected, firms were less likely to use electronic networks when they were working with tangible goods ($\beta = -0.26$, $p < 0.002$). However, the relationships between use of networks and greater time pressure ($\beta = 0.10$, $p < 0.11$) and order frequency ($p < 0.40$) were not significant.

With the exception of ease of input description, we made no predictions about whether vulnerability to opportunism would be associated with the use of electronic networks. Contrary to the expectations of Malone and colleagues (1987), firms used networks more when they were acquiring items that were more difficult to describe ($\beta = -0.23$, $p < 0.001$). Networks were also used more to acquire inputs that were more tailored to their firm rather than generic ($\beta = -0.16$, $p < 0.02$). Both findings suggest that firms used electronic networks when most exposed to potential opportunism from suppliers. The findings do not, however, mean that the networks were being used to outsource more under these conditions. Rather, as we show in later analyses, firms seem to use networks to support in-house production when they are vulnerable to opportunism.

Finally, our data clearly support the contention that personal relationships and electronic networks are used as complementary rather than substitutable coordination mechanisms. Even controlling for other factors, firms that use personal relationships to coordinate with a supplier were also more likely to use electronic networks ($\beta = 0.16$, $p < 0.01$).

	Model 1: Predicting Network Use	Model 2: Predicting Outsourcing (all suppliers)	Model 3: Predicting Outsourcing (when main suplier is external)	Model 4: Networks Moderating the Effects of Opportunism
	Dependent Variable			
	Electronic Networks	Outsourced Production	Outsourced Production	Outsourced Production
Intercept	0.16	−0.01	0.41†	−0.03
Publishing vs. Apparel	−0.30†	−0.53†	0.18†	−0.52**
Pharmaceutical vs. Apparel	−0.15	0.28†	0.17	0.25
Advertising vs. Apparel	−0.18	0.24	0.16	0.23
Firm Size	0.06	0.03	0.02	0.04
Supplier Size	0.00	0.19**	0.01	0.18**
Time Pressure	0.10	−0.01	−0.05	−0.02
Order Frequency	0.01	−0.09	−0.02	−0.06
Tangibility	−0.26**	0.31***	0.24**	0.29***
Ease of Input Description	−0.23**	−0.06	0.02	−0.05
Object Certainty	−0.04	−0.02	−0.02	−0.03
Order Predictability	−0.10	0.09	0.05†	0.09
Generic Object	−0.16*	0.10†	0.06	0.11†
Personal Relationships	0.16**	−0.01	−0.02	0.01
Network Use	—	−0.28***	−0.07**	−0.29***
Network Use × Description Ease	—	—	—	0.04
Network Use × Certainty	—	—	—	0.03
Network Use × Predictability	—	—	—	−0.04
Network Use × Generic Object	—	—	—	−0.15*
N	250	250	185	250
Rsq	0.23	0.34	0.24	0.36

Table 3 Predicting use of electronic networks and outsourcing.

Note
Entries are standardized beta coefficients from a multiple regression analysis.
†: $p < 0.10$, *: $p < 0.05$, **: $p < 0.01$, ***: $p < 0.001$.

Research Question 2: Does the Use of Electronic Networks Predict Virtualization?

Model 2 in Table 3 shows results of a regression analysis predicting virtualization (i.e., outsourcing of a key production input) from control, production efficiency, and opportunism variables.

Transaction Cost Predictors of Virtualization. Transaction cost theory posits that organizations outsource most when inputs are more generic, stable, predictable, and easier to describe (and thereby find). Results only weakly supported these predictions. Firms were more likely to outsource more generic inputs ($\beta = 0.10$, $p < 0.10$). In addition, they outsourced tangible inputs more than intangible ones ($\beta = 0.31$, $p < 0.001$). Specifically, the inputs of fabric, trim, chemical raw materials, packaging materials, paper, and the time slots for advertisements[2] were more likely than average to be acquired externally. Garment designs, cutting services, artwork and graphics, stories, color work, television ad distribution, and packaging assembly were more likely to be accomplished in-house. The findings are consistent with transaction cost theory predictions about asset specificity. That is, highly customizable inputs, most of which were intangible (e.g., garment designs and stories) were more likely to be produced in-house. More commodity-like inputs or those that could be supplied to multiple producers (e.g., fabric and paper) were more likely to be outsourced. While transaction cost predictions about asset specificity were supported, other predictions derived from transaction cost theory were not. In particular, order certainty, order predictability, and ease of order description were not associated with outsourced production.

Firms were more likely to outsource when suppliers were larger ($\beta = 0.19$, $p < 0.01$). However, this seems to be an artifact created by the roughly one quarter of the sample for which an in-house division was the major supplier. Presumably, when a firm does in-house production, the in-house division is small relative to other firms in the industry. The relationship between in-house production and supplier size disappears when we include in the analysis only the subset of the sample whose major supplier is an external organization (see Model 3 in Table 3).

Model 2 in Table 3 shows that the more firms used electronic networks, the more they acquired needed inputs from internal sources rather than external ones ($\beta = -0.28$, $p < 0.001$). This is inconsistent with the hypothesis that network usage leads to greater outsourcing.

To determine whether this result was an artifact of the minority of firms whose main supplier was a division of their own company, we redid the analysis using the subset of the sample whose major supplier was a separate company (see Model 3 in Table 3). Here again, the results show that the more heavily firms used electronic networks to coordinate with their main external supplier, the less likely they were to outsource production ($\beta = -0.07$, $p < 0.01$). This finding again suggests that the more firms relied on electronic networks to coordinate the acquisition of important inputs, the more integrated they were with their suppliers.

Vulnerability from Opportunism. We also tested to see if network usage influenced virtualization by moderating the effects of vulnerability to opportunism. The "protection from vulnerability" hypothesis holds that by using electronic networks, firms can reduce their propensity to produce in-house when they are vulnerable to opportunism, because the networks help them lower the costs of identifying and acquiring firm-specific assets in the market and monitoring transactions involving them. Model 4 (Table 3) shows the results of the relevant analysis, predicting the degree of outsourced production from the interaction of electronic networks with each of the independent variables intended to measure vulnerability to opportunism (i.e., ease of input description, object certainty, order predictability, and generic object). The analysis used the full sample of firms.

Only the interaction between electronic network use and object specificity reached significance ($\beta = 0.15$, $p < 0.02$). Figure 1 illustrates the interaction. It shows that when firms did not use electronic networks, they produced firm-specific inputs in-house and outsourced the generic ones, as transaction cost theory would predict. When they used electronic networks, this gap in outsourcing was reduced. However, the reduction was not consistent with the hypothesis that electronic networks protect firms from opportunism. Firms did not use networks to outsource firm-specific assets. Instead, their use was associated with more in-house production overall, and especially for the generic objects they had previously outsourced.

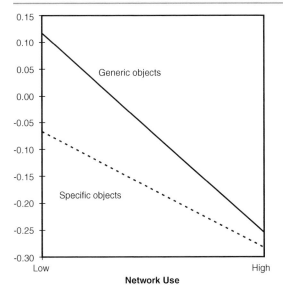

Figure 1 Influence of use of electronic networks and object specificity on degree of outsourcing.

Note

Plot from Model 4 in Table 3, showing the influence of increasing the degree of network use and object specificity by one standard deviation on extent of outsourcing (in standard deviation units). Scores are averaged across the four industries, with all other variables set to their mean levels.

Research Question 3: Effects of Coordination Mode on Transaction Outcomes

Table 4 shows the association of the coordination mechanisms with respondents' judgment of the outcomes of ordering from their primary supplier, controlling for a number of other factors that might lead to use of the coordination mechanisms or directly change business process outcomes. Model 5 in Table 4 shows the analysis for order quality—efficiency of performing orders and their freedom from various types of errors. Model 6 shows the analysis for respondents' satisfaction with their supplier. As can be observed, the significant predictors of the two outcome variables are different.

Order quality. On the variable measuring overall order quality, large firms performed more poorly ($\beta = -.19$, p < .05), especially in terms of terms of the efficiency subscale. When firms were acquiring inputs that were more stable and did not change from order to order (e.g., trim in the apparel industry or printing services in magazine publishing), working with the supplier was both more efficient and error-free ($\beta = .13$, p < .05). In contrast, firms had more trouble with orders that were predictable far in advance ($\beta = -.16$, p < .05) (e.g., fabric in the apparel industry). It is unclear whether this occurs because time lags lead to problems or because firms allow more lead time when they anticipate problems.

Surprisingly, the more firms used electronic networks for coordination, the poorer was the overall quality of the ordering process with their supplier ($\beta = -.17$, p < .05).[3] In contrast, the more they used personal relationships, the better were the outcomes ($\beta = .15$, p < .05). Of particular importance is the finding that these two modes of coordination had a statistically significant interaction on overall order quality ($\beta = .17$, p < .01). Their interaction, plotted in Figure 2, shows that using electronic networks actually degraded the overall quality of the order process when firms

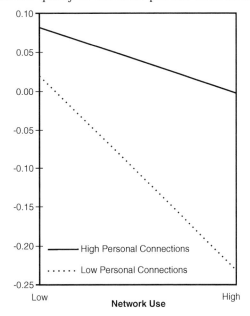

Figure 2 Influence of use of electronic networks and personal relationships on order quality.

Note

Plot from Model 5 in Table 4, showing the influence on order quality (in standard deviation units) on increasing the degree of network use and use of personal relationships by one standard deviation. Scores are averaged across the four industries, with all other variables set to their mean levels.

	Model 5 Predicting Order Quality	Model 6 Predicting Satisfaction
	Dependent Variable	
Predictor	Overall Order Quality	Satisfaction With Major Supplier
Intercept	−0.09	−0.01
Publishing vs. Apparel	−0.21*	0.04
Pharmaceutical vs. Apparel	0.16	0.05
Advertising vs. Apparel	0.32	−0.10
Firm Size	−0.19**	0.07
Supplier Size	0.05	−0.15*
Time Pressure	0.05	0.03
Order Frequency	0.03	0.11†
Tangibility	−0.04	−0.03
Ease of Input Description	0.00	0.00
Object Certainty	0.13*	−0.08
Order Predictability	−0.16*	0.03
Generic Object	0.07	0.01
Outsourced Production	0.14†	−0.06
Personal Relationships	0.15*	0.20***
Electronic Networks	−0.17*	0.07
Network Use × Personal Relationship	0.17**	−0.04
Network use × Outsourced Production	0.06	−0.05
Order Quality		0.37**
N	250	250
Rsq	0.22	0.24

Table 4 Predicting coordination success.

Note
Entries are standardized beta coefficients from a multiple regression analysis.
†: $p < 0.10$, *: $p < 0.05$, **: $p < 0.01$, ***: $p < 0.001$.

failed to supplement network use with personal relationships. However, when personal relationships were also used for coordination, the negative effect of electronic networks was reduced.

Supplier Satisfaction. In terms of the other outcome measure, satisfaction with supplier, respondents were more satisfied with their primary supplier when the supplier was smaller ($\beta = -0.15$, p < 0.05), and if they acquired inputs from it more frequently ($\beta = 0.11$, p < 0.10). They were more satisfied with suppliers who delivered orders more efficiently and with fewer errors ($\beta = 0.37$, p < 0.001). Holding constant the overall quality of the ordering processes, the use of personal contacts was again associated with better outcomes ($\beta = 0.20$, p < 0.001). Respondents were more satisfied with the supplier the more they used personal relationships as a coordination mechanism, while use of electronic networks was unrelated to respondents' satisfaction with suppliers.

Discussion

This article focuses on the choices that firms make when coordinating production: outsourcing key components or producing them in-house and coordinating with their suppliers through personal relationships and electronic networks. Much of the literature on virtual organizations has argued that (a) electronic networks substitute for personal relationships in coordinating production; (b) electronic data networks are a prerequisite for virtualization of production, and the the availability of electronic networks may lead firms to outsource more components of production than they would have otherwise; and (c) use of these networks generally has a positive effect on coordinating production. Our results, summarized in Figure 3 challenge all three assumptions.[4]

First, our data provide evidence that the use of personal relationships and electronic networks are complementary methods of coordination with suppliers rather than competing mechanisms. Firms use personal relationships and electronic networks concurrently to coordinate. The same firms that report using electronic networks heavily also report heavily using personal relationships for coordination. In multivariate regressions, controlling for firm and product characteristics, the existence of personal relationships between a focal firm and a potential supplier is a predictor of their use of electronic networks to coordinate production.

Our regression results further suggest that firms were more likely to use networks to acquire complex, specific, and intangible inputs. Firms appear to be using networks as a means of reducing their vulnerability to opportunism, although not, as discussed below, by coupling network use with greater outsourcing, but by using it to support internal production.

Indeed, our second set of findings is inconsistent with the hypothesis that increased use of electronic networks as a coordination mechanism leads to greater outsourcing of production. Instead we find that the more firms use electronic networks with an external supplier the more they produce their key inputs in-house. That is, increased use of electronic networks is associated with less virtualization and with less use of the market to acquire key production components. This finding is consistent with the growing research literature showing that interorganizational networks are associated more with

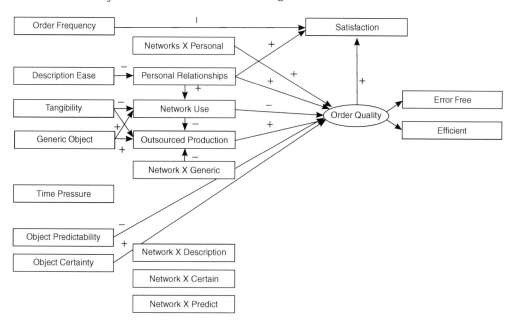

Figure 3 Relationships among product, coordination, and outcome variables.

hierarchical relations than market-based ones (Brousseau 1990, Hart and Estrin 1991, Keen 1988, Malone and Rockart 1993, Steinfield et al. 1995, Streeter et al. 1996).

We are making no claim here that the use of networks leads to in-house production. Indeed, the most plausible interpretation of the findings is that the control and trust that results from ownership encourages firms to invest in greater internal network use. This account of the evolution of firms' use of networks is consistent with trade press reports that firms are using intranets rather than placing strategic company business on the public Internet; if this is indeed a common occurrence, then this use of electronic networks illustrates that firms desire a great degree of control and trust before they are willing to make such investments or share sensitive data. To tease apart causal direction will require longitudinal data.

We were surprised to see that greater use of electronic networks was associated with poorer outcomes when working with suppliers. This finding runs counter to the expectation that the use of information technology is associated with increased efficiency and quality of interorganizational transactions (Davidow and Malone 1992, Malone and Rockart 1993) and with several quantitative case studies that show strong positive effects of using EDI and other interorganizational computer networks (e.g., Kekre and Mudhopadhyay 1992). It is likely, however, that the case studies have overstated the positive impact of using networks by focusing on large firms dealing with high volumes of very routinized transactions. The positive impacts found in these contexts may not generalize to a wider and more heterogeneous sample. The data are consistent with analyses challenging the value of information technology in improving business productivity (e.g., Attewell 1995, Landauer 1996).

It is possible that the relationships we observed in this study between use of electronic networks and errors and inefficiencies in ordering may be transient. Many firms have had little experience using electronic networks with suppliers and are relying on proprietary and ad hoc applications to share data. Because of lack of standards, inexperience, and problems with security, applications that allow data sharing with one supplier do not necessarily allow sharing with others. For example, an international advertising firm that shares files with other offices in their own firm over a proprietary data network was unwilling to send the files to outsiders electronically. They were worried both about the security of their corporate network and about compatibility standards with outsiders. Thus, they used courier services to exchange removable digital media with clients and color separation services they worked with.

Another reason that network use was associated with poorer outcomes such as late deliveries may, paradoxically, be a result of their speed. The instantaneous communications networks may encourage firms to wait until the latest possible moment before shipping, increasing the chances of late delivery and last minute errors. In our interviews in the advertising and magazine publishing industries, precisely this behavior was described by managers. Magazine publishers noted, for example, that using networks allowed advertisers to transmit their copy much closer to scheduled publication. This just-in-time effect increased the likelihood of missed deadlines and mistakes.

In contrast to the poor outcomes associated with use of electronic networks for coordination, greater use of personal relationships was associated with better quality and efficiency in executing transactions. As we have described previously, these personal relationships seemed to be especially important for dealing with nonroutine events. Moreover, it was when firms supplemented use of electronic networks with personal relationships that they achieved the best results with the electronic networks, or at least mitigated some of the potential negative consequences of using networks for coordination. For example, one pharmaceutical company reported heavy use of electronic mail to keep in touch with researchers with whom they had contracted out clinical trials. But they reported that the electronic mail worked only because it was supplemented by periodic face-to-face meetings, in which the principals worked out strategy and monitored performance. Other interviewees reported that it was crucial to know people in the other firms when dealing with the problems that arose in unusual situations (such as rush or special orders) and for dealing with problems that were created by computerized ordering systems (such as orders that had to be delivered by a particular date, even when a customer had failed to schedule a warehouse receiving slot until after the date).

Greater use of personal relationships was associated with greater satisfaction with the client, even after controlling for the relatively objective performance outcomes. To the extent that satisfaction is a proxy for the likelihood of reusing a supplier, personal contact is likely to be associated with repeat business.

Directions for Future Research

Our conclusions must be tempered by several limitations constraining our ability to interpret these data. The major source of ambiguity is the cross-sectional design of this research, which prevents us from determining the causal direction of many of the relationships identified here. Research in this area would benefit from longitudinal data collection.

Given the cross-sectional design, we are unable to offer any quantitative evidence for the direction of causation. This is a particular problem in trying to determine the extent to which electronic network usage influences firms' make-buy decisions. Our finding of a positive association between network usage and in-house production is generally inconsistent with an electronic markets hypothesis. We believe that the most plausible interpretation for this finding is that firms first develop an in-house capability for producing needed inputs, and then subsequently interconnect the related units with intracompany, local, and wide area networks.

The same problem exists in interpreting the positive association between electronic network usage and personal relationships as coordination modes. It may be that the very act of putting in and working with electronic networks causes a greater need for personal coordination. Alternately, preexisting personal relationships used for organizational coordination may also help firms to coordinate electronically.

We do not know the extent to which the results reported here are stable or will change as organizations become more sophisticated in outsourcing and in using electronic networks. Over time, it may be that successful internal network use will encourage firms to extend electronic transactions across boundaries to their closest trading partners—a pattern suggested by the popularity of intranets and extranets as the basis for enterprise networks. It may be that as external networks become more widespread (as they are becoming with the growth of the Internet), the quality and satisfaction problems associated with their use will disappear.

Follow-up research would also benefit from a more detailed examination of the mechanisms by which use of electronic networks and personal relationships influence virtualization and coordination success. In particular, it would be useful to understand better the way that effects of these coordination mechanisms are mediated by flexibility and trust.

Summary

Despite these limitations, the research reported here advances our understanding of the causes and consequences of virtual organization. We have tried to make the case that virtual organization is a matter of degree, rather than a unique type of organization. Ours is one of the few empirical studies of virtual organization and electronic networks that has examined causes and consequences across a broad range of firms in several industries. Contrary to much recent speculation, the research did not find that use of electronic networks for transactions was associated with increased outsourcing, but rather with greater dependence on internal production. Moreover, the use of interpersonal relationships for coordination, which many think of as an alternative to electronic network use, was associated with greater network use. Surprisingly, use of electronic networks was negatively associated with such outcomes as order quality and efficiency, and satisfaction with suppliers, while more reliance on personal linkages was associated with better outcomes and mitigated the negative consequence of using electronic networks.

NOTES

1 We included the relative power of the focal firm vis-à-vis the supplier in preliminary analyses. This variable consisted of the ratio of the standardized focal firm size divided by the standardized supplier firm size. However, because this variable did not approach signifiance in any analysis and because we include focal and supplier firm size as control variables, we have excluded this ratio from the results reported below.

2 In retrospect, the selection of broadcast time slots as an input was a poor choice. While airtime is a key input into a broadcast advertising campaign, since no advertising agencies own radio or TV stations, they must by necessity buy this input from an external organization.

However, our conclusions remained the same when this input was eliminated from analyses.
3 This negative association with electronic network usage was also evident when examining separately the subscales of transaction outcomes—ordering efficiency and percentage of orders with some type of error.
4 Figure 3 summarizes the path analysis implied by the coefficients in Tables 3 and 4. We do not wish, however, to present a full, structural equation model with coefficients for all variables in the model for two reasons. First, we make no strong claims for the causal order of variables represented in Figure 3. Second, we make no strong claim that we have captured all of the important variables that influence the decision to outsource and order quality.

ACKNOWLEDGMENTS

The authors are grateful to the National Science Foundation for the grant (IRI-9408271) which funded this research. They would also like to acknowledge the helpful comments of several anonymous reviewers.

REFERENCES

Attewell, P. 1995. Information technology and the productivity paradox. D. H. Harris, ed. *Organizational Linkages: Understanding the Productivity Paradox*. National Academy Press, Washington, DC.

Bakos, Y. 1991. A strategic analysis of electronic marketplaces. *MIS Quart.* (September) 295–310.

——, E. Brynjolfsson, 1992. When quality matters: Information technology and buyer-supplier relationships. Unpublished manuscript, University of California, Irvine, CA.

Baura, A., C. Kriebel, T. Mukhopadhyay. 1995. Information technologies and business value: An analytic and empirical investigation. *Inform. Systems Res.* **6** (1) 3–23.

Bradley, S. 1993. The role of IT networking in sustaining competitive advantage. S. Bradley, J. Hausman, R. Nolan, eds. *Globalization, Technology and Competition: The Fusion of Computers and Telecommunications in the 1990s*. Harvard Business School Press, Boston, MA. 113–42.

——, J. Hausman, R. Nolan, eds. 1993. *Globalization, Technology and Competition: The Fusion of Computers and Telecommunications in the 1990s*. Harvard Business School Press, Boston, MA.

Brousseau, E. 1990. Information technologies and inter-firm relationships: The spread of inter-organizational telematics systems and its impact on economic structure. Paper presented to the International Telecommunications Society, June meeting, Venice.

Brynjolfsson, E., T. Malone, V. Gurbaxani, A. Kambil. 1994. Does information technology lead to smaller firms? *Management Sci.* **40** (12) 1628–44.

Cash, J. I., B. R. Konsynski. 1985. IS redraws competitive boundaries. *Harvard Bus. Rev.* (March–April) 134–42.

Chan, A. P. 1997. Coordination and control of retailer-supplier transactions: Factors influencing organizational adoption and use of electronic information networks. Unpublished Doctoral Dissertation, Michigan State University, East Lansing, MI.

Choi, S. Y., D. Stahl, A. Whinston. 1997. *The Economics of Electronic Commerce*. Macmillan Technical Publishing, Indianapolis, IN.

Clemons, E. 1993. Information technology and the boundary of the firm: Who wins, who loses, who has to change. S. Bradley, J. Hausman, R. Nolan, eds. *Globalization, Technology and Competition: The Fusion of Computers and Telecommunications in the 1990s*. Harvard Business School Press, Boston, MA. 219–42.

Coase, R. H. 1937. The nature of the firm. *Economica*. N. S. 386–405.

Davidow, W., M. Malone. 1992. *The Virtual Corporation*. Harper Business, New York.

Granovetter, M. 1985. Economic action and social structure: The problem of embeddedness. *Amer. J. Soc.* **91** (3), 481–510.

Gurbaxani, V., S. Wang. 1991. The impact of information systems on organizations and markets. *Comm. ACM* **34** (1) 59–73.

Hart, P., D. Estrin. 1991. Inter-organizational networks, computer integration, and shifts in interdependence: The case of the semiconductor industry. *ACM Trans. Inform. Systems*, **9** (4) 370–98.

Hennart, J. 1993. Explaining the swollen middle: Why most transactions are a mix of "market" and "hierarchy." *Organ. Sci.* **4** (4) 529–47.

Johanson, J., L. G. Mattson. 1987. Inter-organizational relations in industrial systems: A network approach compared with the transactions-cost approach. *Internat. Stud. Management Organ.* **17** (1) 34–48.

Johnston, H. R., M. R. Vitale. 1988. Creating competitive advantage with inter-organizational information systems. *MIS Quart.* (June) 153–65.

Johnston, R., P. Lawrence. 1988. Beyond vertical integration: The rise of the value-adding partnership. *Harvard Bus. Rev.* (July–August) 94–101.

Kalakota, R., A. Whinston. 1997. *Electronic Commerce: A Manager's Guide.* Addison-Wesley, Reading, MA.

Kambil, A. 1991. Information technology and vertical integration: Evidence from the manufacturing sector. M. Guerin-Calvert, S. Wildman, eds. *Electronic Services Networks: A Business and Public Policy Challenge.* Praeger, New York. 22–38.

Keen, P. 1988. *Competing in Time: Using Telecommunications for Competitive Advantage.* Ballinger Press, Cambridge, MA.

——, M. Cummins. 1993. *Networks in Action: Business Choices and Telecommunications Decisions.* Wadsworth, Belmont, CA.

Kekre, S., T. Mudhopadhyay. 1992. Impact of electronic data interchange technology on quality improvement and inventory reduction programs: A field study. *Internat. J. Production Econom.* **28** 265–82.

Kling, R., W. Scacchi. 1982. The web of computing: Computer technology as social organization. *Adv. Comput.* **21** 1–90.

Krackhardt, D. 1992. The strength of strong ties: The importance of philos. N. Nohria, R. Eccles, eds. *Networks and Organizations: Structure, Form, and Action.* Harvard Business School Press, Boston, MA, 216–39.

Kpres, D. M. 1990. Corporate culture and economic theory. J. E. Alt, K. A. Shesple, eds. *Perspectives on Positive Political Economy* Cambridge University Press, New York. 90–143.

Landauer, T. 1996. *The Trouble with Computers.* MIT Press, Cambridge, MA.

Lawrence, P., J. Lorsch. 1967. *Organizations and Environment.* Addison-Wesley, Reading, MA.

Lorenz, E. 1989. Neither friends nor strangers: Informal networks of subcontracting in French industry. D. Gambetta, ed. *Trust: Making and Breaking of Cooperative Relations.* Basil Blackwell, Oxford: 194–210.

Malone, T. 1987. Modeling coordination in organizations and markets. *Management Sci.* **33** 1317–32.

——, J. Rockart. 1993. How will information technology reshape organizations? Computers as coordination technology. S. Bradley J. Hausman, R. Nolan, eds. *Globalization, Technology and Competition: The Fusion of Computers and Telecommunications in the 1990s.* Harvard Business School Press, Boston, MA. 37–56.

Malone, T., J. Yates, R. Benjamin. 1987. Electronic markets and electronic hierarchies: Effects of information technology on market structure and corporate strategies. *Comm. ACM* **30** (6) 484–97.

——, ——, ——. 1989. The logic of electronic markets. *Harvard Bus. Rev.* (May–June) 166–171.

Miller, D., E. Clemons, M. Row. 1993. Information technology and the global virtual corporation. S. Bradley, J. Hausman, R. Nolan, eds. *Globalization, Technology and Competition: The Fusion of Computers and Telecommunications in the 1990s:* (283–307). Harvard Business School Press, Boston, MA.

Ouchi, W. G. 1980. Markets, bureaucracies, and clans. *Admin. Sci. Quart.* **25** 129–41.

Perrow, C. 1986. *Complex Organizations: A Critical Essay*, 3rd. ed. Random House, New York.

Porter, M. E. 1980. *Competitive Strategy: Techniques for Analyzing Industries and Competitors.* The Free Press, New York.

——, V. E. Millar. 1985. How information gives you competitive advantage. *Harvard Bus. Rev.* (May–June) 149–60.

Powell, W. 1987. Hybrid organizational arrangements: New form or transitional development. *California Management Rev.* **30** 66–87.

——. 1990. Neither market nor hierarchy: Networked forms of organization. B. Staw, L. Cummings, eds. *Research in Organizational Behavior*, vol. 12. JAI Press, Greenwich, CT. 295–336.

Saunders, C., P. Hart. 1993. Electronic data interchange across organizational boundaries: Building a theory of motivation and implementation. Working Paper #7–93, Decision and Information Systems, Florida Atlantic University, Boca Raton, FL.

Steinfield, C., L. Caby. 1993. Strategic organizational applications of videotex among varying network configurations. *Telematics and Informatics* **10** (2) 119–29.

——, ——. 1997. Changer les relations dans la société de l'information: Les effets des infrastructures de l'information sur les relations entre usagers et professionels. *Reseaux.* **84** (July–August) 47–65. (in French). Published in English as Changing relationships in the information society: The effect of information infrastructures on relations among business users. *Trends in Communication.* **3** 93–115.

——, ——, P. Vialle. 1993. Internationalization of the firm and the impacts of videotex networks. *J. Inform. Tech.* **7** 213–22.

——, R. Kraut, A. Plummer. 1995. The impact of interorganizational networks on buyer-seller relationships. *J. Comput. Mediated Comm.* (online). **1** (3). Available at <http://www.ascusc.org/jcmc/vol1/issue3/steinfld.html>.

Stinchcombe, A. 1990. *Information and Organizations.* University of California Press, Berkeley.

Streeter, L. A., R. E., Kraut, H. C. Lucas, L. Caby. 1996. The impact of national data networks on firm performance and market structure. *Comm. ACM.* **39** 62–73.

Suchman, L., E. Wynn. 1984. Procedures and problems in the office. *Office: Techn. and People* **2** (2) 133–54.

Tornatsky, L., K. Klein. 1982. Innovation characteristics and innovation-adopting implementation: A meta-analysis of findings. *IEEE Trans. Engrg Management* **EM-29** 28–45.

Tushman, M., D. Nadler. 1978. Information processing as an integrating concept in organizational design. *Acad. Management Rev.* 613–624.

Uzzi, B. 1997. Social structure and competition in interfirm networks: The paradox of embeddedness. *Admin. Sci. Quart.* **42** (March) 35–67.

Weber, J. 1995. Just get it to the stores on time. *Bus. Week* (March 6) 66–7.

Wildman, S., M. Guerin-Calvert. 1991. Electronic services networks: Functions, structures, and public policy. M. Guerin-Calvert, S. Wildman, eds. *Electronic Services Network: A Business and Public Policy Challenge.* Praeger, New York. 3–21.

Williamson, O. 1975. *Markets and Hierarchies: Analysis and Antitrust Implications.* Free Press, New York.

——. 1985. *The Economic Institutions of Capitalism.* Free Press, New York.

——. 1991. Comparative economic organization: The analysis of discrete structure alternatives. *Admin. Sci. Quart.* **36** (2) 269–96.

——. 1996. Economics organization: The case for candor. *Acad. Management Rev.* **21** (1) 48–57.

Zucker, L. 1986. Production of trust: Institutional sources of economic structure: 1840–1920. B. Staw, L. Cummings, eds. *Research in Organizational Behavior,* vol. 8. JAI Press, Greenwich, CT. 53–111.

Accepted by Peter Monge and Geraldine De Sanctis; received June 1997. This paper has been with the authors for three revisions.

Appendix

Scale Items

Supplier Size (alpha = 0.73)

Approximately how many employees does this supplier have?

1 Fewer than 20
2 Between 20 and 99
3 Between 100 and 499
4 500 or more

Would you say that this supplier is one of the ten largest sources in the United States for [this product or service], a major source but not in the top ten sources, an average source, or a smaller than average source?

1 Top ten
2 Major, but not top ten
3 Average
4 Smaller than average
5 Not in this country

Time Pressure (alpha = 0.68)

To meet schedules related to this product or service, we need to use every available minute effectively.

In developing, producing, and distributing this product or service we have to meet tight deadlines.

Ease of Description

It is difficult to describe the [product or service] we routinely acquire.*

Object Certainty (alpha = 0.68)

How much does the availability of the [product or service] change over the course of a year?
How much does the [product or service] your company needs change from order to order?
How much does the unit price of the [product or service] change from order to order?

Object Predictability (alpha = 0.79)

For a typical [product or service], how far in advance can you predict what [product or service] your company will need?

1. Within a week of when you'll need it
2. Within a month
3. Within three months
4. Within six months
5. More than six months from when you'll need it

How far in advance can you predict the quantity of [product or service] your company will need?

1. Within a week of when you'll need it
2. Within a month
3. Within three months
4. Within six months
5. More than six months from when you'll need it

Order Frequency

How often do you acquire [the product or service] from this supplier?

1. At least monthly
2. At least every three months
3. At least every six months
4. At least every 12 months
5. Less frequently than every 12 months

Generic Object (alpha = 0.65)

My firm is about the only one that uses [this particular product or service].
The [product or service] we get from this supplier has certain features that are only useful to us.
Many other companies use the same types of [products or services] when they develop, make, and distribute products or services].
The [products or services] we get from this supplier are fairly standard for the industry.

Outsourced Production (alpha = 0.86)

The next set of questions asks about your company's dealings with one supplier of [good or service]. Think of an important supplier you've worked with in the past 12 months. This supplier can be a person or department in your firm or an outside individual or another company. Is this supplier:

An individual or department within your firm?
A partly or fully owned subsidiary of your firm?
Another firm which is your partner in a joint venture?
Outsiders (individuals you don't employ or suppliers in which your firm has no ownership stake)?
What percentage of [the input] does your firm get from in-house sources?

Coordination Through Electronic Networks (alpha = 0.93)

An electronic connection is one that connects your computers with each other and with computers from outside your company. When we refer to electronic connections here we mean any type of computer communication that exchanges information with this supplier. We include modem connections, local area networks, online data, electronic mail, electronic data interchange, Lotus Notes, or the Internet.
Using a five-point scale where "1" is very important and "5" is not at all important, please tell me how important electronic connections are to the following items, or tell me if you don't use electronic connections at all for this purpose. How about...?

a   In selecting this supplier
b   Developing the specification of the [product or service] your company orders
c   The negotiation of agreements to acquire [the product or service]
d   Getting [the product or service]

e Monitoring the quality of the [products and services] you receive
f Fixing problems after you have ordered [the product or service]

Coordination Through Personal Relationships (alpha = 0.85)

Now I'm going to ask you about the importance of personal relationships in selecting and working with this supplier. When I refer to "personal relationships" I am including any type of personal connection or personal knowledge between your firm and people in the supplier's organization. Using a five-point scale where "1" is very important and "5" is not important, please tell me how important are personal relationships in ...?

a Helping your company select this supplier
b Developing the specification of the [product or service] your company order[s] from this supplier
c The negotiation of agreements to acquire [the product or service]
d Getting [the product or service]
e Monitoring the quality of the [products and services] you receive
f Fixing problems after you have ordered [the product or service]

Satisfaction with Supplier (alpha = 0.89)

Using a five-point scale where "1" is extremely satisfied and "5" is not at all satisfied, how satisfied are you with these aspects?

a The supplier's ability to handle a rush order
b The supplier's ability to handle an out-of-the-ordinary order
c That you have identified the best supplier
d That your company can easily specify exactly what you want from your supplier
e The terms your company is able to negotiate
f The efficiency with which your company gets [the product or service]
g Your company's ability to monitor and evaluate [the product or service] quality
h Your company's ability to fix problems with this supplier

Using the same five-point scale, please tell me how satisfied are you with this supplier overall.

Overall Order Quality (alpha = 0.65)

Subscale: Order Quality (alpha = 0.72) Now I would like to ask you about the quality of the orders you have placed with this supplier over the past year.

What percentage of these orders arrived after your targeted delivery date?
What percentage of these orders had an error of any sort?
What percentage of these orders did not meet your quality standards?

Subscale: Order Efficiency (alpha = 0.48) When you are processing a typical order for [this product or service], approximately how many different people in your firm would be involved with it, from the time it is placed to the time it is actually received?

During a typical cycle, from the time you think you need [the product or service] to the time the order is delivered, how many times would someone in your firm communicate with the supplier?

Approximately how much time would elapse from when your company places a typical order for [this product or service] to when you would actually receive it?

Note: *For attitude statements, most items were represented as a five-point Likert scale, where 5 meant strongly agree with the statement and 1 meant strongly disagree with the statement.

PART FOUR

The impact of IT on markets

INTRODUCTION TO PART FOUR

A much cited paper by Malone et al. (1987) can arguably be credited with launching the stream of IS research that is the focus of Part Four – electronic marketplaces. That paper argued, in part, that "electronic hierarchies" (the interorganizational systems of the preceding section) would evolve into "electronic markets". Some IS researchers continue to view electronic marketplaces as instances or special cases of interorganizational systems. Viewed in that way, they hardly warrant a separate section in this volume. But most analysts of the phenomenon argue that electronic marketplaces constitute a separate unit of analysis. Whereas electronic hierarchies represent dyadic one-to-one or one-to-many connections between a focal buyer (or suppliers) and its supplier(s) (or buyer(s)), electronic marketplaces create networks of many-to-many connections between buyers and sellers. Electronic marketplaces are usually launched by incumbent intermediaries in an industry (e.g. wholesales/distributors) or by consortia of buyers and/or sellers, but they are rarely the initiatives of *individual* buyer or supplier companies, as is common with the interorganizational systems discussed in the previous section. The papers in this section focus on electronic markets as interorganizational networks of buyers, sellers, and an IT-supported intermediary.

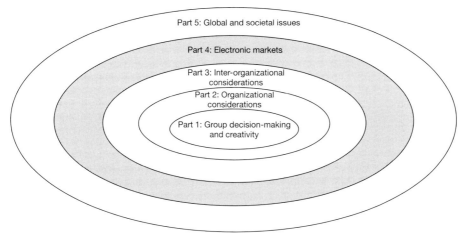

Figure 4.1 Electronic markets within the broadening focus of Information Systems research considerations.

IS research on electronic marketplaces also generally differs from interorganizational systems research in theoretical perspective. Whereas interorganizational systems have often been viewed from a strategic management perspective, electronic marketplaces are generally explored through the lens of transaction cost theory, the conceptual basis for the theory of electronic markets (cf., Malone, 1987 and others). In comparison to research on interorganizational systems, the literature on electronic

INTRODUCTION TO PART FOUR

	Lee (1998)[1]	Weber (1999)[2]	Kumar, van Dissel and Bielli (1998)[3]	Kambil and van Heck (1998)[4]
Research method	Descriptive single case study, quantitative and qualitative data analysis.	Simulation comparison of open outcry and electronic order matching, 5000 trial runs of 250 buy and sell orders.	Holistic explanatory single case study.	Descriptive multiple-case study – two successful electronic markets, two unsuccessful – in a single industry.
Focus and perspective	General economic perspective on impacts of AUCNET, an electronic market for trading used cars among dealers in Japan; focus on effects on prices and explanation for effects.	Impacts on speed of trade completion and transactions costs of Cantor Financial Futures Exchange, an electronic exchange, compared to its open outcry rival in Chicago; market theory.	Collapse in usage of SPRINTEL, an interorganizational system in Prato, Italy, after the end of a "no charging" pricing regime, analyzed through three theoretical perspectives: techno-economic rationality, conflict and power, trust-based rationalism.	Success of electronic markets in the Dutch flower industry viewed through process stakeholder perspective of exchange process (e.g. search, payments, product representation, dispute resolution).
Technology	Electronic auction system: catalog of automobiles with inspection reports and photos; during the auction, members bid remotely using joysticks.	Electronic futures market characterized abstractly in terms of its features for data entry, price determination, price information availability, etc.	Telecommunication network and online databases based on videotext technology.	Electronic auctions with various features for product representation, communication, etc.
Findings and contribution	Prices were higher, contrary to expectations, because AUCNET's inspections were viewed as a surrogate for quality; sellers' market power increased; results do not imply that electronic markets necessarily increase prices, but rather that they can create new opportunities for win-win benefits for buyers and sellers.	Electronic exchange is faster and a third less costly; traditionally, exchange markets have supported the middle of the value chain (e.g. back-room processing), but new electronic markets provide broader support for the value chain; the benefits suggest that acceptance of electronic markets will increase.	Economic theory cannot explain system failure or failure to adopt more market-like modes of exchange; findings suggest the need for caution when applying economic theories outside the cultural contexts in which they were developed; basing system development solely on technical-economic rationality can lead to failure; success may require a win-win approach to benefit sharing.	IT can enable increased efficiency; IT changes information available to different parties, affecting their acceptance; no new IT-enabled market will succeed if any player is worse off; provides a useful guide to the developers of electronic markets.

Table 4.1 Comparison of chapters in Part 4.

Notes
1. Lee, H. G. (1998) "Do electronic marketplaces lower the price of goods?" *Communications of the ACM*, 41(1), January, 73–80.
2. Weber, B. W. (1999) "Next-generation trading in futures markets: A comparison of open outcry and order matching systems", *Journal of Management Information Systems*, 16(2), Fall, 29–45.
3. Kumar, K., van Dissel, H. G. and Bielli, P. (1998) "Trust, technology and transaction costs: Can theories transcend culture in a globlized world?" [This article is an extended version of "The Merchant of Prato – Revisited: Toward a third rationlity of information systems"], *MIS Quarterly*, 22(2), June, 199–226.
4. Kambil, A. and van Heck, E. (1998) "Reengineering the Dutch flower auctions: A framework for analyzing exchange organizations", *Information Systems Research*, 9(1), March, 1–19.

marketplaces tends to place much more emphasis on technological features. Indeed, a review of the economics literature on business-to-business electronic commerce highlights the design features of auctions as a promising area of research (Kauffman, 2001).

Methodological differences also characterize the two bodies of literature. The relative scarcity of electronic marketplaces makes it challenging to mount survey research similar to Palmer and Markus (2000) and Kraut, *et al.* (1999) in the previous section. (Although electronic marketplaces proliferated during the dot-com bubble, most industries today host only a handful of viable electronic markets.) Consequently, most studies are single cases (with varied research approaches); of the studies featured in this section, that by Kambil and van Heck is unique in its comparison of multiple electronic marketplaces in the same industry.

The first two selections are both single case studies that compare particular electronic marketplaces with their traditional counterparts. But there the resemblance ends. Lee's is a descriptive case study of a Japanese business-to-business enterprise that auctions used cars to dealers. The author uses both quantitative and qualitative data to explain why, contrary to economic theory, this electronic marketplace results in higher prices for the traded goods. Published in *Communications of the ACM*, which favors short articles for a more general audience, this article does not describe the author's research method.

The article by Weber similarly compares electronic to traditional markets, in this case, electronic futures exchanges, exemplified by the Cantor Financial Futures Exchange, with traditional open outcry markets, such as the Chicago Mercantile Exchange. Electronic futures markets are designed with price and time priority matching rules that are intended to reduce trading delays and improve sales prices relative to current practices; naturally, research is needed to determine whether these markets achieve their intended effects. Weber addresses this question by using the technique of simulation.

As a pair, these papers suggest the following questions:

- What role do numbers play in Lee's "descriptive" case study? How is this different from the role of numbers in Weber's case study?
- Given that Weber's primary method is simulation, what is the role of textual description in his study?
- What are the pros and cons of simulation as a research method? Why is simulation particularly useful for Weber's research purposes? What design choices did Weber have to make in crafting his simulation?
- Would simulation have been an appropriate method to answer Lee's research question? Why or why not?
- In your opinion, are the findings of these studies equally credible? Why or why not?

One might imagine that AUCNET was the only electronic marketplace for used car trading in Japan when Lee did his study. But when Kambil and van Heck studied the Dutch flower industry, a number of electronic marketplaces had been established – some eventually successful, others not. Although the total number of such marketplace experiments was not sufficient to permit a cross-sectional survey of the sort designed by Palmer and Markus (2000) and Kraut, *et al.* (1999), Kambil and van Heck were able to use the multiple case research design (Yin, 1999) to examine the question of electronic marketplace success versus failure. The conceptual framework for this study is a model of exchange processes against which each case is described and analyzed. Comparisons of the successful cases with the failures yield observations in the form of propositions capable of being tested in future research.

- What are the pros and cons of single case versus multiple case studies? Under what conditions are single case studies most appropriate? (Consider Yin, 1999.)
- How does Kambil and van Heck's research design fit their research question? Put differently, could they have answered the question of why some electronic marketplaces succeed and others fail using Lee's research approach?
- What do you think about the use of propositions as a way of summarizing research findings? Are the propositions posed by Kambil and van Heck the ones you would use to summarize their findings? How would you design a research project to test these propositions?
- The article by Lee mentions the unsuccessful "Slide Auction" implemented to reduce the capacity constraints of traditional Japanese used car markets. Apply the theoretical framework of Kambil and van Heck to explain why AUCNET succeeded whereas the Slide Auction failed.

Like the articles by Lee and Weber, the third article in this section, by Kumar et al., examines a single case study. Unlike Lee and Weber, the authors analyze their single case with multiple theoretical perspectives. Kumar et al. take issue with economic transaction cost theory – the dominant theoretical perspective in IS studies of electronic marketplaces – as inapplicable to traditional economic cultures such as that of the Prato textile district in Italy. They argue that transaction cost theory is unable to explain why use of SPRINTEL electronic marketplace collapsed after usage fees were imposed. In the authors' analysis, electronic marketplaces do not support the establishment and maintenance of interorganizational trust in a society where most business partners are family members or former employees.

- Contrast the research approaches of multiple case comparisons, as in Kambil and van Heck, and multiple theoretical analyses of a single case, as in Kumar et al.. How does Kumar et al.'s approach fit into the case study research method as discussed by Yin (1999)?
- The studies of Kambil and van Heck, and Kumar et al. both examined the question of electronic marketplace success versus failure. In the former instance the comparison is across cases; in the latter instance, the comparison is within cases, over time. What are the pros and cons of each approach?
- Suppose you conclude that electronic marketplace pricing (not just trust) was a key factor in SPRINTEL's failure. How would you want to redesign the Kambil and van Heck study to incorporate the investigation of pricing (for participation in the auctions) as a possible explanation for the failures of Vidifleur and VBA Sample Based Auction *versus* the successful Holland Supply Bank and Tele Flower Auction System?
- Compare and contrast the Kumar et al. article in this section with the published version that appeared in *MIS Quarterly* in 1998. What in addition do you take from the current version as compared to the previously published version? Why do you think the *MISQ* editors preferred the shorter version, other than the obvious reasons of space limitations? Would you take issue with them were you an author of the paper?

Finally, the four studies in this section address electronic marketplaces for very different products in very different industries in very different countries. Imagine that you wanted to conduct a study of electronic marketplaces that cuts across products, industries, and countries.

- What kinds of research design choices would you have to make? Which methods would be most appropriate and why? Which methods would not be applicable?

- Could you use Kambil and van Heck's process stakeholder model for all products and industries? If not, why not?
- Which explanatory concepts would you have to add to accommodate contextual diversity? Which would you have to leave out to keep the study manageable?
- Could you incorporate multiple theoretical perspectives in such a study? If so, how?

REFERENCES

Kambil, A. and van Heck, E. (1998) "Reengineering the Dutch flower auctions: A framework for analyzing exchange organizations", *Information Systems Research*, 9(1), March, 1–19.

Kauffman, R. J. and Walden, E. A. (2001) "Economics and electronic commerce: Survey and directions for research", *International Journal of Electronic Commerce*, 5(4), 5–116.

Kraut, R., Steinfield, C., Chan, A. P., Butler, B. and Hoag, A. (1999) "Coordination and virtualization: The role of electronic networks and personal relationships", *Organization Science*, 10(6), 722–40.

Kumar, K., van Dissel, H. G. and Bielli, P. (1998) "The Merchant of Prato – revisited: Toward a third rationality of information systems", *MIS Quarterly*, 22(2), June, 199–226.

Lee, H. G. (1998) "Do electronic marketplaces lower the price of goods?", *Communications of the ACM*, 41(1), January, 73–80.

Malone, T., Yates, J. and Benjamin, R. (1987) "Electronic markets and electronic hierarchies: Effects of information technology on market structure and corporate strategies", *Communications of the ACM*, 30(6), 484–97.

Palmer, J. W. and Markus, M. L. (2000) "The performance impacts of quick response and strategic alignment in specialty retailing", *Information Systems Research*, 11(3), September, 241–59.

Weber, B. W. (1999) "Next-generation trading in futures markets: A comparison of open outcry and order matching systems", *Journal of Management Information Systems*, 16(2), Fall, 29–45.

Yin, R. K. (1999) *Case Study Research: Design and Method*, third edition, Thousand Oaks, CA: Sage.

Do Electronic Marketplaces Lower the Price of Goods?

Ho Geun Lee

An assistant professor in the Department of Business Administration at Yonsei University in Seoul, Korea, hlee@base.yonsei.ac.kr

The efficiency of electronic means for commerce is sometimes countered by increased product cost, as demonstrated in this case involving an auction system for used cars in Japan.

Abstract

Electronic marketplaces have become increasingly popular alternatives to traditional forms of commerce [8, 9]. This increase in popularity has led many to predict that one effect will be to lower the market price of goods. This reduced price hypothesis was proposed by Bakos in his seminal article on electronic marketplaces [2]. Buyers in market-intermediated transactions have to bear search costs to obtain information about the prices and product offerings of sellers. High search costs of buyers enable sellers to maintain prices substantially above their marginal costs and result in allocational inefficiencies in market transactions. Electronic market systems can reduce the search costs that buyers must incur to acquire information about seller prices and product offerings, thus enabling buyers to locate suppliers that better match their needs. The lowered search costs allow buyers to look at more product offerings and make it difficult for sellers to sustain high prices. The reduced price hypothesis predicts that buyers will enjoy lower product prices as a result of the increased competition among sellers in electronic marketplaces.

An industry case study is presented in this article to demonstrate that the prices of goods traded through electronic marketplaces can actually be higher than those of products sold in traditional markets. AUCNET is an electronic marketplace introduced to reduce search costs of buyers for usedcar transactions in Japan. The average contract price of secondhand cars sold through AUCNET is much higher than that of traditional, non-electronic markets. The industry case study suggests that the analysis of electronic market impacts on product prices should take into account economic factors beyond the buyers' search costs. This article investigates why the product prices in AUCNET are higher than those of traditional markets by examining other economic variables than the buyer's search costs.

The Japanese Auto-Auction Market

Japanese consumers generally purchase secondhand cars from licensed used-car dealers: A complex web of title registration and regulations makes direct trading of secondhand cars between individuals difficult. Avoiding risks of hidden defects and securing financial loans also encourage Japanese consumers to deal with reliable used-car dealers. New car dealers typically sell trade-ins to used-car dealers rather than reselling them to consumers. Besides trade-ins, new car dealers often register blocks of new cars as used cars and sell them to used-car dealers in an attempt to meet sales quotas at the end of a selling period.

Retail used-car demand is becoming increasingly differentiated in Japan, and used-car dealerships are specializing in late model cars, sports cars, sport-utility vehicles, or even a particular make and model of automobile. Direct sales between used-car dealers are limited. Dealers are not inclined to rely on their competitors' inventories, particularly those selling to the same market segment, although relationships sometimes develop between dealers selling non-competing lines. Most urban dealers, if the car desired by a client is not in their inventory, typically go to an auto auction site to locate additional product inventory.

In auto auctions, buyers and sellers assemble at a central auction site, cars are brought onto the auction floor one at a time and most buyers personally inspect the cars prior to the auction. The auctioneer starts with a low price and continues to increase the bid price until the highest bid is registered. Most large auto auctions use a POS (point-of-sales) system, introduced in the late 1970s, in which buyers press a POS button to register their bids instead of raising their hands.

There are 144 auto auction markets in Japan, which serve as intermediaries for used-car transactions between dealerships. The number of cars offered to auto auctions increased at about 11% annually over five years (see Table 1). Slightly more than half of the cars offered to the market are sold: In 1995, over 3.58 million cars were brought to the auto auction markets and around 2 million cars (55% of total car offerings) were sold at an aggregate contract value of ¥1,482 billion (U.S. $15 billion).

AUCNET

AUCNET was introduced in 1985 by an entrepreneurial used-car dealer, who foresaw that the redesign of the auto auction business using computers and advanced communication technologies would significantly improve the market where efficiency of used-car transactions [12]. The AUCNET system is a centralized, online wholesale market where cars are sold using video images, character-based data and a standardized inspector rating.

Buyers and sellers in AUCNET remain at their respective businesses without physically traveling to auction sites. Sellers must have their vehicles inspected by AUCNET mechanics, who evaluate cars and summarize their quality rank with a single number between 1 and 10. The inspection results, together with photos of the car (exterior and interior shots), are entered into the AUCNET central computer to create an auction catalog that is transmitted to all potential buyers prior to the electronic auction. Dealers interested in a specific car listed in the catalog can preview its image by keying its lot number in the catalog. During the electronic auction, buyers bid by pressing the button on the top of computer joysticks, which will increase the current bid by ¥3,000 (U.S. $30). A car sold through AUCNET remains at the seller's location until the transaction is completed. Then, a transport company delivers the sold car directly to the buyer's location. Member dealers are linked to AUCNET's central host computer through an electronic network, which integrates satellite links with terrestrial transmission lines (see Figure 1). Satellite links are used to broadcast car images from AUCNET to member dealers (one-way broadband communication), while text-based data transmission between dealers and the AUCNET host computer uses ground lines (two-way narrowband communications).

AUCNET has carved out a niche in the top end of the wholesale used-car market using computer and telecommunication technologies. AUCNET started its service in 1986 with 560 dealer members. Since then, AUCNET's throughput (listed cars) has increased at an annual compound growth rate of 26%. AUCNET listed more than 1 million cars and sold over 486,000 cars from 1986 through 1995 (see Table 1). In 1995 AUCNET sold 95,778 cars out of 206,312 vehicles listed in its system. As a result, AUCNET has become the largest among 144 auto auctions, with 5.75% of market share (the second largest one, USS Kyushu, recorded 3.80% of market share in 1995). The membership network among dealers has expanded at an annual growth rate of about 11% over five years, reaching 4,150 (about 20% of dealerships in Japan) at the end of 1995.

		1991	1992	1993	1994	1995
Auto Auction (total)	cars offered ('000)	2,328	2,558	2,906	3,175	3,589
	cars sold ('000)	1,263	1,511	1,593	1,794	1,973
	contract value (billion¥)	999	1,237	1,249	1,402	1,482
AUCNET	cars offered ('000) (market share)	114 (4.91%)	128 (5.03%)	147 (5.07%)	165 (5.18%)	206 (5.75%)
	cars sold ('000)	50	62	70	80	96
	contract value (billion ¥)	76	97	106	121	143
	dealer members	2,724	2,973	3,228	3,624	4,154

Table 1 Total auto auction market and AUCNET performance.

Note
(¥100 ≅ U.S. $1)

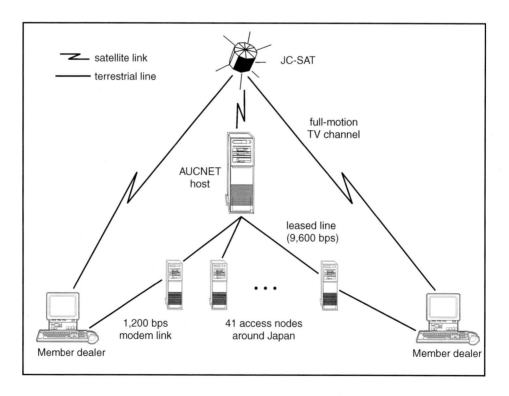

Figure 1 AUCNET network.

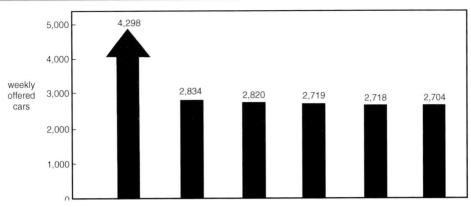

Figure 2 Number of weekly car offerings for auto auctions.
Note
*EA: electronic auction   POS: point-of-sales auction (face-to-face auction)

Higher Prices of Cars in AUCNET

AUCNET was built by a used-car dealer who realized the use of computer and communication technologies for auto auctions would reduce the time involved in searching for cars. Attending physical auto auctions is a time-consuming process for most buyers. Because there is no precise schedule for when certain cars will be sold, a buyer may need to attend a traditional auction the entire day to bid on only one or two cars. Since used-car dealers usually engage in sales activities themselves, they lose sales opportunities while attending traditional auctions.

AUCNET's strength over traditional auto auctions is its ability to help buyers search for cars by efficiently distributing product and price information. Since the auction catalog, which contains information about the products to be auctioned as well as their auction schedules, is distributed in advance through the satellite network, used-car buyers in AUCNET are well informed about what kinds of products are available in the market. Buyers can also limit their auction time to only the cars they are interested in buying. When a car desired by a client is not in their stock, dealers can download the data and images of offered cars through the AUCNET network, show the information to the client, and include the car in their bidding list upon the client's request.

AUCNET also provides member dealers with a crucial market intelligence service: post-transaction information. At any time, dealers can access a database that displays information on the five most recent transactions of the same model, including their quality characteristics and prices paid by buyers (minimum, maximum, and average price paid). This post-trading information keeps buyers well informed about the potential market price of goods with specific characteristics of interest, thus making it difficult for sellers to gain a margin significantly above the market prices.

Traditional auto auctions in Japan have limits in the number of car offerings because they are located in metropolitan areas where it is becoming increasingly difficult and costly to secure additional parking spaces for cars to be auctioned. The car storage constraint restricts the number of cars that can be offered to buyers. Figure 2 compares a weekly car offering in the top five traditional auto auctions with that of AUCNET. The large traditional auto auctions offer a relatively smaller number of cars (compared to AUCNET) in their

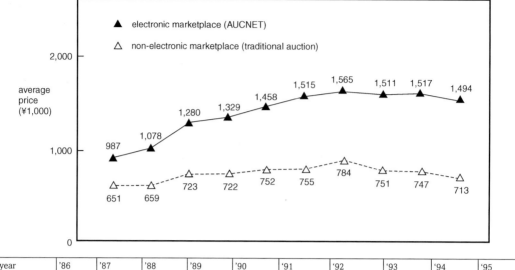

Figure 3 Average contract price per car.

Note
(¥100 ≅ U.S.$1)

weekly POS auctions: there is no single auction site that so dominates the market because of their similar parking capacities. AUCNET created the largest auto auction market without owning a single parking space, listing around 4,300 cars every week in 1995. Furthermore, AUCNET can easily increase the number of car offerings without extending its physical infrastructure, being able to accommodate an expected annual growth rate of 15% over the next five years. This contrasts with traditional auto auctions whose parking capacities limit the growth of car offerings. As a result, buyers in AUCNET enjoy wider vehicle choices than are available in regional auto auctions, thereby reducing their search costs to locate products that best suit their needs in terms of prices and qualities.

According to the reduced price hypothesis, the significant reduction of buyer search costs will lead to lowered prices of cars in AUCNET. On the contrary, the market prices of secondhand cars in AUCNET have turned out to be much higher than those of traditional face-to-face auction markets. Figure 3 compares the average car price in AUCNET with that of traditional non-electronic auction markets over the past 10 years. When AUCNET started its service in 1986, its average contact price per car was 50% higher than that of traditional marketplaces. Since then, the price gap has continuously widened and there are no signs of the price difference narrowing. In 1995, the AUCNET price was more than double that of traditional marketplaces. This result suggests the reduced search cost of buyers is not a single and major economic factor influencing the price of goods in electronic marketplaces, at least in the used-car wholesale market.

Wholesale Auction Market Characteristics

The reduced price hypothesis has a solid theoretical grounding and has been empirically supported in many industries. For instance, the effective distribution of price information in NASDAQ and SEAQ wiped out excess profits once enjoyed by financial dealers.[1] The tremendous growth of the Internet and the World-Wide Web is bringing significant changes to the economics of marketing and is likely to lower prices for retail consumer products. A variety of electronic intermediary services available on the Internet, such as Yahoo!, EINet Galaxy, Lycos and InfoSeek, enable consumers to navigate extensive databases of Web sites and to locate their most preferred vendors, thus significantly reducing search costs of individual consumers or buying firms [10]. The reduction of the search costs makes the market more competitive by increasing the price competition among suppliers.

Furthermore, the Internet often allows manufacturers to bypass existing market intermediaries by internalizing activities that have been traditionally performed by market intermediaries [3]. Market intermediaries add costs (middleman margins) to the industry value chain that result in a higher final price of products or services. If part of the savings resulting from this bypass is transferable to consumers in the form of price reduction, electronic marketplaces may more than compensate for the costs of searching and matching, and could become a preferred alternative to existing intermediary markets.[2] The recent proliferation of suppliers' virtual storefronts, where consumers can directly buy goods over the Internet, supports the proposition that the Internet enables vendors and customers to leap over intermediaries and that it potentially reduces the prices of goods.

AUCNET significantly reduced the search costs that buyers had to bear to locate the most preferred car in the market. AUCNET did not add a new intermediary layer in the vertical chain of used-car trading but rather created an alternative auction intermediary incurring similar transaction costs (consignment fees and contract fees of AUCNET are similar to the industry average). Then why are prices of cars in AUCNET higher than those in traditional and non-electronic marketplaces? In order to address this question, it is necessary to discuss characteristics of wholesale auction markets, together with accompanying features of their electronic marketplaces.

Internet-based retail electronic marketplaces do not include the function of discovering the market price of goods, although they have potential to influence retail prices by increasing competition among suppliers [7]. Products sold in electronic retail markets are mostly standardized and mass-produced, and thus products from one supplier can be assumed to be identical. These systems typically employ posted-off pricing, where producers offer asking prices and customers decide how many items to buy at relatively fixed prices. Retail electronic marketplaces focus on facilitating comparison of a wide variety of products by consumers who purchase goods based on price tags and brand names.

This contrasts with electronic marketplaces for wholesale auction markets, one of the major functions of which is to determine the market price of goods. Sellers who join the wholesale auction markets have fixed quantities for supply without price tags: sellers are not price makers but price takers, although they have a certain level of reserve prices. The electronic market systems determine the market price of goods by tuning supplies and demands within their markets. Buyers who purchase goods in these markets are not the ultimate consumers but dealers who resell their purchased items to their clients. Since qualities of offered products by suppliers widely vary (even products from the same supplier differ in quality from time to time), descriptions of the product quality are essential to buyers who regularly join the market to purchase goods at the wholesale level.

Finally, traders completing transactions in wholesale auction markets are subject to institutional rules that govern and regulate transactions made within the auction markets [5,6]. Agreement over the governing rules can be facilitated because the members meet frequently and deal in a restricted range of goods. The enforcement of the rules is possible because the opportunity to trade on the exchange itself is of great value: the withholding of permission to trade is a sanction sufficiently severe to ensure compliance for most member traders. When the transaction facilities are scattered and owned by a vast number of people, as in the case of various retail online shopping systems over the Internet, the establishment and administration of a private institutional rule would be very difficult. Those operating in these markets have to depend, therefore, on the legal system at the State level.

Reasons for Higher Prices in AUCNET

The unique features of wholesale auction markets have made the single economic variable of the buyer search costs insufficient to explain the impact of electronic market adoptions on product prices. We suggest that the following three institutional and economic factors offset the impact of the reduced buyer search costs and eventually contribute to higher prices of cars traded through AUCNET.

Relatively Newer Secondhand Cars. Before the advent of AUCNET, traditional auto auction markets implemented a Slide Auction in an attempt to overcome their limits in the number of car offerings resulting from the parking capacity restrictions. The Slide Auction, which was intended to separate the transportation of cars from physical auction processes, held auctions by using 35mm color slides shown to buyers present in auction sites for bidding. The Slide Auction ended in failure because buyers were unable to judge the quality of cars offered through the slides. Used-car dealers, who had grown accustomed to inspections of cars at auction sites, did not trust that the color slides adequately represented the product.

The failure of the Slide Auction suggests that buyers will encounter new transaction risks if they join electronic marketplaces. Unlike traditional auto auction markets, electronic marketplaces do not allow a buyer to "kick the tires" to formulate personal assessments of used-car qualities. Buyers using electronic marketplaces have to place bids based on electronic information alone without physically inspecting the cars, and thus face risks of incomplete and distorted information by sellers. Secondhand vehicles transactions are a classic example of a market with asymmetric information because there is always a great possibility that sellers may not reveal hidden defects of cars.[3] Since the quality of secondhand cars offered to wholesale markets widely varies, the use of electronic networks for market transaction can further magnify the information asymmetry phenomenon. If electronic marketplaces fail to assure qualities of products, buyers are not likely to adopt the new online trading services.

AUCNET established a new institutional policy, together with a rigorous car inspection process, to reduce potential transaction risks resulting from uncertain product quality. Used-car sellers must have their vehicles inspected by AUCNET mechanics. The inspection results are summarized in a single number between 1 and 10 (10 indicates a new car, and a car rated 5 or 6 could be resold to the consumer without additional work). A car whose inspection rate is lower than 4 cannot be listed in AUCNET. The consignment contract requires that AUCNET members abide by AUCNET's decisions in such matters. As a result, vehicles sold in AUCNET are of higher quality than those traded in traditional auto auctions where there is no such restriction for car offerings and any car can be put into the auction. The quality difference between the two markets is reflected in the average age and mileage of vehicles traded: cars sold in AUCNET are 2.5 years old with 35,000km of mileage, while cars traded in traditional auto auctions are 4.0 years old and have average mileage of 55,000km.[4] Because of the quality risks associated with online trading, AUCNET had to employ unique institutional policies and tight inspection procedures to reduce transaction risks of traders. These new institutional policies and rules eventually resulted in transactions of higher quality cars, which receive better prices than those sold in traditional auto auction markets.

Increased Market Power of Sellers. Sellers in wholesale markets establish reserve prices for transactions because they do not have perfect information about the consequences of their actions in markets. The reserve price plays a role as sequentially rational rules for suppliers under incomplete market information [11]. Sellers who bring their cars to traditional auto auctions are often forced to accept prices lower than their reserve prices since the transportation costs of bringing the unsold products back home are high.

The separation of transportation of cars from the auction process in AUCNET enables used-car sellers to keep their reserve prices relatively high because unsold products do not incur return costs.

This is partly reflected in AUCNET's lower contract rates than traditional auto auctions, as shown in Figure 4. The percentage of sold cars to total car offerings in AUCNET was lower (by 9.5% on average) than that in non-electronic marketplaces during the past five years. In 1995, for instance, 55.5% of cars registered in traditional auto auctions were sold while only 46.4% of cars listed in AUCNET were contracted through online auctions, indicating that sellers in AUCNET are less fearful of unsold cars and are likely to preserve their higher asking prices. The decoupling of car transportation from the auction process has increased the market power of sellers by making the market process more transparent for suppliers and has eventually contributed to higher prices of secondhand cars in AUCNET.

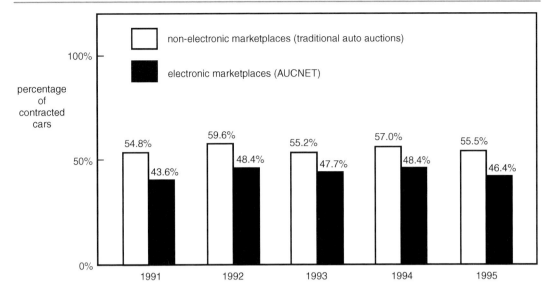

Figure 4 Percentage of sold cars to car offerings.

Buyer Externality. Used-car suppliers in wholesale markets receive prices determined by market demands. Sellers who take their orders to a less active and less liquid market face economic penalties of lower prices [4]. Buyer externality implies that benefits realized by individual sellers increase as more buyers join the bidding. The number of prospective buyers to whom sellers can expose their products in weekly auctions is confined to the maximum number of bidder seats. Figure 5 contrasts the weekly bidding capacity of AUCNET with those of the top five non-electronic auction markets. The largest one of the traditional auction markets—USS Kyushu—has 1,000 buyer seats for bidding and others are equipped with a bidding capacity ranging from 400 to 850 seats.

AUCNET has significantly increased the buyer externality for auto auctions. Sellers in AUCNET can present their supplies to over 4,000 potential buyers, compared to less than 1,000 buyers in

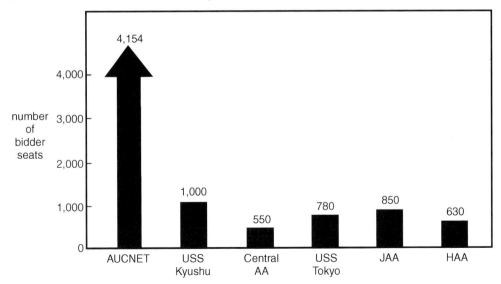

Figure 5 Bidder seat capacities for weekly auctions.

traditional auctions. As AUCNET decreases the search costs of buyers by eliminating the need for a parking capacity for cars to be auctioned, it simultaneously reduces the search costs of sellers by relaxing the constraint on bidding capacities of face-to-face auctions. Furthermore, the number of potential bidders in AUCNET will increase as AUCNET enlarges its dealer member network. The increased buyer externality in AUCNET enables sellers to expose products to buyers who value them most and are willing to pay higher prices.[5]

Conclusion

AUCNET has reengineered the traditional market transactions by separating transportation of cars from auction processes. This IT-enabled market process innovation has created risks of insecure product qualities as buyers have to make purchasing decisions without physically inspecting vehicles. To minimize the potential of quality risks, AUCNET focuses on relatively newer secondhand cars. The difference in average car qualities is quite likely the prime reason why the contract prices of cars in AUCNET are higher than those in traditional and non-electronic auto auctions. However, managers in AUCNET are confident in saying that even cars of similar quality can receive a slightly higher price in AUCNET than in traditional auto auctions because sellers can preserve their asking prices while being able to expose their products to a wider range of buyers. The higher contract prices have made AUCNET attractive to many sellers and have contributed to increasing the number of cars listed. This in turn has attracted more buyers as AUCNET has offered more purchase choices. Buyers are willing to pay the premium (a slightly higher price) because they not only avoid an immense waste of time spent on attending physical auctions but also easily locate a vehicle that best matches their preferences.

It should be noted that our findings in this article provide a supplementary line to the reduced price hypothesis rather than offering an opposite hypothesis. The reduced price hypothesis is established based upon markets where there are more buyers than sellers, and suggests that electronic market systems may wipe out excess profits enjoyed by a few monopolistic sellers in price-competitive markets and increase buyer welfare by lowering the market prices of goods. Thus, it indicates that those sellers may try to delay electronic market adoptions or try to control their development because their market power is likely to decrease when electronic marketplaces emerge. The implication of this research is that, when electronic marketplaces are introduced to markets with large numbers of buyers and sellers, they can improve the welfare of both the sellers and the buyers by increasing market efficiency. New market intermediaries in such markets can build electronic marketplaces that benefit both buyers and sellers (win-win market), rather than transferring welfare from sellers to buyers (lose-win market).

NOTES

1 NASDAQ (National Association of Security Dealers Automated Quotation) and SEAQ (Stock Exchange Automatic Quotation) are quote-driven systems in the U.S. stock market and the London stock exchange. Both systems display dealers' quoted prices on a widely distributed electronic billboard system so that investors can execute transactions at the best dealer bid-offer quote.
2 Bypassing intermediaries, however, is difficult when traditional intermediaries have channel power strong enough to force producers to abandon efforts to bypass them fully. A supplier would be risking loss of the majority of its business with traditional intermediaries if intermediaries retaliate against the firm for bypassing them. For instance, Air France once planned to allow corporate customers to completely bypass travel agencies through direct online seat reservations. Air France later canceled this plan, fearing travel agencies would retaliate by giving more business to competing airlines [10].
3 Akerlof [1] presents secondhand car transactions as an example of a market with asymmetric information. It would be very costly for a buyer of a secondhand car to accurately determine its true quality. There is no guarantee of full disclosure of the seller's knowledge of the car's history and quality during the transaction, particularly if the vehicle is a "lemon" the seller is eager to sell.
4 The average age and mileage of traded cars are estimates of PROTO Corp., a used-car magazine company that collects transaction data from used-car auction markets and publishes weekly and monthly auto auction results.
5 A linear regression model is used to test relations between the maximum bidder capacity and the average contract price in traditional auction markets. The result shows a statistically significant correlation between the number of bidder seats (independent variable) and the average price (dependent variable), thus supporting the existence of the buyer externality even in the traditional auto auction markets.

REFERENCES

1. Akerlof, G. A. The market for "lemons": Qualitative uncertainty and the market mechanism. *Q. J. of Economics 84* (Aug. 1970), 488–500.
2. Bakos, J. A strategic analysis of electronic marketplaces. *MIS Q. 15*, 3 (Sept. 1991), 295–310.
3. Benjamin, R. and Wigand, R. Electronic markets and virtual value chains on the information superhighway. *Sloan Management Review 36*, 2 (Winter 1995), 62–72.
4. Clemons, E. K. and Weber, B.W. Alternative securities trading systems: Tests and regulatory implications of the adoption of technology. *Information Systems Research 7*, 2 (1996), 163–88.
5. Coase, R. H. *The Firm, the Market, and the Law*. University of Chicago Press, 1988.
6. Lee, H. G. and Clark, T. Market process reengineering through electronic market systems: Opportunities and challenges. *J. Management Info. Systems 13*, 3 (Winter 1996), 113–36.
7. Lee, H. G. and Clark, T. Impacts of electronic marketplace on transaction cost and market structure. *Int. J. Electronic Commerce 1*, 1 (Fall 1996), 127–49.
8. Malone, T., Yates, J., and Benjamin, R. Electronic markets and electronic hierarchies. *Commun. ACM 30*, 6 (June 1987), 484–97.
9. Rayport, J. F. and Sviokla, J. J. Managing in the marketspace. *Harvard Business Rev. 72*, 6 (Nov.–Dec. 1994), 141–50.
10. Sarkar, M. B., Bulter, B. and Steinfield, C. Intermediaries and cybermediaries: A continuing role for mediating players in the electronic marketplace. *J. Computer-Mediated Communication 1*, 3 (1996); <http://www.usc.edu/dept/annenberg/journal.html>.
11. Stigler, G. J. Public regulation of the securities markets. *J. Business 37* (Apr. 1964), 117–34.
12. Warbelow, A. and Kokuryo, J. AUCNET: TV auction network system. Harvard Business School Case Study, 9–190–001, July 1989.

Next-generation Trading in Futures Markets: A Comparison of Open Outcry and Order Matching Systems

Bruce W. Weber

Abstract

The introduction of new screen-based systems for trading securities and futures contracts has led to the emergence of a "market for markets," and exchanges, broker-dealer firms, and market data vendors are competing to offer trade execution services that will attract customers and trading volumes. This competition is favored in the United States by regulatory bodies such as the SEC and the CFTC, which have taken steps such as encouraging the listing of equity options on multiple exchanges and approving the applications of screen-based systems for designation as contract markets. This paper examines the design of one screen-based futures market, the Cantor Financial Futures Exchange (CX), and describes its capabilities relative to the rival, floor-based market in Chicago. In comparison to traditional open-outcry mechanisms, the CX order-matching system maintains strict first in–first out time priority among submitted orders. Using a simple simulation model, we see that order matching leads to faster completion of desired trades and about a one-third reduction in transactions costs.

Key words and phrases: electronic futures trading, screen-based trading, trading automation.

This paper compares "open outcry," an established and widely used method for trading futures contracts on a market or exchange floor with a screen-based alternative, electronic "order matching." In a number of markets today, automation reliably handles trading functions, including order routing, quote display, price determination, and trade execution. And yet, research into the relative merits of alternative market structures has not yielded a conclusive answer to the question, *What trading mechanism maximizes participants' satisfaction and minimizes transactions costs?* Today, floor markets and screen-based markets, with different trading mechanisms, coexist and both operate and provide good levels of liquidity to participants [9, 11].

Studies have shown that well-designed trading automation can be valuable to investors and traders in markets [4, 5, 12]. For example, the introduction of the SEAQ screen-based market system as part of the London Stock Exchange's 1986 Big Bang market reforms improved the quality of the LSE market [3] and played a part in increasing trading volumes from $280 million a day in 1985 to $4.1 billion a day in 1994. In the same period, bid–ask spreads (an important trading cost) for FTSE 100 stocks fell from 1.0 percent to 0.8 percent, and commissions shrank from 0.33 percent to 0.17 percent. Thus, the cost of a round-trip investment (purchase and subsequent sale) fell to 1.14 percent (= 0.8 percent + 2 × 0.17 percent) from 1.66 percent. Comparing SEAQ to the floor, London's electronic market proved to be more open and competitive than the floor market and led to lower transactions costs for investors. Similarly, the introduction of the Nasdaq screen market in 1971 to replace the OTC "pink sheets" led to a reduction of the average bid–ask spread in a 174-stock sample to 40.3 cents from 48.7 cents [8]. The best explanation for the reduction in trading costs from SEAQ and Nasdaq is that market makers' quotes were made visible on a computer terminal, which forced dealers to compete by posting more aggressive bid and offer quotes, thus narrowing the bid–ask spread.

Today, alternative venues and market mechanisms exist for the trading of U.S. Treasury securities futures contracts.[1] From the introduction of Treasury futures in 1977 until 1998, trading occurred exclusively in an open-outcry market structure on the floor of the Chicago Board of Trade (CBOT). A competing market, the Cantor Financial Futures Exchange (CX), opened in 1998 and provides screenbased order matching with two distinctive innovations compared with open-outcry markets: (1) first in–first out (FIFO) time priority among orders at a particular price, and (2) a "clearing time" period and "exclusive time" period for the providers of the best bid and ask quotes in the market.[2] The significance of this market, and the presence of alternative trading venues with very different trading mechanisms, highlights the importance of careful study of the impact of trading alternatives.

This paper describes developments in electronic markets for futures trading and then details a model of order arrival and trading that will be used to compare open-outcry trading with a FIFO-based order-matching alternative. It is found that the use of an order book, imposing time and price priority, does indeed improve market quality and customer satisfaction, and does indeed reduce trading costs for investors. Moreover, as explored in a later section, these results hold true even under the most conservative of assumptions concerning the behavior of market participants after the introduction of electronic alternatives.

Environment for the Electronic Futures Trading

More sophisticated strategies and investors desire for more direct market access and greater control over trading are leading to growing interest in screen-based trading systems [10]. In Europe, futures trading has moved rapidly to screen-based markets (see Table 1). The rise of these screen-based markets, and the erosion of activity in Europe's open-outcry floor markets surprised many observers.

While a number of factors may have contributed to the success of these screen-based markets, one contributing factor may be the attractiveness of their order-matching trading mechanisms relative to the open-outcry markets they displaced. In order matching, limit orders[3] placed by market participants can be maintained in time priority, unlike open outcry, in which shouted orders that are tied at the same price are eligible for execution.

To assess order-matching markets, and the implications of their broader adoption for traders, we examine the U.S. Treasury futures market, which has the greatest trading volume of any futures market worldwide. With a model of order arrival and trading, market quality under open-outcry rules is compared with order-matching rules.

The U.S. Treasury Futures Market: Current Order Handling and the Potential for Screen-Based Trading

The U.S. Treasury securities market is the world's largest fixed-income securities market and is one of the largest and most sophisticated financial markets. The U.S. Treasury debt outstanding was $5.4 trillion as of fourth quarter 1997. The treasury market is intended to facilitate the distribution of the U.S. national debt through as efficient a mechanism as possible. Treasury securities are issued in regularly scheduled auctions. Participating directly in the auctions are about thirty-six primary dealers who bid for securities in monthly and quarterly refinancings. There can be as many as 156 separate auctions per year, and recently about $2 trillion in Treasury securities have been auctioned per year [1]. After issuance, Treasury

Market	Recent developmens
Marche à Terme International de France (Matif) Paris-based futures market opened in 1985.	On June 2, 1998, less than two months after the April 8, 1998 introduction of its Nouveau Systeme de Coation-Version Future (NSC-VF) system, floor trading was ended for Matif's interest rate futures contracts. • Matif had planned for a longer period of parallel operation of its open-outcry pit market and NSC-VF. In 1997, *after-hours* trading of Matif contracts on the Globex system accounted for just 9 percent of the exchange's total volumes. • In the day after the NSC's launch, however, floor trading fell to 30 percent of the total. After a month, pit trading had dwindled to 10 percent, down from over 90 percent, and the population of floor trades fell from 400 to 100. • Later in 1998, floor trading ended in options and physical commodities futures, and aws moved to electronic trading exclusively.
Deutsche Terminbörse (DTB) screen-based future market of the Frankfurt-headquarted Deutsche Börse. Opened January 26, 1990. Renamed Eurex in 1998.	From 1991 to 1997, over *70 percent* of trading volume in the German goverment 10-year bond future (Bund) occured on the floor of the London International Financial Futures Exchange (LIFFE). The other 20–30 percent took place on the DTB. • In 1997, the Bund was the world's third most actively traded futures contract, after the U.S. T-Bond (CBOT) and the 3-month Eurodollar contracts (CME). • In the first quarter of 1998, trading volumes on the Deutsche Terminbörse (DTB), the screen-based market of the Frankfurt-headquartered Deutsche Börse began to exceed the LIFFE's volumes for the first time. • In April 1998, the DTB had a 70 percent share of trading in the Bund, and the DTB's share reached 82% on Wednesday April 22, when it traded 430, 187 Bund contracts. • The period of competition between the LIFFE and DTB markets led to reduction in exchange trading fees per contract to about $0.45 on LIFFE and to about $0.25 on the DTB.

Table 1 Recent developments in screen-based futures trading in Europe.

Sources: "Frankfurt Exchange Overtales LIFFE," *Financial Times* (June 3, 1998), p. 17; "DTB Bund Volumes Overtake Chicago," *Financial News* (London) (April 27, 1998), p. 15; "Matif set to End Open Outcry," *Financial Times* (May 14, 1998), p. 20; "Matif Brokers Ponder Future," *Futures* (June 1998), pp. 14–16.

securities trade in an active cash or secondary market. Investors hold treasury securities because of their high credit quality and because of the market's vast liquidity [6].

The Treasury Futures Market

Derivatives of treasury securities are actively traded. In May 1999, the Chicago Board of Trade (CBOT) traded 9.6 million Treasury bond futures contracts, making it the most active futures contract in the world. T-Bond futures accounted for over 39 percent of the total 24.5 million futures and options contracts traded in May 1999 at the CBOT. The value of Treasury futures contracts traded on an average day is about $60 billion and represents about 40 percent of cash market volume.

U.S. Treasury bond and note futures are widely used by a number of different groups to transfer risk and to manage the risks of holding Treasury and other fixed-income securities. For instance, firms that originate mortgages (i.e., lend at fixed rates to home buyers) will hold and trade futures contracts to hedge the interest-rate risk of their mortgage positions. Other common uses of futures on treasury securities in investment management are [2]:

- *Lock in a purchase price:* If an investment manager anticipates positive cash inflows that will be used to purchase fixed-income securities in the future and is concerned about the possibility of higher prices, he or she can buy Treasury futures with delivery months near the time of the anticipated cash flows. This establishes a maximum purchase price.
- *Safeguard investment value:* By selling Treasury futures, an investment manager can lock in attractive selling prices and preserve the value of a portfolio or a security against possible price falls.
- *Cross-hedge:* U.S. Treasury securities prices and yields are the benchmark against which most other fixed-income instruments are compared. Treasury futures can be used to control risk and enhance the returns from non–U.S. government securities. For instance, an investment-grade corporate bond maturing in 10 years could have its yield quoted as 85 basis points over the benchmark Treasury security, which in this case is the 10 year treasury note. As a result of its benchmark characteristics, Treasury futures are useful risk-management tools for corporate bonds, mortgage-backed securities, agency securities, Eurobonds, non–U.S. government and private-sector bonds, and other fixed-income instruments.
- *Enhance returns:* Treasury futures are used to increase exposure to changing rates, allowing investors to profit from anticipated interest-rate moves and to enhance overall returns.
- *Fine-tune positions:* Treasury futures are used to adjust positions or to fine-tune risk-management strategies. By buying or selling futures, a fund manager can lengthen or reduce the duration of a portfolio, thereby increasing or decreasing sensitivity to interest-rate changes.
- *Profit from shifts in the yield curve:* Investors can construct trades based on the differences in interest-rate movements at different points on the yield curve. For instance, an investor that expects the yield curve to steepen, making short maturity instruments gain in value relative to long dated instruments, can sell bond futures and buy two-year-note futures contracts.

These strategies are beneficial because they give investors a chance to customize the risk and return characteristics of their portfolio and to adjust positions quickly to reflect their outlook on interest rates and the market.

Current Open-Outcry Practices

To trade most futures contracts, including Treasury security futures, a market participant must have an account with a broker or Futures Commission Merchant (FCM). As figure 1 indicates, the FCM receiving the customer's order will phone a floor broker with a booth on the appropriate futures exchange. The clerk will write down the order and time-stamp it, and then either pass the order ticket to a runner who will deliver it to the firm's trader in the pit or flash it to the trader using hand signals. Once the order is executed in the pit, information on the filled order is sent back to the booth and relayed to the customer. The process may take from several seconds to several minutes, depending on the complexity of the order and the level of activity in the pit [10].

In recent years, the major futures markets have installed order-routing systems to deliver orders electronically to the FCM booth or to the trading pit. Two systems, TOPS and COMET, are used in the CBOT markets. While such systems eliminate

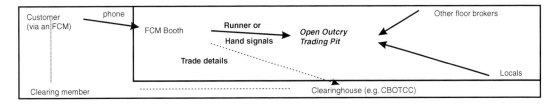

Figure 1 Open-outcry market flows.

Note
A customer's order is phoned to the booth of the Future Commission Merchant (FCM) along the wall of the market floor. The instructions are written on an order ticket and then either flashed with hand signals to a broker with the FCM firm or a runner delivers a ticket to the broker in the pit.

paperwork and speed the transmission of information, they essentially automate existing practices. TOPS transmits the customer's order in electronic form from a remote trading desk to the FCM's booth, or, if it is for 10 contracts or less, directly to an "electronic clerk" (EC) terminal in the trading pit. COMET terminals are in place in most FCM booths and are used to direct orders received over the telephone to an EC terminal in the appropriate pit. Once an order arrives at the EC terminal, it is routed to the appropriate trader, who fills the order and sends the fill information back to the FCM booth and the customer. In addition, several floor-based markets have screen-based systems for after-hours trading. These systems and their sponsors include: Project A (CBOT), Globex (Chicago Mercantile Exchange), and Access (New York Mercantile Exchange). All are screen-based mechanisms for trading derivatives contracts after normal trading hours. To date, few exchanges have run trading systems in direct competition during the open hours of their floor markets.

Electronic Trading: The CX Example

In electronic futures markets, participants also have accounts with a broker or a clearing bank, who backs their trades financially. Instead of relaying information through the broker floor staff, the investor can monitor quoted prices on a screen and can enter buy or sell orders directly into the market [7]. In the CX market, there are two ways participants can access the market (see figure 2): They can directly enter their own orders via their workstation keyboard or they can phone a CX "terminal operator" (TO), who inputs orders into the system and can rapidly modify customer orders. The TO provides a clerical function, exercises no trading discretion, and does not accept "not held orders."

The CX trading system is designed to provide a competitive market and to handle peak trading volumes efficiently. The CX currently does not maintain an order book or a "deck." The CX screen simply shows the best bid and best offer quotes with their size. Bids or offers are deleted from the screen after a better-priced order arrives. For example, bids to buy at 115 are deleted when a better bid of $115^1/_{32}$ or greater arrives. In some markets, these less competitive bids and offers are retained in an order book that may be visible to the market. Quoted sizes on the CX screen could be the aggregate of several orders at that price, or a single order.

The Cantor Exchange was developed and is operated by Cantor Fitzgerald, the leading interdealer broker in the U.S. Treasury market. The CX is owned by the New York Board of Trade (NYBOT) and its members. The NYBOT is responsible for self-regulatory oversight and clearing of all trades executed on CX.

Compared with open-outcry markets, the CX imposes different pricing and trade-allocation priorities, which are intended to create incentives to provide liquidity and place orders at attractive prices. The CX screen displays orders anonymously, showing bids and offers with size, and the last trade price, but no information about the counterparties is displayed. A trade in the system occurs when an order that was entered and displayed is "hit" (sold to) or "lifted" (bought from). The trader that is the "aggressor"—that is, who actively initiates a trade by hitting or lifting a displayed quote—will pay a transaction fee (see Table 2).

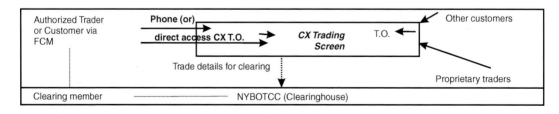

Figure 2 CX market flows.

Note

A customer's order is phoned to a CX terminal operator and displayed on a trading screen for other market participants to react to and trade with. Because traders on the CX enter orders in the market directly or through a TO, a layer of interaction and information transmission that occurs on the floor is eliminated.

CBOT	CX
65,000 sq. ft. Floor opened February 1997 at cost of $182 million. Holds 8,000 people. Formed as an agricultural products market in 1848.	Screen-based, interactive matching market opened in September 1998.
Users contact floor brokers directly or indirectly through an FCM. Floor broker transmits order into trading pit for execution.	Users enter orders directly or contact TOs, who enter their orders for immediate display or execution. The step of moving the order into the trading pit is eliminated.
Prices determined by open outcry in designated trading pits.	Price determined electronically on screen using an order-matching algorithm.
Price data available to the public immediately after trade. Indicative price quotes available from service providers.	Firm bid and offer quotes with size shown on screen. Trade price and quantity information provided live at the time of execution.
Contract delivery months: March, June, September, December	Contract delivery months: March, June, September, December
Trading hours: 7:20 A.M. –2:00 P.M., Chicago time	Trading hours: 7:30 A.M.–5:30 P.M., New York time

Other differences between the CX market and open outcry markets:

Anonymity. The source of an order may be revealed or inferred in an open-outcry trading pit. The identity of a CX order is not revealed during trading and is not revealed during clearing and settlement because the NYBOT clearinghouse steps in and is counterparty to the two sides of each trades.

Transparency. The CX will provide firm and executable bid and offer quotes with sizes that are visible to all market participants. Open-outcry markets do not provide the same degree of pretrade transparency.

Table 2 Comparison of market characteristics and traded contracts on CBOT and CX.

A Simulation Comparison of Open Outcry and Electronic Order Matching

Earlier we saw how electronic futures trading was growing in importance worldwide. We then looked at the current status of the U.S. Treasury futures market, and at CX, an electronic system for trading treasuries futures contracts. This section develops a simulation model to demonstrate the potential improvements available to these markets from the use of electronic trading systems.

In the CX, and in most electronic order-matching systems, orders are filled according to price and time priority. That is, a market order to sell will be matched with the best (highest price) buy order (bid) that arrived the earliest. Similarly, arriving buy orders execute at the lowest available offer price and are matched with the earliest arriving sell order(s) at that price. In an open-outcry market, sell (buy) orders must be traded at the highest shouted bid (offer) quote, but time priority is not maintained at a particular price level. As a result, any of the traders in a pit that are shouting the same bid or offer quote could fill the next order.

To compare open-outcry and electronic order matching, a simulation model was developed in Crystal Ball 4.0, an Excel spreadsheet add-in. Simulation has provided insights into other market-structure questions, when institutional details precluded obtaining results in closed form [4]. In each run of the simulation, a sequence of 250 buy and sell orders is entered into the market. Orders are either market orders or limit orders. Averaged over 5,000 trial runs of the simulation, the sequence of 250 orders in the model generated 112.5 trades, which reflects about 30 minutes of trading activity during a typical day in the CBOT T-Bond futures market. About 10 percent of the arriving orders are limit orders that ultimately do not execute.

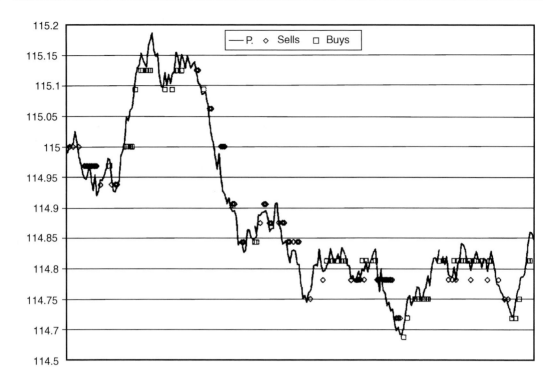

Figure 3 One simulation run of 250 orders, showing P^*, and the sell market order-initiated (sells) trades and the buyer-initiated (buys) trades.

Note
The first part of the run that buying orders with higher reservation prices than the current market offer quote depleted the sell orders and led to price increases. In the last part of the run, selling initiated by orders with lower reservation prices drove trading prices down.

Modeling Assumptions

Each simulated order in the model has an associated quantity (a uniformly distributed number of contracts from 1 to 50), and a reservation price that is sampled from a lognormal distribution around a simulated equilibrium value for the futures contract. This value, P^*, itself follows a white-noise random walk over the simulation period with the 250 interevent returns i.i.d. $N(0, ó^2)$, where $ó^2 = 0.015$ percent. The resulting high–low ranges in the simulation averaged 38 cents, or 0.33 percent of the contract's value, which is consistent with the actual market's ranges during half-hour periods in the trading day.[4]

The simulated market processes successive orders according to the following logic:

- A buyer with a reservation price
  — that is *greater* than the current lowest offer price buys using a market order the size of his or her order, or the size of the current offer quote, whichever is less.
  — that is *less than* the current lowest offer will enter that order as a limit order into the market. If the limit price is greater than the current highest bid, a new improved bid is quoted in the market.
  — that is at the *same price* as the bid has an increased quantity available to buy at the bid. If the reservation price is below the current bid, the order is canceled.
- Similarly, for sellers, their orders will hit the bid if their reservation price is less than or equal to the bid quote, or they will provide an offer quote that could join the current offer quote, or improve on it by offering to sell at a lower price level.

Since reservation prices are distributed around the P^* value, buying and selling will lead to market price changes that will keep quote and trade prices in line with P^*. Figure 3 shows an example of the evolution of P^* and market prices over the course of one simulation run. The price increment in the market is $1/32$ of a point, as is used in the actual Treasury market. The initial price is 115, which reflects the approximate current price levels of Treasury bond futures contracts.

The distinction in order handling across the two markets is that the electronic order-matching system holds the orders in time priority, and in the simulation of the open-outcry mechanism, when several orders are tied at the same bid or offer quote price, the simulation randomly assigns which of the orders will execute the next trade. This distinction could lead to different order-placement strategies, but, for now, we will hold constant the background conditions in the simulated market and examine the impact of the order-matching mechanism on an individual market participant. (The decision to hold order-placement strategy constant is the most conservative strategy available; that is, if the use of the order book improves market quality for investors even without changes or adaptation in their behavior, it would be reasonable to assume that after learning and adaptation these benefits would only increase.)

Model of an Individual Trading Strategy with Measurable Trading Costs

To compare trading costs, we model a hypothetical trader who seeks to open a long position (buy a futures contract) and later close (sell) it. This is done without loss of generality, and the results would be equivalent if the focal trading strategy were to go short by selling and later cover with a buy order. The approach the modeled trader uses is to place a limit buy order for one contract at the 100th order in the simulation sequence, and close out the position using a sell limit order at the 200th order in the sequence. The buy order is placed to establish a new best bid in the market, and the sell order is placed one price increment better than the offer quote. For example, if the market quote is 115 $3/32$ bid, the arriving buy order is placed at 115 $4/32$. If, at the point of the 200th order in the simulation, the ask quote is 115$6/32$, the sell order will be placed at 115$5/32$.

For both orders, the simulated trader will only wait for the arrival of ten orders, and if his or her limit price was not reached the simulated trader will cancel it and buy or sell with a market order. The advantage of this strategy is that it enables users to "earn" the bid–ask spread when their limit orders execute. However, after a period of ten orders, the user will "pay" the spread and buy or sell with a market order at the time of the 111th or 211th order. The strategy described is realistic and allows for trading costs and limit order execution probabilities to be computed.

Because other processes in the simulation were held constant except for the results achieved by the hypothetical simulated trader, many market measures are the same for both mechanisms. The overall market conditions are shown in Table 3.

The bid–ask spread was about 7 cents or 2.2 ticks (a tick is $1/32$ of $1), which is consistent with market-quality data from the actual market. The high–low range of prices over the simulation of 250 order arrivals averaged about 38 cents with a standard deviation of 13 cents. This is about 33 basis points ($^{0.38}/_{115}$ = 0.33 percent) of variation on average in the simulation period. A longer window would lead to a wider price range but would not change the results substantially. Comparisons of the two trading mechanisms are shown in Table 4.

Market means	Results (same under two market designs) and sample standard deviations
(Initial price = $115) High	$115.185 (0.164)
Low	$114.810 (0.165)
Price range	37.8 cents (13.0)
Bid–ask spread	6.84 cents (2.07)
Number of trades (resulting from 250 orders)	1125.5 (7.3)
Orders contained in best bid quote and ask quote	3.20 (0.94)
Quantity of contracts contained in best bid and ask quotes	44.4 (13.7)

Table 3 Overall market characteristics (5,000 simulation trials).

Means for individual trader (buy: order #100–#110, and sell: order #200–#210)	Open outcry	Order-matching system	Difference (std. error)*
Round-trip trading costs (difference between buying and selling prices)	2.326 cents	1.555 cents	0.771 cents (0.0005)
Average wait until order filled: limit orders only	5.56 orders	4.20 orders	1.36 (0.029)
Average wait until order filled: all	7.67 orders	6.45 orders	1.10 (0.025)
Percentage of orders filled as limit orders	45.1	55.0	9.9 (0.6)
Percentage of orders converted to market orders	54.9	45.0	9.9 (0.6)

Table 4 Comparisons of Market quality and costs (5,000 simulation trials).

Note
*F-test significant at 0.001 level.

Trading costs (average difference between the buying price and selling price) incurred by the individual trader are 2.3 cents under open-outcry trading and fall to 1.6 cents with order matching. Trading costs are positive because the trader's market orders pay the bid–ask spread, and because the trader's limit orders are subject to adverse selection [7, 11], since he or she is assumed to have no privileged information about P^*.

The trading cost reduction of 33 percent under order matching is mainly due to more of the individual's orders executing as limit orders, 55 percent, compared with the open-outcry market, 45.1 percent. Limit orders that execute reduce trading costs, and the CX's FIFO priority enables the individual's orders to execute before others at the price they established. In the open outcry, random assignment means that a later-arriving order could execute before the individual's buy or sell order.

A second important improvement in the order-matching environment is that traders' limit orders that establish a new best bid or new best ask quote execute more rapidly than in open outcry, where a later-arriving order could be filled first. Under order-matching rules, limit orders that executed were filled after the arrival of 4.2 subsequent orders on average, compared with an average wait for 5.6 orders to arrive for an open-outcry execution. The overall averages were 7.7 orders and 6.5 orders of delay and are somewhat greater due to the market orders used

	Price		Quantity	
BID (buy)	OFFER (sell)	BUY (Bids)	SELL (Offers)
* 120.01	—		100 x

Figure 4 A buy order arrives.

Note
After trading opens, an authorized trader (AT A enters a buy order for the September Treasury bond contract by placing a bid at $120^1/_{32}$ for 100 contracts. On the CX screen, this will be displayed as "120.01". An asterisk (*) is placed next to the price to indicate that it is newly arrived. The $120^1/_{32}$ bid for 100 is the "first best bid" because it was the first posted on the Cantor System and provided the highest bid at the time.

	Price		Quantity	
BID (buy)	OFFER (sell)	BUY (Bids)	SELL (Offers)
120.01	120.02*		100 x 150

Figure 5 A sell order arrives.

Note
As in figure 4, AT B enters an offer to sell 150 contracts at $120^2/_{32}$.

at the time of the eleventh order arriving after the limit order was placed.

Upon reflection, these results appear reasonable. In particular, if time priority is *not* maintained, then any limit order at a given price has a greater chance of "bumping up against" the ten-order limit. That is, if time priority is not maintained, then the variance in execution time increases, and, if other traders can trade ahead of a customer's limit order, this increases the number of limit orders that do not execute within ten trades. This, in turn, increases the number of limit orders that are "canceled" and converted into market orders. This explains the principal simulation results under open-outcry trading:

- More limit orders are converted to costly market orders.
- More customer orders pay the bid–ask spread.
- Fewer customer limit orders are available to provide price improvement to the market.

As a result of a fairly small enhancement, FIFO order matching in an electronic futures market, market users derive substantial benefits. Round-trip trading costs are reduced by 33 percent with order matching, and the delay between placing a limit order and its execution falls by 16 percent.

Other CX Features

In addition to its FIFO order-matching algorithm, the CX market provides a clearing time period and an exclusive time period for the providers of the best bid and ask quotes in the market. These features also distinguish it from open-outcry trading. The duration of these time periods will be set by the CX to create the appropriate level of incentives to place orders. The more liquid and actively traded the contract, the shorter these time periods are likely to be.

- *Clearing time* is a period in which the AT (a CX acronym for authorized trader) that placed the first best bid or offer has an exclusive right to respond to a contra offer or bid that has just arrived. During this time, the AT whose bid or offer is showing can trade against the newly arrived contra-side order. If the AT does not respond, then other ATs can trade with the order. *For instance, if two bids for 10 contracts each are displayed 120–01, and an order to sell 10 at 120–02 arrives from another AT, then the first of the two bidders has the duration of the clearing time to decide how much, if any, of the 10 contracts he or she wants to buy at 120–02.*
- *Exclusive time* is given to the AT who placed the first best bid or offer, and whose order was just hit or taken by a contra order. After the AT's order was traded, exclusive time begins and gives the AT the chance to do any additional

quantity that the aggressor has to trade. *For example, if two bids for 10 contracts each are displayed 120–01, and a sell order for 20 arrives from another AT who is interested in selling an additional 30 (for a total sale of 50), then the first of the two bidders has the duration of the Exclusive Time to decide how much, if any, of the 30 additional contracts he wants to buy at 120–01.*

A third time period, "execution time," is not fixed by the CX but determined by the sequence of orders and trades that occur. During execution time, the price and current volume transacted flash on the screen, indicating that a trade is being "worked up" to a larger quantity. Once the buying interest or selling interest at that price is exhausted, the execution time ends and the price and quantity traded no longer flash.

Examples of CX Orders and Trades

When an AT enters an order or phones a CX TO, the AT must specify: the code for the customer account or the proprietary order, the contract delivery month, the order type (buy or sell), the quantity, and the price. Figures 4 and 5 illustrate the operation of the CX market system.

After the counterpart offer quote arrives from AT *B*, "clearing time" begins. During clearing time, AT *A*, by virtue of having placed the first best bid, is given an exclusive right to respond to the offer quote that just arrived. AT *A* has several alternatives during clearing time:

1. AT *A* can do nothing, and the market remains quoted as in figure 5, or any of numbers 2 through 5 below.
2. *A* can be the "aggressor," and lift 100 of the 150 contracts offered. A trade of 100 occurs and flashes on the screen as "TAK 100," indicating that 100 contracts have been "taken" or bought (figure 6). Both ATs fill out tickets, and the TO passes the trade data on for

Price		Quantity	
BID (buy)	OFFER (sell)	BUY (Bids)	SELL (Offers)
120.—	120.02		×**TAK100**
BID (buy)	OFFER (sell)	BUY (Bids)	SELL (Offers)
120.—	120.02		x50

Figure 6

	Price		Quantity
BID (buy)	OFFER (sell)	BUY (Bids)	SELL (Offers)
120.—	120.02		×TAK 150

Figure 7

transmission to the CCC, which forwards it via TIPS (Trade Input Processing System, a clearing system used by the NYBOT) to the respective clearing members, who accept or reject it within thirty minutes of posting.

3 Or A can be the aggressor and take the entire offer. The screen will flash the completed trade of 150 contracts at $120^2/_{32}$. The "exclusive time" begins, and A and B have exclusive rights to trade more with each other or with others who wish to buy or sell at $120^2/_{32}$. If both A and B decline to trade any additional quantity, the trade price and size will stop flashing (figure 7).

4 If A wants to buy more, and B wants to sell more, A has the exclusive right to buy more at that price. If B offers to sell another 100 contracts, and A accepts, the screen will appear as in figure 8.

5 If B declines to sell more, then C, another AT who wants to sell, can offer to sell an additional quantity to A.

At any time, except during the execution time, any account may improve upon A's bid or B's offer. In that case, the improved quote shows on the screen and the other is automatically removed. For instance, if C offers to sell 100 at $100^1/_{32}$, then B's offer will be removed and A has the right to trade with C.

Notice that the trade algorithm did not require the participants initially to display the full size of their orders. Both features, price protection and the ability to restrict the display to a size that will not cause market impact, should improve the quality of the market.

	Price		Quantity
BID (buy)	OFFER (sell)	BUY (Bids)	SELL (Offers)
120.—	120.02		×Tak 250

Figure 8

Portfolio decision making →	Implementation/ trading →	Posttrade Processing →	Position accounting and risk management
Research, decision support Investment analytics Cash-flow needs Liability matching	Real-time market data Order handling Trading	Resolution of out-trades, errors, etc. Clearing Margining: original and variation Settlement and delivery	Position analysis Risk-adjusted return on capital Value at risk

Figure 9 A simplified value chain for institutional investors in fixed-income securities depicting the activities at each of four stages.

Note
Information technology is increasingly used to integrate and streamline the linkages between these activities.

Conclusions

A number of new screen-based trading systems for derivatives trading such as CX are based on price and time priority matching algorithms. Because the most competitively priced orders in the system are filled first, such systems provide an incentive to place market-improving quotes, which lead to trading-cost advantages and reductions in trading delays compared with current practices. The CX is evidence that a "market for markets" has emerged, which will improve the trading choices available to market participants and the quality of the market.

In addition to trading enhancements, screen-based markets provide operational improvements. A value chain for an institutional investor in fixed-income securities is depicted in figure 9.

Traditionally, exchange markets have supported the *middle* of the value chain, that is, the trading and trade-processing functions. Separate technologies are used by fund managers to support portfolio decision making and to manage risk. Electronic markets, such as Eurex and NSC in Europe, and the CX however, offer the capability of integrating investors' portfolio systems with the placement of orders into the market. And, with an electronic price and clearing feeds, investors can have real-time position accounting in order to manage risks more effectively.

The economic advantages of order matching were analyzed here and shown to be favorable. This can partly explain market users' willingness to adopt a number of screen-based markets that were in competition with established open-outcry market.

For exchange officials, the implications of order matching's benefits and the good response to screen trading by market participants in Europe indicate that:

1. Traders want direct access to the trading and discovery process and benefit from the trading priorities that can be enforced in screen-based markets.
2. Financial markets will move rapidly to "better" venues when there are benefits to an alternative mechanism for trading.
3. Exchanges must compete by reducing costs and trading fees and by implementing rules and systems that meet the needs of investor-customers, even when these changes may erode privileges enjoyed by member-firm intermediaries.

Although trading volumes on the CX in the first months after its September 8, 1998 launch were modest, by April 1999, about 10,000 contracts a day were trading, or about 2 percent of the 450,000 contracts trading a day on the CBOT. As liquidity develops and as the advantages of screen-based mechanisms for trading become more widely recognized, the CX and other screen-based markets will challenge the dominance of many of today's established open-outcry futures markets. Established markets that do not respond with improvements will see trading activity quickly won over by screen-based order-matching rivals.

NOTES

1. A futures contract is an agreement to purchase or sell a commodity for delivery at a specified time in the future (e.g., December 1999) at a price that is determined at the time of the purchase of the contract in a futures market. In the case of the U.S. T-Bond, it is a contract for $100,000 face value of 8 percent coupon T-Bonds. Higher interest rates will lead to lower futures prices.
2. A bid is a firm indication of willingness to buy at a stated price. An ask or an offer is a firm indication of willingness to sell at a given price.
3. Participants in a futures market submit buy and sell orders usually through a broker. An order can be a *market order*, an instruction to buy or sell at the best available price in the market at that moment, or a *limit order*, which sets a *limit price* as a upper bound on the most they will pay to buy, or a lower bound on what they will sell for.
4. A section of the CBOT web site, <http://www.cbot.com/mplex/quotes/>, provides times and sale data that can be used to determine high-low ranges for short intraday trading periods.

ACKNOWLEDGMENTS

Phil Ginsberg and John Eley of Cantor Fitzgerald provided thorough descriptions of the CX's trading mechanism and its attractions for market participant. Mike Uretsky provided useful comments on an earlier version of the paper. Eric K. Clemons likewise provided useful comments on a later draft.

REFERENCES

1. Bartolini, L. and Cottarelli, C. Designing effective auctions for Treasury securities. *Current Issues in Economics and Finance* (Federal Reserve Bank of New York) (July 1997), 1–6.
2. Bortz, G. Does the treasury bond futures market destabilize the treasury bond cash market? *Journal of Futures Markets, 4*, 1 (Spring 1984), 14–24.
3. Clemons, E. K. and Weber, B. W. London's big bang: a case study of information technology, competitive impact, and organizational change. *Journal of Management Information Systems, 6*, 4 (1990), 41–60.
4. Clemons, E. K. and Weber, B. W. Alternative securities trading systems: tests and regulatory implications of the adoption of technology. *Information Systems Research* (June 1996), 163–188.
5. Domowitz, I. and Steil, B. Automation, trading costs, and the structure of the securities trading industry. Working Paper, Royal Institute of International Affairs, London, February 1998.
6. Fleming, M. The around the clock market for U.S. treasury securities. *FRBNY Economic Policy Review* (July 1997), 9–32.
7. Grunbichler, A., Longstaff, F. and Schwartz, E. Electronic screen trading and the transmission of information: an empirical

examination. *Journal of Financial Intermediation, 3* (1994), 166–87.

8. Hamilton, J. Marketplace organization and marketability: Nasdaq, the Stock Exchange, and the national market system. *Journal of Finance, 33* (March 1978), 487–503.

9. Hamilton, J. Electronic market linkages and the distribution of order flow: the case of off-board trading of NYSE-listed stocks. In H. Lucas, Jr., and R. Schwartz (eds), *The Challenge of Information Technology for the Securities Markets: Liquidity, Volatility, and Global Trading.* Homewood, IL: Dow Jones-Irwin, 1989.

10. Massimb, M. and Phelps, B. Electronic trading, market structure and liquidity. *Financial Analysts Journal* (January–February 1994), 39–50.

11. Schwartz, R.A. *Reshaping the Equities Markets: A Guide for the 1990s.* Chicago: Business One Irwin, 1993.

12. Schwartz, R. A. and Weber, B. W. Next-generation securities market systems: an experimental investigation of quote-driven and order-driven trading. *Journal of Management Information Systems, 14*, 2 (Fall 1997), 57–79.

Trust, Technology and Transaction Costs: Can Theories Transcend Culture in a Globalized World?[1]

Kuldeep Kumar, Han G. van Dissel and Paola Bielli

Rotterdam School of Management, Department of Decision and Information Sciences,
  Erasmus University, Rotterdam, the Netherlands
Università Luigi Bocconi, Information Systems Department, Milan, Italy

Abstract

The failure of SPRINTEL, an inter-organizational information system in Prato (Italy) raises a number of interesting questions with regard to the technical-economic and socio-political perspectives that currently dominate the information-systems/information technology literature. These questions underscore the importance of developing additional theoretical perspectives to help us better understand the role of information systems in organizations. In this article we reflect upon these questions and their theoretical foundations in the context of a case study. The case study describes the implementation, usage and outcome of an inter-organizational information system in the industrial district of Prato. An analysis is made of the extent to which the technical–economic and socio–political perspectives are sufficient to explain the failure of this information system. The outcome of the analysis shows that these two perspectives are insufficient to provide an explanation. Based upon literatures from a variety of sources, we develop a third, complementary perspective. Like Kling (1980)'s socio-political perspective, this perspective is also an interactionist perspective. However, instead of focusing on politics and conflict as the primary interaction mode, it focuses on collaboration and cooperation as the key to understanding interaction processes. This perspective introduces a third rationality in which trust, social capital, and collaborative relationships become the key concepts for interpretation.

Introduction

  The entire history of Italy and Europe leads to Prato.
    (Malaparte, 1994, p. 61, our translation)

Theories commonly used to understand and explain management phenomenon are primarily based upon a technical–economic rationality.[2] Within the information-systems/information technology (IS/IT) domain, Kling (1980) recognizes two distinct rationalities.[3] He calls the first rationality the *System Rationalism* perspective on IS. The central concept of this rationality is that all actors/stakeholders in an organization subscribe to the same economic goal of maximizing the organization's economic efficiency and effectiveness through technology. It focuses on the narrowly bounded world of computer use in which the computer user is a central actor and emphasizes the beneficial or positive role that computerized technologies play in the organizational life (Kling, 1980, p. 63). He also identifies a second rationality or the *Segmented Institutionalism* perspective, which is emerging in IS literature. Unlike the first rationality, the second rationality is not technocentric, that is, it does not presume a technological imperative or economic rationality in human behavior, but instead opens the door to the investigation of human and social phenomena in the management of information systems. In this rationality, organizations are considered to be forums for political activity where actors are engaged in conflict, intrigue, and negotiation based on their private interests (Kling, 1980, pp. 63–65). Thus, individuals and sub-units may work to achieve their own objectives while sometimes adversely affecting the overall objectives in their organization. Consequently, power and politics become the key concepts while the interplay of conflicting objectives and operation of supposedly "non–rational" choice processes determine the consequences of technology (Markus and Robey, 1983).

The purpose of this article is to contribute a third rationality – that of *relationships* and *trust*. Like the second perspective, it also goes beyond technocentric and economic considerations, but unlike the second perspective, the relationship and trust perspective sees more than just conflict and politics in organizations. It recognizes that while issues of politics, exercise of power, and conflict inherent in organizations go beyond the rational decision making perspective, so does the existence of relationships and trust. Trust, thus, provides us with a third way of studying the role of information technology, both in and between organizations.

In this paper, we will bring out and develop the third rationality in the following way. First, we offer a brief discussion of the traditional economic theories, where we identify utility maximization, self-interest, and opportunism as the premises underlying these theories. This discussion will show that these assumptions also underlie Kling's first two rationalities – system rationalism and segmented institutionalism. Next, traditional technical-economic theories, as found in the organizational, inter-organizational systems, and IT implementation literatures, will be used to develop predictions about the implementation and role of information technology in inter-organizational situations. Finally, by subjecting the predictions of the traditional technical–economic theories to an empirical test in a real-world setting, the paper will expose the limitations of these theories and thus motivate the need for additional or different explanations. The real-world setting for the test is remarkable in the sense that in this particular setting trust and relationships, rather than self-interest, opportunism, and conflict are the predominant values underlying socio-economic behavior; hence, this setting is valuable for testing the adequacy of traditional technical-economic theories in providing valid, general explanations. The refutation of the predictions of the technical–economic rationality, and by extension those of Kling's first and second rationalities, will lead to a search for a trust-based rationality or the third rationality.

Research Methodology

Empirical tests of socio-economic theories are usually conducted in the same cultures in which these theories were conceived and developed. As theories, especially social and economic theories, are the products of their native culture, they are likely to be consistent with the dominant socio-economic assumptions of the society in which they were conceived and developed. Thus, field tests of a theory in native surroundings are likely to encounter benign conditions consistent with the underlying assumptions of the theory. The technical–economic theories underlying the first two rationalities are no exception. These theories were conceived and developed in western, mainly Western-European and North American, contexts and have reached their apogee in North American management thought. Thus, they are based upon assumptions and premises that are commonly accepted in these cultures. Moreover, as most empirical observations in testing these theories are usually made in a western context, the observations are less likely to refute the predictions of these theories.

However, when theories and their predictions are used in cultures other than the one in which they were originally developed, they are confronted with "alien" conditions that could severely test the limits of these theories and their applicability.[4] To the extent to which a theory survives such a test, it is considered "robust." From such confrontations and tests, there sometimes arises a need for a re-examination of the premises underlying the theory; this, in turn, could lead to a reformulation or extension of the theory, and occasionally even its rejection and replacement.

In this paper, we present a story of such a confrontation and test of traditional technical-economic theory. The story deals with the development, implementation, and subsequent failure of an interorganizational system (IOS) in the *Prato* industrial district in Tuscany, Italy. With its rich history, unique organizational structure, and its continuing viability and success, this district has been the focus of much attention from management theorists (e.g. Casson and Pannicia, 1995; Inzerilli, 1990; Piore and Sabel, 1984; Ritaine, 1990). This attention has resulted in detailed accounts and extensive socio-economic analyses of the current inter-organizational model in Prato. The growing interest in inter-organizational forms of collaboration has generated a variety of theoretical arguments and empirical evidence highlighting the importance of IT and positing a central role of IOS in managing such organizations (Clemons and Row, 1991; Malone, et al., 1987; Reekers and Smithson, 1995; Kumar and van Dissel, 1996). Further, previously published accounts of the story of Prato interpreted through technical-economic lenses already exist in the management literature (Johnston and Lawrence, 1988; Malone and Rockart, 1991). Thus, not only does this story provide an "alien" context for testing the technical–economic theories traditionally used in the IOS literature, the story makes it also easier to test these theories by confronting the previous accounts with subsequent events.

To frame this empirical test, we start with the hypothetico-deductive logic of inquiry proposed by Lee (1989, 1991). The first step in the this logic employs the technical–economic perspectives of Williamson (1975, 1985)'s theory of Transaction Cost Economics (TCE), Porter (1985)'s theory of competitive advantage, Miles and Snow (1986)'s concept of Dynamic Networks, and theories of IS implementation as starting points for identifying a set of propositions[5] about the introduction, implementation, and success or failure of IT-based inter-organizational systems in *Prato*. The development of these propositions is further supported by previously published accounts (by Johnston and Lawrence, 1988; Malone and Rockart, 1991) of the role of IT in *Prato*. These accounts take these traditional theories to their logical conclusion, by offering not only predictions but also observations of the existence of strong and worthwhile contributions by IT-based systems in *Prato*.

Next, we confront these propositions and predictions with, and thus test them against, our observations of subsequent events from the field. Orlikowski (1993, p. 311) suggests that "in order to produce accurate and useful results, the complexities of the organizational context have to be incorporated into an understanding of the phenomenon, rather than be simplified or ignored." We used two strategies for data collection to develop these contextual observations. First, interviews were conducted with key actors at the *SPRINT*[6] offices and in the district of *Prato*. Second, we used secondary sources such as historical accounts, newspaper articles, and internal documents of the *SPRINT* IOS project.

The contradictions identified between the predicted events and subsequently observed events set the stage for a re-examination of the premises underlying the initial theories. This re-examination is a precursor to theory reinterpretation and re-formulation. Lee (1991) suggests conducting an interpretive investigation to identify the assumptions behind the initial positivist theory in order to explain what the theory apparently did not address, or addressed incorrectly. Thus, the subsequent analysis attempts to interpret the subjective meanings held by human subjects that may have contributed to their observed behavior in the *Prato* industrial district. First, we collected data so as to develop a rich description of the *Prato* phenomenon. Second, we used "new" (to the IS literature) interpretive theories from existing literature, including the theories of relationships and bonds in industrial networks, of the concept of trust, and of inter-cultural differences, together with the collected data to interpret and develop our own understanding of the observed behaviors. Thus a strategy of inductive theory discovery coupled with the interpretation of our observations through the lenses of "new" pre-existing theories allowed for the develop(ment) of a theoretical account of the phenomenon while simultaneously grounding the account in empirical observations or data (Martin and Turner, 1986). Finally, based upon this interpretation, we suggest a reformulation and extension to the theoretical technical–economic perspective by proposing a complementary trust and relationship based rationality for understanding inter-organizational behavior in the instance of *Prato*.

The above research methodology also provides the framework for structuring the underlying arguments of the paper, and thus can be used as a map or guide for its perusal. Part I of the paper presents the technical–economic theories and the propositions and predictions that we deduce from these theories. On the basis of the collected data and literature we develop a rich historical, social, and organizational context in which the story of *Prato* takes place. Part I ends with a test of these predictions by confronting them with observations of events from the field. The result of this confrontation is that we find these theories lacking in their explanations. The inadequacy of the traditional theories is used to motivate the re-examination of the assumptions of these theories. In Part II of the paper we present new additional theories from the literature which we use to re-interpret the Prato phenomenon. This reinterpretation leads us to formulate extensions to the theory in the last section of the paper.

The story of Prato provides the backdrop against which this methodological logic plays out. As most traditional stories end, this will conclude with a "moral of the story."

Part I: A Story of Technical-Economic Rationality

The Technical–Economic Rationality

> .... many neoclassical economists have come to believe that the economic method they have discovered provides them with the tools for constructing something approaching a universal science of man. ... These economists believe that they are right in a deeper epistemological sense as well: through their economic methodology, they have unlocked a fundamental truth about human nature that will allow them to explain virtually all aspects of human behavior.
>
> (Fukuyama, 1995, p. 17)

A story normally makes sense only in the context of the cultural setting that produces it. When a story is recounted within a familiar context, there is usually no need to make the underlying assumptions behind its rationality explicit. However, when the same story is narrated in an "alien" context, in order for the story to have meaning for its new audience, it becomes necessary to make the underlying assumptions of the story explicit. For example, a story about the inevitability of a Japanese soldier's committing *seppuku*[7] in face of defeat usually does not make much sense to western audiences. Only when the Japanese notions of honor, duty, and shame becomes clear, the new audience can understand the logic of the soldier's action and the poignancy of the situation.

The same observation applies to understanding interorganizational information systems (IOS). Traditionally, the IOS literature relies on the concepts of economic rationality to explain the emergence, structure, and behavior of IOS phenomena. (e.g. Clemons and Row, 1987, 1991; Malone, et al., 1987; Reekers and Smithson, 1995). The economic rationality of most theories describing IOS would be intuitively obvious to the readers of the business and economic literature. Thus the following summary of economic rationality is a minimal description intended only as a background for the story, and not a comprehensive analysis of the economic argument.

> "The first principle of Economics is that every agent is actuated only by self interest."
>
> (Fukuyama, 1995, p. 18).

One premise underlying technical-economic rationality is that organizations and individuals are economically rational actors whose primary purpose is to maximize their respective economic utilities.[8] In their quest for utility maximization, these actors take a zero-sum view of the economic pie and an adversarial view of other actors (e.g. customers, suppliers, competitors) in their environment. Under conditions of direct competition, this adversarial view naturally leads to aggressive competitive strategies. Even in interactions which are supposedly based upon the expectations of mutual benefits, such as collaborative relationships and/or supplier–customer interactions, this focus on self-interest is presumed to lead to opportunistic acts. The assumption of opportunism suggests that under conditions of environmental complexity and uncertainty, and assumptions of bounded rationality and information asymmetry, economic actors may act to further their personal interests even at the cost of harming those with whom they are supposed to have collaborative and mutually beneficial dealings.

The result is intra- and interorganizational arrangements and governance structures designed to guard against the presumed opportunistic behavior, while at the same time maximizing the economic value for the actors. Transaction cost theory (Williamson, 1981; 1985), which is frequently used in the IOS literature to explain the emergence and structure of inter–organizational systems, is concerned with the minimization of a cost function consisting of transaction costs[9] and production costs, and the containment of risks which arise due to the opportunistic behavior[10] of the parties to the interaction. Interorganizational information systems are said to reduce coordination costs by decreasing the search costs for potential suppliers and partners, and by providing information based–mechanisms to reduce uncertainty and information asymmetry, thereby reducing the costs of drafting and monitoring the implementation of transaction contracts. As a result, IOS, by reducing the incentives for vertical integration, are considered to induce to a move towards market arrangements (Malone, et al., 1986; Clemons and Row, 1987).

Similarly, the concept of competitive advantage (Porter and Millar, 1985), frequently used to describe the emergence and impacts of IOS, emphasizes the importance of gaining power over the firm's suppliers and customers while attempting to contain the power of competitors (e.g., Johnston and Vitale, 1988). Also organizational theories such as resource dependency theory (Pfeffer and Salancik, 1978) suggest that in the drive to optimize its self–interest the objective of a firm is to minimize its dependence on other firms and to maximize the dependence of other firms on itself (Reekers and Smithson, 1995). In both these cases, interorganizational systems are considered to be the instruments which, by locking–in customers and dominating the suppliers, increase the firm's power over them.

At this point it should be noted that the above mentioned concepts of utility maximization, self-interest, opportunistic behavior, and the exercise of power underlie both the first and the second rationalities of information systems (Kling, 1980). In the case of *system rationalism*, or Kling's first rationality, these concepts operate at the inter-organization or system level; i.e. the organization or the system as a whole maximizes its utility, acts opportunistically vis-à-vis other organizations, and exercises power in relation to its competitors, suppliers, or customers. In the case of the second rationality, or *segmented institutionalism*, these concepts apply at the individual or sub-unit level and operate at the intra-organization level, resulting in exercise of power, politics and conflict between individuals and organizational sub-units within the organization. In either case the key principles are those of self-interest, opportunism, and utility maximization, albeit for actors at different levels.

A second premise of technical-economic theories is the existence of an ideal market (suppliers or customers) consisting of homogeneous, faceless, actors acting independently with no memory of the past transactions, and with little or no historical context. While the suppliers are considered as interchangeable in terms of their products and services, their relationships with their equally homogeneous and therefore virtually indistinguishable customers are considered to be governed by impersonal price mechanisms (Cunningham and Tynan, 1993). In the traditional IOS literature, these assumptions in turn lead to the argument that information technology, by increasing the amount of market information available, increases the choice of supply sources and thus hastens the move towards markets governed primarily by price mechanisms (Malone, et al., 1987).

These two premises, taken together, provide the background for the seminal article by Miles and Snow (1986) in which they assert that information technology and computer-based information systems will play a central role in enabling a new emerging organizational form – the "*dynamic network*." They identify (pp. 64–65) four key characteristics of a dynamic network. First is the vertical disaggregation of business functions. Miles and Snow suggest that different business functions typically conducted in a single organisation will be performed by independent firms in a network. Second, since each function is not part of a single organisation, brokers in the network will perform the role of assembling the business group. Third, market mechanisms will play an important regulatory role. Contracts and payment for results, rather than direct supervision and control through planning and progress reports are likely to be used. Finally, Miles and Snow postulate the role of full-disclosure information systems. They suggest that broad-access computerized information systems, which are continuously updated and thus allow participants to verify mutually and instantaneously each other's contributions, are likely to become substitutes for lengthy trust-building processes.

Following Miles and Snow's lead on the role of IT/IS in predicting a move away from vertical hierarchies, the IOS literature has traditionally focused on an IT induced move towards markets and intermediate forms of organisation (e.g., Malone, et al., 1987; Clemons and Row, 1987, 1991). This in turn, underlies Johnston and Lawrence (1988)'s conceptualization of a value-added partnership (VAP). VAPs are used to describe a set of independent companies that work closely together to manage the flow of goods and services along the entire value-added chain. Unlike Miles and Snow, Johnston and Lawrence (1988, p. 94) do recognize that "VAPs are not, however, necessarily technology driven. They may emerge as the result of computerized links between companies or they may exist before the technical links have been made. Computers simply make it easier to communicate, share information, and respond quickly to shifts in demand." Thus, from a theoretical perspective, IT and computer-based systems are either considered central to the concept of dynamic networks (Miles and Snow, 1986), or are considered instrumental in increasing their effectiveness, coordination, and efficiency (Johnston and Lawrence, 1988).

Johnston and Lawrence (1988) use the example of Prato to illustrate the concept of VAPs and the role of IT in such partnerships. In the context of Prato they state: ... the players in the Italian textile VAPs are eager to share information and cooperation. In recent years they have developed computer systems that rush information from partner to partner. The technology enhances coordination and boosts the speed and quality of responses to the market. (Johnston and Lawrence, 1988, p. 97)

Malone and Rockart (1991, p. 95) further added that the operation of the network of small textile firms in Prato "was coordinated in part by electronic connections between them." As the following story will show these assertions by Johnston and Lawrence (1988) and Malone and Rockart (1991) about the role of information technology in Prato turned out to be somewhat premature.

Prato – The Historical Context

It was an old saying in Prato that if a man cared to look beneath the foundation of the city walls, he would find there a tuft of wool. Certainly from the twelfth century until the present day the city's fortunes have waxed and waned with the cloth–trade ...

(Origo 1957, p. 55)

The emergence and growth of the Pratesian textile industry, and the antecedents of the initial acceptance and subsequent decline of IT-based inter-organizational systems in Prato, can best be understood in the context of its historical roots. As the above quote illustrates, the roots of the textile industry in Prato can be traced to the end of the middle ages. Prato is situated in a socially and economically homogeneous geographical area in Tuscany (North-Central Italy) located thirty kilometers North-West of Florence (see Figure 1). Historical accounts from the thirteenth and fourteenth century[11] indicate the existence of a nascent wool industry in this region. This industry had its origins in the customs of the *mezzadri* (share-cropper farmers) who complemented their farming incomes by sheep shearing and wool processing (Ritaine, 1990).

Figure 1 Location of Prato in Italy.

Over time, these modest beginnings developed into an extensive network of a large number of small firms. Each firm in the network specialized in one or a few steps in the process of converting wool to cloth, while the network as a whole was organized and commercialized by the merchants of Prato. The introduction of the first spinning machines by Giovanni Battista Mazzoni in the nineteenth century helped launch the industrial revolution in Prato. The consequent increase in demand for inputs very quickly outstripped the local supply of wool. The merchants of Prato found an innovative solution to this problem. The textile business in Prato was soon relying on re–cycling cloth remnants and old clothes from as far away as the United States for raw materials. In the latter half of the nineteenth century, the textile production in Prato was based so much on recycled cloth materials that the city was commonly known as the *citta' degli stracci* (city of rags).

The period between the two world wars brought a short period of consolidation. By the 1930s, driven by overseas demand for inexpensive cloth from the Asian markets, and the need for large amounts of standardized military clothing for the Italian armed forces, about 80 percent of the production (and textile employment) was concentrated in around 30 large integrated textile mills (*lanifici*). Under Mussolini's regime, this trend towards centralization was further reinforced by the fascist ideological drive towards state-led large scale industrialization to solve the economic crisis resulting from the crash of 1929, and by nationalist protectionism under the guise of patriotism, the so called 'autarchic' policy. The extent of this drive towards centralization is illustrated by the fact that, around 1934, the public sector of the economy was larger in Italy than in any other Western capitalistic country (Procacci, 1973, p. 430).

However, in the Pratesian time–scale of centuries, this period of concentration was a relatively short-lived aberration. With the restoration of democracy in 1948, individual initiative, or better, family based initiative prevailed again. After the end of the Second World War, the declining overseas demand, the loss of the military as the major customer, a trend towards greater product variety, and the increasing labor costs and labor militancy led to vertical disaggregation of the large integrated firms. This trend was further reinforced by an increase in production in developing countries with low labor costs. The loss of state (government) support was also a major factor in this break-up of large textile firms. The Pratesean companies were forced to reduce their costs through the fragmentation of the production units (Gandolfi, 1988). In most cases, either the owners of the large firms sold off major parts of the production processes to their former employees thereby spawning a number of small firms, or the former employees, using their skills and training acquired in these firms, set themselves up in the business (Johnston and Lawrence, 1988). The indicators of production fragmentation reported by Ritaine (1990) clearly show the depth of this phenomenon. The factory loom to domestic loom ratio went from 4:1 in 1949 to 1:6 in 1954.

Over the years, despite economic recessions, both at home and in the global markets, the Prato system of production through a network of small specialized firms has shown remarkable resilience and has continued to flourish. Prato survived the world recession in the '70s because of strong differentiation through the introduction of innovative textiles at reasonable prices and high levels of service (Gandolfi, 1988). The latest available statistics from a *Financial Times* survey (Graham, 1994) indicate the existence of approximately 8500 small firms, of which fully 47 percent employ fewer than 10 people, while a further 40 percent employ between 10 and 50 people. The figures from this survey also indicate that Prato is the single biggest agglomeration of textile manufacturing facilities in Europe with a turnover close to ItL 6,500 billion (approx. US$ 4.5 billion).

Today Prato caters essentially to the middle and higher end of the fashion industry (Buxton, 1988). No longer dealing with recycling rags,[12] the textile business in Prato, in addition to the traditional wool, now deals with a variety of both man–made and noble (natural) fibers. The production processes thus need to be sophisticated enough to work with a wide diversity of fashion needs which are supplied through infinite variety in combining different yarns and of producing small batches of custom fabrics. In any one year, the Prato companies may turn out 70,000 types of different materials. These are usually produced in small runs for specialized niche markets (Graham, 1994).

Finally, in response to the rapidly changing whims of the fashion market, the small production runs, and the consequent demands for flexibility in the design and production processes, the small entrepreneurial firms continue to adopt sophisticated and innovative production technologies. Firms focusing on a specific phase in the production cycle have a deep knowledge of this production process, and therefore are able to identify potential improvements, and are willing to assess and adopt useful innovations. The use of CAD (computer-aided design), CA/IM (computer-aided and integrated manufacturing), and numerically controlled machines and looms is wide spread in this area (Gandolfi, 1988).

The high number of specialist firms in Prato creates a fertile spawning ground for new ideas in design and technique. Further, Prato is said by those who work there to be an "open factory" in which firms can go freely into each other's units (Buxton, 1988) thus allowing easy sharing and transfer of innovations. While the "bigger" small firms normally take the lead in adopting newer sophisticated technologies, the diffusion of knowledge about these innovations, and their rapid adoption and assimilation to overcome the innovation gap are further facilitated by the unique organizational model prevalent in Prato.

Prato: The Organizational Model

> Despite the grand technical and management innovations which have come about through the centuries, the organizational model invented by Medieval merchants is substantially still operative, and evidently, thoroughly suited to the mentality of the Prato operators.
>
> (Balestri, 1994)

The essential nature of the Pratesian industrial network can be understood if one imagines a production and logistics–based value chain[13] (*filiera tessile*) where individual steps in the production cycle are made up of independent firms. From an organizational point of view, this means the de–verticalization and decentralization of all the major functions of the organization, from purchasing, to individual production steps and substeps, to marketing, each of which is performed by independent firms (Inzerilli, 1990). Moreover, the competing units of production are further specialized in only one or two of the different phases of production, e.g. sorting, carding, spinning, dyeing, weaving, printing, finishing, etc. Figure 2 depicts the *filiera tessile*.

However, unlike the value chain conceptualized by Porter (1980), this value chain is not a static phenomenon. The chain comes together to produce and deliver a particular, usually specialized order, and dissolves once the delivery is complete. This is not to say, that another value chain with the same actors may not be reconstituted in the same sequence to deliver another order[14]; it means, instead, that the particular value chain has meaning and existence only for the duration of order fulfillment. Furthermore, at any given time, an individual firm may work for, and therefore be a member of, different dynamically constructed value chains.

The primary catalyst of this network is the unique and traditional role of the typical Pratesian entrepreneur, the *impannatore*.[15] The *impannatore*, is described in the literature as an "independent master broker" (Johnston and Lawrence, 1988), a "head firm" (Inzerilli, 1990), a "commercial intermediary" (Ritaine, 1990), or a "pure entrepreneur" (Casson and Pannicia, 1995). Ritaine describes the *impannatore* as an entrepreneur without a firm. He is the quintessential medieval merchant[16] with an overview of the external markets and the internal district. He obtains orders and distributes work to firms in the production process. He never ties up his own capital, and thus can adapt his activity to market fluctuations without risks. On the other hand, Inzerilli (1990) also points out that the firm performing this role may, at times, perform one or more steps of the production cycle itself. Sometimes, the *impannatore* may also have the role of planning and coordinating between different steps of the production and logistics cycle (Casson and Pannicia, 1995). However, the *impannatore* does not have formal contractual power and thus is not be able to exercise any significant formal control over the firms that constitute the dynamic value chain. Among themselves the *impannatori* compete on the basis of price, innovation and service (e.g. order delivery deadlines).

The coordination in the chain is primarily achieved by horizontal communication between the adjacent parts of the chain and through a lesser extent by the flow of information to and from the impannatore who "owns" the order. Thus the *filiera tessile* can almost be considered to be a self–organizing dynamic value-chain in which production materials and information flow directly from one firm (and one step in the production process) to the next with only minimal interference or

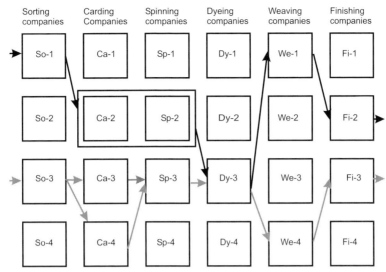

Figure 2 Example of two *filiere tessili* in Prato.

Note
The first chain passes through company So-1, Ca-2 and Sp-2 (a company that masters both processes), Dy-3, We-1 and Fi-2 before the order is delivered to the customer. The second chain passes through So-3, is then spit between Ca-3 and Ca-4, and continues its route via Dy-3, We-4 and Fi-3. Each of the chains representing an order is "owned" by an *impannatore* (defined below).

control by the *impannatore*. It is very common for the *impannatore* to communicate only with the first and last actor in the chain, and to communicate with others only to track order progress and in case of problems or exceptions.

Inzerilli (1989, 1990) outlines a number of factors that contribute to the continuing viability and thriving nature of the Prato organizational model. First, from a straight economic/functional perspective, this organization provides economies of scale, the flexibility to meet highly specialized and variable market demand, and an ability to maximize capacity utilization of the system as a whole (for an extensive discussion explaining these advantages, see Inzerilli, 1989, 1990). There are also a number of factors external to the above organizational model that further enhance its viability. These include legislation providing preferential loans and tax incentives to small firms, labor legislation exempting small firms from union representation and requirements for specific working conditions, and public services such as public transportation, industrial worker-training, industrial land and price control, and availability of water (important to cloth making) provided by the local governments. The close proximity of residential and industrial areas may also have facilitated the mobilization of family resources thereby contributing further to the resource base.

Ritaine (1990) and Inzerilli (1990) mention that the underground economy (*the sommerso*) may have also played a role in the growth of the model. Ritaine suggests that this underground economy may include non–declared domestic work hours at home, non–accounted overtime, non–declaration of income and activity, personal use of business assets, and other forms of tax evasion[17]. Being predominantly family businesses, they tend to report as little profits as possible, while both directly related and unrelated costs are charged to the business. In the rich years the firms lavishly support all family needs, while in years of recession the family supports the firm with non–declared hours and wages and personal assets. However, Ritaine points out that while the underground economy, by allowing the accumulation of capital resources, may have facilitated competitiveness, it is not sufficient to explain the competitiveness completely.

Finally, to explain the continuing viability of Prato, Balestri (1994), Casson and Pannicini (1995), and Inzerilli (1990) also identify the existence of a variety of local institutional factors, such as professional and employers' associations, powerful *Cassa di risparmio* (savings and loans) and local commercial banks, and the local communist party which provide sources of support and a rich resource–base for financing, worker training, and information and knowledge accumulation and dissemination. The meeting halls of the local associations and the communist party also provide both informal and formal meeting places and forums for communication and deliberations among the firms.[18]

Prato: A Technological Experiment

> In the name of God, the 12th. of February 1395. Pay at useance, by this first of exchange, to Giovanni Asopardo £ 306 13s. 4d. Barcelonesi, which are for 400 florins received here from Bartolomeo Garzoni, at 15s. 4d. per florin. Pay and charge to our account there and reply. God keep you. Francesco and Andrea, greetings from Genoa. Accepted March 13.
>
> Example of one of the more than five thousand bills of exchange found. (Origo, 1957, p. 147)

As the above letter from Francesco Datini illustrates, ultimately the business of Prato did rely on the transmission of instructions and information between different parties (in fact, given the great number of letters, bills of exchange, and other documents found in *Palazzo Datini*, one may wonder if Datini spent a great part of his day processing information). Thus the arguments in the IOS literature about the role of information and information technology in shaping the inter–organizational phenomenon are also likely to be applicable to Prato.

> Participants in the network agree upon a general structure of payment for value added and then hook themselves together in a continuously updated information system …
> (Miles and Snow, 1986, p. 65)

By the middle of 1980s, conditions seemed ripe for the introduction of IT/IS in Prato. A great number of vertically disaggregated small firms were already interconnected in a cohesive industrial network. The general structure of payments and mutual obligations had been established through tradition and through a system of agreements brought about by a combination of professional and employer associations with roots in the Medieval guilds, and through the brokering role of the *impannatore*. The existence of these associations, along with the support provided by the local governments and local financial institutions assured the availability of a support infrastructure and of resources. Further, as demonstrated by their use of sophisticated technologies such as CAD/CAM systems, the network itself consisted of firms which had shown themselves receptive to accepting and using innovative production technology directly supporting their core business. Finally, the Italian government through ENEA,[19] and the European Union through its funding were anxious to promote innovative uses of information technology. The only thing that remained was for these firms to "hook themselves together in a continuously updated information system" (Miles and Snow, 1986, p. 65).

The catalyst for this hooking of themselves together was the *SPRINT*[20] program, funded and nurtured by ENEA, the European Union, the Tuscany Regional government, local governments and financial institutions, the Florence Chamber of Commerce, unions, and the various professional and industrial associations representing the network of firms. The mission of *SPRINT* was the creation of a telecommunications infrastructure for the Prato industrial district, activities to promote awareness and to increase the diffusion of information and communication technologies in the district, and the design and implementation of information and communication technology based services useful for the district.

The *SPRINT* consortium went about its mission in a thorough and professional manner. In October 1982, they commissioned a Milan-based IT and telecommunications consulting company, *Reseau*, to perform a requirements study and to develop the concept further. During 1983–86, following the classic systems development life cycle paradigm, *Reseau* first conducted a study of the existing information flows connecting the economic actors in the districts (i.e., the manufacturing firms, brokers, transport firms, financial institutions, local government organizations, industrial and professional associations, etc.). The purpose of this study was to define the information and communication needs of the firms in the industrial network.

The output of this study was a map containing the structure, nature and quantities of information flows in the district. Particular attention was paid to analyzing the information that was communicated between the various actors and the means of communication. The study clearly showed that most of the information communicated between the parties was originated by and used for day-to-day activities of the firms such as placing orders, technical and design data for production, arranging for transportation, and the transfer of funds for payments. The use of external information such as data on potential markets, new technologies, and on foreign competitors was perceived as less urgent. However, the small production firms did show some interest in having access to such external data which were, until then, in the exclusive province of the larger firms.

The study also showed that, while the firms used information technology for CAD/CAM and administrative and record keeping purposes, information for current operations (orders, transportation, status inquiries, payments, etc.) was exchanged primarily through a variety of personal interactions, telephone calls, fax, and sometimes mail. External information was acquired mainly through traditional channels such as agents, fairs and expositions, and trade magazines.

On the basis of this study, and in keeping with its mission of promoting the use of information technology, the *SPRINT* consortium concluded that information and communication technology support would be appropriate for both the communication flows between the actors in the textile district and for the acquisition and dissemination of external information. With regard to internal information flows, the creation of a telecommunications network called *SPRINTEL* connecting individual manufacturing firms to each other, and to other actors such as banks, transport companies, and financial institutions was proposed and approved in September 1984. Subsequently, a technical committee was appointed in 1985. This committee included representatives of all parties involved and was headed by *Reseau*. Its objective was to develop a more detailed design and to develop, implement, and manage a pilot experiment. This design included a common, centralized purchasing and order-entry function through which the participating firms could place orders, arrange for transportation, check order status, and make arrangements for electronic payments through the local financial institutions. It was felt that the on-line monitoring of orders through the *filiera tessile* and the monitoring and control of work-in-process, raw materials, and finished goods inventories through electronic communication would increase the overall efficiency of the district.

Three main characteristics of information and communication technology were considered as key to achieving the system benefits: compression of time and space among the actors in the district, flexibility of being able to exchange various types of information among the actors, and maintaining business relationships by supporting information transfers and organizational networking between the actors independent of direct contacts between the participants (Bielli and Giacomelli, 1995). The last was especially noteworthy in the light of Miles and Snow's assertion that in dynamic networks, "Broad-access computerized information systems are used as substitutes for lengthy trust–building processes based on experience" (Miles and Snow, 1986, p. 65) To promote the use of IT in the district, an information brokering center was planned for scanning information from both traditional and electronic (mainly on-line databases) external sources.

In keeping with a risk-averse implementation strategy, the proposed system was developed with a relatively simple technical base. Each firm participating in the network received a *videotext* terminal connected through a public telephone network to two servers located in the *SPRINT* offices. These servers, in turn, performed the function of gateways to the *Videotel* host in Milan. The online databases providing external information were also to be accessed through the servers in *SPRINT* offices. The *videotext* technology was chosen for three reasons: first, it was considered to be an already tested and tried technology exemplified by the success of the French Minitel services (see e.g., Cats–Baril and Jelassi, 1994) and thus carried little technological risk; second, at that time the Italian telephone company (SIP) was investing heavily in this technology and was promoting the use of its *Videotel Nazionale* services; and finally, the technology itself was considered to be user friendly and low cost in terms of both hardware and software.

Next, given the high profile of the *SPRINT* experiment, it was important to ensure high levels of participation in the system. Thus the *SPRINT* consortium made a special effort to market the system. Direct marketers were hired to knock at doors and convince the Prato entrepreneurs to participate. Training programs were set up to train designated people from the firms in using the services of *SPRINT*. A three-year pilot testing program, consisting of 300 participating actors with some 450 terminals, beginning in June 1987 and ending in 1990, was used to fine tune the system. Finally, the first 300 *videotext* terminals and the information services on the network were initially offered free of charge, both to the manufacturing firms, and to other actors in the network such as banks, the chamber of commerce, the local municipality, trade unions, and the industrial associations. From June 1988 onwards additional terminals could be hired for a nominal fee of only ItL 7,900 (approx. US$ 3.50) per month.

By the end of the pilot the system seemed to be highly successful and 500 firms were on the waiting list for getting onto the system.[21] This was considered as a good premise for the full blown rollout of the project. *SPRINT* decided to increase the access to additional external data sources. Access to two major Italian data sources (CERVED, providing information on corporate balance sheets, and ISTAT, providing statistical data) and two major foreign hosts (DIALOG and ECHO) was provided. Further, to stimulate use, a unit within *SPRINT* acted as an "information broker" by helping formulate the information queries, searching the databases for information requested by the firms, and communicating the results of the database-search to the firms through paper-based reports.

At this point, *SPRINT* also proposed to start selling some of its services to become, at least partially, self-sustaining. It intended to do so by charging part of its costs to the information providers (industrial and crafts associations, municipalities, banks, and chambers of commerce). The proposal to bill the information providers was also influenced by the fact that by this time most of the external moneys (i.e. European Union funds) had been depleted, and no new external funding seemed to be on the horizon. Consequently, *SPRINT* intended to bill the information providers for a small part of the cost of the system and allowed them in turn to bill the firms which used their information services. This resulted in information providers offering their paper-based services electronically for the same price, while new additional services were offered for free. The projected cost of the external information services to the end-user firms was in the neighborhood of US$ 6–7 per month per connection. *SPRINT,* however continued to subsidize the difference between the real cost of the services and the minimal fees charged of the information providers. The internal messaging and communication services to the individual manufacturing firms were to continue to remain free.

The pilot tests had also identified some problems. The user firms considered the system sometimes too slow, and sometimes technical failures reduced the system's availability. Consequently, powerful new computers were installed at the local *SPRINT* offices. The connection to the *Videotel* center at Milan was eliminated and these new computers took over as local *Videotel* hosts of the network. It was expected that faster computers, coupled with local hosts, would reduce the technical and speed problems.

After the pilot, the firms started complaining that the services were not worth the time and money. *SPRINT's* response was to increase the sources of external information and increase the variety of information services. Despite these efforts, the total number firms connected to *SPRINTEL* started to decline in the 1991–93 period. From approximately 440 connections in September 1992, the number of connections steadily declined to approximately 70 connections in December 1993.

SPRINT reacted to this decline by attempting to minimize costs by reducing some services, while maintaining the information brokering service, and further increasing the information sources in an attempt to make the information services more valuable. Thus, it attempted to stem the tide of desertions by providing additional information and more powerful technology. Despite these attempts, the steady decline continued. By the beginning of 1994 it was clear that the *SPRINTEL* experiment had failed. In October 1994, as a result of continuing decrease in the use of the remaining services, and a new round of reductions in external financing, the *SPRINTEL* experiment was formally terminated.

Examining Implementation Failure: A Technical–Economic Perspective

As the above account shows, the introduction and implementation of information technology–based systems in Prato was a textbook example of building and implementing IS in the established technical–economic tradition. In this tradition, the literature on IS implementation identifies a number of technical economic factors which are said to have a bearing on the success or failure of information systems (e.g., Lucas, 1975; Swanson, 1988). In the context of Prato, most of these factors were overwhelmingly positive.

First, there is the technical–economic factor that technology should serve a valid purpose. Theoretically at least, the situation in Prato seemed to be ideal for the introduction and use of an IOS. The literature on IOS has long argued for the enabling and beneficial role of IT in interorganizational alliances (e.g., Malone, et al., 1986; Clemons and Row, 1991; Miles and Snow, 1986; Snow, et al., 1992). Moreover, Johnston and Lawrence's (1988, p. 97) observation that, as early as 1988, IT was already "enhance(ing) co-ordination and boost(ing) the speed and quality of responses to the market" in Prato, added the *imprimatur* of practical wisdom on top of these theoretical arguments. While it is true that the requirements analysis process done in Prato discovered that the need for external information such as data on potential markets and on foreign competitors was perceived as less urgent, small production firms did show some interest in having access to such external data. From an information processing perspective it can be argued that better and additional information on the markets and competition, by reducing uncertainty in demand, should have provided added value. Thus, both theoretically and practically, information technology was supposed to serve a valid and important purpose in the industrial network of Prato.

Second, successful implementation requires the support of key stakeholders; in the instance of Prato, such support was readily forthcoming from industrial and craft associations,[22] chambers of commerce, local and regional governments, and financial institutions. Moreover, while there were some discussions among members of the associations about whether to continue with the project, no reported evidence of jockeying for power and consequent conflict among these stakeholders was suggested in any of the accounts. It can be argued that the major sponsors and beneficiaries of *SPRINTEL* were the larger firms and financial institutions, while the smaller firms did not have as much to gain from the system. However even if this imbalance existed, the reported evidence does not seem to suggest any major power plays or conflict which could have scuttled the system. Thus, Kling's (1980) *segmented institutionalism* perspective and Markus and Robey's (1983) *sociopolitical conflict* perspective are also not likely to be major factors explaining the failure of the *SPRINTEL* project.

Third, there is the factor that successful development and implementation requires adequate resources; generous funding from ENEA, the European Union, and various participants in *SPRINT* ensured this adequacy.

Fourth, users who are receptive to technology and innovation increase the probability of success; through their adoption and use of innovative and sophisticated production technologies, the firms in the Prato case had demonstrated their willingness to acquire and use new technologies.

Fifth, a careful requirements analysis and specification process are considered to be a prerequisite for successful systems; the detailed requirements study conducted by Reseau ensured this.

Sixth, in order to minimize risks of technological failure as far as possible, the systems should be implemented using pre-tested technology; the use of proven *videotext* technology illustrates this.

Seventh, the implementation process should include sufficient user training; *SPRINT* provided such training.

Finally, a pilot test phase provides a risk–averse strategy for shaking down any remaining problems, fine-tuning the system to the user's needs, and in general increasing the acceptability of the system; *SPRINT* conducted a pilot test involving some 300 firms and as a result fine-tuned the system by providing additional computing power and increasing the amount of information and services provided by the system.

Unexplained Questions and Issues

Despite such strong theoretical advantages and a positive implementation climate, the use of IT-based information systems could not be sustained over time in Prato. The firms were willing to participate in the system only as long as it was free of charge. The decline in system usage started when the use-charges, however minimal, were proposed for some of the services. About this time, the firms and through them the crafts and trade associations started complaining that the services were not worthwhile. From a technical–

economic perspective we may argue that the use of the system declined because the participating firms did not find the system worthwhile and were therefore unwilling to pay for it. This raises the question as to why the firms did not find the system worthwhile. Both economic theory as interpreted by IOS researchers, and observations by Johnston and Lawrence (1988) provide strong arguments for the benefits of information technology in a network structure such as exists in Prato. The low worth attached by the Prato firms to the IT-system runs counter to these arguments. It is, therefore, important to investigate the possible reasons for the seemingly insignificant value attached to the information system. Related questions are, why, given the perceived low marginal value by the district firms, was the system initiated in the first place, and what were the reasons behind the acceptance of the system during the earlier phases?

Beside the low perceived marginal value, a second phenomenon, also counter to the traditional economic arguments, was mentioned earlier. The break-up of *lanifici* into a network of specialized production units after World War II and the key role of *impannatori* are consistent with Miles and Snow's (1986) concepts of vertical disaggregation and the role of brokers in dynamic networks. In addition, Miles and Snow had suggested that in a dynamic network, the major functions would be held together by market mechanisms rather than plans and controls. However, Inzerilli (1990) points out that the reliance on markets and contracts as regulation mechanisms could be inappropriate for the organizational model in Prato. He suggests that pure market regulation would involve high transaction costs in such a volatile system: "Given the variability of markets in which this system normally operates, the continuously changing web of interdependencies among firms, and the functional specialization of each firm, one should expect very high transaction costs due, among other things, to uncertainty, specificity of investments, idiosyncratic tasks, and asymmetrics of information." (Inzerilli, 1990, p.12).

This obviously was not the case in Prato. Thus, Miles and Snow's assertion about the increasing reliance on markets, contracts and price mechanisms in dynamic networks needs further examination.

Finally, Miles and Snow also suggested that broad-access computerized information systems would become substitutes for lengthy trust-building processes (Miles and Snow, 1986, p. 65). In light of the subsequent rejection of IOS in Prato, this statement too would require further scrutiny. We are, thus, left with anomalies that technical–economic arguments alone can not adequately explain.

Part II: An Interpretation

Over the past generations economic thought has been dominated by neoclassical or free market economists … . We can think of neoclassical economics as being, say, eighty percent correct; it has uncovered important truths about the nature of money and markets because its fundamental model of rational, self-interested human behaviour is correct about eighty percent of the time. But there is missing twenty percent of human behaviour about which neoclassical economics can only give a poor account. As Adam Smith well understood, economic life is deeply embedded in social life, and it can not be understood apart from the customs, morals, and habits of the society in which it occurs. In short, it can not be divorced from culture.

(Fukuyama, 1995, p. 13)

In contract to the positivist technical–economic perspective, an interpretive perspective on this story relies upon complementary, but different, theoretical bases for examining the IT experience in Prato. We have abstracted three such bases from literature. First, the concept of "Industrial Networks", originally proposed by the Industrial Marketing and Purchasing (IMP) group at the Stockholm School of Economics, and further developed at the Universities of Uppsala in Sweden and Lancaster University in the United Kingdom, will provide the underlying structural foundation for this interpretation. This concept uses long-term, stable inter–organizational relationships as a basis for examining industry-wide networks exemplified by the *filiera tessile* in Prato. The applicability of the industrial network concepts to interorganizational systems has recently been recognized in the IOS literature

(Cunningham and Tynan, 1993; Reekers and Smithson, 1995). Moreover, from an organizational theory perspective, Casson and Pannicia (1995) have used the concept of industrial networks to examine the current organizational model in Prato and to compare it to the industrial system in the nineteenth century South Wales.

Second, we will introduce the concept of trust as an alternate basis for understanding interorganizational relationships. In the first part of the story, we found rational–economic concepts of opportunism and transaction costs to be insufficient for explaining the rejection of information technology in Prato. Literature in organizational theory identifies the role of trust as counterbalancing the tendency towards opportunism (Thorelli, 1986; Ghoshal and Moran, 1996). In the following discussion we will use the concept of trust to develop an alternate explanation for the events in Prato.

Finally, we will examine some cultural differences between Italian and American societies to develop additional explanations for the events in Prato. Trompenaars (1993) interprets culture as the meaning that societies attach to, among other things, relationships between people and to time. This view of culture, which focuses on relationships, complements the Industrial Network view. The cultural perspective is further developed by using Hofstede (1980)'s characterization of cross-cultural differences and Wall (1994)'s conceptualization of task-oriented and relationship-oriented societies.

Industrial Networks

A network is a model or a metaphor which describes a number, usually a large number of entities which are connected. In case of industrial, as opposed to say social, communication, or electrical networks, the entities are actors involved in the processes which convert resources to finished goods and services for consumption by end users whether they be individuals or organizations (Axelsson and Easton, 1992). The conceptual foundations of industrial networks can be traced back to sociological and anthropological theories concerned with "network relations" (Tichy, et al., 1979). It originally found its expression in the work of the IMP Group (Håkansson, 1982). At times, variously known as the *"Nordic School,"* the *"Interactionist Approach,"* and the *"New Institutionalism,"* it has evolved into what is currently known under the umbrella term "The Industrial Network" approach (Axelsson and Easton, 1992). Only those aspects of the approach that are relevant to our arguments are summarized below.

The concept of networks is based upon the premise that the basic models of classical and neoclassical economic theories do not fully reflect the reality of business-to-business relationships. Such relationships are considered to be the norm, rather than the exception, in many interorganizational transactions (Ford, 1992). Relationships among firms are the *sine qua non* of the industrial network approach: "The existence of relationships, many of them stable and durable, among firms engaged in economic exchange provides a compelling reason for using interorganizational relationships as a research perspective" (Easton, 1992, p. 3).

Johanson and Mattsson (1987) distinguish between relationships and interactions. *Relationships*, in general, are cumulative over time, stable, and long-term in nature. As the history clearly shows, in Prato, the pattern of relationships has evolved over centuries and is the accumulation of past experiences and consequences of future expectations. Relationships, in turn, are comprised of four elements: mutual orientation, or the inclination and expectations to interact with each other; dependence, or the price the individual firms are willing to pay for the benefits the relationship bestows; bonds, or the "tying" between partner firms in a relationship; and the investments each firm has made in the relationship.[23]

Mattsson (1984) indicates that *bonds* can be thought of as having, variously, economic, social, technical, logistical, administrative, informational, and legal dimensions. While economic, technical, informational, and logistical elements of the bond may be self-evident in industrial networks, the relevance and significance of non-economic dimensions such as social or legal dimensions are mainly culture specific. Hamfelt and Lindberg (1987) indicate that the social dimension is characterized by patterns of individual social contacts. They suggest that the salience of the social dimension relative to other dimensions is a function of culture, i.e. commonly held beliefs about the basis of social activity in the network. Furthermore, the manifestation, interpretation, and use of the informational elements of the bond are highly dependent upon the *meaning* attached to information, which in itself is also culturally dependent. Meaning is not something that is transmitted, but it arises and changes in the use of words. Finally, since meaning is inherently social (i.e. intersubjective), common language is a key element in culture. The more subjective elements of the common language that result from direct interactions and tradition (e.g., previous informal agreements and history) are difficult to capture with IT/IS.

Mattsson (1988) suggests that the social aspect of bonds can be a significant factor in the overall strength of inter–organizational relationships. In some cultures, it is possible that social bonds transcend and even replace economic bonds as the *raison d'être* for the relationship. Thus, the level of significance of social bonds in Prato can possibly provide some indications about the failure of *SPRINTEL*.

Relationships form the context in which the interactions or transactions take place. *Interactions*, by contrast, represent the "here and now" of inter-firm behavior and constitute the dynamic aspects of relationships (Easton, 1992). Interactions comprise exchange processes and adaptation processes. *Exchange processes* represent the operational, day-to-day exchanges of an economic, technical, social, or informational nature occurring between firms. *Adaptation* comprises the processes by means of which firms adjust and maintain their relationships by adjusting production, products, routines, and mutual expectations. Thus adaptations represent an ongoing process of investment in relationships. The consequences of adaptation are strengthening of bonds between firms, easier resolution of conflict, confirmation that continuing adaptation is possible, and the development of mutual knowledge on adaptation (Johansson and Mattsson, 1987). Over time, the variety of economic and non–economic exchanges result in the adaptation of both the economic and non-economic dimensions of bonds (Easton and Araujo, 1992). Like bonds, the existence and significance of different types of exchanges, too, are a function of the cultural context of the industrial network. Cultures that value the social aspects of relationships are more likely to engage in social exchanges. A greater level of social exchanges, in turn, is likely to result in further adaptation and strengthening of social relationships. Consequently, a positive feedback cycle exists between social relationships and social exchanges. In Prato this feedback cycle has been going on for over seven centuries.

Revisiting Opportunism: Trust in Relationships

According to Fukuyama (1995, p. 25): "Communities depend on mutual trust and will not arise spontaneously without it. .... Trust does not reside in integrated circuits or fiber optic cables. Although it involves an exchange of information, trust is not reducible to information. A 'virtual firm' can have abundant information coming through network wires about its suppliers and customers. But if they are all crooks and frauds, dealing with them will remain a costly process involving complex contracts and time-consuming enforcement.

Koenig and van Wijk (1994) define trust[24] as an informal mode of control governing mutually identified actors. They suggest that it reduces uncertainty regarding mutual behavior through a process of self-control. The trusting party develops mostly implicit assumptions regarding the trusted party's behavior. Aware of the anticipations held by others regarding its general conduct, the trusted party becomes "trustworthy" if it feels the obligation to fulfill these anticipations. This bidirectional combination of mutual anticipation and obligation yields an effective mode of coordination.

On the other hand, the edifice of traditional transaction cost theory depends upon the universal assumption of a general *lack of trust*, i.e. a belief in the existence of *opportunism* in the context of environmental uncertainty and complexity.[25] Thus, as mentioned above, when confronted with the high variety and complexity of the Pratesian dynamic network, transaction cost theory would predict high transaction costs leading to vertical integration. Furthermore, the traditional IOS literature (e.g. Malone, et al., 1987; Johnston and Lawrence, 1988) would have us rely upon IT and information systems to compensate for this presumed lack of trust by using information to reduce transaction costs. Similarly also Miles and Snow (1986) assumed that computer-based information systems could supplant the need for trust.

However, the continuing segmented nature of the *filiera tessile,* and the low marginal economic value attached to the *SPRINTEL* system by the potential users and their subsequent rejection of it, suggest that transaction costs were not of much consequence in Prato. The so-called "universal assumption" of opportunism, therefore, needs to be re-examined in the context of Prato.

Inzerilli (1989) considers this is a crucial point. He argues that according to Williamson, environmental factors and bounded rationality lead to high transaction costs in the presence of opportunism. On the other hand if opportunism is subject to variation, the impact of environmental uncertainty and complexity, and bounded rationality on transaction costs would also be subject to variations. Williamson himself assumes, albeit implicitly, the variability of opportunism when he suggests that it can be reduced through processes of organizational socialization.

Furthermore, Williamson suggests that the level of opportunism may be determined by the context in which transactions take place. He indicates that while in most cases transactions are treated by the actors as instrumental to benefit maximization, in other contexts the transactions may be valued by itself (Williamson, 1975, p. 39). In the latter case, the reasons for opportunism, and therefore opportunism itself may no longer be relevant. Additionally, Williamson suggests that opportunism is not endemic – indeed he admits that business managers often do act on the basis of trust (Williamson, 1975, p. 109). Williamson's more recent work (Williamson and Ouchi, 1981; Williamson, 1985; Williamson, 1991) recognizes this relationship between trust and opportunism more explicitly.

Bromily and Cummings (1992) argue that trust reduces transaction costs. They suggest that higher levels of trust, not only reduce the cost of monitoring performance, they also eliminate the need for developing detailed contracts and installing control systems. Hill (1990) proposes that a reputation for non-opportunistic behavior leads to reduced transaction costs. While opportunistic actions may yield short-term benefits, they create a longer term cost in terms of a lack of trust which increases future transaction costs. Reputation, of course, is the result of trustworthy behavior in the past. Brenner (1983, p. 95) observes that "Trust is one resource that, by diminishing contract uncertainty, lowers the cost of exchanges in the economy." Arrow (1974, p. 23) goes even further by suggesting that "Trust is an important lubricant of a social system". Finally, Thorelli (1986) suggests that in some cultures trust is a vital supplement to contractual arrangements and may even take their place.

Evidence of trust in the industrial networks in Prato would indicate the existence of low transaction costs, and thus, account for the continuing segmentation of production in the textile chain and the low marginal value attached to the *SPRINTEL* system in Prato. Moreover, earlier, in discussing industrial networks, we presented Mattsson's (1988) suggestion that in some cultures, it is possible that social bonds transcend and even replace economic bonds as the *raison d'être* for the relationship. This is consistent with Williamson's above observation that in some contexts transactions may be valued per se, over and above the economic benefits provided by the transaction.

Fukuyama (1995, p. 26) defines social capital as the capability that arises from the prevalence of trust in a society or parts of it. It differs from other forms of human capital insofar as it is usually created and transmitted through cultural mechanisms such as tradition or historical habit. According to him effective organizations are based on communities of shared ethical values and do not require extensive contractual and legal regulations of their relations. Further, relationships, especially social relationships, are normally built upon a foundation of trust. Thus, an examination of the existence of social capital and strength of social bonds in Prato would in addition also possibly provide explanations for the low transaction costs encountered in Prato.

Origins of trust in relationships in Prato

The role of trust in network organizations, particularly in Italian industrial districts has been emphasized by a number of authors (Inzerilli 1989, 1990; Jarillo 1988; Jarillo and Stevenson 1991; Lazerson, 1988; Ritaine 1990, Lazerson, 1995). According to Erickson (1959) a fundamental mechanism for the development of trust is mutual identification. In Prato a variety of mechanisms for fostering mutual identification exist.

First, as reported by Lazerson (1988), and also confirmed by Johnston and Lawrence (1988), most of the interacting firms in Prato are spawned either by relatives or employees first splitting off as quasi-independent satellite firms usually financed by the family/employer, and then gradually become more independent. The tradition of the extended family being the economic family and the economic family the extended family, goes back to the days of Datini:

> ... famiglia embraced a very wide field. Fuoco, famiglia, parentela – these terms were the terms used to designate not only a men's immediate descendants, but every relative living under the same roof and eating the same bread – aunts and uncles and cousin's and cousin's children, down to the most remote ties of blood. They belonged to the casato, as they had to the Roman gen, and often the term was extended to include even people bound to the family by common economic interests or by dependence, such as partners, employees, and servants.... The family, moreover, was an economic entity, as well as a social one. Its importance, which had once depended on the number of its fighting men, was now chiefly assessed by the strength and variety of its connections at home and abroad, in politics and in trade.
> (Origo, 1957, pp. 181–182).

This tradition has continued in Italy over the years. A study conducted by Rieser and Franchi (1986) in the Modena industrial district in the northern part of Italy showed that 47 percent of business partners were former employees of the firm owners. Lazerson (1988, p. 338) reported that firm owners sometimes even refused to expand their business if they could not find trustworthy partners. Partners considered trustworthy were either members of the extended family, friends, or former employees. The partner's contribution of capital or his specialized skills were no more than secondary considerations. Demattè and Corbetta (1993) confirm these observations and discuss the problem of finding trustworthy partners in greater detail.

Second, the web of relationships that develop within such a family-oriented structure is necessarily diffuse rather than specific. Parsons and Shils (1951) argue that different cultures view relationships as either specific (i.e. they limit relations to specific, limited spheres of activity, e.g. economic) or diffuse (i.e. relations exist concurrently in a variety of spheres of activities). Trompenaars (1994, p 73–91) reports that while relationships are pre-dominantly specific in the North American and West European cultures (the same cultures where traditional opportunism based theories were first developed), they tend to be much more diffuse in Southern European cultures such as Italy.[26] Thus, in Tuscany, and in Prato, relationships between men who do business together, possibly went to school together, and now are also members of the same church, sports and hunting clubs, trade and crafts associations, and even political parties are sooner diffuse than specific. This too encourages mutual identification, and ultimately trust.

However, it also needs to be recognized that diffuse societies are usually closed societies that create "barriers" for outsiders. Barbieri, et al. (1989) suggest that both *ENEA* and *Reseau* were perceived as outsiders by the manufacturing firms in the district of Prato. This greatly reduced their effectiveness. Both organizations had a technical-economic bias and championed the projects without fully understanding the social structure in the district and the social implications (discussed below) of *SPRINTEL*. While the Pratesean firms realized that the projects were utopic, they were prepared to go along, hoping for at least some minimal gains with very limited costs.

Another factor which may have influenced the development of long term bonds and trust in Prato is the Italian attitude towards time. Trompenaars (1994, pp. 107–24) classifies cultures according to their concept of time. In sequential cultures like the United States[27] and North Western Europe, time is viewed as passing in a straight line consisting of a sequence of disparate events; in synchronic cultures, e.g. Italy, and to a much greater extent in Asia, past, present, and future are all interrelated so that expectations of the future, and accumulated experiences of the past, both, shape present action (Nonaka and Takeuchi, 1995, pp. 28–9). Moreover, these accumulations and expectations are not limited to the present generation. Memories are long, and harm done or benefits bestowed onto parents, grandparents, and granduncles, continue to influence the relationships and interactions both now and in future. Consequently, the attitude towards time also means that current actions not only affect current generations, but also contribute to 'building' for the future family. In Prato with the low geographic mobility of its inhabitants, the trust relationships have evolved over generations.

Finally, there is also direct empirical evidence supporting the hypothesis of trust in inter-firm relationships. In a survey of 100 firms in the Tuscany area, Bagnasco and Trigilia (1990) report that, of the 93 firm owners asked to describe their relationships with other firms, 56 defined the relationship as one of "trust" or "co-operation," 33 as "correct market relationships," and only 4 as a "hard and sometimes disloyal market relationship." Of a total of 100 firms in the sample, only 19 reported the use of contingency and long-term contracts, while another 34 reported advance agreements with other firms to undertake specific activities and to meet specific orders, the joint use of technology, or shareholding in other companies. If we discount the 19 legal contracts, most firms were found to rely either on non-contractual agreements or no formal agreements at all. This too supports the reliance on trust mechanisms and social controls as opposed to contractual legal mechanisms.

SPRINTEL failed to the extent that it had its justification in the need to reduce transaction costs. Given the relationship and trust basis of Prato, these costs were low to begin with.

The cultural context

> It is difficult to be Italian, but it is even more difficult to be Tuscanian ... And this is not because in Tuscany we are better or worse than others, Italians or foreigners, but because, thank God, we are different from any other nation: for something which is within us, in our true nature, something different from others. Or maybe because ... we simply do not want to be like the others.
>
> (Malaparte, 1994, p. 7, our translation)

As the above quote from the Italian best-seller book suggests, Tuscanians consider themselves to be a people apart, different not only from the rest of the world, but also from their fellow Italians. The fact is that the Tuscan culture, by and large, is similar to, and can be identified with other societies in Southern Europe and the Mediterranean. On the other hand, it does exhibit certain traits which definitely set it apart from the North American and Northwestern European societies, traits that may add to the explanation of the puzzle in Prato. Three such traits, a diffuse view of relationships, an emphasis on trust as a mode of operation, and a synchronic perspective of time were discussed in the previous section. In this section we identify two additional cultural dimensions which further differentiate the Tuscanian culture.

The first dimension deals with a relationship orientation, as opposed to a task orientation of the culture. Walls (1993) adapts Toennies' (Toennies, 1965) concept of *gesellschaft* and *gemeinschaft* societies to develop the concept of task-oriented versus relation-oriented communities (Walls, 1993, pp. 155–9). Toennies' *gesellschaft* refers to societies defining themselves by impersonal contractual and legal relationships, based more upon mutual need to achieve specific tasks than on kinship. His *gemeinschaft*, on the other hand, is a natural grouping of people based upon kinship and neighborhood, shared culture, and folkways. Walls interprets *gesellschaft* societies as "task-focused societies" with firm (fixed) goals and rules, where relationships are defined only by legal contract, and all other aspects of the relationships are easily jettisoned or altered in pursuit of the firm task-oriented goals. "Relationship-focused" cultures, on the other hand, put their primary emphasis on maintaining long-term, multifaceted relationships, sometimes even at the expense of modifying the goals in order to avoid harm to the relationship (Walls, 1993, p. 157).

Parson and Shil's (1951) concept of "universalist" and "particularist" cultures, interpreted by Trompenaars (1994, pp. 29–46) as "rules-oriented" and "relationship–oriented" cultures can also be directly mapped onto Walls' classification of cultures. Within the business context, these concepts are illustrated by Dore's distinction between the company law model typified by the Western/American corporation, and the community model commonly found in businesses in Japan (Dore, 1987, pp. 53–5). Dore's earlier study of the textile industry in Nishiwaki town of Japan, shows marked similarity to Prato in terms of de-verticalization of the industry, co-ordination through trust, "good-will," and social norms, and the lack of reliance on contracts (Dore, 1983, p. 465). It should however be recognized that the characterization of a culture as either "task–oriented" or "relationship-oriented" is an over simplification, which is indeed the case with all ideal types. In reality cultures which are sustained and survive over time will include both orientations, albeit in different degrees.

Walls goes on to suggest that communication plays very different roles in the two types of cultures. In a task-oriented culture, the only purpose of communication is to further the goals or the tasks of the community. In a relationship oriented culture, communication, in addition to furthering tasks, is also used for maintaining relationships. In the context of the industrial network model discussed above, while a task-oriented society will primarily focus on task-oriented economic and informational exchanges, the exchanges in a relationship-oriented culture are likely to include social, relationship-maintenance aspects as well.

Projection of the self, i.e. communication of social-psychological presence, is key to relationship maintenance. Projection of self requires rich communication. In information systems literature, communication richness and role of media in communication has been examined from three paradigmatic perspectives: information richness theory perspective, interpretivist perspective, and critical social theory perspective (Nwenyama and Lee 119, p. 146). Information Richness Theory (Daft and Lengel, 1986; Fulk, et al., 1987) suggests that mediums of communication differ in the degree to which they convey the senders' and receivers' social-psychological presence. Moreover, richer mediums of communication, such as face-to-face meetings, telephone conversations, and to some extent even written communication through mail and fax, provide greater capability of furthering social presence as compared to traditional electronic telecommunication media such as e-mail and EDI (Short, et al., 1976).[28] Consequently, according to information richness theory "richer media" are likely to better support exchanges involving both tasks and relationships and are likely to be preferred in relationship-oriented societies (Ishii, 1990). Second, in the interpretive tradition, Lee (1994, p. 144, 154) observes that communication richness emerges through the interaction between the organizational context and the communication medium. Zack (1993) suggests that context even more than content is a crucial factor in choosing a communication mode. Lee goes on to suggest that the preferred "medium is one that *becomes* best or appropriate over time, through its interactions with its users and through the user's adaptation or re-invention of the medium to suit their own purposes" (Lee 1994, p. 155).

In Prato, the rejection of EDI-based communication in favor of other more personal modes of communication is consistent with both these theoretical perspectives. The spatial arrangements in the industrial district of Prato, and the physical proximity of its inhabitants facilitate the usage of "rich" communication media. The *impannatore* and other Pratesian actors often need just to cross the street to communicate, making "walking" a convenient substitute for electronic communication. Furthermore, as the following quote from "The Merchant of Prato" shows, over the centuries, the men of Prato have developed and fine-tuned a variety of mechanisms for communication:

> And on summer evenings the men would sit for hours on the long stone benches beneath the Palazzo Pretorio – a custom which gave rise to a new verb, pancheggiare – while the gossip of the town ran rife. Every man's business was also his neighbor's.
>
> (Origo, 1957, p. 61)

In addition to face-to-face meetings in a variety of diffuse contexts, the custom of *lo struscio* and *pancheggiare*,[29] still persists in modern Prato. Furthermore, the growing proliferation of cellular phones suggests that Pratesians are willing to take advantage of modern communication methods as long as they do not unduly limit the richness of expression needed for social exchanges and relationship maintenance. The reader may want to contrast the sterility of a typical EDI message with the following quote from an early fifteenth century letter to Datini: "We went to the fairs and purchased some very fine, perfect, and good wool at the cost of 11 marks. Please God these countries of ours will be at peace, so that trade can flow once more" (Origo, 1957, p. 7).

Finally in the critical social theory (CST) perspective Ngwenyama and Lee (1997, p.152) suggest that: "In CST communication richness involves not only understanding what the speaker or writer means, but testing the validity claims associated with the action type enacted by the speaker or writer." While such a test of validity claims will be difficult in the case of a typical EDI message, it is much more likely in the types of social exchange existing in Prato.

The last dimension of cross-cultural difference deals with what Hofstede (1980) calls the level of "masculinity" in the society. Masculinity is associated with assertiveness, focus on advancement and earning power, achievements in terms of recognition and wealth, exhibitionism, and dominance. Hofstede (1980, p. 279) reports much higher levels of masculinity in the Italian society as compared to the North American and Northwestern European nations. Contemporary accounts of Tuscany and Prato emphasize the level of conspicuous consumption and display of wealth in the modern Pratesian society (Malaparte, 1994). Prato is one of the wealthiest towns in Italy and has for instance the highest car ownership in the country (Buxton, 1988, p. 10). The masculinity of the Pratesians is illustrated by the following quote:

> [The people from Prato] ... are good workers, businessmen, job creators and their heart is bigger than their own hand: they spend all the money they get; and they are good people as long as they are poor workers, but they suddenly become mean, when they make money – any way they make it.
> (Malaparte, 1994, p. 56, our translation)

This tendency towards exhibitionism could be an explanation for the initial adoption and early growth of *SPRINTEL* despite its seemingly low marginal economic value to the Prato firms. The 1980s were heady days for Prato. Scholars from the world over were studying Prato, trying to describe and understand the success of the "newly discovered" Italian model of organization. Even prestigious American journals like *Harvard Business Review* and the *Scientific American* had mentioned and profiled Prato. The Italian government through ENEA, the European Union through its grants, and various other private and public institutions were willing to invest in the "Grand Technological Experiment" in Prato. The Pratesians justifiably felt that they were creating a new and shining example for Italy, if not for the rest of the world. This initial enthusiasm was further compounded by the Italian equivalent of the "keeping up with the Joneses" syndrome, or the *Giovanni* effect. Firms which did not get a *SPRINTEL* connection in the initial round of pilot tests wanted a connection to keep up with the next door *Giovanni's* who already had one. Thus, as long as someone else was paying for it, *SPRINTEL* provided an ideal outlet for self-aggrandization. It is only when the minimal charges were proposed that the low marginal economic worth of the IOS became evident and the steady decline in enrollments started.

Another example of masculinity in the context of this story was the non-use of a service offered by *SPRINT* to help sell redundant inventories. The companies in Prato could offer their overstocked inventories to other firms via *SPRINTEL*. On the basis of sound economic reasoning during the requirements definition phase, it was thought that such a service should be in great demand. However, for the Pratesian entrepreneur, acknowledging the existence of redundant inventories would have meant a public confession of an inaccurate market forecast. Thus, almost nobody utilized this service.

Finally, one last argument for the decline of *SPRINTEL* may be found in Pratesian entrepreneur's reliance on the *sommerso* or the underground economy. Earlier we mentioned that the underground economy including non-declared domestic work hours at home, non-accounted for overtime, non-declaration of income and activity, personal use of business assets, and other forms of tax evasion could have played a role in the capital formation and growth of Prato firms. An IOS like *SPRINTEL,* which formalizes all information flows and directs them away from the direct control of the firm, would undoubtedly be considered a threat to this underground economy.

Conclusions and moral of the story

The objective of this paper was to examine the limitations of the applicability of technical-economic theories currently dominant in IS and management literature and through this examination extend our understanding of the role of information technology within and across organizations. To explore these limits, we used these theories in an attempt to explain the story of the implementation, use, and subsequent rejection of an IT-based IOS in the Prato textile district in Italy. In conducting this investigation we followed the methodological logic proposed by Lee (1991). Following this logic we first formulated propositions based upon traditional technical–economic theories about the implementation and use of information technology in Prato. These propositions were confronted with the sequence of events in Prato. In this confrontation, we found the technical–economic theories inadequate in explaining the events. Next, in the second part of the story, we developed an interpretation of the story using a combination of theoretical perspectives from pre-existing literatures on trust, relationships, and inter-organizational networks, and information about Prato collected from literature and interviews in the Prato district. These propositions, observations, and revised interpretations are summarized in Appendix A.

So what did we learn from this examination and interpretation? This story has three key conclusions or "morals." The "first moral" is that we need to use caution in applying economic theories and their predictions to cultures which are different from the US and Northwestern European cultures. While the logical structure of the underlying theories might very well be sound, their applications, such as those found in the traditional IOS literature, may rest upon premises which are valid mainly in their originating cultural context. In the technical–economic story presented in Part I, the traditional transaction cost theory, theories derived from it, and their predictions vis-à-vis the role of IOS in network organizations were based upon the premise of self-interest and opportunism in the context of a faceless market. These assumptions may hold in the United States, and to a lesser extent in the economies of Northwestern Europe. These are the same societies where these economic theories and arguments were originally developed and have now become commonly accepted. However, when the so-called "universal assumptions" of opportunism and faceless markets are replaced by the notions of trust and long-established multifaceted relationships, and when the parties at both ends of the transaction have known faces and recognizable personalities, as in the case of Prato, the conclusions originally derived from these theories are no longer tenable. Thus the arguments that computer-based systems substitute for trust-based relationships (Miles and Snow, 1986, p. 65) and that a reduction in transaction costs hasten a move towards markets (Malone, et al., 1988) or induces a move to the middle (Clemons and Row, 1991) are at best problematic in the context of Prato.[30]

The second "moral" or conclusion deals with how information systems are conceived and developed. Many of the methods and techniques for analyzing information systems requirements, and the management of factors critical to the success of these systems have been developed on the basis of the traditional technical–economic rationality. However, as our analysis shows, these methods and factors provide at best only a necessary but not a sufficient facet in the success or failure of information systems. Thus, these traditional development approaches need to be augmented with additional strategies which, as a precursor to development, examine the existing patterns of culture, relationships, and trust (or distrust) in the development situation, and take them into account for devising a development and implementation strategy. The design principles proposed by Ciborra (1993), his use of neotransaction cost theory based IS design exercises, and his notion of a contractual IS development paradigm may provide some early pointers to such methodologies. Similarly, a reinterpretation, in the light of the notions of relationship and trust, of methodologies such as the Soft Systems Methodology (Checkland and Scholes, 1990), could also provide the beginnings of such development and implementation methodologies.

At the end of the day, the reality of business is about people, where involvement, interaction, making meaning and contribution are critical to business success. In supply chains, materials, knowledge and costs rarely move up or down the chain without human intervention and cooperation. Since most business processes involve people, transactions are inherently impacted by the interaction between people and their environment. At their turn, the effectiveness of these interactions is often determined by the relationships and trust, the company cultures, prior experiences, etc. of the individuals involved. Neither the economic-technical rationality, nor the social–political rationality adequately explains these aspects of the usage of ICT in organizations.

The notions of trust, relationships, and cooperation lead to the third or final contribution of our story. These concepts, by introducing a third complementary perspective, add to Kling's (1980) theoretical perspectives in IS literature, the "*System Rationalism*" and "*Segmented Institutionalism*" perspectives. As discussed in part I, both of Kling's perspectives are rooted in the premise of utility and self-interest maximization under conditions of opportunism – a premise which may or may not hold in settings other than the one in which the economic rationality was conceived. In such settings other modes of behavior may prevail.

As this story shows, the existence of trust and cooperation, in addition to power and politics, provides a third way of studying the role of information technology in organizations. The focus on self-interest and opportunism underlying the first and second rationalities is likely to create a *win–lose* view of business transactions and relationships. This view results in jockeying for position and power both across and within organizations, leading to a relentless quest for competitive advantage. Thus conducting business becomes synonymous with waging "war" (Tzu, 1995) and IT strategy with wielding IT as a strategic weapon (McFarlan, 1984). A third rationality which recognizes the existence of trust and relationships is likely to help identify and create win–win strategies and thus lead to cooperative strategies and the concept of IT for collaborative advantage (Moss-Kantor, 1994, Kumar and van Dissel, 1996).

Like Kling's segmented institutionalism, this new perspective is also an interactionist perspective and refers to social phenomena. However the third rationality, instead of focusing on political conflict and power plays as the primary interaction mode, focuses on collaboration and cooperation as the key interaction processes. The third rationality calls for an analysis of existing levels of trust and relationships in studying the role of information technology in organizations, and suggests an examination of the impact of IT on trust and cooperation. Trust-related concepts of responsibility, benevolence, fairplay, and altruism supplement (rather than supplant) the previous concepts of self-interest, betrayal, and conflict. Thus, this perspective introduces a *third rationality* in which trust, social capital, and collaborative relationships are introduced as the key concepts. It is our belief that this third rationality will not only provide yet another way of examining the role of information technology in organizations, it will provide guidance for developing a new generation of collaborative technologies for an increasingly interdependent, multicultural, globalized world.

NOTES

1 This article is an extended version of: The Merchant of Prato – *revisited*: towards A Third Rationality of Information Systems, 1998, *MIS Quarterly*, June, pp. 199–226.
2 The economic rationality commonly used in the IS interorganizational systems literature is described at the beginning of Part I, A Story of Technical–Economic Rationality.
3 For details of the two rationalities see Kling (1980) and Kling and Scacchi (1982).
4 A theory can be considered to be a "thought artefact" constructed by humans with the purpose of describing and explaining phenomenon. Following Simon (1981, pp. 16–17)'s logic of "Limits to adaptation" in his treatise "*Sciences of the Artificial*," an artefact usually functions smoothly in benign or "home" conditions for which it was originally designed. It is only when it is exposed to extreme conditions, possibly in an alien environment, do its limits become visible. Similarly, the limits to theories are likely to become apparent once they are transplanted and used in a culture other than their native culture.
5 A summary of the propositions, observations, and extensions to theory is presented in Appendix A.
6 SPRINT is the acronym of the *Sistema Prato Innovazione Tecnologica* consortium that managed the IOS and is not to be confused with the well-known US telecom operator.
7 Japanese ritual suicide in face of defeat, failure or shame; colloquialism: *hara-kiri,* a practice dating back to the times of Shoguns and Samurai. A more recent incarnation of this practice is the suicide by a Japanese banking executive who believed that he had brought dishonor to his bank in the recent Japanese banking scandal.
8 The concept of utility could include economic as well as social-psychological aspects of utility.
9 Transaction costs consist of costs incurred in searching for the best supplier/partner/customer, the cost of establishing a supposedly "tamper-proof" contract, and the costs of monitoring and enforcing the implementation of the contract.
10 One of the central assumptions of traditional transaction cost economics is the belief that the other party in a relationship is not to be trusted, and the risk of opportunism is high. Opportunism is defined by Williamson (1985, p. 47) as "*self-interest seeking with guile.*" However, as will be discussed in Part II, recent extensions to transaction cost theory (e.g. Ghoshal and Moran, 1996), and even Williamson himself, recognize that opportunism is not a constant, and under certain circumstances the existence of trust and moral obligations may counter and even supercede the need to control opportunism.
11 Through a curious coincidence, what is known about the early textile industry in Prato is based upon the

information systems (records) of the fourteenth-century textile merchant *Francesco di Marco Datini* (1335–1419). Extensive records of Datini's business and social dealings were discovered in the *Palazzo Datini* in the nineteenth century (1870). These records consisting of some 140,000 letters and a great number of business documents form the basis of historical treatises describing the business, political, and social life of that period (Avigdor 1961, Origo 1957). Famous medieval merchants such as Datini, immortalized in Origo's classic book *The Merchant of Prato*, have now become the symbols of the spirit of Prato.

12 The ragtrade now consists of only 10 percent of the overall business in Prato (Graham, 1994).

13 Given its structural connotations, the term "value chain" is probably not appropriate for capturing the dynamic nature of this phenomenon. In this article we explore the use of other concepts, such as "value added partnerships" (Johnston and Lawrence, 1988) and "dynamic networks" suggested by Miles and Snow (1986). Inzerilli (1990) in his description of such networks has used the term "corporation." Inzerilli's concept of corporation is derived from the Italian *corporazioni*, the guilds of masters and craftsmen in the medieval cities. However, in English the term corporation evokes images of relative stability which are even less suitable. In the context of this story a more appropriate term may be the Italian term *filiera tessile* or the thread (chain) of textile processing. But, given its general acceptance and usage, we will continue to use the terms "value chain" and *fileria tessile* interchangeably.

14 Actually, it is quite likely that in most cases repeated chains may be constituted from the same set of actors. The chains are usually formed based upon the mutual knowledge and prior experience with actors in the chain - and therefore are likely to include, substantially, the same actors as long as they are able to supply the required manufacturing capabilities and capacity at competitive prices.

15 From *panno* (cloth). The word *impannatore* literally translates to the cloth-man.

16 See footnote 10 on Datini.

17 Tax evasion is called "*elusione*" in Italian. It refers to "barely" legal ways to avoid taxes. Recently the International Monetary Fund (IMF) guesstimated the size of the Italian underground economy at 15 percent of the gross national product. In comparison, the IMF guesstimate for the other Western European countries was 2.3 percent on average.

18 From the 1970s onwards, the so-called Eurocommunism of the Italian communist party has had very little in common with communism in the former Soviet Union. Massimo Salvadori, a famous journalist of *La Stampa*, one of Italy's biggest newspapers, expressed this very well (26, February 1988): "To what extent are you still communists" he asked, "If you are still communists, what then is the substance of your communism? If not, isn't it becoming time to resolve this useless misunderstanding?" The integration of the communist party within the Italian political system is exemplified by the fact that when a Pope dies, communist organizations hang the red flag at half mast. Since the fall of the communist regime in the former Soviet Union, the Italian communist party has changed its name to *Partito Democratico di Sinestra* (Democratic Party of the Left) to resolve the misunderstanding. Overwhelmingly, the Pratesian entrepreneurs, their extended families, and their employees, were members of the communist party. The Pratesian and the Tuscanian culture was often called the red-culture (Lazerson, 1988; Malaparte 1994). A similar role was performed by the Christian Democrats (*Democrazia Christiana*) in the industrial districts of Italy's north east (Veneto).

19 *ENEA* was the former state-owned agency for nuclear energy in Italy. After the ban on nuclear research, it was looking for a new 'mission,' and was assigned the role of the agency in-charge of technological innovation.

20 The SPRINT consortium was founded in 1983. Its activities covered three main areas: telematics, infratechnologies and energy. Here we concentrate only on the telematics area. Infratechnologies covered production automation, CA/IM and robotics and energy aimed at reducing the energy costs in production.

21 As stressed in an interview conducted by P. Manacorda with Marco Capponcelli, the director of SPRINT, in the newpaper "*Il sole -24 ore*" (21, November 1989).

22 It should be noted that the associations were not able to formally impose anything on the Pratesean firms. They were able to propose innovations, like infrastructures, but had little leverage in forcing the firms to adopt them.

23 While, each of these elements can be shown to have some relevance in context of Prato, the concept of bonds is key to understanding the cultural dimension of relationships in Prato and is therefore discussed further. For details of the other elements see Easton (1992, pp. 8–11).

24 Hosmer (1995) provides a more formal and scholarly definition of trust. This definition synthesizes the conceptualization of trust from two intellectual traditions – organizational theory and philosophical ethics. Hosmer (1995, p. 399) states: "Trust is the expectation by one person, group, or firm of ethically justifiable behavior – that is morally correct decisions and actions based upon ethical principals of analysis – on the part of other person, group, or firm in a joint endeavor or economic exchange." It should however be noted that Hosner misses the concept of felt obligation and trustworthiness.

25 See Ghoshal and Moran (1996, p. 17–20) for an extensive discussion of the role of opportunism in Williamson's transaction costs economics.

26 In North America and Northwestern Europe this specificity is reinforced through societies that are based on

the individual and the state as two equal pillars. Their success is considered in the light of the rational–impersonal interaction between these micro and macro components. Furthermore, the macro element is also characterized by "bigness," e.g. big government or big businesses. On the other hand, the Italian society is based on the concept of small social groups, i.e. the extended family, while the reaction of the individual towards the state is essentially adversarial (Ward, 1990, Chapter I: "La Società", p. 23).

27 The United States is considered to be the ultimate sequential culture. The sequential nature of the US culture is evident in phrases such as "what have you done for me lately?" and "you are only as good as your last quarter" which are part of the common business and day-to-day lexicon. The comparatively short term gain taking view in the American culture is said to promote opportunism as opposed to trust-based relationships. Empirical support for the connection between the view of time and trust and co-operation comes from game theory. In a game with a finite number of plays (i.e. a short–term time horizon) opportunistic behavior predominates and the co-operating or trusting actor always looses. However, Friedland (1990) found that in games with infinite number of plays the chances for such loss are substantially reduced.

28 While Short, et al. (1976) discussed electronic media and e-mail in general, and did not specifically mention EDI, EDI with its minimal, highly structured data interchange, is likely to have even less social presence than e-mail.

29 From *panca,* a bench. The socially embedded practice of bench sitting mainly done by the town's elders. It is closely related to the custom of *lo struscio* – the stroll before dinner by the youth (everybody younger than 40). In many Italian provincial cities every evening people slowly stroll around the town square to "see and be seen." At this occasion, informally, all sorts of information is exchanged. In many Italian provincial cities and towns these customs continue with the ostensible reason to enjoy fresh air, while at the same time serving as an important forum for exchanging the day's social and business gossip and news.

30 In the case where no prior history of trust and relationships exist and few mechanisms for mutual identification are available, for example in the case of global open-electronic commerce, it is possible that IT/IS to a certain extent may provide trust substitutes such as trusted third parties.

ACKNOWLEDGMENTS

The authors gratefully ackowledge the help of the former senior editor of MISQ, Allen Lee. We further acknowledge the helpful comments and insights of Dr Giorgio Inzerilli in preparation of the manuscript. Finally we would like to thank Giorgia Valentini and Stefano Giacomelli for help in collecting the data.

REFERENCES

Arrow, K. J. (1976) *Limits of Organization.* Norton, New York.

Avigdor, E. (1961) *L'industria tessile a Prato.* Feltrinelli, Milano.

Axelsson, B. and Easton, G. (eds) (1992) *Industrial networks: A New View of Reality,* Routledge, London.

Balestri, A. (1994) *The Textile Industry of Prato: An Overview.* Centro Studi UIP, 13 October 1994.

Barbieri, G. P., Bellini, N., Giordani, M. G.; Magnatti, P., Pasquini, F. and Scarpitti, L. (1989) *La Valutazione di Interventi di Modernizzazione Tecnologica a Carattere Territoriale: Una Prima Applicazione Operativa.* Società di Studi Economici S.p.A., Bologna, Laboratorio di Politica Industrial (internal report).

Bielli, P. and Giacomelli, S. (1995) Reinforcing Networks of SMEs in Italian Districts: The Role of Information and Communication Technologies. Paper presented at: *The First CEMS (Community of European Management Schools) Conference,* Vienna, Austria.

Brenner, R. (1983) *History – the Human Gamble.* University of Chicago Press, Chicago.

Bromily, P. and Cummings, L. L. (1992) Transaction Costs in Organizations with Trust. *Working paper # 28. Strategic Management Research Center,* University of Minnesota, Minn.

Buxton, A. (1988) Italy's Prato Textile Industry. *Textile Outlook International,* The Economist Intelligence Outlook, January, pp. 8–18.

Casson, M. and Pannicia, I. (1995) Networks and Industrial Districts: A Comparison of Northern Italy and South Wales. In *Proceedings of the Workshop on Industrial Networks,* Blanc, G. (ed.), European Science Programme (EMOT), Groupe HEC, Jouy–en–Josas, France, 3–4 February.

Cats–Baril, W. L. and Jelassi, T. (1994) The French Videotext system Minitel: A successful implementation of a national information technology infrastructure. *MIS Quarterly*, 18, (1), 1–20.

Checkland, P. and Scholes, J. (1990) *Soft Systems Methodology in Action*. John Wiley and Sons, Chichester, UK.

Ciborra, C. U. (1993) *Teams Markets and Systems: Business Innovation and Information Technology*. Cambridge University Press, Cambridge.

Clemons, E. K. and Row, M. (1987) Structural differences among firms: A potential source of competitive advantage in the application of information technology. In *Proceedings of the 8th International Conference on Information Systems*, San Diego, 1–3 December.

Clemons, E. K. and Row, M. C. (1991) Sustaining IT Advantage: The role of structural differences. *MIS Quarterly*, 15 (3) 275–92.

Coleman, J. S. (1984) Introducing social structure into economic analysis. *American Economic Review*, 74, 84–8.

Cunningham and Tynan (1993) Electronic trading, interorganisational systems and the nature of buyer–seller relationships: The need for a network perspective. *International Journal of Information Management*, 13, 3–28.

Demattè, C. and Corbetta, G. (1993) *I processi di transizione delle imprese familiari*. Collana 'Studi e Ricerche', 15, Mediocredito Lombardia, Milano.

Dore, R. (1983) Goodwill and the spirit of market capitalism. *British Journal of Sociology*, 23, 459–82.

Dore, R. (1987) *Taking Japan Seriously: A confucian Perspective on Leading Economic Issues*, Stanford University Press, Berkeley.

Easton, G. (1992) Industrial networks: A review. In: *Industrial networks: A New View of Reality*, Axelsson, B. and Easton, G. (eds), Routledge, London, pp. 1–27.

Easton, G. and Araujo, L. (1992) Non–economic exchange in industrial networks. In *Industrial networks: A New View of Reality*, Axelsson, B. and Easton, G. (eds), Routledge, London, pp. 62–84.

Erikson, E. (1959) Identity and the Life Cycle. *Psychological Issues*, 1, (1).

Ford, D. (ed.) (1990) *Understanding Business Markets, Interaction, Relationships, and Networks*. Academic Press, London.

Friedland, N. (1990) Attribution of Control as a Determinant of Cooperation in Exchange Interactions. *Journal of Applied Social Psychology*, 20, 303–20.

Fukuyama, F. (1995) *Trust: The Social Virtues and the Creation of Prosperity*. Hamish Hamilton, London.

Fulk, J, Steinfield, C. W., Schmitz, J. and Power, G. J. (1987) A Social Information Processing Model of Media Use in Organizations. *Communications Research*, 14, 520–52.

Gandolfi, V. (1988) *Area sistema: internazionalizzazione e reti telematiche*, Franco Angeli, Milano.

Ghoshal, S. and Moran, P. (1996) Bad for Practice: A Critique of Transaction Cost Theory. *Academy of Management Review*, 21 (1), pp 13–47

Graham, R. (1994) Prato Textile Industry: Remarkable Resilience. *Financial Times Survey*, 25 October 1994.

Granovetter, M. (1985) Economic Action and Social Structure: The Problem of Embeddedness. *American Journal of Sociology*, 91, pp. 481–510

Håkansson, H. (ed.) (1982) *International Marketing and Purchasing of Industrial Goods*. John Wiley and Sons, Chichester, UK.

Hamfelt, C. and Lindberg, A. K. (1987) Technological Development and the Individual's Contact Network. In Håkansson, H. (ed.), *Industrial Technological Development: A Network Approach*, Croom Helm, London 1987.

Hill, C.W.L. (1990) Co-operation, opportunism, and the invisible hand: Implications for transaction cost theory. *Academy of Management Review*, pp. 500–13.

Hofstede, G. (1980) *Culture's Consequences: International Differences in Work–Related Values*. Sage Publications, London.

Hosmer, L.T. (1995) Trust: The connecting link between organization theory and philosophical ethics. *Academy of Management Review*, 20 (2), 379–403.

Inzerilli, G. (1989) Transaction costs, opportunism, and social control: A reply to Lazerson. *Management Report Series*, No. 50, March, Rotterdam School of Management, Erasmus University, Rotterdam, The Netherlands.

Inzerilli, G. (1990) The Italian alternative: Flexible organization and social management.

International Studies of Management and Organization, 20 (4), 6–21, (Winter).

Ishii, H. (1993) Cross–Cultural Communication and CSCW. In Harasim, L.M. (ed.), *Global Networks: Computers and International Communication,* The MIT Press, Cambridge, Mass.

Jarillo, C. (1988) On strategic networks. *Strategic Management Journal,* 9, 31–41.

Jarillo, C. and Stevenson, H. (1991) Co-operative Strategies – The Payoffs and the Pitfalls. *Long Range Planning Review,* 24 (1), 64–70.

Johansson, J. and Mattson, G. (1987): Inter-organizational relations in industrial systems: A network approach compared with the transaction–cost approach. In *International Studies of Management and Organization,* pp. 34–48.

Johnston, R. and Lawrence, P. R. (1988) Beyond vertical integration – the rise of the value–added partnership. *Harvard Business Review,* pp. 94–101, (July–August).

Johnston, R. H. and Vitale, M. R. (1988) Creating competitive advantage with inter–organizational systems. *MIS Quarterly,* 12 (2), 153–65.

Kling, R. (1980) Social analysis of computing: Theoretical perspectives in recent empirical research. *ACM Computing Surveys,* 12 (1), 61–110.

Kling, R. and Scacchi, W. (1982) Web of computing: computer technology as social organization. In Yovits, M.C. (ed.), *Advances in Computers,* 21, Academic Press, pp. 1–90.

Koenig, C. and Wijk, G. van (1994) Inter-organizational collaboration: Beyond contracts. In *Workshop on Schools of Thought in Strategic Management: Beyond Fragmentation?* ERASM/Erasmus University, 14–15 December (internal report).

Kumar, K. and Dissel, H. G. van (1996) Sustainable collaboration: Managing conflict and collaboration in inter-organizational systems. *MIS Quarterly,* 20 (3), 279–300.

Lazerson, M. H. (1988) Organizational growth of small firms: An outcome of markets and hierarchies? *American Sociological Review,* 53, 330–52.

Lazerson, M. H. (1995) A new phoenix? Modern putting-out in the modena knitwear industry. *Administrative Science Quarterly,* 40, 34–59.

Lee, A. S. (1989), A scientific methodology for MIS case studies, *MIS-Quarterly,* 13 (1) 33–50.

Lee, A. S. (1991) Integrating positivist and interpretive approaches to organizational research, *Organization Science,* 2 (4), 342–65.

Lee, A. S. (1994) Electronic mail as a medium for rich communication: An empirical investigation using hermeneutic interpretation. *MIS Quarterly,* 18 (2), 143–57.

Lorenzoni, G. (1987) Costellazione di Imprese e Processi di Sviluppo. *Sviluppo e Organizzazione,* 102, 59–72.

Lucas, H.C. (1975) *Why Information Systems Fail?* Columbia University Press, New York, NY, 1975.

Malaparte, C. (1994) *Maledetti Toscani.* Leonardo, Milano

Malone, T. W., Yates, J. and Benjamin, R.I. (1987) Electronic markets and electronic hierarchies. *Communications of the ACM,* 30 (6), 484–97.

Malone, T. W. and Rockart, J.F. (1991) Computers, Networks, and the Corporation. *Scientific American,* pp. 92–9, (September).

Markus, L. and Robey, D. (1983) The organizational validity of management information systems. *Human Relations,* 36, 203–26.

Martin, P. Y. and Turner, B.A. (1986) Grounded theory and organizational research. *The Journal of Applied Behavioural Science,* 22 (2) 141–57.

Mattsson, L. G. (1984) An application of a network approach to marketing: Defending and changing market positions. In Dholakia, N. and Arndt, J. (eds), *Changing the Course of Marketing, Alternative Paradigms for Widening Marketing Theory,* JAI Press, Greenwich, Conn.

Mattsson, L. G. (1988) Interaction strategies: A network approach. In *Proceedings of the AMA Marketing Educators Conference,* Summer, San Francisco, CA.

McFarlan, F. W. (1984) Information technology changes the way you compete. *Harvard Business Review,* pp. 98–103, (May–June).

Miles, R. and Snow, C. (1986) Organizations – new concepts for new forms. *California Management Review,* 28 (3), 62–73.

Moss-Kantor, R. (1994) Collaborative advantage: The art of alliances. *Harvard Business Review,* pp. 96-108, (July–August).

Nonaka, I. and Takeuchi, H. (1995) *The Knowledge Creating Company.* Oxford University Press, New York.

Ojelanki, N. and Lee, A. (1997) Communication richness in electronic mail, critical social theory and the contextuality of meaning. *MIS Quarterly*, 21 (2), 145–59.

Origo, I. (1957): *The Merchant of Prato: Daily Life in a Medieval Italian City*. Penguin Books, London.

Orlikowski, W. J. (1993); Case tools as organizational change: Investigating incremental and radical changes in systems development *MIS Quarterly*, 17 (3), 309–40.

Osta, J. van (1992). Het labyrint van de politiek. In *Italie and Italie. Cultuurhistorische hoofdstukken uit het naoorlogse Italie*, Meulenhoff, Amsterdam, pp. 9–32 (in Dutch).

Parsons, T. and Shils, E.A. (1951) *Toward a General Theory of Action*. Harvard University Press, Cambridge, Mass.

Pfeffer, J. and Salancik, G. (1978): *The External Control of Organizations: A Resource Dependence Perspective*. Harper & Row, New York, NY.

Porter, M.E. (1985) *Competitive Advantage: Creating and Sustaining Superior Performance*. Free Press, New York, NY.

Porter, M.E. and Millar, V. (1985) How information technology gives you competitive advantages. *Harvard Business Review*, 63 (4) 149–60.

Procacci, G. (1973) *History of the Italian People*. Penquin Books, London.

Piore, M. and Sabel, C. (1984) *The Second Industrial Divide: Possibilities for Prosperity*. Basic Books, New York.

Reekers, N. and Smithson, S. (1995) The impact of electronic data interchange of interorganizational relationships: Integrating theoretical perspectives. In Nunemaker, J. F. and Sprague, R. H. (eds), *Proceedings of the 28th Annual Hawaii International Conference on System Science*, Vol. IV, pp. 757–66.

Rieser, V. and Franchi, M. (1986) *Innovazione technologica e mutamento organizzativo nell'impresa artigiana: Una ricerca sull'artigianto metalmechanicco di produzione nella provincia di Modena*. Unpublished manuscript reported by Inzerilli (1990), on file with other survey results with the Confederazione Nazionale dell' Artigianato of Modena.

Ritaine, E. (1990) Prato: An extreme case of diffuse industrialization. *International Studies of Management and Organization*, 20 (4), 61–76 (Winter).

Short, J. Williams, and Christie, B. (1976) *The Social Psychology of TeleCommunications*. Wiley, London.

Simon, H.A. (1981); *The Sciences of the Artificial*. 2nd edition, MIT Press, Cambridge, Mass.

Snow, C.C., Miles, R.E. and Coleman, H.J. (1992) Managing 21st century network organizations. *Organizational Dynamics*, pp. 5–20, (Winter).

Swanson, E.B. (1988) *Information Systems Implementation*. Irwin, Homewood, Ill.

Tichy, M.N., Tushman, M.L. and Fombrun, C. (1979) Social network analysis for organizations. *Academy of Management Review*, 4, 507–19.

Thorelli, H.B. (1986) Networks: Between markets and hierarchies. *Strategic Management Journal*, 7, 37–51.

Toennies, F. (1965) *Community and Society*. Harper & Row, New York, 1965.

Tzu, Sun (1995) *The Art of War for Executives*. D.G. Krause (ed.), Nicholas Brealey, London.

Trompenaars, F. (1993) *Riding the Waves of Culture: Understanding Cutural Diversity in Business*. The Economist Books, London.

Van Maanen, J. (1988) *Tales From the Field*, University of Chicago Press, Chicago, Ill.

Walls, J. (1993) Global networking for local development: Task focus and relationship focus in cross–cultural communication. In Harasim, L.M. (ed.), *Global Networks: Computers and International Communication*, The MIT Press, Cambridge, Mass.

Ward, W. (1990) *Getting It Right in Italy – A Manual for the 1990s*. Bloomsbury, London.

Williamson, O.E. (1975) *Markets and Hierarchies: Analysis and Anti–trust Implications*. Free Press, New York.

Williamson, O.E. (1981) The economics of organization: The transaction cost approach. *American Journal of Sociology*, 87 (2), 233–61.

Williamson, O.E. (1985) *The Economic Institutions of Capitalism: Firms, Markets, Relational Contracting*. Free Press, New York, 1985.

Williamson, O.E. (1991) Comparative economic analysis: The analysis of discrete structural alternatives. *Administrative Science Quarterly*, 36, 269–96.

Williamson, O.E. and Ouchi, W.G. (1981) The markets and hierarchies program of research: Origins, implications, prospects. In: Joyce, W. and Van der Ven (eds), *Organisational Design*, Wiley, New York, pp. 347–70.

Zack, M.H. (1993) Interactivity and communication mode choice in ongoing management groups. *Information Systems Research*, 4 (3), 207–39.

Appendix A

Proposition from techno-economic perspective	Observations	Revised explanations based upon alternative interpretations
Value of IS/IT in Inter-organizational relationships: IS/IT, by reducing transaction costs, contributes to interorganizational relationships. Thus, in an inter-organizational context IS/IT would have high added value.	Though initially there was interest in the system, once minimal use charges were proposed, membership in the system fell precipitously and ultimately the SPRINTEL was cancelled. Members observed that it was "not worth the money."	• Trust, by reducing opportunism, had already minimized transaction costs, and thus the added value of IS/IT in reducing transaction costs was minimal. • Culturally, the Pratesians are more relationship oriented, and prefer richer relationship oriented communication. SPRINTEL provided primarily task-oriented communication and was poor from a relationship maintenance perspective. Thus it may actually have provided negative value. • Initial acceptence of SPRINTEL could be explained by the relatively high levels of "masculinity," Giovanni effect, and tendency to exhibitionism
The key characteristics of Dynamic Networks: Organizations become vertically disaggregated. Brokers have a key role in bringing together the parties in the network. Markets provide the key co-ordination mode. IS/IT based systems substitute for trust-building processes.	• The large firms, which had been built up during the period between the World Wars, at the end of WWII broke up into small firms, each firm specializing in one or a few steps of the production process. • *Impannatori* in Prato perform the role of brokers. • Markets seem only to have a limited role within the Prato network. Moreover, the network structure and the high variability of the large number of orders in Prato would suggest high transaction costs and therefore vertical re-aggregation which was also not observed in Prato. • IS/IT-based systems ultimately failed to substitute for relationships in the Pratesian Network.	• Consistent with the theory. In addition, historically, Pratesian entrepreneurs had always been accustomed to a disaggregated organizational model. • Consistent with the theory. In addition, historically the role of *impanatore* has existed in Prato for a long time. • Trust and relationship based coordination mechanisms substitute for both vertical hierarchies and market-based coordination mechanisms. Transaction costs are lowered through trust. • The cultural, historical context of Prato, and the existence of trust and relationships turned out to be stronger than the potential of information technology in substituting for these trust-based relationships.
Factors for IT/IS implementation success: The following factors are considered critical to the success of IS/IT implementations: • Technology should serve a perceived need. • Support from key stakeholders. • Adequate resources for development and implementation. • Use of experienced developers. • Detailed analysis of requirements and specification prior to implementation. • Risk averse implementation strategy including a pilot test. • Use of proven technologies. • Users receptive to technological innovations. • User training prior to implementation.	• All these factors were present in the case of Prato. In the short run there was a high adoption level of the IOS technology; however, in later stages the use of the technology declined very rapidly and the whole project was abandoned.	• The technical–economic success factors provide at best necessary conditions but are not sufficient guarantors of success. • Perceived need was not a real need – see above discussion about real value of IS/IT in the Prato situation. • Given the close-knit Prato culture, Reseau, the consulting firm that developed the system, and even ENEA who sponsored the system, were seen as outsiders, and their solutions were in the end not easily accepted because they were based on wrong (technical–economic) assumptions. • Thus, in addition to the rational success factors described in traditional implementation literature, a variety of culture specific factors could influence the ultimate success or failure of a system.

Reengineering the Dutch Flower Auctions: A Framework for Analyzing Exchange Organizations

Ajit Kambil and Eric van Heck

Information Systems Department, Stern School of Business, New York University, 44 West 4th Street, New York, New York 10012, akambil@stern.nyu.edu
Department of Decision and Information Sciences, Rotterdam School of Management, Erasmus University Rotterdam, P.O. Box 1738, 3000 DR Rotterdam, The Netherlands, eheck@fac.fbk.eur.nl

Abstract

This paper specifies a generalizable model of exchange processes and develops a process-stakeholder analysis framework to evaluate alternative market designs. This framework is applied to analyze a number of information technology initiatives in the Dutch flower markets. The Dutch flower auctions are the world's leading centers for trading cut flowers and potted plants. We undertake a cross-case analysis and apply our framework to analyse successes and failures in the introduction of new IT-based trading mechanisms in these markets. Based on our study, we develop a number of testable propositions on: the separation of physical and informational processes in trading, the responses of stakeholders to changes in available information due to IT initiatives, and economic and incentive conditions required for adoption of new trading processes. Finally, our detailed cases illustrate the institutional and incentive constraints, and complexities encountered in the introduction of new electronic markets.
(*Electronic Markets; Transaction Costs; Reengineering; Technology Adoption*)

1. Introduction

Information technology enables new ways of coordinating the exchange of goods and services. Improved communications and processing systems, such as the Internet and video conferencing, allow individuals and organizations to radically re-engineer existing trading processes enabling new forms of electronic commerce and markets.

Markets are real or virtual meeting places where buyers, sellers, and intermediaries meet to exchange or transfer property rights from one party to another. Typically, buyers and sellers confront many different uncertainties and risks in trading. These include the inability to forecast or plan for the future given the traders' bounded rationality and market uncertainty, combined with the potential for opportunism given the divergent interests of different market participants (Williamson 1975, 1985). Markets and other governance mechanisms provide specific information processing capabilities to exchange parties and serve to generate information and trust to reduce the uncertainty and risks inherent in the exchange process (Kambil 1992). To reduce uncertainty and risk, markets as institutions provide specific routines and procedures for coordination, and for the information and guarantees that enable the buyers and sellers to undertake the exchange.

Prior research on the effects of information technology (IT) on exchange organizations and processes typically applied transaction costs and agency theory to predict shifts from hierarchies toward market or other intermediate forms of organizations (Bakos 1991a, Johnston and Lawrence 1988, Malone et al. 1987, Powell 1990). A central argument of these articles was that information technology would improve communications, searches, monitoring, and information-sorting capabilities to reduce transaction costs and allow purchasers to take advantage of production economies available in markets. A critical drawback of this analysis was the definition of markets in abstract economic terms (i.e., markets coordinate economic activity through a price mechanism) without consideration for differences in market organization. For example, some different market types include direct search markets where partners directly seek each other out (e.g., a typical retail fruit and vegetable market), brokered markets where exchange parties employ agents to seek compatible partners (e.g., private placement offerings by investment banks), dealer markets where dealers as intermediaries hold, buy, and sell product inventories (e.g., NASDAQ), and auction markets where traders transact directly through a centralized intermediary (e.g., Sotheby's art auctions). Each of these mechanisms organizes the trading process and related information processing activities in different ways. Thus, we can expect the impact and role of IT can vary across types.

More recently, information systems researchers have examined the impact of information technology on market institutions. Konsynski et al. (1989, 1990, 1992) provided a number of descriptive case studies of electronic markets providing a rich case base for research. Clemons and Weber (1990, 1991) examined the effects of computerization on the London Stock Exchange, and Lee (1993) provided a technology to support order matching of complex products. These studies begin to consider the importance of different market types and institutional histories. Despite the increasing number of cases on electronic markets, a common language and categorization of key processes in a trading system is lacking. Such a categorization is a precondition to building a framework to systematically evaluate and compare the structure and benefits of different markets for various stakeholders. Building on transaction costs and information-processing models (Galbraith 1974, Tushman and Nadler 1978), this paper provides both a model of key exchange processes and a process-stakeholder analysis framework for comparing across different market structures. Such a framework is important to managers who must select and design new trading models. It is also important to researchers who want to develop models of how information technology affects exchange organizations and create theories that can be generalized across cases. We illustrate and apply this framework to evaluate different electronic commerce initiatives in the Dutch flower markets. The framework for analysis developed in this paper can apply to the analysis of other market structures.

2. Exchange: A Process Stakeholder Analysis

Exchange is a critical subject of study in marketing, economics, finance, and information systems. Building on these disciplines, information systems researchers have typically adopted a number of outcome dimensions to characterize and evaluate markets. For example, Clemons and Weber (1990) characterize financial markets in terms of liquidity, volatility, and transparency. Liquidity is the ability of the market to absorb large orders without significant price changes, indicating the depth and extent of market participation by buyers and sellers. Volatility refers to the variance in day-to-day volumes and prices, and transparency refers to keeping the widest group of buyers informed on current prices and allowing them access to trading. Highly liquid markets with low volatility and high transparency provide stable forums for exchange.

While these outcome characterizations of different market institutions are important, they are not easily generalizable across settings. For example, liquidity is a key indicator of market quality for financial markets but it is an inappropriate measure in the case of Dutch flower auctions. First, the size of any one flower order is limited, so it is hard to evaluate the impact on price of large orders. Second, the flowers are sold from the auction to the buyer without further resale among the buyers. Common outcome characterizations such as liquidity do not apply to all market exchange relations. Outcome characterizations also do not help managers evaluate the potential impact of a new information technology innovation on the operations of a trading system, or the costs and benefits to different stakeholders.

A model of key exchange processes is necessary to provide a reference point for comparing the organization of different market institutions. Traditional process classifications of exchange include search, bargaining and negotiation, monitoring, and enforcement processes (Coase 1960, Dahlman 1979).[1] However, this level of classification does not provide sufficient detail for the comparison of alternative market structures. What is required is a classification of exchange processes that is generalizable, focuses attention on salient dimensions of an exchange relation and provides sufficiently detailed information for effective classification of exchange features into independent process categories. We adapt work by Kambil (1992) to build such a classification scheme grounded in transactions cost and information-processing theories.

2.1. A Model of Exchange Processes

Kambil (1992) identifies ten distinct processes that can operate in an exchange relation, providing a basis for a finer grained analysis of market structure and costs. This categorization also recognizes that basic trade processes exist within a wider context of support processes that implement dispute resolution, and other mechanisms to ensure that counterparties to trades will meet their obligations. We adopt, refine, and extend his model into five trade processes (search valuation, logistics, payments and settlements, authentication) and five trade context processes (communications and computing, product representation, legitimation, influence, and dispute resolution). The basic trade processes are distinct processes required in all transactions of goods and services. The trade context processes facilitate and enable or reduce the costs of or "frictions" in the basic processes. Figure 1 illustrates these processes, which are defined and discussed in Tables 1 and 2.

The ten process categories described in Tables 1 and 2 provide parties to an exchange relation with the information processing, coordination, and influence mechanisms required to mitigate uncertainty and exchange-related risks. While we cannot prove that our categorizations define all processes related to exchange, we propose that these categories define the key processes underlying trades. The categories allow systematic identification of processes affected by a specific administrative or technological innovation and enable us to systematically examine and represent the impacts of information technology across interdependent processes and stakeholders in an electronic market.

The implementation and operation of the above processes and systems create transaction costs for different stakeholders to the exchange. Transaction costs and the complexity of these processes increase as products and production become more complex, and as the market environment becomes more uncertain. Transaction costs also increase as exchange parties establish more complex mechanisms to defuse opportunism risks. Traditional economics postulate that firms will choose to organize exchange relations so that they achieve the lowest possible combined transaction and production costs. This would suggest the trading mechanisms with the lowest sum of transaction costs across all the processes will evolve and dominate the market. However, real-market institutions are complex, constituting a meeting point for stakeholders with both convergent and divergent interests. Innovations by a specific stakeholder may reduce overall transactions costs but can increase the costs to other parties. This can result in the failure to adopt a specific technology or administrative innovation in the marketplace. Thus we propose the need for a process-stakeholder analysis to systematically evaluate the effects of technological innovations on specific processes as well as costs and benefits that arise from the innovation on different stakeholders.

Below we develop and illustrate the utility of a process-stakeholder analysis in focusing attention on critical features of trading mechanisms, by using it to systematically represent and evaluate the effects of various information technology initiatives in the Dutch flower auctions (DFA).

3. The Dutch Flower Auctions

The Netherlands is the world's leading producer and distributor of cut flowers and potted plants. This industry consists of about 11,000 growers and nearly 5,000 buyers. Growers typically are family businesses, while buyers represent both large and small wholesalers and retailers. The Dutch

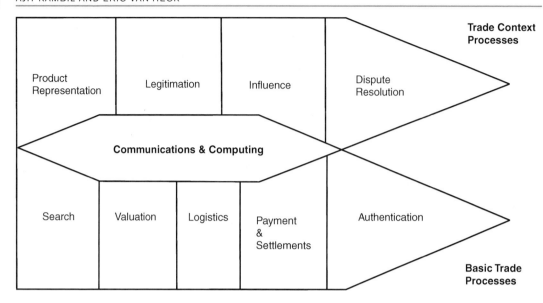

Figure 1 Exchange processes.

Note
Reprinted by permission from: "Doing Buiness in the Wired World," Ajit Kambil, *IEE Computer*, May 1997. © IEEE 1997.

dominate the world export markets, with 59% world market share for cut flowers and 48% world market share for potted plants in 1996. The Dutch flower auctions established and owned by grower cooperatives play a vital role in Holland's leadership of this industry by providing efficient centers for price discovery, and the exchange of flowers between buyers and sellers. The world's two largest flower auctions are the Flower Auction Aalsmeer (VBA) and the Flower Auction Holland (BVH) in Holland. These auctions trade over 30 million flowers and conduct nearly 60,000 transactions daily, generating together over $2.4 billion U.S. in annual sales of cut flower and potted products from the Netherlands and other producers such as Israel, Kenya, and Zimbabwe. Annual sales of the seven Dutch flower auctions exceeded $2.9 billion U.S. in 1996.

The flower auctions use the "Dutch" auction method for price determination. This method uses a clock, with the clock hand starting at a high price determined by the auctioneer and dropping until a buyer stops the clock by pushing a button to bid for a lot at the price determined by the clock hand. This method was invented by a Dutch cauliflower grower in the 1870s to trade agricultural products. It reduces the time spent by growers at markets and frees them from the task of bidding and price determination, thereby allowing them to focus on production. It also reduces the time buyers spend in bidding or bargaining for goods.

The flower auctions provide a central location for the meeting of buyers, with suppliers allowing for efficiencies in quality control, logistics, and product redistribution. Flower Auction Aalsmeer (VBA) near Schiphol airport and the Flower Auction Holland (BVH) near Rotterdam are located close to major transportation facilities and have some of the largest commercial buildings in the world. Both auctions spread across the equivalent of about 100 soccer fields, each with the capacity to host 2,000 buyers. At the VBA the flower auctions occur in different auction rooms under 13 different clocks. The computerized clock in the auction hall provides buyers with information on the producer, product, unit of currency, quality, and minimum purchase quantity. The flowers are transported through the auction hall and shown below the clock to buyers. When the buyer stops the clock, the auctioneer uses the intercom to ask the buyer how many units of a lot he or she will purchase. The buyer provides the bid and, when the bid is accepted, the clock is reset and the process is restarted to clear the remaining inventory. The

Process	Description of Basic Trade Processes
1 Search	The information gathering and evaluation process undertaken by buyers and sellers to identify opportunities. A trading opportunity consists of a supplier willing to offer a good and a buyer interested in purchasing the good. Thus search processes identify potential suppliers, goods, and customers willing to purchase the goods. Different market mechanisms implement varied methods and venues to identify trading opportunities. Different search methods allocate different search costs to buyers, sellers, or intermediaries.
2 Valuation (Price Negotiation)	The process and methods of negotiating and discovering a purchase or sale prices for a product. A variety of different price discovery and bidding processes exist that differentially attributes costs to buyers, sellers and intermediaries. For example, various auction methods are specified in Davis and Holt (Davis and Holt 1993) for finding prices. Price discovery mechanisms can also be biased to shift surplus from the trade to specific transaction stakeholders.
3 Logistics	The processes specifying and coordinating the actual delivery of goods from the seller to buyer. Again, in a trade the direct costs of delivery may be assumed differently by different players.
4 Payment and Settlements	These processes define the terms and methods of payment and ensure the settlement of payments in the exchange. Third parties often provide the infrastructures for exchange.
5 Authentication	This is a core set of activites used to verify the quality and features of the product offered, the authenticity of the trading parties, and monitor conformance to the contract or agreement among parties. Authentication may involve third parties who provide credit check, notarization, and other services to reduce the uncertainty of buyers and sellers. Autheniticating the quality of a product prior to purchase mitigates against adverse selection risks (Akerlof 1970, Klein et al. 1978). Monitoring processes in a complex transaction work to ensure that the trade is successfully completed and mitigate against moral hazard risks. Third parties to the transaction may play a critical role in monitoring transactions, ensuring nonrepudiation and competion of agreements.

Table 1 Description of basic trade processes.

auctioneer may introduce a new minimum quantity to ensure the entire lot clears. Each clock can handle nearly 1,000 transactions an hour (approximately one transaction in four seconds). On average, 15 million flowers are traded daily at the VBA in 30,000 transactions.

We investigated the use of information technology in these auctions over the span of three years, collecting data through interviews with auction officials and participants and through review of all relevant published reports and archival material. Next we used the process categories identified in the previous section to organize the information and systematically evaluate or compare the impacts of specific IT innovations and process changes on different stakeholders in the DFA. This process-stakeholder analysis was used to identify changes in benefits and costs of different exchange processes, and the relative value of these innovations to different stakeholders. A case method was used because it enables rich data and "reality" to be captured in greater detail than other methods. The case method also permits us to consider a greater number of explanatory variables. We then undertook a cross-case analysis on multiple IT initiatives in the Dutch flower markets to generate propositions that explain the success or failure of specific innovations and more general impacts of new information technologies on markets. While we acknowledge that a single case cannot be generalized into a theory, a cross-case analysis is a useful step in theory construction by identifying patterns that provide testable propositions (Eisenhardt 1989, Yin 1981).

Process	Description of Trade Context Processes
6 Communications and Computing	Communications and computing processes underlie and bind all trading processes. New communications capabilities in terms of richer media, faster transfer speeds, improved ease of use and lower infrastructure costs transform coordination capabilities within and across trade processes. Similarly, improved processing, storage, input-output, and software technologies, further transform the coordination and decision making capabilities of stakeholders in each process. These transformations change the specific transaction costs and opportunity costs perceived by buyers, sellers, and intermediaries.
7 Product Repesentation	Product representation processes determine how the product attributes are specified to the buyer or other third parties. To what extent is there a standardized product and quality specification language? Both the buyers' and sellers' costs are reduced by informational standards or well-specified languages to represent the key quality, functional and aesthetic attributes of the product. If the product is well represented, the uncertainly of the buyer is lower. The costs and methods of implementing a uniform product representation scheme vary in different market mechanisms.
8 Legitimation	This process is used to validate the trade or exchange agreement. It defines how bids, or agreements are recognized as valid and binding commitments by exchange parties. In different trading situations, different procedures and rituals are undertaken to legitimate a transaction. Often this is done by announcing the agreement, and participants to the agreement to a broader audience through a formal process.
9 Influence Structures and Processes	These processes implement mechanisms to enforce obligations or penalties to reduce opportunism risks. Kambil (1992) defines two distinct influence processes developed from social exchange and transactions cost theories. These are: incentives and sanctions, and credible commitments. Incentives and sanctions include performance rewards for completing a contract on time, or penalties for misrepresenting products, or denial of access to an exchange network. A reputational system defines a special class of incentives and sanctions. Actions taken to enchance and create a good reputation serve to create trust in the party. When a social system establishes a reputation as a valued asset to sellers or buyers, opportunistic behaviour that can sully the reputation becomes more costly. Investment in credible commitments (Williamson 1985) or asset-specialized investments by both exchange parties creates hostages such that the loss of the relationship will create a significant cost for both parties. This serves to signal commitment to the relationship by both parties and mitigates against opportunism risks.The types of commitments, the importance of reputation, and institutional authority to create incentives and sanctions can serve to reduce risks and the costs associated with other processes.
10 Dispute Resolution	All exchanges occur in a broader legal or institutional context which provide various processes for resolving disputes among parties and structures decision rights in the event of conflict. Firms may elect to directly resolve disputes through negotiation, commit to third-party arbitration or court ordering of disputes. The latter is typically the most expensive alternative. Dispute-resolution processes adopted in an exchange relation, and the allocation of decision rights to structure solutions to disputes are crucial to reducing risks for buyers and sellers. When goods are exchanged, disputes can be resolved by replacement of a defective product, and warranty mechanisms.

Table 2 Description of trade context processes.

3.1. An Evaluation of the Traditional Dutch Flower Auction

Utilizing the previous process categories, Table 3 summarizes the key features, strengths, and weaknesses of the traditional Dutch flower auction. This provides a baseline for comparision and evaluation of process changes enabled by information technology.

The Dutch flower auctions provide buyers and sellers with a highly efficient infrastructure to support exchange. As summarized in Table 3, they provide efficient search, communication, and product representation capabilities for highly varied products. The auctions are also very efficient in determining price, enabling approximately 1,000 transactions per clock per hour on a low-value good. The auction method eliminates problems of haggling and reduces bargaining costs that are otherwise associated with the trade of nonstandard and unusual-value products (Milgrom and Roberts 1990). By operating at a high clock speed, encouraging competition among buyers, and setting a fixed time by which to complete a transaction (or the clock goes to zero), the auction enables efficient trading. The hub and spoke operation and the logistics within the auction hall allow the timely transfer of a large volume of flowers from suppliers to buyers. By intermediating all buyer-grower exchanges, the auction also provides efficient quality control, settlement, and dispute resolution mechanisms that mitigate against opportunism risks encountered by buyers and sellers. These features reduce transaction costs to growers and buyers.

However, as summarized in Table 3, there are disadvantages of this specific model of trading. First, buyers must be physically present at the correct location and correct time to bid on a specific product. In addition, it is hard to simultaneously search other markets to estimate prices. Today, large buying organizations provide cellular phones to their buyers to coordinate purchases across auctions. In response, the auctions are beginning to coordinate trading similar products at the same times in different auction houses. This will allow the large buyers to find the best price across auctions. The auction structure also inhibits growers from understanding the true preferences of buyers, as growers are separated from the sale of their produce by the auction.

Other disadvantages for specific stakeholders are understood when we consider the auction's ownership and incentive structure. A number of auction rules favor growers, who are the auction owners, over buyers. First, the clocks move at a high speed. Higher speeds reduce the decision time available to buyers. This results in higher bid prices. The auctions have experimented with different clock speeds and it is widely known that faster clock speeds result in higher prices. However, the auctions are unwilling to fully disclose the results of their experiments and selection of auction clock speeds. Second, the auction rules and service costs to buyers for processing trades favors trading in smaller lots instead of purchases of large lots. This ensures that no one party can purchase the entire lot without competition. It also increases the cognitive complexity and competition confronted by buyers. Buyers must now purchase products and adjust purchasing decisions to availability across multiple clock auctions. This imperfect information and low transparency about available inventories favors growers, although more empirical work needs to be done to discover the magnitude of the growers' advantage.

3.2. Challenges to the Traditional Dutch Auctions

In the early 1990s the Dutch flower industry and auctions faced a number of challenges. First, rapid growth in the flower industry created increasing demands for logistics support, space in the auctions and complaints about nearby traffic congestion. As the two major auctions were close to their limits in terms of complexity, capacity, and room to expand, increases in capacity would require changes in auction logistics or investment in new facilities. A second problem confronted by the industry was the consolidation of buyers into larger entities. Buyers for supermarkets and large retail store chains wanted to purchase larger lot sizes and coordinate purchases across different markets to take advantage of the best price. Some large retailers, like Marks and Spencer, were bypassing the Dutch auctions and their commissions to source directly from growers in Spain and other countries. Third, the success of the Dutch auctions as the world's leading flower auctions and entry point into downstream logistics and distribution led to an increased flow of foreign cut

Exchange Process	Implementation	Strengths	Weaknesses
1 Search	All trading opportunities are found at the auctions.	Search costs are reduced as the centralized auction aggregates major trading opportunities in one location.	The buyers have to come to the auction halls, and sellers have to deliver products to the auction hall.
2 Valuation	The Dutch auction clock method.	Very efficient for small lot trading. All transactions occur in a fixed time slot leading to 1,000 transactions per hour.	Auction prices for specific products decrease during the day. To assure fairness, assignment of auction sequence for different growers is done by lottery. Clock speeds and small lots stretch the cognitive capabilities of buyers favoring higher prices for growers.
3 Logistics	The auction facilities are a central hub providing logistics support for transferring products from sellers to buyers. These costs are shared among all buyers.	Very efficient transfer of product from seller to buyers. The centralized auction trading allows specialization, transportation, and other resources.	Packaging costs are incurred multiple times—for transport to and from the auction. Multiple handling of flowers can damage them.
4 Payments and Settlements	The auctions provide systems for order tracking, payment, and settlements with efficient one-day settlement periods.	Shared costs among auction participants and economies of scale in these systems.	
5 Authentication	The auction house authenticates participants and grades the quality of the products. It is also responsible for tracking and ensuring delivery of orders to buyers.	Very efficient for large numbers trading. The auction reduces counter-party risks, product, and related authentication costs for buyers and sellers.	Buyers and growers perceive quality grades as too broad, artificially inflating valuation of products at the lower end of quality rating.
6 Communication and Computing	Simple visual communications of competitors, product, price, and other trading information on the clock. A simple computerized system for communicating bids.	Efficient and low costs for shared communications infrastructure.	Requires synchronous communitions for trading, and colocation of parties to the limited trading floor. Growers do not know final demand patterns for products.
7 Product Representation	The product is represented by itself, and simple codes identifying grower and a gross level quality grade.	Buyers can directly inspect the product in the auction hall if they want to do so.	Buyers and growers perceive quality grades as too broad. The cost of visual inspection.

Exchange Process	Implementation	Strenths	Weaknesses
8 Legitimation	The auction intermediary is responsible for recording bids and legitimating transactions.		
9 Influence	The auction can exclude growers or buyers who do not meet various criteria.	The auctions, by centralizing and requiring the product to be delivered prior to sale, minimizes opportunism risks.	Auction rules tend to favor the growers.
10 Dispute Resolution	The auctions provide for various arbitration mechanisms and rules for resolving problems. This reduces dispute resolution costs.		

Table 3 Features of the traditional Dutch flower auctions.

flowers to the Dutch flower auctions. In 1994, 15% of flowers traded at the largest Dutch Flower auctions were imported from and re-exported to foreign countries (sometimes back to the country of origin). This increased foreign competition to Dutch growers, and in 1993 the import of foreign rose stems to the two largest auctions reduced average prices by 40% during the winter season. Subsequently, the Dutch grower cooperatives voted to limit foreign access to the major auctions in 1994. These challenges led to a number of successful and unsuccessful information technology-enabled transformations of trading in the industry. These initiatives are critically examined below, using a process-stakeholder analysis, to illustrate major process changes and benefits or costs of the initiatives to different stakeholders.

4. IT-Enabled Dutch Auction Transformations: A Process-Stakeholder Analysis

Four major information technology-enabled initiatives were undertaken to transform the Dutch flower auctions and respond to the above challenges. Two unsuccessful initiatives were the BVH's Holland Vidifleur system and the VBA's Sample-Based Auction (SBA) system. These initiatives promised to simplify auction hall logistics and add extra transaction capacity to the auction. Two successful initiatives were the Holland Supply Bank (HSB) system and the Tele Flower Auction System (TFA). The former system responded to the need of large buyers, and the latter provided a new auction for foreign produce. While the auctions have used information technology extensively to support order processing, inventory, and logistics, they primarily supported or automated existing auction processes having impacts typically limited to the auction intermediary (Van Heck and Groen 1994). The IT-enabled initiatives discussed below fundamentally changed the implementation or conduct of one or more specific auction processes, and the way in which buyers, auctions, and sellers interact. Summary information on the initiatives is given in the Table 4.

4.1. The Vidifleur Initiative

The BVH auction implemented the Vidifleur initiative in response to the growing capacity and logistics requirements of the auction. Vidifleur intended to use video auctioning to decouple the price determination and logistics mechanisms and to allow buyers to trade from outside the auction hall. When the product arrived at the auction a picture was taken, digitized, and stored in auction computers. These computers transferred the picture for display to a screen in the auction hall, where buyers could bid for the product based on the image of the product. Buyers were also able to bid for and look at the flowers on computer screens in their private auction offices. The computers in the private office provided a screen-based representation of the clock which was synchronized with the clock in the auction hall.

The Vidifleur experiment initiated by the BVH information technology staff (Spooner and Copeland, 1992),[2] was implemented with minimal changes to the original auction clock process. Experimenters converted one of three auction clocks in a room to have video screens for product display around the clock. The auction also moved the real product under the auction clock, providing a second visual display of the product. An auction room solely dedicated to this experiment was unavailable because of the demand for limited auction facilities, and altering existing logistics processes was deemed too expensive for the experiment.

Auctioneers expected this remote video auctioning to provide buyers with better information, as they could work from the comfort of their own offices and also access their office computers for purchasing, order, sales, and local inventory information. The auctioneers also knew that buyers often tended to select goods from specific sets of growers rather than inspect the product in great detail and expand the selection of producers. This buyer behaviour suggested that grower reputations played a substantial role in shaping purchases, and that the visual inspection of flowers prior to purchase was less important to the process. Despite these expectations, buyer reaction to screen-based trading was negative and led to the termination of the experiment in late 1991. Table 5 summarizes the areas of key process changes, benefit expectations and actual results of the initiative. We focus only on buyers and the auction intermediary, as no major process changes occur for sellers.

As highlighted in Table 5, buyers cited three main reasons for not adopting the new system. First, the clock-based trading system provided no new efficiencies for buyers. Second, the quality of the auction hall video display was perceived as poor, and trading from outside the auction hall created an informational disadvantage. In floor-based trading the buyers could observe each other, and the reactions of other major buyers (from supermarket chains, etc.) to specific bids. This important nonprice information was incorporated into the decision making of the buyers. Video-based auctioning on a computer was a limiting medium, not rich enough to capture this information. Thus buyers had limited advantage in trading from their offices, when there was also a floor-based auction which provided the traders with more information. Indeed, buyers who tried to trade from offices complained that they could not feel the "tension" in the marketplace. Third, at the back of each auction hall is a coffee shop

	Traditional DFA	Holland Vidifleur System	VBA Sample Based Auction	Holland Supply Bank	Tele Flower Auction System
Start (End) Year	1890	1991(1991)	1994 (1994)	1993	1993
Product	Flowers and Potted Plants	Potted Plants	Poted Plants	Flowers and Potted Plants	Flowers
Sellers	Growers	Growers	Growers	Growers	Growers
Intermediary	Flower Auctions	Flower Auction Holland (BVH)	Flower Auction Aalsmeer (VBA)	Flower Auction Holland (BVH)	East African Flowers (EAF)
Buyers	Wholesalers	Wholesalers	Wholesalers	Wholesalers	Wholesalers
Valuation	Dutch Auction Clock	Dutch Auction Clock	Dutch Auction Clock	Negotiated Posted Offer	Dutch Auction Clock

Table 4 Dutch flower auction initiatives.

Exchange Process	Auction Intermediary	Buyer
1 Search	No change.	No Change
2 Valuation	No change.	No change.
3 Logistics	Expected: Auction hall would be eliminated allowing for cheaper and more frequent transactions and new space for clocks.	No change.
	Acual: No change during the experiment due to potential disruptions to the current process. Real product appeared in the hall synchronized to clock.	
4 Payments and Settlements	No change	
5 Authentication	No change in the quality grading process.	In a full-scale system, buyers would find it harder to visually inspect and authenticate quality.
6 Communication and Computing	Shift to electronic communications. Electronic representation of the clock as well as the product.	Expected: Electronic communications could substitute for buyer presence in the auction halls.
		Reality: Electronic communications does not capture nonprice information, such as other bidders' expressions and reactions to specific bids. It reduces social interaction opportunities.
7 Product Representation	Expected: Electronic flower clock and video representation of flowers would not adversely affect the sale.	Expected: Buyers would make decisions from their offices using computers rather than the auction hall.
	Result: Negative buyer reaction to video quality.	Result: Negative reaction. Buyer perceive the video quality as poor.
8 Legitimation	No change.	No change.
9 Influence	No change.	No change.
10 Dispute Resolution	No change.	No change.
Net Benefits	None.	Negative.

Table 5 Process-stakeholder impacts of Vidifleur.

where buyers interact informally and share information about the market. Again, access to the social interaction and information was more difficult through screen-based trading.

4.2. The Aalsmeer Sample-Based Auction

In contrast to the BVH, the Aalsmeer auction began a sample-based auction for trading potted plants (Griffioen 1994a, 1994b; Van Heck and Groen 1994). In this model, growers send a sample of the product to the auction house along with information on available inventory. During the auction the sample represents the entire inventory available to buyers who can bid for the product and specify product packaging and delivery requirements. Growers then package the product as specified and deliver it the next day to the buyer location in the auction complex or to other buyer warehouses. Growers, buyers and auctions used electronic data interchange to share all the information required in this process.[3] This trading model reduces the number of times a product is handled, reducing overall packaging costs and damage.

The different actors—the auctions, growers, and buyers—expected a number of different benefits. First, by uncoupling logistics and price determination, the auctions and growers expected the number of transactions per hour to increase. In reality the number of transactions per hour decreased, as buyers had to specify terms of delivery. Second, while the auctions expected 45% of the supply of potted plants to be transacted in the sample-based auction, only 10% of the product was transacted this way. Thus, sample-based auctions also did not effectively reduce storage requirements at the auction. After numerous attempts to increase the volume of sample-based auctions, they were discontinued in late 1994. Table 6 summarizes results of the initiative. Only areas of change are discussed below.

The sample-based auction failed to meet expectations for many reasons. First, the incentives and benefits to buyers and growers (in particular) did not change substantially to encourage their participation in this market. Specifically, growers received no extra compensation for modifying packaging and delivery practices to suit the customer. Second, the growers perceived that they got lower prices in a slower auction. To overcome this disadvantage growers would break the same product into different sample lots so that it would be priced multiple times during the auction, hoping it would lead to higher prices. Third, the auction rules initially did not provide incentives to buyers by supporting transactions on large lots. Instead, the auction maintained rules to favor transactions in small lots. Thus, an insufficient number of buyers and sellers initially adopted this new form of trading (Van Heck and Ribbers 1996).

In response to this failure, the auction undertook various rule changes beginning in March 1994. First, the auction established a price floor of 70% of average price of a flower type in the most recent five days to reduce volatility and downside risks. Second, the *representative* lot was auctioned first. This sample lot typically received a higher price than following lots of the same type and quality. Buyers believed the first sample to always be of best quality within any quality rating. Third, the lot size was increased to three stapelwagens[4] from one. The growers were also forced to increase the number of plants offered in any one transaction, so that they did not game the system. To meet the needs of large buyers, buyers were permitted to buy either one or three stapelwagens per transaction. To prevent gaming, the grower for any type of flower was only allowed one auction per category of product for price determination. These rule changes temporarily stabilized the market, but the lack of growers and buyers adopting the method led to its demise in late 1994.

4.3. The Holland Supply Bank: Image Representations and Negotiated Trading

To respond to the needs of larger buyers, the BVH and Aalsmeer auctions have diversified to create a brokered market for flowers and potted plants. In the "be-middelingsbureau" (BB), or Mediation Office, an auction employee acts as an agent for the growers and negotiates between growers and buyers in a forward market. Prices, product specifications, amount of lots, and delivery specifications are specified in a contract which is legitimized and monitored by the mediation office. In 1994, about 18% of flowers and potted plants were traded this way. In 1996, this amount increased to 22%. The mediation office is useful for the sale of large lots to large buyers, like supermarkets, for the occasions market.

The mediation office, as an honest broker, reduces the costs of search, communications, and bargaining for buyers and sellers. It also provides a mechanism to legitimate the transaction and resolve disputes in the event the contract is not met. Image databases of product types and inventory are especially useful for transactions in the mediation office. Wholesale buyers find that the pictures of the product are also very useful marketing tools to downstream retailers. Electronic data interchange is used for communications of orders and the coordination of settlements, and delivery. This results in fewer errors and greater transaction efficiency.

Table 7 summarizes the major impacts of the Holland Supply Bank on different stakeholders.

Exchange Process	Growers	Auctions	Buyers
1 Search	No change.	No change.	No change.
2 Valuation	Expected no change. Result: Classify into smaller lot sizes so same product is sampled and effectively priced multiple times. Price is on average 10% less on remainder than sample lot. This results in less revenue.	Expected no change. Auction process is slowed down due to the need to specify packing information and buyer gaming. Lower commissions on remainder for sample lot. 70% of previous five days' average price as a price floor.	Greater uncertainty about product quality leads to discounting of nonsample lots.
3 Logistics	Increased burden from needing to customize package to buyer specifications.	Lower logistics costs.	More efficient delivery of the product.
4 Payments and Settlements	No change.	No change.	No change.
5 Authentication	No change.	Minor change ensuring product is properly delivered.	Inability to authenticate quality ahead of time. Must verify product is properly received.
6 Communication and Computing	Minor change: More use of EDI to send coordinination information.	Minor change: More use of EDI to send coordination information.	Minor change: More use of EDI to send coordination information.
7 Product Representation	Enabled to select a sample lot that is of higher quality than the remaining product.	Sample lot substitutes for actual transaction lot.	Buyers cannot visually evaluate the whole offer. Must decide to buy based on the sample lot.
8 Legitimation	No change.	No change.	No change.
9 Influence	More difficult due to more complex transaction.	No change.	No change.
10 Dispute Resolution	No change.	No change.	No change.
Net Benefits	Negative.	Negative.	Negative.

Table 6 Process-stakeholder impacts of the Sample Based Auction.

4.4. The Tele Flower Auction System

Since 1993, the Dutch flower industry has been concerned about the increasing flow of foreign flower products into the Dutch flower auctions and downstream logistics and distribution systems. These imports were having an adverse effect on the margins of Dutch growers and the prices of key commodity products like roses and carnations. After a referendum in September 1994, the growers, who are the owners of the Dutch auctions, decided to ban foreign grower participation in the auctions during the summer. These efforts to reduce foreign access to the traditional Dutch auctions led buyer organizations and foreign growers to announce the creation of competing auctions. In particular, the East African Flower (EAF) importers association responded to

Exchange Process	Growers	Auctions	Buyers
1 Search	Benefit: The HSB permits growers to search for buyers.	Benefit: Efficient image database combined with broker matching system.	Benefit: Large buyers' search costs are reduced.
2 Valuation	Expected no change. Benefit: Slighty higher prices with less uncertainty.	Mediation process is less costly than logistics-intensive process.	Cost: slightly higher prices but reduced uncertainty.
3 Logistics	Benefit: Efficient product transfer in bulk to buyer.	Benefits: Less logistics in the auction hall.	Benefits: Direct delivery from seller with no repacking costs.
4 Payments and Settlements	As specified in the contract.	Managed by mediation office.	As specified in the contract.
5 Authentication	No change.	Costs: Quality managers evaluate sample, and assign a quality grade. Monitor and ensure delivery.	Cost: Inability to authenticate quality ahead of time.
6 Communication and Computing	Minor change: More use of EDI to send coordination information.	Minor change: More use of EDI to send coordination information.	Minor change: More use of EDI to send coordination information.
7 Product Representation	Cost: Growers send data and sample lot. Growers perceive quality grades as broad and don't reflect true quality.	Cost: Quality managers create a digital image of sample lot to present to buyers in addition to the rating.	Cost: Buyers cannot visually evaluate the offer and perceive quality grades are broad. Benefit: However, images are useful for downstream trade.
8 Legitimation	No change.	Mediation office.	No change.
9 Influence	No change.	No change.	No change.
10 Dispute Resolution	No change.	Mediation office.	No change.
Net Benefits	Positive–higher prices and less uncertainty.	Positive–less logistics requirements.	Positive—ability to transact for large lots into the future.

Table 7 Process-stakeholder impacts of the Holland Supply Bank.

this ban by developing the Tele Flower Auction (TFA) (Van Heck et al. 1997).

TFA is a completely electronic auction that enables buyers to trade at a distance. Decoupling the logistics and price discovery processes, the buyers can bid for different flowers via their personal computer (PC) screens (Bos 1995, 1996; Eras 1995; Van Vliet 1994). Using ISDN services of the Dutch PTT, each buyer's PC is connected from his or her office to a fully computerized Dutch auction clock. The PC provides information on the producer, product, unit of currency, quality, and minimum purchase quantity. For added convenience, the PC also provides the buyer with textual information on the available flowers, specific lots for sale, and two digital images of each lot. One image gives an overview of the lot to inspect ripeness of flowers, and the other presents details of the flower bud, to evaluate its size and diagnose potential diseases. The buyer can then use the PC's custom software to flag interesting lots, so that the computer can alert the buyer when it is time to auction the specific lot. On the PC screen the buyer sees the Dutch auction clock. The clock hand starts at a high price and drops until a buyer stops the clock by pushing the space bar at the keyboard of the PC. The auctioneer then asks the buyer via an open telephone connection how many flowers of the lot he or she will buy. When the buyer provides the amount, the auctioneer legitimates the transaction. The clock is then reset, and the process begins for the next lot until the remainder of the product is sold.

The auction logistics are substantially simpler than the traditional DFA. The EAF receives and stores flowers and plants from growers in a distribution warehouse facility in the nearby town of Amstelveen. The flowers from the Amstelveen facility are transported to the wholesaler's address directly by transporters of EAF. In an agreement with the VBA, the EAF was also permitted to deliver products to the buyer's rented facilities at the VBA. This way the auction minimizes most internal logistics operations and is able to deliver flowers to buyers in a timely way. The major process changes relative to the traditional Dutch auction are summarized in Table 8.

Overall, TFA was an immediate success. Since its implementation in March 1995, it has grown from auctioning the product of 2 main East African growers to include products from 40 growers in Africa, Spain, Colombia, France, India, and Israel. By the end of the year, the TFA had nearly 160 buyers (about half are located in the Aalsmeer area; the rest are scattered throughout the Netherlands). The TFA was estimated to be the fourth largest flower auction in Holland by the end of 1996.

Why were buyers willing to accept this new auction? First, it uniquely satisfied the buyers' needs to inexpensively purchase foreign flowers and include them in their existing distribution chains. Second, buyers we interviewed said the flowers they purchased from the TFA often exceeded their quality expectations as determined by the quality grades. Thus, the TFA created a reputation for setting high quality standards, enhancing trust in the auction and minimizing adverse selection. If necessary buyers close to the warehouse location could physically inspect the product. Buyers were also impressed with the computer data, the quality of the products, and the speed of delivery after their purchases were completed. The buyers also felt they had a better overview of the auction, as compared with the traditional Dutch auction.

5. Lessons Learned from IT Innovations in the Dutch Flower Auctions

Based on our research we did a cross-case analysis to identify factors that influence the successful or unsuccessful use of information technology in markets. We use this analysis to develop a series of observations on the market impacts of IT as well as the success and failure of IT-based initiatives in markets. These observations stated as propositions, and the supporting analysis and discussion are given below.

PROPOSITION 1. *The application of information technologies to trading can enable increased efficiency and separation of informational and physical trading processes. This in turn will permit more varied forms of trading, customized to different user requirements.*

The four cases illustrate the use of information technology to separate the informational and physical trading processes. In all cases the valuation and logistic processes are increasingly de-coupled from one another in time and space. For example, the Holland Supply Bank illustrates the implementation of the mediation bureau to broker trading of large futures contracts between large buyers and different growers separating logistics and information processes. Similarly, in the Tele Flower auction, the buyers are able to bid for products from remote locations. The buyers can customize the system by specifying preferences and receiving alerts when the product they want is being auctioned.

These observations also lend support to Malone et al. (1987), who note that information technology enables personalized markets where software agents would support trading customized to individuals and firm preferences. While the new flower auctions are not as fully advanced, they achieve a better fit between the transaction preferences of larger buyers and the organization of exchange processes by overcoming the capacity limits and colocation requirements for floor-based trading. These new trading mechanisms better suit the unique requirements of large versus small buyers.

Exchange Process	Growers	Auctions	Buyers
1 Search	No change from DFA.	Benefit: Efficient databases and online system for presenting trade opportunities.	Benefit: Up-date online database reduces search costs for specific products. Alerts inform buyers of opportunities.
2 Valuation	No change.	Very efficient screen-based auction clock process.	Benefit: Screen trading enables better oversight and concentration on bidding.
3 Logistics	Benefit: Efficient product transfer to Amstelveen Warehouses.	Benefits: No auction hall logistics are necessary. Less errors due to simplicity of a decoupled system. Cost: Transport to buyer locations.	Benefits: Faster delivery from auction due to decoupling of logistics and price discovery.
4 Payments and Settlements	24-hour payment and settlement clearing.	Managed by EAF.	24-hours payment and settlement clearing.
5 Authentication	No change.	Costs: quality managers evaluate sample, and assign a strict quality grade. Monitor and ensure delivery.	Benefit: Auction quality control reduces risks—may exceed buyer expectations. Cost: Inability to authenticate quality ahead of time.
6 Communication and Computing	Simplified phone and data communications.	Simplified phone and data communications.	Text representations and digital images enable simplified communications.
7 Product Representation	Cost: Growers perceive quality grades as broad and don't reflect true quality.	Cost: Text representation of product in addition to the quality rating.	Cost: Buyers cannot directly evaluate the offered product. Benefits: Screen-based representations enable remote trading.
8 Legitiwation	Benefit: Growers don't have to be an EAF member, enabling foreign imports.	EAF is primary authority.	No change.
9 Influence	Growers have little influence on auction policies.	EAF organization defines the market rules.	Buyers do not have direct influence on auction rules.
10 Dispute Resolution	No change.	No change: Arbitration by the auction reduces costs.	No change.
Net Benefits	Positive: for excluded foreign growers.	Positive: lower cost auction model.	Positive: product access and remote buying.

Table 8 Process-stakeholder impacts of the Tele Flower.

The observations across cases illustrate that information technology substantially reduces coordination costs and enhances communications capabilities. These changes allow transacting parties to separate in space (e.g., TFA bidding process) and time (e.g., HSB delivery logistics) the informational and physical components of trading processes. Indeed, the informational and physical components within each of the ten exchange processes defined in this paper could be decoupled from one another and provided by different actors without the necessity for colocation. Over time this will permit the creation of more distributed and customized market institutions for the trade of physical goods.

PROPOSITION 2. *Information technology-enabled process innovations change the information available to different stakeholders. When different stakeholders perceive these changes reduce the relevant information or trust, they will act to either increase the information available or discount prices to account for increased risks.*

Market organizations, like other governance systems, are information processing systems that represent a consensus among different competing stake-holders on organizing the interdependent processes identified in § 2. These processes serve to generate or organize information, or to reduce uncertainties and mitigate opportunism risks faced by different stake-holders (Kambil 1992). The prior literature on markets and technology generally presumes that information technology applications enable the provision of more relevant information to decision makers (Bakos 1991b, Brynjolfsson et al. 1994, Gurbaxani and Whang 1991, Malone et al. 1987). However, the cases above illustrate that IT can actually reduce the relevant information available to decision makers. The application of information technology to communication and product representation processes had major impacts on relevant information available for price discovery, authentication, and logistics processes. When information or trust as perceived by specific stakeholders was reduced in these processes, they or other interested stakeholders acted to increase the information and trust, or reduced prices to discount for new risks.

For example, in the SBA, the buyers chose to discount the prices bid for non-sample lots by nearly 10% because they could no longer authenticate quality by visual inspection. Similarly, growers in the SBA decreased lot sizes or reclassified their product so that the same product was displayed multiple times. The growers' objective was to have multiple new bidding cycles for price determination, hoping that multiple bidding competitions for smaller lots among buyers would generate new demand information leading to higher prices. Both of these responses reduced the attractiveness and efficiency of the sample-based auction, leading to its ultimate demise.

While the Holland Supply Bank and the Tele Flower auction reduced the ability of real-time authentication of the product, they provided new benefits to the different stakeholders or created trust in new ways. In the former, parties were able to trade in a way that reduced uncertainty and generated information about future availability of demand, prices and goods. In the latter, to create trust among buyers in the new auction process and establish a reputation for quality, the Tele Flower auction ensured that the product matched or exceeded buyers' expectations given a specific quality rating. While we cannot establish that the product was traded at a lower price than when it was visually inspected, the tight quality control that assures a product matches or exceeds its quality ratings leaves it open to being discounted to the price level of the quality rating rather than the price of the flower given perfect quality information.

The above observations are consistent with prior research on the governance of exchange relations. First, the need for information seeking to reduce uncertainty is well known. The generation of trust (the expectation of future nonopportunistic behavior) and preservation of good reputations is also important to reduce transaction costs or "friction" in exchange (Arrow 1974, Fukuyama 1995). Third, social exchange theorists have identified the importance of equity and fairness (Blau 1964, Homans 1961). The cases illustrate that reductions in information available due to an IT-enabled process discontinuity creates new information seeking or trust-generation costs. Market designers can respond to these requirements by allowing more information gathering by different stakeholders (buyers and sellers) or create trust and reputations for fairness. Both these choices lead to differential levels and allocations of costs across stakeholders. These costs and their distribution across stakeholders must be considered by designers of new IT-enabled market mechanisms if the new market is to be adopted.

PROPOSITION 3. *Market organizations are the meeting point for multiple stakeholders: buyers, sellers, and intermediaries with conflicting incentives. Given existing or competing market alternatives, no new IT-based initiative is likely to succeed if any key stakeholder is worse off after the IT- enabled innovation.*

The process-stakeholder analysis framework developed in § 2 of this paper was applied to map process changes and their impacts on different stakeholders. In the two cases of failure, the application of the framework clearly identified either the grower or the buyer was worse off from the innovation. For example, in the SBA the grower incurred new packaging costs and logistics costs. In the Vidifleur auction, the buyers did not perceive a new benefit from the system. The video quality was poor, authentication of quality less convenient, and trading online did not provide all the information available in the auction hall.

In contrast, the Holland Supply Bank and the Tele Flower auctions created new advantages for all stakeholders who participated. In the Holland Supply Bank, buyers were able to order large quantities of product and specify a contract for delivery on a future date, reducing the growers' uncertainty about future price levels. The Tele Flower auctions provided Dutch buyers with a source of foreign flowers, foreign growers with a trading location that could take advantage of the Dutch wholesaler's demand for flowers, and downstream logistics facilities.

These cases illustrate the need to carefully design and implement electronic markets and to include adequate incentives to different stakeholders for participation. All stakeholders must be better off with the new system rather than the old system, otherwise the new market will fail and revert back to the prior system if this alternative remains available. The process-stakeholder analysis framework can be used by market designers to evaluate the consequences of their designs on processes and stakeholders to fulfill the above conditions.

PROPOSITION 4. *Biases in trading will depend on the ownership of the market and the availability of cost-effective trading alternatives. As owners of existing markets may not benefit from electronic trading, this incentive incompatibility will impede the transition to new electronic markets. A solution to this problem is to change market ownership or create new sources of competition.*

Why did the traditional Dutch markets fail to successfully adopt an electronic trading system? Why did the leading auctions allow the Tele Flower auction to reframe competition among the auctions? In part, the various electronic initiatives of the traditional auctions did not fulfill compelling needs among the different stakeholders. In addition, the IT-based trading initiatives were embedded in a specific organizational context and set of assumptions that limited the initiatives.

Historically, the Dutch grower cooperatives have owned the auctions and have been the dominant influence on the use of technology in the auctions. The well-organized growers who own the auction intermediary established a consensus on the trading rules and organization that best favored specific grower groups—from clock speeds to auction mechanism.

The technology-enabled initiatives of the existing auctions, with the exception of the Holland Supply Bank, generally assumed the existing valuation model and auction structure. As auction officials mentioned to us in interviews prior to the Tele Flower initiative, it was unlikely that any other form of trading or price discovery mechanism would be approved by the owners other than the Dutch auction. The innovations were thus embedded within the context and framework of the traditional auction model. Converting this model to an IT-based representation had the potential for a number of advantages to sellers. For example, by having a very fast auction clock, and trading in small lots in the traditional auction, the sellers found they could get higher prices. Electronic trading that preserved these features benefits the sellers but creates more costs for larger buyers, who are increasingly important in the market. Other price discovery mechanisms and trading rules, including larger lots, can be more efficient for large buyers.

As buyers become more powerful, and move to bypass the auctions or source foreign flowers, the existing consensus auction arrangement is being renegotiated. For example, by not participating in the sample auction, the buyers were able to influence the restructuring of the sample-based auctions. Second, the Holland Supply Bank brokered market represents another new arrangement for large buyers. Finally, the Tele Flower auction was the EAF and the buyers' response to restrictive foreign flower trading rules in the traditional Dutch flower auctions. The latter was a departure from renegotiating the consensus around a specific auction to creating a competing structure. This structure was created because the traditional auctions were forced to restrict the sale of foreign flowers by their owners.

The emergence of Tele Flower is especially instructive. Despite the advice of the traditional Dutch Auction officials, the owners voted to deny access to foreign growers. The subsequent creation of the Tele Flower auction resulted in the loss of a key revenue stream to the existing auctions and its owners. Conforming to the existing auction models limited innovation by the traditional Dutch auctions. In contrast, information technology allowed the EAF to coordinate and implement trading processes without colocation and at lower costs than setting up a facility similar to the traditional Dutch auctions. As buyers and foreign growers implement and continue to innovate in developing alternate markets, they will, most likely, increasingly reshape the access, price determination and trading rules to their favor. The emergence of Tele Flower could probably have been avoided if the buyers had an ownership stake and greater influence in the traditional Dutch auctions. This would have led to less biased or restrictive trading practices. We expect that information technology will enable new trading patterns that are less biased in favor of Dutch growers. These new trading models (e.g., clearinghouse auctions, brokered markets) could better serve different buyer segments.

In summary, current owners of the auctions may not have sufficient incentives to adopt new or more efficient auction mechanisms. The capacity of existing market structures to adapt to new electronic and non-electronic innovations will be determined by the relative power and capabilities of different parties to the trade, the current ownership structures, and potential of competitive entry. In the Dutch auctions, we see an increasing shift of power toward buyers as new sources of supply become available, buyers become larger, and technology enables new low-cost trading mechanisms leading to competing markets.

6. Conclusions

This paper makes four key contributions to the literature on information technology and markets. First, we identify a series of distinct processes that underlie exchange relations. In contrast to the traditional case studies of electronic markets, this categorization provides a useful basis for examining how technology innovations affect specific trading processes. Second, we propose and illustrate the use of the process-stakeholder analysis framework for comparing different forms of trading, and evaluate the impacts on different market participants. This is a useful guide for designers of new market systems as well as those studying markets to evaluate or explain the successes and failures of IT-based initiatives in new markets. Third, the development of a detailed case study of market change and a cross-case analysis of IT initiatives in the Dutch flower markets result in a series of testable propositions. The propositions address the separation of physical and informational processes in trading, the responses of stakeholders to changes in information in markets due to IT initiatives, and economic and incentive conditions required for adoption of new trading processes. Finally, we propose a more complex and subtle view of markets than typically analyzed in the MIS literature.

While there are a number of teaching and research cases on electronic markets, these are either unsuitable for generating new theories (Kambil and Short 1994) or articles analytically apply transactions cost or network externality models to emergent markets (Bakos 1991b, Lee 1996). While some of the analysis provides conditions for market adoption, they shed little light on why different actors failed to adopt new market forms. In contrast, the rich cases in this study illustrate that markets represent a socially constructed consensus that is constantly renegotiated among stakeholders with different levels of power. Information technology, by enabling new low-cost trading mechanisms and changing information available to stakeholders, is one factor that can alter the relative costs and benefits to actors, or the power of actors leading to a renegotiation of the consensus. When the renegotiation fails, the new market form is not adopted. As new information technologies make feasible new models of electronic commerce, our analysis suggests that designers of new markets should carefully manage the context in which they apply their technologies by constructing suitable incentives and processes to arrive at a consensus.

Finally, the paper provides new tools and directions for inquiry for managers and researchers of electronic commerce. First, the process-stakeholder framework allows managers and researchers to systematically characterize process changes and the allocation of costs and benefits from electronic commerce. Second, the cases highlight new questions for research. As the Internet evolves to a powerful and reliable infrastructure for electronic commerce, Dutch auctions become more important as a trading mechanism. However, there is little published *empirical* research on the Dutch auction mechanism and the effects of clock speed, and other information variables on equilibrium prices, buyer strategies, and the distribution of benefits. New development tools (Java, etc.) will enable researchers to develop simulations and test market designs over the Internet to develop a richer understanding of individual behaviors in electronic markets.[5]

NOTES

1. Coase (1960) states for "a market transaction it is necessary to discover who it is one wishes to deal with, inform people that one wishes to deal and on what terms, conduct negotiations leading up to a bargain, to draw up a contract, to undertake the inspection needed to make sure that terms of the contract are being observed and so on." This specifies a minimum of three sequential activities: search to gather information and determine products and parties to transactions; bargaining and negotiations to determine the terms of exchange; and the construction and adoption of monitoring, policing, and enforcement mechanism to ensure parties will satisfactorily perform the exchange.
2. Spooner and Copeland (1992) provide a teaching case of IT initiatives at BVH (previously called Flower Auction Westland) in which they discuss the motivation and technology used to implement the Vidifleur experiment.
3. Beginning in 1992, growers and auctions adopted electronic data interchange to communicate product, order, and transaction information to each other. EDI reduces double entry of data into systems and paperwork, and reduces errors. Growers who use EDI can submit data on shipments until 3:00 p.m. instead of 12:00 p.m. for paper transfer.
4. Stapelwagens are a fixed-size cart used to transport flowers through the auction complex.
5. We thank Hank Lucas, Ted Stohr, and the anonymous reviewers of *Information Systems Research* for their comments and suggestions on previous versions of this manuscript.

REFERENCES

Akerlof, G. A., "The Market for 'Lemons': Qualitative Uncertainty and the Market Mechanism," *Quarterly J. Economics*, 84 (1970), 488–500.

Arrow, K. J., *Limits of Organization*, Norton, New York, 1974.

Bakos, J. Y. "Information Links and Electronic Marketplaces: The Role of Interorganizational Information Systems in Vertical Market," *J. Management Information Systems*, 8, 2 (1991a), 31–52.

_____, "A Strategic Analysis of Electronic Marketplaces," *MIS Quarterly*, September (1991b), 107–19.

Blau, P., *Exchange and Power in Social Life*, John Wiley & Sons, New York, 1964.

Bos, J., "De Klok met Afstandsbediening," *Groot Handelsblad*, November (1995), 22–3.

_____, "Een Jaar TFA: De Innovatieve Noodsprong," *Groot Handelsblad*, March (1996), 8–11.

Brynjolfsson, E., T. W. Malone, V. Gurbaxani, and A. Kambil, "Does Information Technology Lead to Smaller Firms?", *Management Sci.*, 40, 12 (1994), 1628–44.

Clemons, E. K. and B. Weber, "London's Big Bang: A Case Study of Information Technology, Competitive Impact, and Organizational Change," *J. Management Information Systems*, 6 (1990), 41–60.

_____ and _____, "Evaluating the Prospects for Alternative Securities Markets," in *Proc. Twelfth International Conf. on Information Systems*, New York, 1991.

Coase, R. H., "The Problem of Social Cost," *J. Law and Economics*, 3 (1960), 1–44.

Coase, R. H., "The Nature of the Firm," *Economica*, 4 (1937), 386–405.

Daft, R. L. and R. H. Lengel, "Organizational Information Processing Requirements, Media Richness and Structural Requirements," *Management Sci.*, 32, 5 (1986), 554–71.

Dahlman, C. J., "The Problem of Externality," *J. Law and Economics,* 22, 1 (1979), 141–462.

Davis, D. and C. Holt, *Experimental Economics*, Princeton University Press, Princeton, NJ, 1993.

Eisenhardt, K. M., "Building Theories from Case Study Research," *Acad. Management Rev.*, 14, 4 (1989), 532–50.

Eras, P., "Tele Flower Auction Veilt via Telefoon en Beeldscherm," *Telecommagazine*, 10, 10 (1995), 40–43.

Fukuyama, F., *Trust—The Social Virtues and the Creation of Prosperity,* The Free Press, New York, 1995.

Galbraith, J. R., "Organization Design: An Information Processing View," 4, 3 (1974), 28–36.

Griffioen, A., "Goede Start Informatieveilen in Aalsmeer," *Vakblad voor de Bloemisterij*, 2 (1994a), 9.

_____, "Informatieveilen Moet de Stagnatie Overwinnen," *Vakblad voor de Bloemisterij,* 7, (1994b), 56–7.

Gurbaxani, V. and S. Whang, "The Impact of Information Systems on Organizations and Markets," *Comm. ACM*, 34, 1 (1991), 59–73.

Homans, G., *Social Behavior: Its Elementary Forms*, Harcourt, Brace and World, New York, 1961.

Johnston, R. and P. R. Lawrence, "Beyond Vertical Integration—The Rise of the Value-Adding Partnership," *Harvard Business Rev.,* 66, 4 (1998), 94–101.

Kambil, A., "Electronic Integration: Designing Information Technology Mediated Exchange Relations and Networks," Doctoral Dissertation, Sloan School of Management, Massachusetts Institute of Technology, Cambridge, MA, 1992.

_____ and J. E. Short, "Electronic Integration and Business Network Redesign," *J. Management Information Systems*, 10, 4 (1994), 59–83.

Klein, B., R. Crawford, and A. Alchian, "Vertical Integration, Appropriable Rents and the Competitive Contracting Process," *J. Law and Economics*, 21 (1978), 297–326.

Konsynski, B., "Thinking of Linking: Managerial Perspectives on Electronic Linkages Across Organizations," in *Proceedings of Scientific Research on EDI*, R. J. Streng, C. F. Ekering, E. van Heck, and J. F. Schultz (Eds.), Samson Bedrijfs Informatie, Alphen aan den Rijn, Netherlands, 1992.

_____ and A. Warbelow, "American Gem Market System," Case Study, 9–190–001, Harvard Business School, Cambridge, MA, 1989.

_____ and _____, "Cooperating to Compete: Modelling Interorganizational Interchange," Harvard Business School, Cambridge, MA, 1990.

_____, _____, and J. Kokuryo, "AucNet", Case Study, 9–190–001, Harvard Business School, Cambridge, MA, 1989.

Lee, H. G., "Intelligent Electronic Markets for Commodity Auction: An Integrated Approach of Economic Theory and Social Choice Theory," Ph.D. Dissertation, University of Texas, Austin, TX, 1993.

_____, "Electronic Brokerage and Electronic Auctions: The Impact of IT on Market Structures," in *Proc. Hawaii International Conf. on System Sciences*, Honolulu, Hawaii, 1996.

Malone, T. W., J. Yates, and R. I. Benjamin, "Electronic Markets and Electronic Hierarchies," *Comm. ACM*, 30 June (1987), 484–97.

Milgrom, P. and D. J. Roberts, "Bargaining Costs, Influence Costs, and the Organization of Economic Activity," in J. Alte and K. Shepsle (Eds.), *Perspectives on Positive Political Economy*, Cambridge University Press, Cambridge, England, 1990.

Powell, W. W., "Neither Market nor Hierarchy: Network Forms of Organization," *Organizational Behavior*, 12 (1990), 295–336.

Spooner, I. and D. Copeland, "Flower Auction Westland: The Cosmos Project," Case Study, University of Western Ontario, Canada, 1992.

Tushman, M. L. and D. A. Nadler, "Information Processing as an Integrating Concept in Organization Design," *Acad. Management Rev.*, July (1978), 613–24.

Van Heck, E. and T. P. Groen, "Towards an Electronic Flower Auction?," in *Proc. Management Studies and the Agri-business 1994: Management of Agri-chains*, Wageningen Agricultural University, Wageningen, 1994.

_____ and P. M. A. Ribbers, "Economic Effects of Electronic Markets," Discussion Paper, No. 9669, Center for Economic Research, Tilburg University, The Netherlands, July, 1996.

_____, E. Van Damme, J. Kleijnen, and P. Ribbers, "New Entrants and the Role of Information Technology, The Case of the Tele Flower Auction in the Netherlands," in *Proc. Hawaii International Conf. on System Sciences*, Vol. 3, IEEE Computer Society Press, Los Alamitos, CA, 1997.

Van Vliet, C., "EAF wil Met Importveiling Centrum Bloemenhandel in Nederland Houden," *Vakblad voor de Bloemisterij*, 49, (1994), 40–41.

Williamson, O. E., *Markets and Hierarchies: Analysis and Antitrust Implications*, The Free Press, New York, 1975.

———, *The Economic Institutions of Capitalism: Firms, Markets and Relational Contracting*, Free Press, New York, 1985.

Yin, R. K., "The Case Study Crisis: Some Answers," *Admin. Sci. Quarterly*, 26 (1981), 58–61.

Jiro Kokuryo, Associate Editor. This paper was received on April 11, 1995 *and has been with the authors* 10 *months for* 1 *revision.*

PART FIVE

Global and societal issues

INTRODUCTION TO PART FIVE

This section of the book deals with wider, societal impacts of information and communication technologies (ICT), and the recent trend towards globalization. In some ways, the metaphor of the universe applies to the field of Information Systems (IS). Like our universe, the field seems to be expanding at an ever-increasing rate. As an illustration, this growing interest in the wider aspects of IS parallels the developments that have taken place in the International Federation for Information Processing (IFIP) Technical Committee 9 (TC9) – The Relationship between Computers and Society <http://www.ifip.org/bulletin/bulltcs/tc9_aim.htm>. Established in 1976, TC9's stated aim was: "To influence the applications of computers with respect to individuals, groups, institutions and society".

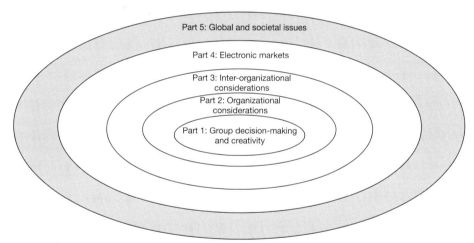

Figure 5.1 Global and societal issues within the broadening focus of Information Systems research considerations.

In some respects, the establishment of TC9 may be seen to have arisen from the sociotechnical school of thought, which had its genesis in the work of the Tavistock Institute in the 1950s. While by no means the focus of the majority of research in the IS field internationally, there has been significant interest in sociotechnical systems in the UK (e.g. Land, 1996; Mumford, 1999) and in Scandinavia (e.g., Iivari and Lyytinen, 1999; see also *Scandinavian Journal of Information Systems*) *ab initio*. Two Working Groups (WGs) were formed in the year following TC9's launch: WG9.1 – Computers and Work, and WG9.2 – Social Accountability. WG9.1 was formed to deal with such issues as the impact of IT on employment, job content, working conditions, career patterns, stress levels and the like, while WG9.2 was formed to raise awareness of the social consequences of IT, for example in relation to ethical considerations and the impact of IT on the general public. An excellent source of articles on the social aspects of ICT is the journal founded and edited by the late Rob Kling: *The Information Society* <http://www.indiana.edu/~tisj/>.[1]

	Walsham (2002)[1]	Watson, Kelly, Galliers and Brancheau (1997)[2]	Drori and Jang (2003)[3]	Andersen and Kraemer (1995)[4]
Research method	Two interpretative case studies based on previously published work by the author and others, utilizing an alternative theoretical lens.	Secondary data analysis, based primarily on previous survey research.	Hypothesis testing using ordinary least squares regressions, utilizing measures of national economic development, the level of activity in international markets, and the extent of democracy.	Secondary data analysis of previous research by the authors and other researchers, synthesized into cross-national comparisons.
Focus and perspective	Cross-cultural contradiction and conflict in multi-national software development teams. Use of structuration theory in analysing cross-cultural issues in software development and technology transfer.	International comparisons of key IS management issues, national and international surveys of IS executives.	Identification of factors shaping the global digital divide; the contention being that national ideological commitment to social development, rather than social complexity *per se*, is at the heart of world-wide trends towards greater adoption of IT.	A comparison of the fundamentally different approaches by governments in the USA and Scandinavia in the introduction and use of IT, and automation more generally, in the public sector.
Technology	Insurance application system development and geographical information system.	No specific IT.	No specific IT.	Examples include citizen information, automated services delivery, MIS and detention monitors.
Outcomes, findings, impacts	1. Theoretical: Application of structuration theory provides a richer analysis than Hofstede-type studies, inc. surfacing conflict, sub-cultural issues, power relations and the dynamic nature of culture; also, globalization does not lead to homogenization. 2. Practical: Need for practitioners to be highly sensitive to cultural differences when working in cross-cultural contexts, preparing for such collaboration through prior training and discussion.	1. Theoretical: Revised framework for cross-national key issues studies. 2. Impact of national culture and relative economic development on management issues facing IS executives.	The level of economic complexity has no significant effect on the 1991–5 change in a nation's level of IT sophistication. Net sophistication scores have increased faster in non-Western countries than in Western countries, although there is a sharp divide between the two groups in their tendency to upgrade IT usage. Economic openness and democracy have significant effect; change in IT sophistication.	Differences stem from different perceptions of beneficence of IT, the need to "re-engineer", and the role of government in anticipating and mitigating the effects of IT. USA: individualistic approaches by governments; small scale; ad hoc/reactive. Scandinavia: communal approaches, at all levels of government; large scale; proactive.

Table 5.1 Comparisons of chapters in Part 5.

Notes
1. Walsham, G. (2002) "Cross-cultural software production and use: a structuration analysis", *MIS Quarterly*, 26(4), December, 359–80.
2. Watson, R. T., Kelly, G. G., Galliers, R. D and Brancheau, J. C. (1997) "Key issues in information systems management: An international perspective", *Journal of Management Information Systems,* 13(4), Spring, 91–116.
3. Drori, G. S. and Jang, Y. S. (2003) "The global digital divide: A sociological assessment of trends and causes", *Social Science Computer Review*, 21(2), Summer, 144–61.
4. Andersen, K. V. and Kraemer, K. L. (1995) "Information technology and transitions in the public service: A comparison of Scandinavia and the United States", *European Journal of Information Systems,* 4(1), February, 51–63.

There was something of a hiatus thereafter with WG9.3 and WG9.4 not being formed until a decade or so later – in 1988 and 1989. WG9.3 was something of an extension of WG9.1, focusing on the then new phenomenon of home working, with employees being linked to their employing organization by ICT, thereby being able to work from home rather than the office. However, it was not until WG9.4 was formed that we saw evidence of a growing group of IS researchers concerned with the impacts of IT at the national, and indeed international, level. The focus of WG9.4 was most specifically on developing countries, with one of its goals being "to develop a consciousness amongst [IS] professionals, policy makers and [the] public on social implications of computers in developing nations". A further concern was to develop "culturally adapted" IS for local contexts.[2] So, with the establishment of WG9.5 – Applications and Social Implications of Virtual Worlds – one year later, in 1989, we can see, towards the end of the 1980s, a burgeoning in interest in these wider, societal aspects of ICT.[3] In the years that followed, the whole issue of ICT and globalization emerged as key (see, for example, the work of Beck, Castells, Giddens and Walsham listed in the References at the conclusion of this introduction).

So, what does the extension of the boundaries of the field of IS, to include societal and global issues, mean for the research approaches that might be used in such contexts? Exploring this question is the rationale for this section of the book. We have chosen, for the purposes of illustration, articles that deal with aspects of globalization and cross-country comparisons. The first article, by Walsham, deals with the phenomenon of off-shoring; in other words, the increasing use being made by companies of multinational virtual software development teams. The second, by Watson, *et al.*, is concerned with the different issues perceived to be the most critical facing IS executives in different countries. The third, by Drori and Jang, deals with an issue of considerable social concern – that of the so-called digital divide – in this case, in relation to the differing experiences and opportunities confronting countries on either side of this divide: the technological 'haves' and the technological 'have nots', in other words. Finally, we have an article by Andersen and Kraemer that provides us with a comparative insight into the way in which ICT has been used to transform how public services are made available to the citizens from two quite different regions of the globe: North America (specifically the USA) and Scandinavia. The four articles are by no means, and nor are they meant to offer, a comprehensive treatment of the literature on ICT and societal and global issues. No four articles could hope to. What they do provide is an insight into the kinds of topic that are becoming increasingly prevalent in the IS research literature, and the kinds of research approach being adopted in this particular area of interest. As indicated in Table 5.1 we see the influence of Giddens, and the use of structuration theory emerging in IS research (see, for example, Giddens, 1979, 1984; Orlikowski, 1992; Jones, 1999). We also see the use being made of secondary data analysis in order to provide comparisons between different contexts and cultures. Additionally, we see the use being made of economic and social indices, brought together in a causal model, in a manner common in the Social Sciences. Additional research approaches common in this area of IS research include Actor Network Theory (ANT), Critical Social Theory (CST), theories of power relations, among others (see, for example, the work of Latour, Habermas and Foucault listed in the References).

The first article in this section is by Walsham. In it, he investigates off-shore software development and use – an increasingly prevalent phenomenon (cf., Lacity and Willcocks, 2001) as companies seek to take advantage of the high skill base, but lower costs in such countries as Ireland, India, China, and Eastern Europe (see, e.g. Trauth, 2000; Walsham, 2001). This paper was chosen for this collection since it provides interesting and useful insights into issues of cross-cultural communication, especially in virtual teams (see, for example, Sambamurthy and Jarvenpaa, 2002). As indicated, Walsham's approach is to draw on structuration theory (Giddens, 1979, 1984)[4] as a means of developing a theoretical basis for the study, which is then illustrated by means of two cross-cultural case studies – a Jamaican–Indian collaboration associated with software development, and a US–Indian collaboration involving technology transfer, in this instance, a geographical information system.

The paper is a development of an earlier working paper that is available at: <http://www.jims.cam.ac.uk/research/working_papers/abstract_01/abstract_01_f.html> and is not only of interest for the focus and content of the research, but also the manner in which it illustrates the application of structuration theory in analyzing cross-cultural working and ICT in terms of structural issues, culture, cross-cultural contradiction and conflict, and the management of change.

Points for discussion and reflection include:

- Consider the potential dangers, specifically with an eye ethical considerations, in analyzing different cultural traits and characteristics, in terms of stereotyping. How might the researcher deal with such concerns?
- Consider the use of structuration theory in unpacking cross-cultural communication issues of the kind considered in this paper. What are its strengths and weaknesses? What other research approaches might be used?
- In thinking about and discussing the above point, you may wish to consider as alternatives, the case study method (either positivist or interpretive), critical social theory and ethnographic approaches (e.g. Myers and Avison, 2002), and actor-network theory (e.g. Latour, 1999, 2004; Monteiro, 2000, 2004). With regard to the latter, to what extent is technology an 'actor' in the two cases?
- Read the working paper referenced above and consider the development of the argument into the form published here. What aspects of the earlier paper have been developed further? What aspects have been deleted? Why do you think this is so?
- Consider the role of culture in the two cases. How would you define culture in this context? What impact does culture have on the transfer of IT between countries? (see also, Shore and Venkatachalam, 1996; Galliers, 2003; Westrup, et al., 2003).

Turning now to the second paper in this section, by Watson and colleagues, this deals with the range of management issues perceived by CIOs to be key for their organizations, in different parts of the world. A reason for including this paper in the book is that it provides clear evidence that issues facing organizations may well differ from one country or region to another, and through time. Cultural, economic and technological development, and political and legal reasons for these differences between countries and/or regions are discussed. Such evidence as is contained in the paper may be helpful to researchers when providing a rationale for their work. Another reason for choosing to include this paper in the book is that it uses secondary data analysis of surveys previously undertaken by a number of different authors.

Points for discussion and reflection include:

- A number of previous surveys are referenced in the paper. What are the strengths and weaknesses of the survey method in this context?
- Consider the Watson, et al. paper together with that by Palvia and Palvia (1996). What are the relative strengths and weaknesses of the two, from both an epistemological and ontological standpoint?
- One of the questions raised in relation to the Walsham article dealt with the question of culture. In most IS articles dealing with this subject matter (including the Walsham and the Watson, et al. articles), Hofstede's (e.g. 1980, 1995) work is cited. What are the limitations of Hofstede's analysis?
- In their final remarks, Watson, et al. indicate that they "would like to see a new approach that could provide a firmer basis for comparative key issues studies". How would you take up this challenge?

The third article in this section is by Drori and Jang. In it, they provide a sociological assessment[5] of trends in and causes of the global digital divide. Concerns regarding the technological 'haves' and 'have nots' has grown in the IS academy in recent years (see, for example, Krishna and Madon, 2003), both within countries and between the developed and the developing world. More recently, the idea that the introduction of ICT may be a boon to development has been questioned, however. Authors such as Wade (2004) question whether the unintended outcome of such well-intentioned policies may be a new form of dependency. So, a reason for choosing the Drori and Jang article for inclusion in this book relates to the fact that the digital divide is a key policy issue – and there have been recent calls for policy issues to be brought into IS discourse (e.g. Galliers, 1999: Jarvenpaa and Tiller, 1999). Drori and Jang "compare the effects of ... various social national conditions on the pervasiveness of IT". Somewhat in contrast to the findings of Watson, et al. (1997) and Palvia and Palvia (1996), Drori and Jang find that the determinants of "IT connectedness" relate more to cultural considerations than to economic and political characteristics of nations. As such, "the global digital divide is more a product of networking into global society than ... a mere reflection of local economic capabilities." Thus, another reason for its inclusion relates to the differing conclusions drawn by these authors as compared to others.

Points for discussion and reflection include:

- Compare the findings drawn by Drori and Jang to Watson, et al. (and Palvia and Palvia, 1996). To what do you attribute the different conclusions? Is there any way in which the different research approaches adopted may have biased the data and/or the analysis?
- In what other IS research contexts might a sociological approach be appropriate? In what circumstances would this style of research be inappropriate in the IS field?
- The paper might be described as using a positivist epistemology. What place might there be to complement this with more of an interpretivist approach (cf. Walsham, 1993)? In considering this question, recall that Burrell and Morgan (1979), among others, talk of paradigmatic incommensurability, while Mingers (2001) argues for a pluralist methodology. Consider these apparently contrasting perspectives.
- Consider taking a power/knowledge approach (cf., Foucault, 2003; Willcocks, 2004) to the subject matter of this article. What different conclusions might one draw from such an analysis?

Our final article is by Andersen and Kraemer. This deals with the use of IT in the public services in two quite different contexts: Scandinavia and the United States. There has been comparatively less IS research carried out in the public sector than in the corporate world, and this is one of the reasons why this article is included here. But there are other reasons too. One relates to the consideration of policy issues referred to in the above discussion on the Drori and Jang article. Another relates to the fundamentally different approaches adopted in the USA and in Scandinavia – the more individualistic approach of the Americans vis á vis the more communal approach of the Scandinavians. In addition, and while the authors talk of automation, the examples cited in the paper make it clear that there is more to transformation of service on the back of IT than the mere automation of that which already exists.

Points for discussion and reflection include:

- Consider the empirical evidence provided by and the conclusions drawn by the authors. To what extent would you say that they had established a strong case for their argument?
- How would you describe their research approach? Is it positivist, or interpretivist, or a combination of both? Relate your argument to the penultimate point for discussion and reflection in relation to the preceding paper.
- Were you to carry out a similar study, what alternative approaches might you consider in order to undertake a rigorous analysis of the experiences in the two regions? How might a sociological approach have added weight to your argument? What of Critical Social Theory (cf., Habermas, 1972)?
- Much has been made of national culture in this section of the book. How might a cultural comparison have been conducted in this case? Considering ANT, compare the role of non-human actors in Scandinavia and the USA.

NOTES

1. See, for example, Kling (2000).
2. For a recent example of the work of WG9.4, see Krishna and Madon (2003).
3. Further TC9 WGs have followed: WG9.6 – IT Mis-use and the Law (1990), WG9.7 – History of Computing (1992), and WG9.8 – Women and IT (2001).
4. As Walsham (2002: 362) notes, "Giddens himself makes little direct reference to information technology in his development of [structuration] theory". For an excellent review of structuration theory in the information systems field, see Jones (1999). For examples of the incorporation of information systems within structuration theory's throretical framework, see, e.g., DeSanctis and Poole (1994) and Orlikowski (1992). Walsham (1993) himself, draws on structuration theory in his book *Interpreting Information Systems in Organizations* (see, in particular pp.80–6).
5. For further reading on a sociological approach to the study of ICT, refer to, e.g., Sassen (2004).

REFERENCES

Beck, U. (2000) *What is Globalization?* Cambridge: Polity Press.

Burrell, G. and Morgan, G. (1979) *Sociological Paradigms and Organizational Analysis*, London: Heinemann Educational Books.

Castells, M. (1996) *The Rise of the Network Society*, Oxford: Blackwell.

Castells, M. (2001) *The Internet Galaxy: Reflections on the Internet, Business, and Society*, Oxford: Oxford University Press.

Currie, W. L. (2000) *The Global Information Society*, Chichester: Wiley.

Currie, W. L. and Galliers, R. D. (eds) (1999) *Rethinking Management Information Systems: An Interdisciplinary Perspective*, Oxford: Oxford University Press.

DeSanctis, G. and Poole, M. S. (1994) "Capturing the complexity of advanced technology using adaptive Structuration Theory", *Organization Science*, 5(2), 121–47.

Foucault, M. (2003) *'Society Must Be Defended'. Lectures at the Collége de France 1975–6*, New York, NY: Picador. (Original version in French, 1977.)

Galliers, R. D. (1999) "Towards the integration of e-business, knowledge management and policy considerations within an information systems strategy framework", *Journal of Strategic Information Systems*, 8(3), September, 229–34.

Galliers, R. D. (2003) "Information systems in global organizations: Unpacking 'culture'", in S. Krishna and S. Madon (eds), *The Digital Challenge: Information Technology in the Development Context*, Aldershot: Ashgate Publishing Ltd, 90–9.

Galliers, R. D., Lediner, D. E. and Baker, B. S. H. (eds) (1999) *Strategic Information Management: Challenges and Strategies in Managing Information Systems*, second edition, Oxford: Butterworth-Heinemann.

Giddens, A. (1979) *Central Problems in Social Theory*, Basingstoke: Macmillan.

Giddens, A. (1984) *The Constitution of Society*, Cambridge: Polity Press.

Habermas, J. (1972) *Knowledge and Human Interests*, London: Heinemann.

Hofstede, G. (1980) *Culture's Consequences: International Differences in Work-Related Values*, Beverly Hills, CA: Sage.

Hofstede, G. (1995) "Managerial values", in T. Jackson (ed.), *Cross-Cultural Management*, Oxford: Butterworth-Heinemann, 150–65.

Iivari, J. and Lyytinen, K. (1999) "Research on information systems development in Scandinavia: Unity in plurality", in W. L. Currie and R. D. Galliers (eds), *Rethinking Management Information Systems: An Interdisciplinary Perspective*, Oxford: Oxford University Press, 57–102.

Jarvenpaa, S. L. and Tiller, E. H. (1999) "Integrating market, technology, and policy opportunities in e-business strategy", *Journal of Strategic Information Systems*, 8(3), September, 235–49.

Jones, M. (1999) "Structuration Theory", in W. L. Currie and R. D. Galliers (eds), *Rethinking Management Information Systems: An Interdisciplinary Perspective*, Oxford: Oxford University Press, 103–35.

Klein, H. K. and Huynh, M. Q. (2004) "The critical social theory of Jürgen Habermas and its implications for IS research", in J. Mingers and L. Willcocks (eds), *Social Theory and Philosophy for Information Systems*, Chichester: Wiley, 157–237.

Krishna, S. and Madon, S. (eds) (2003) *The Digital Challenge: Information Technology in the Development Context*, Aldershot: Ashgate Publishing Ltd.

Lacity, M. C. and Willcocks, L. P. (2001) *Global Information Technology Outsourcing*, Chichester: Wiley.

Land, F. (1996) "The new alchemist: How to transmute base organizations into corporations of gleaming gold", *Journal of Strategic Information Systems*, 5(1), March, 5–17.

Latour, B. (1999) *Pandora's Hope: Essays on the Reality of Science Studies*, Cambridge, MA: Harvard University Press.

Latour, B. (2004) "On using ANT for studying information systems: A (somewhat) Socratic dialogue", in C. Avgerou, et al. (eds), *The Social Study of Information and Communication Technology: Innovation, Actors, and Contexts*, Oxford: Oxford University Press, 62–76.

Mingers, J. (2001) "Combining IS research methods: Towards a pluralist methodology", *Information Systems Research*, 12(3), 240–59.

Mingers, J. and Willcocks, L. (2004) *Social Theory and Philosophy for Information Systems*, Chichester: Wiley.

Monteiro, E. (2000) "Actor-network theory and information infrastructure", in C. U. Ciborra and Associates, *From Control to Drift: The Dynamics of Corporate Information Infrastructures*, Oxford: Oxford University Press.

Monteiro, E. (2004) "Actor network theory and cultural aspects of interpretative studies", in C. Avgerou, et al. (eds), *The Social Study of Information and Communication Technology: Innovation, Actors, and Contexts*, Oxford: Oxford University Press, 129–39.

Mumford, E. (1999) "Routinisation, re-engineering, and socio-technical design: Changing ideas on the organisation of work", in W. L. Currie and R. D. Galliers (eds), *Rethinking Management Information Systems: An Interdisciplinary Perspective*, Oxford: Oxford University Press, 28–44.

Orlikowski, W. (1992) "The duality of technology: Rethinking the concept of technology in organizations", *Organization Science*, 3(3), 398–427.

Palvia, P. C. and Palvia, S. C. (1996) "Understanding the global information technology environment: Representative world issues", in P. Palvia, S. Palvia and E. M. Roche (eds), *Global Information Technology and Systems Management: Key Issues and Trends*, Maretta, GA: Ivy League Publishing, 3–30.

Palvia, P., Palvia, S. and Roche, E. M. (eds) (1996) *Global Information Technology and Systems Management: Key Issues and Trends*, Maretta, GA: Ivy League Publishing.

Robertson, R. (1992) *Globalization: Social Theory and Global Culture*, London: Sage.

Sambamurthy, V. and Jarvenpaa, S. (eds) (2002) Special Issue on "Trust in the Digital Economy", *Journal of Strategic Information Systems*, 11 (3–4), December, 183–344.

Sassen, S. (2004) "Towards a sociology of information technology", in C. Avgerou, et al. (eds), *The Social Study of Information and Communication Technology: Innovation, Actors, and Contexts*, Oxford: Oxford University Press, 77–99.

Shore, B. and Venkatachalam, A. R. (1996) "Role of national culture in the transfer of information technology", *Journal of Strategic Information Systems*, 5(1), March, 19–35.

Trauth, E. M. (2000) *The Culture of an Information Economy: Influences and Impacts in the Republic of Ireland*, Dordrecht, NL: Kluwer Academic Publishers.

Wade, R. H. (2004) "Bridging the digital divide: New route to development or new form of dependency?" In C. Avgerou, et al. (eds), *The Social Study of Information and Communication Technology: Innovation, Actors, and Contexts*, Oxford: Oxford University Press, 185–206.

Walsham, G. (1993) *Interpreting Information Systems in Organizations*, Chichester: Wiley.

Walsham, G. (1993) *Making a World of Difference: IT in a Global Context*, Chichester: Wiley.

Westrup, C., Al Jaghoub, El Sayed, H. and Liu, W. (2003) "Taking culture seriously: ICTs culture and development", in S. Krishna and S. Madon (eds), *The Digital Challenge: Information Technology in the Development Context*, Aldershot: Ashgate Publishing Ltd, 13–27.

Willcocks, L. P. (2004) "Foucault, power/knowledge and information systems: Reconstructing the present", in J. Mingers and L. Willcocks (eds), *Social Theory and Philosophy for Information Systems*, Chichester: Wiley, 238–96.

Cross-cultural Software Production and Use: A Structurational Analysis[1]

Geoff Walsham

Judge Institute of Management University of Cambridge Trumpington Street Cambridge CB2 1AG United Kingdom g.walsham@jims.cam.ac.uk

Abstract

This paper focuses on cross-cultural software production and use, which is increasingly common in today's more globalized world. A theoretical basis for analysis is developed, using concepts drawn from structuration theory. The theory is illustrated using two cross-cultural case studies. It is argued that structurational analysis provides a deeper examination of cross-cultural working and IS than is found in the current literature, which is dominated by Hofstede-type studies. In particular, the theoretical approach can be used to analyze cross-cultural conflict and contradiction, cultural heterogeneity, detailed work patterns, and the dynamic nature of culture. The paper contributes to the growing body of literature that emphasizes the essential role of cross-cultural understanding in contemporary society.

Keywords: globalization, cross-cultural work, structuration theory, software development, technology transfer

ISRL Categories: AI0114, AI0703, BD0101, BD05, EL05, EL07, EL09

Introduction

There has been much debate over the last decade about the major social transformations taking place in the world such as the increasing interconnectedness of different societies, the compression of time and space, and an intensification of consciousness of the world as a whole (Robertson 1992). Such changes are often labeled with the term globalization, although the precise nature of this phenomenon is highly complex on closer examination. For example, Beck (2000) distinguishes between *globality*, the change in consciousness of the world as a single entity, and *globalism*, the ideology of neoliberalism which argues that the world market eliminates or supplants the importance of local political action.

Despite the complexity of the globalization phenomenon, all commentators would agree that information and communication technologies (ICTs) are deeply implicated in the changes that are taking place through their ability to enable new modes of work, communication, and organization across time and space. For example, the influential work of Castells (1996, 1997, 1998) argues that we are in the "information age" where information generation, processing, and transformation are fundamental to societal functioning and societal change, and where ICTs enable the pervasive expansion of networking throughout the social structure.

However, does globalization, and the related spread of ICTs, imply that the world is becoming a homogeneous arena for global business and global attitudes, with differences between organizations and societies disappearing? There are many authors who take exception to this conclusion. For example, Robertson (1992) discussed the way in which imported themes are *indigenized* in particular societies with local culture constraining receptivity to some ideas rather than others, and adapting them in specific ways. He cited Japan as a good example of these *glocalization* processes. While accepting the idea of time-space compression facilitated by ICTs, Robertson argued that one of its main consequences is an exacerbation of collisions between global, societal, and communal attitudes. Similarly, Appadurai (1997), coming from a non-Western background, argued against the global homogenization thesis on the grounds that different societies will appropriate the "materials of modernity" differently depending on their specific geographies, histories, and languages. Walsham (2001) developed a related argument, with a specific focus on the role of ICTs, concluding that global diversity needs to be a key focus when developing and using such technologies.

If these latter arguments are broadly correct, then working with ICTs in and across different cultures should prove to be problematic, in that there will be different views of the relevance, applicability, and value of particular modes of working and use of ICTs, which may produce conflict. For example, technology transfer from one society to another involves the importing of that technology into an "alien" cultural context where its value may not be similarly perceived to that in its original host culture. Similarly, cross-cultural communication through ICTs, or cross-cultural information systems (IS) development teams, are likely to confront issues of incongruence of values and attitudes.

The purpose of this paper is to examine a particular topic within the area of cross-cultural working and ICTs, namely that of software production and use; in particular, where the software is not developed in and for a specific cultural group. A primary goal is to develop a theoretical basis for analysis of this area. Key elements of this basis, which draws on structuration theory, are described in the next section of the paper. In order to illustrate the theoretical basis and its value in analyzing real situations, the subsequent sections draw on the field data from two published case studies of cross-cultural software development and application.

There is an extensive literature on cross-cultural working and IS, and the penultimate section of the paper reviews key elements of this literature, and shows how the analysis of this paper makes a new contribution. In particular, it will be argued that the structurational analysis enables a more sophisticated and detailed consideration of issues in cross-cultural software production under four specific headings: cross-cultural contradiction and conflict; cultural heterogeneity; detailed work patterns in different cultures; and the dynamic, emergent nature of culture. The final section of the paper will summarize some theoretical and practical implications.

Structuration Theory, Culture and IS

The theoretical basis for this paper draws on structuration theory (Giddens 1979, 1984). This theory has been highly influential in sociology and the social sciences generally since Giddens first developed the ideas some 20 years ago. In addition, the theory has received considerable attention in the IS field (for a good review, see Jones 1998). The focus here, however, will be on how structuration theory can offer a new way of looking at cross-cultural working and information systems. The rest of this section develops this analysis. A summary of key points is provided in Table 1.

Structuration theory is described by Giddens as an "ontology of social life" or, in other words, a description of the nature of human action and social organization. At the heart of the theory is the attempt to treat human action and social structure as a duality rather than a dualism. In other words, rather than seeing human action taking place within the context of the "outside" constraints of social structure (a dualism), action and structure are seen as two aspects of the same whole (a duality). This device is achieved in part by a careful redefinition of the meaning of structure. Giddens defines structure as:

Structure	• Structure as memory traces in the human mind
	• Action draws on rules of behavior and ability to deploy resources and, in so doing, produces and reproduces structure
	• Three dimensions of actions/structure: systems of meaning, forms of power relations, sets of norms
	• IS embody systems of meaning, provide resources, and encapsulate norms, and are thus deeply involved in the modalities linking action and structure
Culture	• Conceptualized as shared symbols, norms, and values in a social collectivity such as a country
	• Meaning systems, power relations, behavioral norms not merely in the mind of the person, but often display enough systemness to speak of them being shared
	• But need to recognize intra-cultural variety
Cross-cultural contradiction and conflict	• Conflict is actual struggle between actors and groups
	• Contradiction is potential basis for conflict arising from divisions of interest, e.g., divergent forms of life
	• Conflicts may occur in cross-cultural working if differences affect actors negatively and they are able to act
Reflexivity and change	• Reproduction through processes of routinization
	• But human beings reflexively monitor actions and consequences, creating a basis for social change

Table 1 Structuration theory, culture, and ICTs: some key concepts.

Rules and resources, recursively implicated in the reproduction of social systems. Structure exists only as memory traces, the organic basis of human knowledgeability, and as instantiated in action (1984, p. 377).

The crucial point here is that structure, defined in this way, is seen as rules of behavior and the ability to deploy resources, which exist *in the human mind itself*, rather than as outside constraints. (This distinction is often misunderstood in the IS literature which draws on structuration theory; see Jones 1998.) The actions, therefore, of an individual human being draw on these rules and resources and, in so doing, produce or reproduce structure in the mind. So, for example, a manager who reprimands an employee for arriving late at the workplace is drawing on the concept of the start time of an employee, the rule that the employee should arrive before or at this time, and the perceived ability for the manager to deploy the human resource represented by the employee, and thus to reprimand the employee for being late. In carrying out this action, the manager and the employee have the structure of these rules and resources reinforced in their minds as standards of appropriate behavior.

In order to develop a more detailed analysis of the duality of structure, as defined above, Giddens introduced three dimensions concerned with systems of meaning, forms of power relations, and sets of norms. Human action and structure in the mind are composed, according to structuration theory, of elements of each of these dimensions but, as the example of the manager and the employee above demonstrated, the dimensions are inextricably interlinked. So the power to reprimand is linked to the concept of starting time and the norm of what it means to be late. This may seem obvious, but norms of behavior such as this vary widely between cultures. In our analysis later in the paper, it will be seen that it is precisely some of these differences "in the mind" as to what is appropriate behavior that can cause conflict in cross-cultural working.

Culture, at its most basic level, can be conceptualized as shared symbols, norms, and values in a social collectivity such as a country. In Giddens' terms, systems of meaning, forms of power relations, and norms of behavior have a more widespread currency than *merely* within the mind of one person. Giddens defines these as *structural properties*, namely "structured features of social systems stretching across time and space." He comments that social systems should be regarded as widely variable in the degree of *systemness* that they display, and he says that they rarely have the sort of internal unity which may be found in physical or biological systems. In other words, related to the focus of this paper, national cultures are composed of many different people, each with a complex structure in their mind, none of which can be thought of as fully shared. For example, there will be all sorts of nuance as to how individuals view lateness, even within the same cultural context. Nevertheless, it will be argued in this paper that the structural properties of cultures often display enough systemness for us to speak about shared symbols, norms, and values, while recognizing that there will remain considerable intra-cultural variety.

There have been a number of attempts to incorporate information systems within the theoretical framework of structuration theory (e.g., DeSanctis and Poole 1994; Orlikowski 1992). Giddens himself makes little direct reference to information technology in his development of the theory, so that the IS researcher is left to his or her own devices. This paper draws on the conceptualization in Walsham (1993, p. 64), where he argues that:

> A theoretical view of computer-based information systems in contemporary organizations which arises from structuration theory is that they embody interpretative schemes, provide coordination and control facilities, and encapsulate norms. They are thus deeply implicated in the modalities that link social action and structure, and are drawn on in interaction, thus reinforcing or changing social structures.

In other words, IS are drawn on to provide meaning, to exercise power, and to legitimize actions. They are thus deeply involved in the duality of structure.

There is one further element in structuration theory, which has not been widely referred to in the literature, and certainly not in the IS literature, that is of considerable theoretical value in the study of cross-cultural working. This is Giddens' discussion of conflict and structural contradiction. He defines and discusses these concepts as follows:

> By conflict I mean actual struggle between actors or groups...whereas contradiction is a structural concept.... Conflict and contradiction tend to coincide because contradiction expresses the main "fault lines" in the structural contradiction of societal systems (1984, p. 198).

Conflict is thus real activity, while contradiction can be thought of as the *potential basis* for conflict, arising from structural contradictions within and between social groupings. Giddens elaborates on this:

> contradictions tend to involve divisions of interest between different groupings of categories of people....Contradictions express divergent modes of life and distributions of life chances...If contradiction does not inevitably breed conflict, it is because the conditions not only under which actors are aware of their interests but are able and motivated to act on them are widely variable (1984, pp 198–199).

This theorizing has immediate application to cross-cultural working and IS. Contradictions include "divergent modes of life," which can be taken to include cultural differences. They *may* result in conflict if actors feel that the differences affect them negatively, and they are able and motivated to take positive action of some sort. We will see examples of this in the later empirical material.

Structuration theory appears at first sight to be focused on reproduction of structure in the mind, and broader social structures within societies, through processes of routinization of activity and thus reinforcement of existing structures. However, Giddens also emphasizes human knowledgeability, and the way in which human beings reflexively monitor their own actions, that of others, and consequences, both intended and unintended. The latter provides an example of the basis for social change as well as social stability. If a human being takes action and he or she subsequently views the unintended consequences of this as negative, then it is likely that different action will be taken in similar circumstances in the future, with related changed structure in the mind. The following empirical sections will analyze stability and reproduction, but will also focus on change processes.

Software Production in a Cross-Cultural Team

This section is the first of two designed to illustrate the value of the theoretical basis described above, and focuses on a cross-cultural software development team. Software development in the context of a more globalized world is no longer carried out exclusively within the country that needs it, using citizens from that country, but is increasingly outsourced through nonlocal arrangements such as body-shopping and global software outsourcing (Lacity and Willcocks 2001), and the use of global software teams (Carmel 1999). The case below provides a specific example of this in a Jamaican insurance company, with the cross-cultural element being the extensive involvement of a team of Indian software developers. The description of the case below draws from papers by Barrett and Walsham (1995) and Barrett et al. (1996), but the structurational analysis is new.[2]

Case Description

The case concerns a Jamaican general insurance company, called Abco, which formed part of a broader Jamaican conglomerate, called the Jagis Group. Jamaica is located in the high risk catastrophe region of the Caribbean, but the capital base of general insurers in Jamaica is insufficient for high risk insurance coverage, such as that caused by earthquake and hurricane. Jamaican general insurance companies thus rely on worldwide reinsurers, who underwrite some of these high risks. In 1988, Hurricane Gilbert swept through Jamaica, paralyzing business activities on the island for a couple of months. At Abco, computer records were lost, and claims were made on policies that did not exist on the batch system.

After the hurricane and other world catastrophes, reinsurance not only became a problem to obtain, but reinsurers started to demand better quality information from companies such as Abco on risks and levels of exposure.

Responding to this crisis, the Jagis Group's chairman led an investigation as to how IT/IS could be used to provide superior quality service to clients through improved claims handling, as well as providing reinsurers with the more detailed risk and exposure information that they required. The decision was made to develop a new general insurance information system, called Goras. A leading management consultancy was commissioned to conduct the requirements study and a group software development company, Gtec, was set up within Abco in order to strengthen existing information technology skills. In March 1990, an Indian software expert, Raj, and other experienced Indian software developers were recruited from software houses in India to form the top management group of Gtec.

After the requirements study, bids were invited for the job of carrying out the software development, and Gtec was selected. However, in the initial stages of development, it became clear that additional expertise in insurance systems was needed, and a selected team of Jamaicans from the Jagis Group was seconded to the project as insurance consultants, including Roberts, the MIS manager of Jagis. The initial stages of the project were marked by some enthusiasm, at least by team members at the programmer level. Drawing from their experience on past development projects, Indian developers provided guidance to the Jamaican members on software development issues. There were weekly awards for the "most helpful member" and "project champion," and cash incentives for meeting deadlines. A key developer at Gtec reflected later:

> Looking back at it now, it was well organized. Every Monday, a memo came out specifying the deliverables and bonus structure for the week. There was a bonus on top of your salary if you met deadlines...but it was so hard to make your deadlines....Though teams were compliant, deadlines were rather stringent, if not unreasonable.

As time went by, conflict started to develop between the Indians and the Jamaicans, particularly at the senior and team leader levels. Raj was viewed by the Jamaicans as having an autocratic approach as he would "lay down the law which was not to be questioned." In contrast, the senior Jamaican on the project team, Roberts, viewed an appropriate management style with Jamaicans as being more consensual:

> If there is a problem to be solved, we would sit down and solve it ... It was not a sort of hierarchy ... It was a team effort, meet and discuss each project.

Resentment by the Jamaican software developers at all levels had deeper roots than specific conflicts on management style, since some of the locals believed that Indians were not needed in the first place. A key Gtec developer expressed this sentiment:

> The Abco MIS staff felt the whole project had been taken away from them ... They were the natural group to be utilized to develop a new general insurance system for Abco. Instead [the management consultancy] who were a bag of Indians again were asked to do the functional requirements and the initial design. Later on, Gtec was formed, staffed by Indians in all the senior posts, and responsible for the Goras project ... The Indians had been given power over the Jamaicans.

There are, of course, two sides to these cross-cultural issues. Raj, for example, was critical of the more laid-back attitude the Jamaicans had to deadlines, regarding their formal working hours as being all they were prepared to offer to the project:

> With the Indians, there is no discussion once the deadline is agreed; they will work until 9 p.m. every night, weekends if necessary to have it on my desk at the stipulated time. However, with the Jamaicans, this is not the case. If the worker recognizes that they cannot meet the deadline, they will call me up and give some excuse as to why they need more time ... they expect me to understand and accommodate.

Raj also felt that there were significant cultural differences in the way that project activities were coordinated. In India, that task was handled by the project manager whose job was "walking around and seeing how people are progressing," coordinating and administering activities, while in Jamaica project coordination was seen by him to be inherently problematic. Raj attributed this to Jamaicans' inability to "link hands and do parallel work." To illustrate this point, he offered an analogy of Jamaica's performance at international athletics events:

> They are fantastic runners ... they only miss out on medals at international relay races because at the interchange of the baton, it is dropped or it is passed too late outside the permitted exchange ... there is not training to coordinate and keep things moving.

In contast, a Jamaican member of the software team viewed the Indian approach to coordination as representing an adult–child mentality, related also in his mind to the Indian caste structure.

> The strict deadlines seemed impossble, and I was not used to the interpersonal relations of the closely knit teams ... I was reluctant to fully integrate myself into the environment which was different to what we [Jagis MIS staff] were used to ... It was a school room attitude, with someone senior to me telling me to do as he says ... It was hard to relate to their caste system where hierarchy and status were so important.

These comments relate to differences in deepseated cultural attitudes to hierarchy and authority that were recognized on the Indian side also, but of course with a different emphasis on their merits and demerits. Raj gave his view of Jamaicans' attitudes in these areas as follows:

> Everybody treats everybody as equal, The boss is viewed as a supervisor but at the same time they expect to be treated as equal. If something is due at the end of the month, don't intervene [as the boss] ... the attitude is, "I will tell you if the job is done or not, then we reset the data and keep going ... If you feel performance is bad, then fire me with redundancy pay" ... They don't want a monitoring system ... It is demeaning to them if the boss asks about progress of activities in between tasks.

The above quotes from the case study may be thought to reflect racial stereotyping on the part of some of the Indian and Jamaican software developers and managers.[3] They have been reproduced here to exemplify some of the broader issues and problems, which were interpreted by some participants to have arisen from the different cultural backgrounds of the team members. However, not all members subscibed to these views in a simple way, and the importance of individual diversity and difference within the national groups was recognized. For example, the project approach reflected the personality of Raj, in addition to elements derived from his cultural background, and this did not pass unnoticed, demonstrated by his removal from the role in the later history of the case study, as described below.

But first, how successful was the initial project in the cross-cultural team environment? The development of Goras started in 1990. The original plan envisaged a year for completion, but there were significant delays and major project cost overruns. The acceptance testing done by end users showed substantial inadequacies in the design, but the system was finally delivered by Gtec to Abco in August 1992. After further quality assurance, user testing, and system modification, a first attempt at implementation was made in December 1992. The implementation was not a success. System performance was poor in terms of time taken to carry out tasks, and users were critical of the restricted functionality of the new system, partly due to incomplete data conversion from the old systems.

In January 1993, a new CEO of Gtec was appointed, also an Indian expatriate. Ray stayed on as technical director, "preferring to work on technical issues rather than organizational ones." The responsibility for further development of the Goras system and user acceptance testing and training was switched to the Jagis group, although Gtec continued to make a techinal input. By 1995, the Goras system had still not been fully implemented, but new deadlines were in place for implementation later that year. An increased emphasis had been placed on user involvement. One of the Jagis staff described this involvement:

> Testing started in July [1994] with live data from users. Each module is being tested module-by-module and then issue forms are created which then involve a lot of work on the part of MIS [staff] to implement the required changes.

Five years after project inception, there was general optimism about successful project implementation, but it still remained a promise rather than a reality.

Structurational Analysis

Structure

This subsection analyzes the Abco case using the theory articulated earlier. Key points of the analysis are summarized in Table 2. Structure "in the mind" and its links to action, according to structuration theory, can be analyzed through the dimensions of meaning, power, and norms. Cross-culturl interaction is likely to involve basic differences in these dimensions, and the development of information systems in a cross-cultural team can bring these differences into stark contrast. With respect to meaning, metaphors of teamwork used by Abco and Gtec staff can be used as an illustration. A Jamaican software developer described the Indian's approach as a "school room attitude," linked in the mind of this person to the Indian caste system. In contrast, the Indian project leader used the metaphor of international relay races as a way of illustrating his view that the Jamaicans were incapable of working together in a coordinated way.

Turning to the second structural dimension, the case study shows radically different views of appropriate personal and power relations. The Indian team leader was viewed as autocratic by the Jamaican staff, whereas the senior Jamaican staff member thought that an appropriate management style in Jamaica was consensual. In contrast, the Indian team leader felt that the Jamaicans were too equal to make project monitoring and control effective. Related issues arose with respect to the third structural dimension of norms of behavior, for example, with respect to time deadlines for software projects and a sense of urgency. The Indian team leader was critical that the Jamaicans would go home at the "normal" leaving time, whereas the Indian team members would work evenings and weekends if necessary to meet deadlines.

Structure	• Different meaning systems: metaphor of team-work as a school room attitude or international relay races
	• Different views of appropriate power relations: Indians too autocratic; Jamaicans too equal for project control purposes
	• Different norms of behavior: attitude to time deadlines on software projects
Culture	• Strong degree of systemness in terms of different cultural attitudes of Indian and Jamaican groups
	• But important to note that individual difference also matters
	• Culture of IS development also different in the two national groups: high productivity/strict deadlines versus working closely with end users/application backlog
Cross-cultural contradiction and conflict	• Structural contradiction arising from different cultural backgrounds
	• Resulted in conflict since these affected all participants directly, and they had the ability to act: e.g., to enforce deadlines or to resist them
Reflexivity and change	• Increasing recognition on all sides that cross-cultural issues were important, and needed to be managed
	• Pragmatic actions taken on roles and responsibilities, reflecting changed structure on the part of both Jamaican and Indian participants

Table 2 Jamaica-India software development case: structurational analysis.

Culture

The above analysis, in order to make some general points, has downplayed individual differences within the Jamaican and Indian groups. This can be justified on the grounds that there was some consistency of the responses from within each cultural group which supports the argument that there was a strong degree of systemness operating here. In other words, the indigenous elements of Jamaican and Indian national cultures were sufficiently strong in the minds of the individuals concerned to influence their behavior in a broadly similar way to other members of their own culture and, equally importantly, for this to be perceived as such by members of the other culture. However, as noted in the case description, individuals also matter, and the personality of Raj was given as one example of this.

In addition to the influence of national culture, the word *culture* is often used as a metaphor (Morgan 1986) for shared values and attitudes within a specific organization or other form of social grouping. In the Abco case, Barrett and Walsham (1995) highlighted how the culture of IS development was different in the two countries:

While occupational cultures for Indians and Jamaicans alike originated from software development, the impact of the local work culture at Indian software houses and the insurance company respectively were significantly different. The norms of an Indian software house include high productivity and profitability, the software development being driven from a specification under strict project deadlines. The norms of an insurer's MIS department in Jamaica involve application development by MIS personnel working closely with end users with a backlog of applications being quite acceptable. (p. 30)

Cross-Cultural Contradiction and Conflict

Contradiction reflects differences in structural principles, according to structuration theory, such as those arising from different cultural backgrounds. However, conflict is an actual struggle, and we have seen that significant struggle did indeed take place in the case. It was argued earlier that this is likely to occur, first, if the differences affect actors negatively. With respect to the Jamaicans, they felt the force of the structural contradictions in cultural attitudes in a very direct way through Indian approaches to project monitoring and control, attitudes to deadlines and working hours, and what they viewed as excessively hierarchical approaches. The Indian management team, in particular the overall team leader, viewed these as the right way to approach software development, and the Jamaicans' attitudes as largely negative to the goal of effective project monitoring and control. The second condition for actual conflict to arise along the fault lines of the structural contradictions is that the participants have the ability to act to support their perceived position. The Indian management team had the recognized authority to control the project and to make the rules, such as time deadlines. On the other hand, the Jamaican team members were able to resist in various ways, such as giving reasons why more time was needed for a particular software task. In addition, the removal of Raj from the CEO role in the later history of the project can be taken to reflect the resistance of some of the software team members to his leadership.

Reflexivity and Change

The analysis so far has focused on the way in which structure in the minds of actors in cross-cultural interaction affects the way they think and behave, and the way in which they perceive others from a different culture, which may result in disagreement and conflict. However, as noted in the earlier theoretical section, human beings reflexively monitor actions and their consequences, creating a basis for social change. In other words, structure and culture are not immutable. This can be illustrated in the Jamaica-India software development project, in that there was an increasing recognition on all sides that cross-cultural issues were important and that they needed to be managed effectively. This resulted, in the later years of the project, in various actions being taken to mitigate the problems which had occurred. These actions included shifting the role of Raj away from organizational issues to a primarily technical role, and giving increased responsibility for human issues such as user involvement to the Jamaican MIS group. These actions not only reflected a pragmatic interest in getting a better job done, but also changed attitudes, or structure in the mind in Giddens' terms, on the part of the Jamaican and Indian participants.

Technology Transfer of GIS Software

A second way in which software is involved in cross-cultural interaction is through the transfer of IS across borders to different cultural environments from that in which it was initially developed. This technology transfer phenomenon is not a new one, but it is increasingly common in the context of globalization. For example, major software packages such as enterprise resource planning systems have spread extremely rapidly across much of the world, particularly in large organizations, over the last decade (Davenport 1998). The case described in this section will provide a specific example of the technology transfer of another global technology, namely that of geographical information systems (GIS). In particular, the case looks at the transfer of GIS from the United States to India. The description of the case below draws from the paper by Walsham and Sahay (1999), but the structurational analysis is new.[4]

Case Description

The case concerns attempts to develop and use geographical information systems (GIS) to aid district-level administration in India. In particular, the focus is a set of GIS projects that took place under the umbrella of the Ministry of Environment and Forests (MOEF) of the government of India over the period 1991 through 1996. The technical work to develop the systems was carried out by scientists in a range of institutions, including two remote sensing agencies, three research groups within universities, and three other scientific agencies concerned with forestry, space research, and the study of science and technology in development. The systems were intended to be used by district-level administrators. The MOEF initiated 10 GIS projects in January 1991, in collaboration with the eight scientific institutions, with the aim of examining the potential for using GIS technology to aid wasteland development. Wastelands are categorized as degraded land that can be brought under vegetative cover with reasonable effort, and land that has deteriorated due to lack of appropriate water and soil management.

The initiation of the project in 1991 can be traced back to two earlier events. In 1986, the government of India started the National Wastelands Identification Project, involving the mapping of the distribution of wastelands across the various states of India. Detailed maps were produced on a 1:50,000 scale for 147 selected districts using remote sensing techniques. The existence of these maps provided a basis for considering how to develop and manage these wastelands. The stimulus for the possible application of GIS to this issue was provided by a chance meeting of some GIS experts from Ohio in the United States with Indian government officials, in the context of a general USAID mission to India in 1989. This was followed by a visit of an Indian expert team to see GIS installations in the United States in 1990, and then the eight scientific institutions in India were invited by the MOEF to test the efficacy of GIS in wasteland management, using specific districts as research sites.

Phase I of the projects took place over the period 1991 to 1993, and the staff of the scientific institutions saw the objectives to be primarily technological, involving the production of working GIS systems based on real data from the field sites in their particular districts. The detailed models and systems developed by the institutions tended to reflect their view of themselves as scientific research and development centers. For example, there was a heavy reliance on data obtained by sophisticated remote-sensing techniques, reflecting the nature of the interests of the typical research scientist in these institutions. There was less emphasis on other socio-economic variables relevant to wastelands management, such as population and livestock data. In addition, and of crucial importance to later development of the project, many of the scientists involved in the project saw their institutional mandate to be limited to the development of technology rather than to its transfer to administrators at the district level.

Although the Phase I projects were completed in early 1993, proposals for continuation were not submitted until about a year later, and then only by five of the original eight institutions. This period of transition from Phase I to Phase II was characterized by uncertainty about the objectives and nature of the continuation phase. The project director saw it as involving the transfer of the developed systems to the district level so that they could be used for real management applications. However, the project managers in the scientific institutions did not view their staff skills or resources to be adequate for this task in most cases. The institutions asked for further funding largely to provide more hardware and software, whereas the project director felt that the institutions should concentrate on using the existing equipment and on its transfer to the field.

Eventually, five institutions agreed to terms for Phase II and these continuation projects were authorised by the MOEF. Soon after this, the project director left the MOEF and transferred to another institution, and there was very limited further central direction of the Phase II projects. Despite this lack of coordination from the center, all of the five Phase II projects went ahead, in different ways and with different levels of success in terms of the stated project goals. However, by the end of the project in 1996, although some efforts had been made in some of the sites toward transferring the technology to the district level, there were no actual working systems receiving real use.

Structurational Analysis

At one level, this project can be thought of as another example of a failed technology transfer effort, all too common in the history of aid agencies and their attempts to promote the use of western-origin technologies in Third World contexts. One could argue, for example, of the need for improved training and education, or institutional development. While acknowledging that these may be relevant, the theoretical basis of this paper can be used to analyze more underlying reasons. A principal argument will be that information technologies such as GIS, developed in the western countries, can be thought to reflect and embed western values. These may not be compatible with deeply-held beliefs and attitudes in other cultures such as India. Key points of the analysis in this section are summarized in Table 3.

Structure	• GIS embody systems of meaning, such as the representation of space through maps; provide resources; and encapsulate norms, such as the high value of coordinated activity
	• However, these may clash with the structure in the mind of actors in the different, cultural interest groups
Culture	• [U.S. personnel] GIS as appropriate spatial technology; provides means of deploying financial resources; promotes good development
	• [Indian GIS scientists] GIS as lead-edge technology; provides means of gaining financial resources; is suitable for a scientific institution
	• [District-level administrators] GIS as alien technology; requires them to provide data; but need not affect normal job role
Cross-cultural contradiction and conflict	• Interests not threatened in Phase I
	• Some conflict in interim phase between GIS project director and scientific institutions—some of the latter withdrew
	• Passive resistance in the form of nonuse by district-level administrators in phase II
Reflexivity and change	• Increasing awareness of maps and map-based systems in India
	• Resulting in subtle shifts in perception, but major social change over longer time horizons is made up of such minor shifts
	• Some current evidence of successful use of GIS for land management in India, reflecting changed attitudinal rigidities

Table 3 GIS Technology transfer case: Structurational analysis.

Structure and Culture

As with the case study in the previous section, it is not possible to analyze in detail the individual perceptions and actions of the many project participants. Rather, the analysis here aims to aggregate to the level of groups who can be taken to broadly share similar structure in the mind. Three such groups consist of the U.S. GIS specialists and USAID personnel, the Indian scientists concerned with GIS development, and the Indian district-level administrators. With respect to the three structural dimensions of meaning, power, and norms, the first group took the view that GIS was an appropriate technology to help with spatial issues, that they had the power through financial resources to sponsor its application in India, and that computer-based applications such as this were the right way forward for development in India. The Indian scientists saw GIS as a new lead-edge technology which they wished to learn about, that the USAID-sponsored project was a way to obtain the necessary resources, and that this fitted their mandate as a scientific institution.

Finally, the Indian district-level administrators thought that GIS technology was something outside their experience, that they were required to provide data for the systems, but that the norms of carrying out their own job in the usual way still applied.

There is clear structural contradiction here, and an analysis of this can be sharpened by looking carefully at the technology itself and the way in which it can be thought to embed structural properties in terms of meaning and norms, and to provide political resources. With respect to meaning, GIS are a way of representing space through the explicit device of maps, a common enough concept in western societies. However, India is not a map-based culture. Typical Indians will rarely, if ever, use maps in their daily life. A GIS project leader in the National Informatics Center (NIC), one of the other institutions in India trying to introduce GIS, said:

> The most difficult part of GIS introduction is getting people to think spatially. There is no simple strategy here. A first step would be to

motivate NIC's own people. They must start thinking spatially first.

This remark misstates the core of the issue. It is not that Indians do not think spatially, but that they do not in general use external conceptualizations of space, namely maps, as key aids to spatial awareness. District-level administrators, for example, those concerned with forestry management, are well aware of spatial distributions of trees in their areas. However, they do not normally conceptualize this in terms of maps, whether computer-generated or not.

Sahay (1998) linked Indians' conceptualization of space to fundamental aspects of their identity. He argued that Indians view space as basically "inhere," subjective and inherent to the person, rather than "out-there" as some objective entity. Sahay summarized the lack of fit between GIS technology and these aspects of Indian cultural identity as follows:

> The objective reality depicted in GIS software is interpreted to represent a disconnection of space from place, a relationship that allows interaction between absent others. In contrast, in Indian society, a strong relation is seen to exist between notions of space and place arising out of political, cosmological, religious and social considerations. These differences between subjective considerations and objective reality (of the GIS) seem to contribute to the discomfort which some Indians feel in relating to the notion of a GIS map (p. 181).

Sahay added that the purpose of a GIS reflects a sense of being able to control space and nature through technology. This need to dominate nature is also not a concept that comes naturally for many Indians, who typically see themselves as part of nature rather than standing outside of it.

A second feature of GIS technology can be seen as reflecting an organizational norm in western societies that places a high value on coordinated activity. The multi-layered nature of GIS systems, where data on different characteristics are brought together as overlays in the same map-based system, assumes that management issues will be addressed in a coordinated way. For example, the management of land resources in any country involves a wide range of disciplinary specialities, including agriculture, forestry, wildlife management, and many others. However, in India, these issues have typically been handled in relative isolation by the different agencies involved. Over 20 separate government agencies operate at the district level in India, each dealing with a particular functional area, and reflecting the wider governmental funding structures that are built around departmentally-based schemes. An employee in a non-governmental organization operating at the district level in India described this as follows:

> The main problem is the compartmentalism of activities. Different departments do not speak to each other. There is a problem of attitude, people do not want to do things. The crux of the problem is not technical but that of sustained coaxing. The district level engineer says that he is interested only in dams, the agricultural scientist in soils, the forester in trees. Everyone says that I am fine and no one sits and talks with each other. There is extreme compartmentalization. There is a mental barrier among the people.

This feature of compartmentalism of role in India is not a simple matter of inefficient bureaucratic organizations, but reflects some deeply-held cultural beliefs. Indian society has traditionally been stratified on functional lines with caste as the basic structural feature. Hinduism, the religion of the majority in India, emphasizes a social framework that embodies caste rituals, and these have governed the lives of most Indians for hundreds of years. One of the sacred Hindu texts, the *Bhagavad Gita*, says:

> And to thy duty, even if it be humble, rather than another's, even if it be great. To die in one's duty is life: to live in another's is death.

The compartmentalism of role and activity was a clear feature of the GIS projects. Most of the GIS scientists viewed their goal as producing accurate scientific models for the GIS, which they then expected the district level administrators to use.

The GIS can be viewed, therefore, as embodying systems of meaning such as the representation of space through maps, and encapsulating norms such as the need for coordinated action. The systems were thus aligned to the interests and structures in the mind of the U.S. personnel, and can be thought of as *actors* (Walsham and Sahay 1999) introducing those ideas into an Indian context. Another way of expressing this is that the systems provided a political resource for an attempt to use western ideas in Indian district-level administration. No value judgement is being made in this paper about whether this attempt was a "good thing" or not. The point being made here is that there was a marked structural contradiction between the values embedded in the technology and those in the minds of local actors, particularly the district-level administrators.

Cross-Cultural Contradiction and Conflict

Structural contradiction, according to the theory in this paper, does not necessarily result in conflict. Conditions under which conflict is likely to occur are when actors feel that their interests are affected negatively, and when they are able to act to counter this. The relatively smooth nature of Phase I can be explained in that, although the GIS scientists were not map users themselves in their daily lives, they did not feel their interests threatened by the technology. Indeed, it provided a resource for them to learn about a leading-edge technology, with positive career connotations. Although the district-level administrators were, in some cases, required to provide data for the GIS, this did not compromise their normal way of working. The interim period between Phases I and II did, however, start to manifest some conflict, notably when the GIS scientists felt that they were being asked by the project director to carry out a role which was not theirs, namely working closely with the district-level administrators to implement the systems. Some institutions withdrew from Phase II as a consequence.

Phase II itself saw little overt conflict, despite the stark structural contradictions between the values embedded in the technology and those in the minds of the Indian participants. Yet, there was real potential for some participants to be affected negatively. For example, the district-level staff were having alien systems imposed on them, which they saw as of little value. However, forms of resistance are many and subtle. The district-level staff did not, in general, reject the systems or undertake any form of direct action. Rather, they simply did not use the systems—action in the form of inaction, a type of passive resistance. This provides a nice illustration of what Giddens (1984) calls the "dialectic of control," namely the ways in which the seemingly less powerful manage resources in such a way as to exert control over the more powerful.

Reflexivity and Change

This passive resistance to the GIS on the part of district-level staff can be taken as an example of reproduction of structure, but change is also inherent in the human actors' reflexivity here. India is not a static culture and there is an increasing awareness of maps and map-based systems in India, not least since private Indian software companies in places such as Bangalore have been very successful in selling their services as GIS developers in the world software market. Structures in the mind do change over time, even with respect to such a fundamental issue as the conceptualization of space. Changes in culture are often imperceptible over short time periods, but major social change over longer time horizons is made up of such minor shifts.

As an example of longer-term shifting attitudes in the development and use of GIS in India, Puri (2002) describes ongoing efforts to use GIS for land management in the Indian state of Andhra Pradesh. He argues that some indications of successful use are now discernible, in contrast to the earlier work described by Walsham and Sahay (1999). Puri ascribes the later success to shifts in earlier "attitudinal rigidities," and gives examples of new approaches: GIS scientists assuming ownership of implementation as well as development of systems; increasing consultation with local departments and people; and nodal district agencies managing implementation action plans. Puri's research provides a valuable reminder that longitudinal studies of several years length, as carried out by Walsham and Sahay, may still not be long enough to detect the effect of shifting individual attitudes, or structure in the mind, which can aggregate over time to major shifts in national or subgroup cultures.

Theorizing Cross-Cultural Working and IS

In order to assess the contribution the structurational analysis of this paper can make to the study of cross-cultural software production and use, or more generally to cross-cultural working and information systems, it is necessary to examine the existing literature in this latter domain. A good starting point is the widely-cited work of Hofstede (1980, 1991), which describes cultural difference in terms of scores on five dimensions: power-distance, individualism, masculinity, uncertainty avoidance, and long-term orientation. Myers and Tan (2002) noted that much of the literature concerned with cultural and cross-cultural issues in the IS field has relied on Hofstede's work. They analyzed 36 studies from the cross-cultural IS literature, and noted that 24 of these used some or all of Hofstede's dimensions.

While the work of Hofstede, and that of similar style such as Trompenaars (1993), has the merit of alerting us to the importance of cultural difference, it can also be criticized as rather crude and simplistic. Myers and Tan note that the very concept of *national culture* is problematic on several grounds. These include the heterogeneity within a given nation-state and the difficulty of relating national cultural values to work-related actions and attitudes. They propose that IS researchers should adopt a more dynamic view of culture—one that sees culture as contested, temporal, and emergent. The rest of this section will examine why such issues are important to the study of cross-cultural working and IS, and what the structurational analysis of this paper has to offer. The discussion is organized under the four headings of cross-cultural contradiction and conflict, cultural heterogeneity, detailed work patterns, and the dynamic nature of culture. Key points in this section are outlined in Table 4, summarizing limitations of Hofstede-type studies and related contributions from a structurational analysis.

Cross-Cultural Contradiction and Conflict

Hofstede-type studies describe intercultural differences in the selected aggregate variables, and these can be taken as reflecting *contradictions* between different cultures. However, no analytical tools are provided by such studies as to how to analyze whether, and if so how, such contradictions result in actual *conflict*, physical or otherwise. For example, people from different cultures may coexist quite easily despite such differences, but in other cases the differences seem to cause major difficulties. In trying to analyze possible conflict in cross-cultural working and IS, such as in software production and use, the aggregate national variables are of little use.

The structurational analysis in this paper offers a way of addressing the question of both structural contradiction and conflict. It has been argued that conflicts may occur in cross-cultural working if differences in structures in the mind are perceived to affect actors negatively, and they are able to act to resist or oppose these negative impacts. This was illustrated in the Jamaica-India case by identifying differences in cultural views about approaches to teamwork, forms of appropriate power relations, and attitudes to time deadlines. These contributed to conflict since they affected all participants in the software project directly, and in ways that were largely perceived to be negative. Opposition or resistance was possible, and detailed ways in which this occurred were described in the case.

The GIS case also illustrated the value of a structurational analysis of cross-cultural contradiction and conflict, although in a slightly different way. Three cultural subgroups were identified, with rather different structures in the mind with respect to GIS systems, but no significant conflict occurred in Phase I of the project. This was explained by an analysis of the specific interests of the three groups, which were not negatively affected by the GIS project, although they had different views concerning its merits. However, in Phase II, some resistance did occur, for example when the Project Director wanted the GIS scientists to become involved in local-level implementation, something which they viewed as outside their remit.

Cultural Heterogeneity

By treating the concept of national culture through the use of scores on particular dimensions, as is the case in Hofstede-type studies, the implicit assumption is that national culture shows a strong homogeneity. However, there is much evidence against this view of the world. For

Topic	Hofstede-Type Studies	Structurational Analysis	Examples in Jamaica Case	Examples in GIS Case
Cross-cultural contradiction and conflict	Describe aggregate differences between cultures	Detailed way of relating contradiction and conflict	Differences in cultural views about teamwork, power relations, time deadlines	Three different cultural subgroups with different attitudes to GIS
	But provide no link to conflict		Resulting in conflict since perceived negatively and resistance possible	Resulted in resistance in phase II only, when participants perceived negative consequences
Cultural heterogeneity	No description of heterogeneity	Can be used to analyze differences in cultural sub-groups and even individuals	Some analysis of individual difference related to the Indian project director	Analysis of different attitudes of Indian scientists and district-level administrators from the same national culture
Detailed work patterns	Aggregate cultural variables do not easily translate to effect on work patterns	Meaning systems, power relations, norms already targeted at the detailed work level	Example of approaches to control of subordinates	Example of different ways of representing space
The dynamic nature of culture	Normally treated as static	Can analyze reflexivity and change	Increasing recognition over time of importance of cross-cultural issues	Recent work indicates some shift away from the attitudes that characterized the earlier studies
			Example of negotiated culture	

Table 4 Cross-cultural working and IS: Contribution of different theories.

example, India provides a good counterexample. Its one billion people come from many and varied cultural, racial, and religious backgrounds, speak hundreds of different languages, and exhibit enormous variety at different hierarchical levels within the society. Within western countries, there is an increasing heterogeneity of history and background, not least due to the existence of ethnic subgroups (see, for example, Appadurai 1997).

An interesting example of work in the IS field which goes beyond the simple attribution of national cultural characteristics is that of Korpela and his colleagues (Korpela 1996; Korpela et al 2000). Korpela criticized the approach of taking West Africa, an area equal in size to Europe, as one culture characterized by Hofstede's aggregate variables such as low individualism and a high acceptance of an unequal distribution of power. In contrast, Korpela pointed out that the country of Nigeria, for example, is a colonial creation and contains many different groups with "sharp cultural discontinuities." One such group is the Yoruba people, numbering some 20 million. Although there are differences within this large group itself, Korpela drew on the extensive literature on the Yoruba to highlight five aspects of the Yoruba cultural heritage that are distinctive. The work of Korpela and his colleagues used these characteristics to illuminate complex issues of IT development problems in the health sector in Yorubaland.

So, what does structurational analysis offer to the study of cultural heterogeneity and its impacts on IS? If we look back to the case studies of this paper, such an analysis does not require that cultures are regarded as homogeneous, but rather that one should be looking for a measure of systemness or homogeneity within particular social groupings. A good example is provided by the GIS case study. As we saw earlier, the subcultures of the GIS scientists and the district-level administrators, both composed solely of Indian nationals, had radically different attitudes toward the GIS and their value. For example, the first group viewed the GIS as providing ways for them to work with lead-edge technologies and systems, whereas the second group viewed the GIS as alien technology of little relevance to their role. A structurational analysis opens up the possibility of examining the heterogeneous systems of meaning, power relations, and norms of different social groupings within the same national culture.

The Jamaican case study did not analyze cultural heterogeneity within the two national groups directly, but aspects of it can be seen through the discussion of the role of the initial project director, Raj. His interest in organizational issues was limited, and the quotes from him in the text show his tendency to racial stereotyping of the Jamaican software employees. He was later moved to a role dealing with technical issues, leaving the way open for a new Indian CEO with a rather different management and cross-cultural approach. Space and resource limitations provide a natural barrier to case analyses which treat every project participant as an individual person with a different mixture of attributes, but structurational analysis can, in principle, be used to analyze cultural heterogeneity down to the level of subgroups, or even individuals.

Detailed Work Patterns

A further criticism of the use of Hofstede-type national cultural characteristics as a basis for analysis of cross-cultural working and IS is that there is normally a poor link between these characteristics and detailed work-related attitudes and actions. It is one thing to know how the people of a country score on masculinity or uncertainty avoidance, but another to know how this translates into the details of systems development processes, or attitudes to particular technologies. In terms of cross-cultural working, it is not necessarily the case that similarities in national characteristics imply similar work-related patterns. For example, Khare (1999) describes radical differences between Indian and Japanese work patterns, in areas such as commitment to their organization and attitude to time, despite similarities between India and Japan in terms of their scores on individualism, long-term orientation, and power-distance (Hofstede 1995).

In order to analyze detailed patterns in cross-cultural working, it is necessary to go away from the high level of national characteristics to a more detailed focus on behavior at the micro-level of the group or organization. For example, in the general management literature, Lam (1997) described a fascinating longitudinal study of cross-cultural working between Japanese and British engineers. Her detailed analysis demonstrated how differences in educational background, bases of skills, and approaches to coordination of work resulted in very different attitudes to knowledge sharing by the two cultural groups, and thus major problems in cross-cultural working. In the IS literature, a limited number of authors have carried out cross-cultural studies from this perspective of a detailed analysis of work patterns and attitudes. For example, Trauth (1999, 2000) examined the management of IT workers in an American-Irish cross-cultural work environment as part of a detailed longitudinal study of the information economy in Ireland. Barrett et al (1997) described cross-cultural working on software outsourcing from U.S. to Indian companies, examining detailed work patterns in areas such as forms of partnership and coordination mechanisms.

The structurational analysis described in this paper can offer a valuable theoretical underpinning for studies of this latter type, which otherwise tend to be somewhat anecdotal in nature. Such an analysis, as we have seen, focuses on meaning, power, and norms within particular work groups and how these affect particular work patterns and behavior. For example, in the Jamaica-India case, we saw how the Indian managers of the project were used to hands-on approaches to control subordinates, whereas this was viewed as reflecting an "adult-child" approach by one of the Jamaican participants. In the Indian GIS case, we saw how the different ways of representing space between the U.S. developers and the Indian users resulted in passive resistance to the implementation of the technology. The insights from these studies could not have been obtained by a highlevel analysis of cultural dimensions. It may be possible, in theory, to make a connection between Hofstede-type dimensions and detailed work patterns and attitudes, but such an analysis is not easily found in the literature. A structurational analysis, with its focus on meaning, power, and norms, is already targeted at the detailed work level.

The Dynamic Nature of Culture

A final area of weakness of the cultural dimensions approach to cross-cultural working is that culture is not static. For example, we have seen quite dramatic changes in many societies over the last few decades in areas such as attitudes to gender, the environment, race, sex, family life, and religion. In the context of globalization, with increasing contact between different societies, it is increasingly difficult for any group to remain isolated and uninfluenced by other cultures. Thus, in the domain of cross-cultural working, we need theories that reflect change as well as stability, and that are attuned to shifts in attitudes and actions as well as their continuance.

An example of such work in the cross-cultural management literature is that of Brannen and Salk (2000) on *negotiated culture*. They used the case example of a German-Japanese joint venture to show how the attitudes of the two cultural groups shifted over time as they engaged with each other in collaborative work activities. The groups negotiated a compromise between themselves in areas such as styles of decision making and attitudes to time off on weekends and holidays, resulting in a hybrid culture for both groups. This is not saying that the two groups became homogeneous, but that they both shifted in their attitudes from their initial cultural starting point. In the IS literature, Sahay and Krishna (2000) described a similar process in some ways, although they did not use the term *negotiated culture*. They described a case study of a software outsourcing venture over a period of several years from a Canadian multinational to an Indian software house. At first, cultural contradiction produced some conflict, but the authors argued that, later, the relationship "showed signs of maturing" based on both sides gaining an increased understanding of the other's culture. Again, this did not result in the parties becoming the same in terms of attitudes and values, but it certainly supports the view of workgroup culture being dynamic and emergent, and not derived in a static manner from national cultural characteristics.

Although neither of the above studies used a structurational analysis, this would have provided a theoretical framework within which to embed their analyses. Structuration theory, in addition to analyzing structural reproduction, emphasizes reflexivity on the part of human actors and thus changes in structure in the mind. This was analyzed in the earlier case studies under the heading of reflexivity and change. In the Jamaica-India case, we saw this reflected in an increasing recognition over time of the importance of cross-cultural issues, and the necessity for actions to be taken to address such issues. Job roles were changed, people were moved to different positions, and the India-Jamaica team started to function rather better. The negotiated culture concept fits quite well here.

In the Indian GIS case, longer-term attitudinal changes are needed if people working at the local level, such as district-level officials, are to embrace technologies such as GIS in their day-to-day work, or if GIS scientists are to perceive their role as involving implementation as well as technical development of systems. Although such changes are hard to trace in detail in the complexity of a context such as India, the earlier structurational analysis of the case drew on some recent work to indicate, at least in some areas, a shift away from the attitudinal rigidities which had characterized the earlier reported case studies. Indian culture, as with all other societies, is dynamic and emergent, and a structurational analysis can offer insights on such change processes.

Conclusions

In the more globalized world of the 21st century, working with information and communication technologies is increasingly taking place in a cross-cultural context, but we are short of good theory to analyze such phenomena. A recent article by Goodall (2002) argued that this applies to the cross-cultural management literature more generally, namely that "we are short of both rich descriptions of cross-cultural interaction, and theoretical explanations of the same." The primary contribution of this paper has been to provide such a theoretical basis, drawing from structuration theory, which was used to analyze cross-cultural software production and use. The theorization goes beyond the relatively simplistic

Hofstede-type studies which dominate the IS literature to date. In contrast to such studies, it was shown in the preceding section that a structurational analysis can accommodate elements such as the links between structural contradiction and conflict, cultural heterogeneity, an analysis of detailed work patterns, and the dynamic and emergent nature of culture.

The theory has been illustrated using two empirical examples only, with a focus on software production and use, but it could be used to analyze any case study involving cross-cultural working and IS. Viewed from a more critical perspective, however, any theory illuminates some elements of particular case situations and is relatively silent on others. Structuration theory is no exception, and as noted by Giddens (1984) himself, the use of structuration theory does not preclude the use of other theories in tandem with it. For example, Walsham and Sahay (1999) drew on actor-network theory to analyze elements of the GIS case other than those discussed in this article. In particular, they focused on the detailed processes of human reflexivity, technical adaptation and network building involved in the case. The structurational analysis in this paper can be supplemented with other specific theories, as appropriate to the particular domain of interest.

Moving finally to the issue of IS practice, what conclusions can be offered? The paper lies squarely within the literature which considers that globalization, facilitated by ICTs, is not leading to simple homogeneity of culture and approach. While it has been argued that culture is not static, the relatively enduring nature of cultural norms and values results from processes of reproduction of structure in the mind. Thus, there is a need for practitioners to be highly sensitive to cultural difference when working in a cross-cultural context. Sensitivity to other cultures does not imply the need for practitioners to change their own attitudes and values to those of the other culture. What is needed is some understanding, and ideally empathy, for the attitudes, norms, and values of others. This offers the possibility of mutual respect between cross-cultural partners and the opportunity for a move toward a more negotiated culture of cooperation.

A detailed discussion of ways in which this can be achieved is beyond the scope of the current paper. However, some broad approaches are worth mentioning in conclusion. Cross-cultural education and training can be achieved through such means as reading, formal courses, and on-the-job facilitation. With respect to the latter, open discussions about difficult cross-cultural issues can be valuable starting points to increased understanding in cross-cultural teams. While technologies, such as GIS, have features that reflect their cultural origins, technology has a degree of *interpretive flexibility* (Pinch and Bijker 1987), and can be adapted and used in different ways. For example, Braa (1997) used the metaphor of *cultivation* to describe the process of adapting Scandinavian technologies and approaches to the different context of the development of South African health information systems. In our more globalized world, cross-cultural working is increasingly common, and the information systems field needs to increase its understanding of the problematic issues involved and approaches to resolving them. It is hoped that this paper makes a modest contribution to these goals.

NOTES

1 Michael D. Myers was the accepting senior editor for this paper.
2 Readers should refer to the earlier published material for details of the research methodology and data collection methods. As a member of the research team, the author had access to all the field notes from the study and has chosen quotes from these as appropriate to illustrate the theme of the current paper, and the new theoretical analysis carried out here.
3 A reader of this section may indeed believe that some of the organizational members were engaging in racial or ethnic stereotyping. Regardless of whether this is or is not the case, we need to make it clear that any such stereotyping reflects the values of those particular organizational members. It does not necessarily reflect the values of other organizational members and it does not reflect the values of the researcher who is reporting the orgnizational members' words. Such stereotyping also does not reflect the values of the editorial policy of the journal publishing the research. We believe it is the responsibility of researchers to report, rather than to cleanse or censure, the data that they collect, where such data include the subjective interpretations that are constructed and held by the organizational members themselves. *MIS Quarterly* stands behind the author of this study in reporting his data, although this does not amount to any endorsement of the organizational members' own opinions. — Michael D. Myers, Senior Editor
4 See footnote 2 above.

ACKNOWLEDGMENTS

The author would like to thank the senior editor, Michael Myers, who was particularly helpful in guiding the paper through the review process. He is also grateful to the anonymous referees and associate editor for their helpful and constructive comments on the earlier drafts of the paper.

REFERENCES

Appadurai, A. *Modernity at Large: Cultural Dimensions of Globalization*, Oxford University Press, New Delhi, India, 1997.

Barrett, M., Drummond, A. and Sahay, S. "Exploring the Impact of Cross-Cultural Differences in International Software Development Teams: Indian Expatriates in Jamaica," in *Proceedings of the Fourth European Conference on Information Systems*, J. D. Coelho, W. Konig, H. Krcmar, R. O'Callaghan and M. Saaksjarvi (eds.), Lisbon, Portugal, 1996.

Barrett, M., Sahay, S. and Hinings, B. "The Process of Building GSO Relationships: The Experience of a Multi-National Vendor with Indian Contractors," in *Proceedings of the Eighteenth International Conference on Information Systems*, K. Kumar and J. I. DeGross (eds), Atlanta, GA, 1997.

Barrett, M. and Walsham, G. "Managing IT for Business Innovation: Issues of Culture, Learning and Leadership in a Jamaican Insurance Company," *Journal of Global Information Management* (3:3), 1995, pp. 25-33.

Beck, U. *What is Globalization?*, Polity Press, Cambridge, UK, 2000.

Braa, J. *Use and Design of Information Technology in Third World Contexts with a Focus on the Health Sector: Case Studies from Mongolia and South Africa*, Unpublished Ph.D. Thesis, Department of Informatics, University of Oslo, Oslo, Norway, 1997.

Brannen, M. Y. and Salk, J. E. "Partnering Across Borders: Negotiating Organizational Culture in a German-Japan Joint Venture," *Human Relations* (53:4), 2000, pp. 451-87.

Carmel, E. *Global Software Teams*, Prentice-Hall, Englewood Cliffs, NJ, 1999.

Castells, M. *End of Millennium*, Blackwell, Oxford, UK, 1998.

Castells, M. *The Power of Identity*, Blackwell, Oxford, UK, 1997.

Castells, M. *The Rise of the Network Society*, Blackwell, Oxford, UK, 1996.

Davenport, T. H. "Putting the Enterprise into the Enterprise System," *Harvard Business Review*, July-August 1998, pp. 121-31.

DeSanctis, G. and Poole, M. S. "Capturing the Complexity in Advanced Technology Using Adaptive Structuration Theory," *Organization Science* (5:2), 1994, pp. 121-47.

Giddens, A. *Central Problems in Social Theory*, Macmillan, Basingstoke, UK, 1979.

Giddens, A. *The Constitution of Society*, Polity Press, Cambridge, UK, 1984.

Goodall, K. "Managing to Learn: From Cross-Cultural Theory to Management Education Practice," in *Managing Across Cultures: Issues and Perspectives* (2nd ed.), M. Warner and P. Joynt (eds), International Thompson Business Press, London, 2002, pp. 256-68.

Hofstede, G. *Culture's Consequences: International Differences in Work-Related Values*, Sage, Beverly Hills, CA, 1980.

Hofstede, G. *Cultures and Organizations: Software of the Mind*, McGraw-Hill, New York, 1991.

Hofstede, G. "Managerial Values', in *Cross-Cultural Management*, T. Jackson (ed.), Butterworth-Heinemann, Oxford, 1995, pp. 150-165.

Jones, M. R. "Structuration Theory," in *Rethinking Management Information Systems: An Interdisciplinary Perspective*, W. L. Currie and R. D. Galliers (eds.), Oxford University Press, Oxford, UK, 1998, pp. 103-35.

Khare, A. "Japanese and Indian Work Patterns: A Study of Contrasts," in *Management and Cultural Values: The Indigenization of Organizations in Asia*, H. S. R. Kao, D. Sinha and B. Wilpert (eds), Sage, New Delhi, 1999, pp. 121-36.

Korpela, M. "Traditional Culture or Political Economy? On the Root Causes of Organizational Obstacles of IT in Developing Countries," *Information Technology for Development* (7:1), 1996, pp. 29-42.

Korpela, M., Soriyan, H. A., Olufokunbi, K. C. and Mursu, A. "Made-in-Nigeria Systems Development Methodologies: An Action Research Project in the Health Sector," in *Information Technology in Context: Implementing Systems in the Developing World*, C. Avgerou and G. Walsham (eds), Ashgate Publishing, Aldershot, 2000, pp. 134-52.

Lacity, M. C. and Willcocks, L. P. *Global Information Technology Outsourcing*, Wiley, Chichester, UK, 2001.

Lam, A. "Embedded Firms, Embedded Knowledge: Problems of Collaboration and Knowledge Transfer in Global Cooperative Ventures," *Organization Studies* (18:6), 1997, pp. 973–96.

Morgan, G. *Images of Organization*, Sage, Beverley Hills, CA, 1986.

Myers, M. D. and Tan, F. B. "Beyond Models of National Culture in Information Systems Research," *Journal of Global Information Management* (10:1), 2002, pp. 24–32.

Orlikowski, W. J. "The Duality of Technology: Rethinking the Concept of Technology in Organizations," *Organization Science* (3:3), 1992, pp. 398–427.

Pinch, T. J. and Bijker, W. E. "The Social Construction of Facts and Artifacts," in *The Social Construction of Technological Systems*, W. E. Bijker, T. P. Hughes and T. J. Pinch (eds.), MIT Press, Cambridge, MA, 1987, pp. 17–50.

Puri, S. K. "Building Networks to Support GIS for Land Management in India: Past Learnings and Future Challenges," in *Proceedings of the IFIP WG9.4 Working Conference on ICTs and Socio-Economic Development: Balancing Global and Local Priorities*, S. Krishna and S. Madon (eds), Bangalore, India, May 2002.

Robertson, R. *Globalization: Social Theory and Global Culture*, Sage, London, 1992.

Sahay, S. "Implementing GIS Technology in India: Issues of Time and Space," *Accounting, Management and Information Technologies* (8:2–3), 1998, pp. 147–88.

Sahay, S. and Krishna, S. "A Dialectical Approach to Understand the Nature of Global Software Outsourcing Arrangements," Working Paper, Indian Institute of Management, Bangalore, 2000.

Trauth, E. M. *The Culture of an Information Economy: Influences and Impacts in the Republic of Ireland*, Kluwer Academic Publishers, Dordrecht, Netherlands, 2000.

Trauth, E. M. "Leapfrogging an IT Labor Force: Multinational and Indigenous Perspectives," *Journal of Global Information Management* (7:2), 1999, pp. 22–32.

Trompenaars, F. *Riding the Waves of Culture*, Nicholas Brealey, London, 1993.

Walsham, G. *Interpreting Information Systems in Organizations*, Wiley, Chichester, UK, 1993.

Walsham, G. *Making a World of Difference: IT in a Global Context*, Wiley, Chichester, UK, 2001.

Walsham, G. and Sahay, S. "GIS for District-Level Administration in India: Problems and Opportunities," *MIS Quarterly* (23:1), 1999, pp. 39–66.

ABOUT THE AUTHOR

Geoff Walsham is a professor of Management Studies at the Judge Institute of Management, Cambridge University, UK. His teaching and research is centred on the development, management, and use of computer-based information systems, and the relationship of information and communication technologies to stability and change in organizations and societies. He is particularly interested in the human consequences of computerization in a global context, including both industrialized and developing countries. His publications include *Interpreting Information Systems in Organizations* (Wiley 1993), and *Making a World of Difference: IT in a Global Context* (Wiley, 2001).

Key Issues in Information Systems Management: An International Perspective

Richard T. Watson, Gigi G. Kelly, Robert D. Galliers and James C. Brancheau

Abstract

This study compares and contrasts the findings of recent information systems (IS) management studies in ten nations or regions as well as one U.S. multinational study. It examines the key concerns of IS executives in these areas, focusing on identifying and explaining regional similarities and differences. Internationally, there are substantial differences in key issues. Possible reasons for these differences—cultural, economic development, political/legal environment, and technological status—are discussed. The analysis suggests that national culture and economic development can explain differences in key issues. The paper concludes with a revised framework for key issues studies that will more readily support comparison across time and nations.

Key words and phrases: information systems management, international computing, key issues in information systems.

INFORMATION SYSTEMS (IS) DEPARTMENTS FACE MANY CHALLENGES in today's rapidly changing, highly competitive global environment. One approach to understanding the challenges faced by IS departments is to survey IS executives and managers to elicit what they consider key issues. This method of gathering and reporting IS issues initially started in the United States over a decade ago [14] and has been extended to several other countries.

Watson and Brancheau [37] report the key-issues studies conducted prior to 1991. After analyzing the similarities and differences among these studies, they explicitly note the importance of such studies and advocate additional key-issues studies in other countries. Furthermore, they suggest that the format and issues of the U.S. Society for Information Management (SIM) key issues study [14] be adopted as the baseline, given its prevalent use by researchers. By using a similar method across studies, comparison of results is enhanced.

Since the publication of Watson and Brancheau's paper [37], there has been considerable research activity in this area. IS key-issues studies have been conducted in many countries, including Estonia, the Gulf Cooperative Council, Hong Kong, India, Slovenia, and Taiwan, while previously published key-issues studies have been repeated in Australia, Europe, the United Kingdom, and the United States, and an international key-issues study based on the perceptions of IS managers of U.S.-based multinationals has been reported. This paper replicates and extends Watson and Brancheau's [37] work by analyzing and integrating the most recently reported key-issues research. We confine our attention to research that included a survey instrument similar to those used in the SIM study. We do not deal with studies based on interviews (e.g., [8,34]) because of the inherent problems and limitations of trying to compare interview and survey research findings.

There are two major motivations for this research. The first motivation is to present the key concerns of IS executives and managers worldwide and to determine which concerns are global and which are regional in nature. By assessing the similarities and differences of regional issues, IS executives should be better prepared to manage their increasingly global function. More important, however, there is a need to understand the forces that shape key issues so that managers can predict and plan rather than react. Thus, the second motivation is to present an explanation for differences in key issues based on variables that differ among regions. While a regional study can attempt to explain the causes of key issues in terms of organizational factors (e.g., [36]), a comparative international analysis can consider other variables (e.g., national culture, economic structure, political/legal environment, and technological status) as well.

Research Method

OUR STUDY EMPLOYS A SECONDARY DATA ANALYSIS [33] using the results of recent studies conducted by other MIS scholars in various regions. Secondary data analysis, which permits the researcher to work with existing data, avoids the costs of data collection, builds on existing findings, and permits comparison of data across studies. It has a downside, however. The original data collection and reporting might have been flawed, the data might have been collected for a different purpose, the data may be obsolete, and there may be problems integrating data from different sources [23, 33]. Despite its shortcomings, secondary data analysis is an appropriate methodology for a comparison of existing key-issues data given the high costs and imposing logistics of administering such a study on a global scale.

Key-Issues Studies

The studies—of Australia, Estonia, Europe, the Gulf Cooperative Council, Hong Kong, India, Slovenia, Taiwan, the United Kingdom and the United States—selected for analysis represent the most current and consistent data available. In the case of the United States, where several studies are available for analysis, two were selected. The most recent U.S. SIM Delphi (hereafter referred to as US SIM) study is chosen because of the strength of its approach and because it was the basis for many of the other studies reported here. Also, the U.S. multinational (hereafter US MNC) study is included because its international perspective aligns with the purpose of this paper.

Rather than reproduce the major findings of each study, we elect to summarize each study briefly (see Table 1) by indicating its region, the authors, and other pertinent details. In addition, we established a World Wide Web site **(www.cba.uga.edu/iris)** to report additional details of each key-issues study. This makes these data available for secondary analysis by other researchers.

Combining the Studies for Analysis

Following the work of Watson and Brancheau [37], the most current published US SIM—based issues [25] are used as the basis for a standardized (international) set of issues. Mapping the national issues into a standard international set permits comparison of the studies from Australia, Estonia, Gulf Cooperative Council, Hong Kong, Slovenia, the United Kingdom, and the United States with only minor reinterpretation. Although not explicitly stated, the wording of the European results is very similar to the US SIM study and allows for easy mapping of issues across studies. The studies of Hong Kong and India, although following the methods of the US SIM study, developed unique issues for each country. Mapping these results into international issues is more difficult as a result of differences in the wording used to describe the issues. Nevertheless, careful review of these studies enables reasonably accurate mapping. The studies for Estonia, Gulf Cooperative Council, Hong Kong, India, and Slovenia include their authors' comparison with the US SIM results [5, 25], thus assisting in the mapping of issues across studies. The US MNC study presents some problems because it includes many issues that are not easily aligned with the US SIM base. Many of these issues deal specifically with problems that multinationals might face (e.g., currency restrictions and exchange-rate volatility). Nevertheless, after reading the rationales for each issue and

Country	Source	US SIM study as basis	No of rounds*	Year data collected	Sample size	Response rate (%)
Australia	Pervan[28]	Yes	3	1991	88	29
Estonia	Dexter, Janson, Kiudorf, and Laast-Laas [13]	Yes	3	1991	24	85
Europe	CSC Index [9]	Yes	1	1993	183	—
Gulf Coop. Council	Badri [2]	Yes	1	1990	96	80
Hong Kong	Burn, Saxena, Ma, and Cheung [6]	No	3	1991–92	46	40
India	Palvia and Palvia [26]	No	1	1990	—	—
Slovenia	Dekleva and Zupancic[12]	Yes	4	1992	148	80
Taiwan	Harrison and Farn [20]	Yes	1	1989	94	39
United Kingdom	Galliers, Merali, and Spearing [17]	Yes	1	1992	66	—
US SIM	Niederman, Brancheau, and Wetherbe [25]	Yes	3	1989	104	59
US MNC	Deans, Karwan, Goslar, Ricks, and Toyne [10]	No	1	1988	121	31

Table 1 Summary of Key-issues studies.

Note
* The original key-issues study [14] used a four-round Delphi approach and many researchers continue to follow the Delphi method [3].

exercising judgment, it is possible to map most of the more general issues.

Table 2 lists the international issues after they were mapped into a parsimonious set. A total of fifty-four unique issues were identified across the eleven studies. Issues that were included in only one survey are not shown. Most of these (21 out of 27) are from the US MNC study.

Table 3 presents the median ranking,[1] in ascending order, of the international issues from the results of the eleven studies presented above. In order to be considered, an issue had to be included in at least five of the eleven studies. This eliminated thirty-five of the fifty-six issues (63 percent), thus indicating that more than half of the issues were unique to a particular nation or region, particularly in the US MNC study.

Inspection of Table 3 indicates some general agreement on the importance of issues; however, there is by no means consensus on what matters. Strategic planning, for instance, the issue with the lowest median ranking, appears in the top three issues in six studies, but it is not deemed important in either the Indian or Estonia studies. Since seven of the median rankings are clustered around 5 and 7, it is not possible to discriminate about the importance of these issues. Essentially, there are several issues that are seen as important across the globe, and there is considerable diversity of opinion as to the relative importance of these issues.

An earlier analysis [37] classified key issues along the dimensions of management/technology, planning/control, and internal/external. This classification was derived from work by Brancheau and Wetherbe [5] and Hirschheim et al. [21]. For example, strategic planning is a managerial, planning, and external issue. Table 4 reports the classification for each issue with the 1991 and 1994 rankings.

KEY ISSUES IN INFORMATION SYSTEMS MANAGEMENT

Issue	National surveys including issue	Surveys issue ranked in top 10	Origination nation/year
1 Developing and implementing an information architecture	5	5	US SIM (1986)
2 Making effective use of the data resource	8	7	US SIM (1983)
3 Improving IS strategic planning	9	6	US SIM (1983)
4 Using information systems for competitive advantage	8	7	US SIM (1986)
5 Aligning the IS organization within the enterprise	11	7	US SIM (1983)
6 Improving information security and control	8	6	US SIM (1983)
7 Increasing understanding of IS's role and contribution	7	4	US SIM (1983)
8 Recruiting and developing IS human resources	11	6	US SIM (1983)
9 Facilitating organization learning and use of IT	8	4	US SIM (1983)
10 Improving the effectiveness of software development	11	7	US SIM (1983)
11 Planning and managing telecommunications	8	4	US SIM (1983)
12 Developing and managing distributed systems	5	0	US SIM (1989)
13 Managing the existing applications portfolio	5	2	US SIM (1983)
14 Measuring IS effectiveness and productivity	7	2	US SIM (1983)
15 Implementing decision and executive support systems (expanded to ESS in 1989)	5	1	US SIM (1983)
16 Facilitating and managing end-user computing	8	2	US SIM (1983)
17 Building a responsive IT infrastructure	7	4	US SIM (1989)
18 Improving disaster recovery capabilities	2	1v	US SIM (1989)
19 Determining appropriate IS funding level	5	1	US SIM (1983)
20 Implementing and managing office automation	3	1	US SIM (1983)
21 Reengineering business processes through IT	2	2	Europe(1989) and UK (1992)
22 Education of senior management	3	3	US MNC (1988)
23 Improving data integrity and quality assurance	5	3	US SIM (1986)
24 Instituting cross-functional information systems	2	2	Europe (1989)
25 Changing technology platforms	2	1	Europe (1993)
26 Moving to open systems/standards	3	1	Europe (1993)
27 National communication infrastructure	2	—	US MNC (1988)

Table 2 International issues in IS management.

Over-all rank	International issue	Australia 1991	Estonia 1991	Europe 1992	Gulf Coop. Council 1991	Hong Kong 1990	India 1991	Slovenia 1992	Taiwan 1989	UK '92	US SIM 1990	US MNC 1988	Median ranking	Conf. interval
1	Strategic planning	1	19	12	1	2	18	3	—	2	3	—	3	1–18
2	IS organizational alignment	3	26	3	5	3	13	5	1	14	7	11	5	3–13
3	Information architecture	6	8	4	—	—	—	—	—	7	1	—	6	1–8
3	Competitive advantage	5	23	—	4	6	—	—	2	6	8	7	6	4–8
3	Data as a resource	4	10	7	14	5	—	—	—	3	2	8	6	3–10
3	Human resources	17	16	6	13	1	2	2	4	15	4	13	6	2–15
3	Security and control	19	4	—	2	—	9	—	8	4	19	2	6	2–19
8	Integrating technology	6	12	21	22	7	6	2	—	—	—	—	7	2–22
9	Software development	8	15	11	9	6	7	12	10	5	9	14	9	7–12
9	IS's role and contribution	9	20	—	6	—	1	1	—	11	11	—	9	1–20

Table 3 International issues in IS management: national rankings.

Note
This table covers only issues included in at least 5 of the 11 studies reviewed.

Among the top ten issues identified in this comparative study, most are management-related concerns. There is a pronounced strategic focus on how IS can support the enterprise, which indicates that IS executives require leadership ability as well as technical skills [1]. Furthermore, there is a split between issues of control and planning, which may reflect the balance between these two perspectives that IS managers must maintain. There is also division between external and internal issues. Clearly, IS management is not a simple task; managers must be able to operate simultaneously in a fluctuating nexus of planning/control and external/issues tensions.

Comparison with 1991 International Key-Issues Study

Although differences in regional coverage make it somewhat dangerous to compare the findings of this study with the 1991 study [37], there is merit in at least viewing the comparison at a high level. There is no association between the 1991 and 1994 ranks (Spearman's rho = 0.54, p = 0.11). This could be due to the rapid changes since the prior study or to the different mix of countries in each study. Observe, however, that the top two issues have not changed, and competitive advantage is still tied for third. There are two new issues, security and control and integrating technology, which displace end-user computing and organizational learning, respectively. Note the elevation of security and control (from 14 to 3) and the demotion of IS's role and contribution (from 3 to 9). End-user computing may have lost importance in part because it has been integrated into organizational computing [24].

Explaining Differences in Regional Rankings

As mentioned previously, we believe key-issues research needs to move beyond just listing and commenting on top issues. It needs to advance understanding of what makes an issue prominent. Several factors (see figure 1) potentially influencing information technology (IT) differences between regions are national culture, economic structure, political/legal environment, and technological status [10]. These variables are consistent with, but broader in scope than those

1994 rank	1991 rank	International issue	M/T	P/C	I/E
1	1	Strategic planning	M	P	E
2	2	IS organization alignment	M	C	E
3	5	Information architecture	M	P	I
3	3	Competitive advantage	M	P	E
3	8	Data as a resource	M	C	E
3	6	Human resources	M	C	I
3	14	Security and control	T	C	I
8	12	Integrating technology	T	C	E
9	10	Software development	T	C	I
9	3	IS's role and contribution	M	P	E

Table 4 International issues in IS management: Classification of issues.

Note
Issues were classified as follows: "M/T" indicates management (M) or technology (T) issue, "P/C" indicates planning (P) or control (C) issue, "I/E" indicates internal (I) or external (E) issue.

suggested by Ein Dor, Segev, and Orgad [15] in their framework for global IS research. That framework proposes that national culture, economic factors, and geographic factors influence the development of IT in different countries. Because of their broader scope, we opted to follow Deans and Ricks' [11] model. Although we present a macro view of the influences on these issues, micro factors may also influence key issues (e.g., CIO personality). Furthermore, there is certain to be considerable debate over the interaction between the factors depicted within the box. For example, the United States, United Kingdom, and Australia favor private enterprise (an economic structure) and democracy (a political/legal environment), and both of these effects may be due to

Country	Power distance	Uncertainty avoidance	Individualism	Masculinity	Time orientation
Australia	36	51	90	61	31
Gulf Cooperative Council	80	68	38	53	
Hong Kong	68	29	25	57	96
India	77	40	48	56	56
Slovenia	76	88	27	21	
Taiwan	58	69	17	45	87
United Kingdom	35	35	89	66	25
United States	40	46	91	62	29

Table 5 Dimensions of culture.

Source: Hofstede [22]

national culture. Nevertheless, we suggest this model as a starting point for considering what influences key issues.

National Culture

Several authors [11, 29] have suggested that differences in national culture may explain differences in IS effects. In one situation, two case studies from Southeast Asia are used to demonstrate how Hofstede's [22] ideas can explain variations in systems in different cultures [29].

Hofstede's [22] revision of his seminal work identifies five dimensions of national culture: power distance, uncertainty avoidance, individualism, masculinity, and time orientation (see the appendix for a description of the dimensions). The values for these dimensions are shown in Table 5, with Europe and Estonia excluded: Europe encompasses many cultures, and Estonia was not included in Hofstede's analysis.

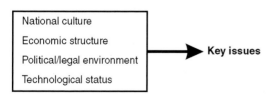

Figure 1 Dimensions influencing key IS issues (from [11]).

Hofstede asserts that the interaction of power distance and uncertainty avoidance results in four dominant organizational forms (see figure 2). Based on Hofstede's measures, the cultures of Australia, the United Kingdom, and the United States are similar, and the difference in these studies should not necessarily be attributed to cultural differences. We could draw similar conclusions for other groups of countries falling in the same quadrant of figure 2. In contrast, differences between countries falling in different quadrants might be explained by cultural differences.

Economic Structure

Another factor to consider is the economic development status of the country or region. Key-issues researchers [27] suggest that, as a nation progresses through different stages of economic and IT development, the relevant key issues should change from infrastructure issues to operational issues, and finally to strategic issues. Table 6 presents gross domestic product (GDP) and GDP per capita for each country. The stage of economic development is based on GDP.

On the basis of economic development, Australia, Europe, Hong Kong, the United Kingdom, and the United States are developed economies. Strategic issues should be of significant interest to this group because of the

Country/region	GDP	Per-capita GDP	Economy
Australia	$375 bililon	$ 20,720	Developed
Estonia	$10 billion	$6,460	Developing
Europe	—	—	Developed
Gulf Cooperative Council	—	—	Developing
Hong Kong	$136 billion	$24,530	Developed
India	$ 1,254 billion	$ 1,360	Developing
Slovenia	$16 billion	$8,110	Developing
Taiwan	$257 billion	$12,070	Developing
United Kingdom	$ 1,045 billion	$17, 980	Developed
United States	$6,738 billion	$25, 800	Developed

Table 6 Economic status for countries/regions.

Source: CIA World Factbook (1995) [7]

advanced nature of IS in these economies. The GCC, India, Slovenia, and Taiwan are developing countries with issues driven more by operational needs. Estonia appears to be a special case within our sample because of its need to create a national IT infrastructure. Economic development coupled with cultural differences may help explain some of the large-scale differences between the studies.

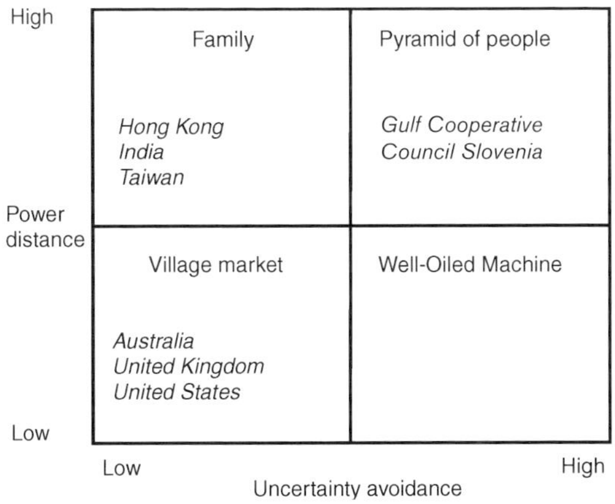

	Low Uncertainty avoidance	High Uncertainty avoidance
High Power distance	Family Hong Kong India Taiwan	Pyramid of people Gulf Cooperative Council Slovenia
Low Power distance	Village market Australia United Kingdom United States	Well-Oiled Machine

Figure 2 National culture and organizational structure (adapted from [22]).

Political/Legal Environment

The political/legal environment can have a significant impact on which IS issues are considered important, as some examples clearly demonstrate. The political transformation in eastern Europe has had a tremendous influence on many aspects of business, including key IS issues such as telecommunications. In India, the government has considerable influence over the deployment of technology. In Hong Kong, the changes anticipated with the 1997 government handover are creating uncertainty. In the United States, the government's policy to convert major elements of the Internet from public to private ownership has made electronic commerce a key issue for many firms. The political/legal dimension is a very prominent factor that can quickly alter which issues are considered critical.

Technological Status

There is considerable diversity in the technological status of many regions. In Estonia, for example, the availability of telephone lines is still very much an issue, and this in turn has a direct impact on telecommunications. Also, the other dimensions (national culture, economic development, and political environment) have an impact on technological status. For example, the GCC countries did not become economic powers until the late 1970s, and it has only been in the last eight years that these countries have invested heavily in technological infrastructure.

Comparing Regional Views Using Correspondence Analysis

Correspondence analysis [19], a visual data analysis method for contingency tables, is an appropriate statistical technique for analyzing regional differences when there are few observations (eleven regions) and four independent variables. It is an exploratory technique for displaying the rows (i.e., issues) and columns (i.e., regions) of a data matrix as points in one- or two-dimensional space. It decomposes the chi-square measure of association of a table into components in a manner similar to that of principal component analysis for continuous data. If two rows or columns have similar profiles, their points appear close together in the perceptual map created by correspondence analysis. Correspondence analysis was used by Watson and Brancheau [37] to display the relationship between key issues and regions. The ranking of issues was transformed into a form suitable for correspondence analysis [37], which was then used to create a perceptual map of the top ten issues and eleven regions.[2]

The eigenvalues generated by correspondence analysis are similar to those obtained in factor analysis. An eigenvalue indicates the amount of variance (inertia in correspondence analysis terms) accounted for by a factor (see Table 7). Slightly more than 65 percent of the variance is explained by the first two eigenvalues, suggesting that a two-dimensional map captures a considerable proportion of the variance.

Another statistic provided by correspondence analysis is the partial contributions to variance for the rows (issues) and column (regions) points. Examination of the issues (see Table 8) reveals that the vertical dimension is mainly explained by integrating technology, competitive advantage, and IS's role and contribution, with respective contributions of 31 percent, 21 percent, and 18 percent. These features represent 70 percent of the variance explained by the vertical dimension (36 percent) or 25 percent (0.36 × 0.70) of the total variance. The horizontal dimension is primarily explained by information architecture (55 percent), or 16 percent of the total variance. Inspection of Table 3 explains why this is so.

Information architecture is ranked relatively high when included in a study, but is missing from many studies.

Examination of the regions (see Table 9) shows that the vertical dimension is mainly explained by Slovenia, US MNC, India, and Taiwan, with respective contributions of 40 percent, 22 percent, 17 percent, and 13 percent. These features represent 92 percent of the variance explained by the vertical dimension (36 percent) or 33 percent of the total variance. The horizontal dimension is mainly explained by Taiwan and Europe, with respective contributions of 24 percent and 22 percent, or 13 percent of the total variance.

One of the advantages of correspondence analysis is the representation of the statistical analysis as a two-dimensional perceptual map. The main purpose of the map is to display graphically those issues that are ranked similarly and those regions in which overall rankings are similar. No relevance can be attached to the particular quadrant in which an issue or region appears. It is the relative vertical or horizontal distance between points that is significant.

Because rankings are missing for some regions, we treat the results cautiously and only point out major differences. Inspection of Figure 3 reveals that on the vertical axis, Slovenia, India, Taiwan, and US MNC (the regions explaining 33 percent of the variance) are quite distant from the remaining regions, which are clustered together in the middle, both vertically and horizontally. As for the top ten key issues, horizontally issue 3 (information architecture, which explains 16 percent of the variance) is quite apart from the rest, possibly, as noted before because six regions did not rank this issue. Vertically, issues 8 and 10 (integrating technology and IS's role and contribution) are close to Slovenia and India because they are relatively more important issues for these nations. Similarly, issues 4 and 7 (competitive advantage and security and control) are clustered near US MNC and Taiwan. It is worth observing that the two U.S. studies are some distance apart. This could be due to differences in the timing and focus of these studies. Broadly, correspondence analysis suggests India, Slovenia, Taiwan, and US MNC differ substantially from the common view because of their attitudes on four issues.

It is useful to compare the perceptual map with a clustering of regions based on national culture and economic development. It is not feasible to cluster on the four explanatory variables because we simply do not have enough data points. Nevertheless, it should be realized that there are strong linkages between these four variables and that economic development captures elements of economic structure, the political/legal system, and technological status. Thus, *village market—developed* (a description used in Figure 4) describes nations that have market economies, democratic systems, and large investments in advanced technologies.

If we use the data from Figure 2 (national culture and organizational structure), Table 6 (stage of economic development), and keep the relative positioning of the perceptual map (Figure 3), we arrive at four clusters (see Figure 4) We cannot say anything about a single region cluster (i.e., family-developed) because we have no points of comparison. Visual inspection indicates that the common culture and stage of economic development explain the relative closeness of Australia, the United Kingdom, and the United States. As mentioned previously, it is the relative vertical or horizontal distance between points on a perceptual map that is significant. Observe that the relative horizontal distance between India and Taiwan is quite small, as it is for GCC and Slovenia. Consequently, the perceptual map provides some graphic support for the notion that national culture and economic development have a major bearing on key issues.

If we examine the issues prominent in the *village market—developed* nations (see Table 3), we see that they center around using strategic planning and data as a resource—reflections of a mature IT economy. This mirrors the nature of these societies, where competitive forces and advanced economies encourage the rapid adoption of new technology. These countries adopted IT many years ago; indeed, they were pioneers in the development of the industry. In contrast, the *pyramid of people-developing* regions, not exposed to the same pressure to compete because decision making is centralized and because there are fewer funds for new technology, have less mature IT economies. Thus, we see many firms in these regions still wrestling with the role of IS in the organization (e.g., IS organizational alignment or IS's role and contribution) and developing human resources.

Regional Analysis

Additional insights are gained from examining how each region compares with the group. A regional analysis also helps to highlight local differences that cannot be attributed to the four dimensions discussed above. We review the regions based on the similarity revealed by the prior analysis and using the categories of figure 4.

Village Market—Developed

The advanced western democracies category, in which markets are the dominant economic form, embraces Australia, Estonia, Europe, the United Kingdom, and the United States. These regions are very similar in terms of economic development, political environment, and technological infrastructure. While there are cultural variations, the market economy is well established in nearly all countries comprising this group, though to differing degrees. We have included Estonia in this category because, although clearly the economy and technological infrastructure are not as developed, it does have a parliamentary democracy and market economy, and the correspondence analysis places it in the vicinity of the members of this group. As a group, the dominant concern is strategic planning and data as a resource. We also observe differences due to local conditions (e.g., reengineering in Europe possibly as a result of continuing economic integration).

US SIM: The US SIM results are not surprising, given the international scope of our analysis. The top issues are strategically oriented, which is to be expected in an advanced economy. Niederman et al. [25] note two important trends: the growing importance of technology infrastructure and the resurgence of internal IS effectiveness issues. The accelerating change in information technology means IS managers have to confront infrastructure issues more frequently. At the same time, IS departments, along with many other corporate functions, are being held to a higher standard of accountability. Downsizing and outsourcing are symptomatic of a highly competitive environment and the requisite greater attention given to organizational effectiveness and the close scrutiny of each unit's organizational contribution.

Dimension	Eigenvalue	Percent
1 (vertical)	0.37	36
2 (horizontal)	0.33	29
3	0.24	15
4	0.18	8
5	0.16	7
6	0.09	2
7	0.06	1
8	0.05	1
9	0.03	0

Table 7 Correspondence analysis eigenvalues.

US MNC: The US MNC study illustrates some of the problems of running an international IS operation. This is particularly evident in the top-ranking key issues (data security, integration of technologies, and the price and quality of telecommunications). The prominence of one issue, end-user computing, is probably due to timing. The study was conducted in 1988 when this was a hot issue. The top-ranked issue, educating senior personnel, is quite distinct. The authors suggest that too many U.S.-based managers assume that what works in the United States will work elsewhere. This both indicates a degree of cultural and technological ignorance that is alarming and has important implications for management well beyond the realm of IS.

Australia: Strategic issues (strategic planning, responsive IT infrastructure, aligning the IS organization with the enterprise) are of top concern in Australia, which is similar to the other well-developed regions in this sample. However, responsive IT infrastructure was ranked higher by Australia than by any other country. The need for integrated technology platforms is recognized as a key component in order to contend in the competitive environment. Open systems have been vaunted as a reality for several years; however, there are still gaps in fulfilling this promise. One reason why IT integration may be of higher importance is that the Australian IS manager, because of trading and historical links, often has more choices than are available in many other countries. Australia and Japan have extensive trade with each other, and Australia has historically had strong business ties with the United Kingdom and United States. Thus, the Australian computer market, as with many other Australian markets, has strong representation from Japanese, U.K., and U.S. firms. The additional choice and competition could make IT integration a more complex task.

An outlier issue, compared with the other studies, is recruiting and developing human resources for IS. This issue was ranked the lowest by Australia. Furthermore, this issue ranked seventeenth in the 1992 study, a very significant decrease from the second place it held in the 1988 Australian study [35]. One reason for this shift may be the current tight job market in Australia resulting in a healthy pool of qualified applicants.

United Kingdom: As with the European results, business process reengineering (BPR) is the top issue in the United Kingdom. As in other advanced regions, strategic issues dominate IS managers' concerns. The low ranking of human resources is similar to that of Australia, but the explanation is different. The researchers suggest that this low ranking is a reflection of "a failure to recognize the importance of staff development for stakeholders ('users' and 'developers') in order to empower

Issue	Vertical (dimension 1)	Horizontal (dimension 2)
Strategic planning	0.06	0.04
IS organization alignment	0.00	0.03
Information architecture	0.02	0.55
Competitive advantage	0.21	0.06
Data as a resource	0.08	0.14
Human resources	0.01	0.03
Security and control	0.13	0.12
Integrating technology	0.31	0.00
Software development	0.00	0.01
IS's role and contribution	0.18	0.01

Table 8 Partial contribution of issues.

Region	Vertical (dimension 1)	Horizontal (dimension 2)
Australia	0.01	0.05
Estonia	0.00	0.05
Europe	0.00	0.22
GCC	0.00	0.06
Hong Kong	0.01	0.00
India	0.17	0.11
Slovenia	0.40	0.03
Taiwan	0.13	0.24
United Kingdom	0.05	0.04
United States	0.02	0.11
US MNC	0.22	0.08

Table 9 Partial contributions of regions.

them to become more effective actors in the utilization of IT in shaping the outcomes of the IS development process" [17]. Furthermore, it is suggested that U.K. managers tend to hold a narrow focus regarding managing IS. U.K. executives appear to have some frustration with information technology. Outsourcing is considered an answer in many cases, but this only puts IT out of sight ... for the time being.

Technology infrastructure was ranked twenty-first, which is an outlier compared with Australia, Europe, and US SIM (2, 7, and 6, respectively). In explaining the low ranking by U.K. managers, Galliers et al. suggest that there is perhaps a naive expectation among U.K. IS managers that open systems and international standards will limit the significance of this issue.

Europe: The more recent European results indicate that BPR is the top issue for European IS executives This issue is missing from all other surveys except the United Kingdom, which is also a more recent survey. Timing of a survey appears to be critical when a new issue rapidly invades the IS and business communities. In addition, building computer systems to support cross-departmental boundaries was a new issue that surfaced in the European survey. This issue is clearly linked with the reengineering focus that is the top concern. Given the goals of the European Union, it is evident that radical changes may be required to support new business alliances and procedures. Organizations need to redesign standard operating procedures to operate in the context of a European rather than a domestic market. Thus, we have the confluence of the current popularity of BPR in the IS press and the need to adapt to a new environment.

Estonia: The key IS issue in Estonia is the planning and implementing of telecommunications systems. The existing telecommunication system is unreliable, resulting in data being transported on disks. Although developed countries, such as the United States, United Kingdom, and Hong Kong, rated telecommunications as an issue (10, 12 and 9, respectively), Estonia's concerns are considerably more basic, because it is concerned with the lack of stable national telecommunications networks. Such networks (taken for granted in developed nations) provide a foundation for building organizational networks. It is currently impractical to link intra- and interorganizational sites. Estonia, like many eastern European countries, has recently experienced massive political and economic changes, which have a direct impact on IS and will continue to do so. Strategic issues such as strategic planning, IS for competitive advantage, and IS organizational alignment ranked nineteenth, twenty-third, and twenty-sixth, respectively, clearly and uniquely distinguish Estonia's IS issues from those of the other countries and regions included in this analysis.

Family—Developing

The developing eastern nations of Taiwan and India, while falling into the same cultural/economic grouping, are vastly different in terms of economic development. This difference is reflected in the prominent key issues. Taiwan, like advanced western economies, shows greater concern for strategic issues and India is wrestling with understanding IS's contribution and human resource issues.

Taiwan: The Taiwanese survey differed the greatest in format from the US SIM survey. It is believed that this departure led to some significant differences in the reported key issues. However, it is interesting to note that two new issues were added: establishing and/or maintaining effective communications with end users, and establishing

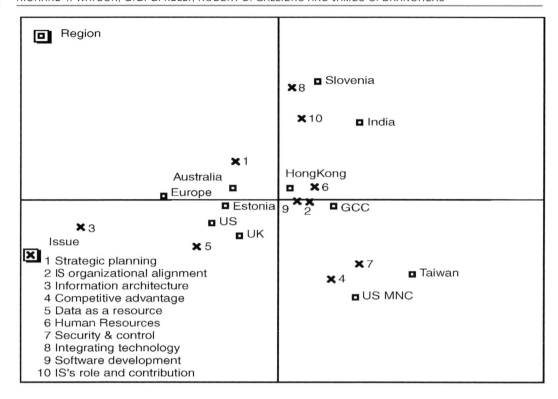

Figure 3 Perceptual map of regions and key issues.

and/or maintaining effective communications with top management (3 and 5, respectively). The emphasis on effective communications has not been explicitly identified in the US SIM survey, although the concept is embedded in many of the issues. By listing this issue separately, the significance of communication for Taiwanese IS managers is highlighted. Would these issues receive as high a ranking if included in the other surveys? This is one of many questions raised in this analysis that can only be answered by further research.

India: India, along with Slovenia, ranked understanding the role and contribution of IS as its top issue, with human resources and personnel for IS as the number two issue. In developing countries, it is not uncommon that there is a lack of knowledge regarding the potential uses of new technologies. It is understandable that human resources and personnel would be a concern for developing countries, given the relative newness of IT for these nations. Education will be required to address these top two issues in both India and Slovenia. A marked difference between India and Slovenia is the issue of strategic planning (eighteenth versus third). IS managers in India appear to be more focused on operational issues with little concern for longer-term issues.

Pyramid of People—Developing

Although these two countries fall into the same cultural/economic quadrant, they are quite dissimilar in many respects. The GCC is an oil province steeped in Arabic culture. Slovenia, an emerging democracy, is a former Yugoslavian republic. Not surprisingly, the key issues differ considerably and appear to be due to local conditions.

Gulf Cooperative Council: The key concerns of the GCC illustrate a mixture of issues. The top issue is improving strategic planning, followed by improving information security and control. This combination correlates with the U.K. ranking of these issues (second and fourth). The data do not provide any clear reasoning for this match. As the world's largest oil-producing region, the GCC is a

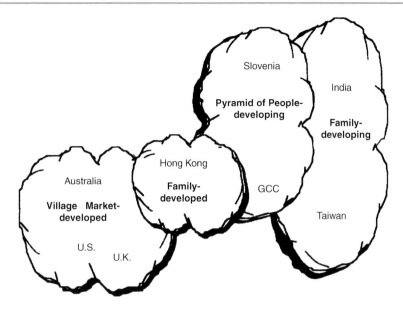

Figure 4 Clustering by national culture and economic development based on the perceptual map.

rich area that is fully embracing information technology. Because the GCC is still in the early stage of development, security and control are an important issue. Furthermore, organizational learning and the use of IS technologies are important as IT revolutionizes the way firms operate. The issues for the GCC appear to be unique with little similarities to the other studies.

Slovenia: The top issues of Slovenia were previously discussed in comparison with India. Although MIS strategic planning was identified as a top issue, other supporting strategic issues such as data as a resource, IS for competitive advantage, and IS portfolio and organizational structure are absent from issues identified in Slovenia. Although it is sensitive to MIS strategic planning, Slovenia as a developing country is primarily concerned with operational issues.

Family—Developed

The only region in this group is Hong Kong, and the issues reflect a very local situation—forthcoming reabsorption into China.

Hong Kong: Retaining, recruiting, and training personnel is a greater issue in Hong Kong than in any other region, ranking as the top issue. Justification for this position is given to the timing of the survey and the political climate that existed in 1990 after the 1989 Beijing incident and with the reintegration of the territory into mainland China in 1997. Unlike Australia, where there may be an oversupply of IS skills, Hong Kong suffers from an unskilled IS work force. Furthermore, many of the skilled personnel have emigrated to other countries. India and Slovenia also face the challenges of finding and retaining qualified IS personnel. Given the international nature of IS and the ready transfer of skills to other countries, it is not surprising to find that Hong Kong's IS professionals are highly mobile. IS professionals, unlike accountants or lawyers, do not face local certification and registration rules (e.g., CPA and bar exams) which impede their movement.

Limitations

KEY-ISSUES STUDIES, WHOSE POPULARITY HAS GROWN RAPIDLY in the past several years, provide insights into the management of IS when the findings for different regions are compared and contrasted. However, any analysis must be done with the understanding that there are several methodological concerns. First, although many studies are based on the US SIM survey, most researchers make modifications when conducting their specific research to take account of local concerns. Second, the period for each of these studies varies from 1988 to 1992. Although there is only a maximum of four years' difference, this can have an impact on the results. Because of the delay between compilation of results and their publication, new and emerging issues may appear only on the most recent regional surveys. For instance, BPR is a phenomenon that has become a key IS issue for developed nations. Although this issue is missing from our analysis, we would be remiss if we did not point out that it currently appears to be paramount in many of IS executives' and managers' agendas throughout the western world, as the U.K. results indicate. Consequently, timing differences contribute to the difficulty of comparing studies because the lack of consideration of some issues results in missing data, with subsequent problems for statistical analysis. Third, the composition of the respondents to the survey can vary from IS executives to general business managers. Finally, methods for analyzing the results of the surveys vary and are reflected in the interpretations and conclusions reported in the regional studies. Despite these limitations, we believe the data provide useful evidence for explaining why some IS management concerns are global and others regional in nature.

Implications and Conclusions

Implications for Practice

Our analysis demonstrates that the substantial differences in key concerns may be due to national culture and economic development. This indicates that IS managers need to be extremely cautious when moving to a culture or economy with which they are unfamiliar. Their experience and expertise may be insufficiently developed for a region whose problems are significantly different from their own. For example, IS managers who have struggled to plan an information architecture in their home country may have insufficient experience to undertake human resources issues in another culture in a different phase of economic development. This suggests that organizations need to assess carefully the particular skill requirements of a region before making an appointment. Those who are successful in one region may not necessarily be best equipped to tackle problems in other parts of the world because they lack relevant exposure to its culture or economic status. Furthermore, the prominence of executive education in the US MNC study implies that appropriate experience may not be enough. Managers will also need education in the cultural, social, political, and economic system of their assigned region before taking up an appointment.

Implications for Research

As we have noted previously, some key issues appear to emerge quickly. The sudden prominence of BPR is somewhat disconcerting because it suggests IS managers may be too willing to respond to a current hot topic, and their attention may be too easily diverted from fundamental, long-term issues. The IS profession is notable for its *fashion swings*. In the last few years the hot topics have included outsourcing, business process reengineering, and client/server. These terms describe fundamental issues that should be of continuing concern to IS managers. Outsourcing can be viewed as a relabeling of a rudimentary but somewhat complex management choice—make or buy. Client/server can be seen as another possibility for minimizing data storage and processing costs. Since these are major costs for any IS department, they should be a continuing issue rather than subject to the vagaries of IS fashion. On the other hand, if the issues become too broad, they may become meaningless to practitioners. Therefore, a careful balance is required in developing an appropriate issue set.

One cause for the perceived rapid emergence of new key issues may be embedded in the US SIM method. There is a large number of distinct issues in this list as well as some indistinct and overlapping issues. Some issues refer to broad general problems, others refer to more narrow and specific concerns. The inconsistent breadth and depth of the issues is an artifact of the US SIM approach of simply asking IS executives what is currently important. The approach has not changed significantly since its first use in 1983 by Dickson et al. [14]. Many researchers have fine-tuned the method, but none has thoroughly reviewed it. We suggest it may be time for rethinking the key issues framework.

Revising or redesigning the key-issues framework is not a trivial task. There are two major avenues open for exploration. One approach is to begin with the international issues listed in Table

1. The broad issues might be used to group the more narrow issues in a two-tier framework. For example, improving the effectiveness of software development (a broad issue) might subsume issues dealing with CASE and object-oriented technology (more narrow issues dealing with software development). Aligning the IS organization goals with those of the enterprise (a broad issue) might embrace issues dealing with structural alignment, strategic planning and competitive advantage (more narrow issues dealing with goal alignment). The two-tier nature of the framework would recognize that broad enduring issues are more stable in time, but that specific aspects of each broad issue come and go. A sufficient number of broad issues would need to be created to cover all the issues in Table 1. Advantages of the "revise" approach include grounding future research in past empirical work and maintaining a reasonable degree of continuity with previous studies. Disadvantages include the lack of a formal theoretical base and the difficulty of classifying "hot new issues" that might cut across several broad issues.

Another approach is to redesign the framework from the ground up. The general idea would be to find a sufficiently relevant theoretical model on which to base a new key-issues framework. Several ideas are introduced here but other directions are also possible. One direction is to build a framework around role theory through careful examination of the roles that IS managers and executives play. There are both an applicable theoretical base [16] and some relevant empirical work with IS executives [1, 4, 30]. A related direction is to examine Sambamurthy and Zmud's [31, 32] empirical work on managerial IT competencies. This work is aimed at identifying the basic capabilities and skills (common to all organizations) needed to effectively leverage IT investments in support of business activities. This work would need to be expanded to apply to smaller enterprises and international contexts, but it might serve as a useful starting point. An additional direction would be to go back to the reference literature in the organizational sciences in search of a suitable theoretical model dealing with general management practices. Advantages of the "redesign" approach include the possibility that the framework be complete, consistent, parsimonious, and both regionally and temporally stable. Disadvantages include the lack of continuity with previous studies and the danger that the issues might become so abstract that they would cease to have meaning to IS managers and executives, thus breaking an important link to practice.

A fundamental issue that needs to be addressed is the tradeoff between a study within a region that attempts to address the particular concerns of that locality and the objective of comparative studies across cultures. To maintain their relevance, local studies need to reflect the issues of their region and focus on variables within that society to explain the causes of key issues. Comparative studies, which need a common base of issues so that variables between societies can be examined, may neglect local concerns. One of the limitations of this study is that it relies on secondary data collected from studies focusing on local issues, and this is reflected in missing observations and a flexible interpretation of the wording of issues in different studies. A possible solution is a research protocol with dual objectives—a focus on the specific and on the general. There is a core set of issues with common wording in all studies that is augmented with additional issues to cover the local situation.

In discussing the limitations, we indicated that the timing of the various studies complicated the problem of comparison. New issues can emerge quickly (e.g., electronic commerce) and are often missing from older key-issues studies. The simple solution is to remove the time variable from the equation. We need to establish a group of scholars in different regions who agree to conduct simultaneously a key-issues study in each region.

We feel the time has come to carefully reconsider the key-issues approach so that we develop a methodology that is less susceptible to fluctuations in IS fashions because it addresses the fundamentals of IS management. There should be less need for adaptation to local conditions so that we then might have a single instrument for many regional studies. Obviously, the comparison of international key issues will be simpler and more accurate if the same set of fundamental issues is addressed in each region. In addition, such an instrument will also be valuable for the comparison of key issues within a single country over a period.

Final Remarks

This comparative study of key issues gives us some understanding of the regional diversity of problems faced by IS managers. We have put forward several explanations for this diversity. First, cultural, economic, political/legal, and technological factors may influence regional key issues. We offer some preliminary evidence that these factors are viable causes of diversity. Second, we also suggest that because of the speed with which new issues emerge, timing of data collection can explain differences. Finally, we suggest the very nature of the methodology encourages diversity because it focuses on current problems rather than fundamental issues. Consequently, we urge our colleagues who conduct key-issues studies to consider redesigning the methodology before undertaking their next study. We would like to see a new approach that could provide a firmer basis for comparative key issues studies.

NOTES

1 Because rankings are ordinal data, the median was used to indicate the central tendency of each issue across nations. An approximate 95 percent confidence interval for the median ranking was calculated using the method described by Gibbons [18].
2 The values reported by the correspondence analysis should be treated with due caution given the mapping of similar, but not identical, issues to a common set for a cross-study comparison and missing values in the data set.

REFERENCES

1 Applegate, L. M. and Elam, J. J. New information systems leaders: a changing role in a changing world. *MIS Quarterly, 16*, 4 (December 1992), 469–90.
2 Badri, M. A. Critical issues in information systems management: an international perspective. *International Journal of Information Management, 12* (1992), 179–91.
3 Bass, B. M. *Organizational Decision Making*. Homewood, IL: Irwin, 1983
4 Benjamin, R. I.; Dickinson, C.; and Rockart, J. F. The changing role of the corporate information systems officer. *MIS Quarterly, 9*, 3 (September 1985), 177–88.
5 Brancheau, J. C. and Wetherbe, J. C. Key issues in information systems management. *MIS Quarterly, 11*, 1 (March 1987), 23–45.
6 Burn, J.; Saxena, K. B. C.; Ma, L.; and Cheung, H. K. Critical issues of IS management in Hong Kong: a cultural comparison. *Journal of Global Information Management, 1*, 4 (September 1992), 28–37.
7 CIA. *The World Fact Book*, 1995. <www.odci.gov/cia/publications/95fact>.
8 Clark, T.D. Corporate systems management: an overview and research perspective. *Communications of the ACM, 35*, 2 (February 1992), 60–75.
9 *Critical Issues of Information Systems Management for 1993. The Sixth Annual Survey of I/S Management Issues*. MA: CSC Index, 1993.
10 Deans, C. P.; Karwan, K. R.; Goslar, M. D.; Ricks. D. A.; and Toyne, B. Key international IS issues in the U.S.-based multinational corporations. *Journal of Management Information Systems, 7*, 4 (1991), 27–50.
11 Deans, C. P. and Ricks, D. A. MIS research: a model for incorporating the international dimension. *Journal of High Technology Management Review, 2*, 1 (Spring, 1991), 57–81.
12 Dekleva, S. and Zupancic, J. Key issues in information systems management: a delphi study in Slovenia. In J. I. DeGross, R. P. Bostrom and D. Robey (eds), *Proceedings of the Fourteenth International Conference on Information Systems*, Orlando, Florida, December 5, 1993. New York: ACM, 1993, pp. 301–14
13 Dexter, A. S.; Janson, M. A.; Kiudorf, E.; and Laast-Laas, J. Key information technology issues in Estonia. *Journal of Strategic Information Systems, 2*, 2 (June 1993), 139–52.
14 Dickson, G. W.; Leitheiser, R. L.; Wetherbe, J. C.; and Nechis, M. Key information systems issues for the 1980's. *MIS Quarterly, 8*, 3 (September 1984), 135–48.
15 Ein-Dor, P.; Segev, E.; and Orgad, M. The effect of national culture on IS: implications for international information systems. *Journal of Global Information Management, 1*, 1 (Winter, 1993), 33–44.
16 Galletta, D. F. and Heckman, R. L., Jr. A role theory perspective on end-user development. *Information Systems Research, 1*, 2 (1990), 168–87.

17. Galliers, R. D.; Merali, Y.; and Spearing, L. Coping with information technology? How British executives perceive the key information systems management issues in the mid-1990s. *Journal of Information Technology, 9*, 3 (1994), 223–38.
18. Gibbons, J. D. *Nonparametric Statistics. An Introduction.* Newbury Park, CA: Sage, 1993.
19. Greenacre, M. J. *Theory and Applications of Correspondence Analysis* London, Academic Press, 1984.
20. Harrison, W. L. and Farn, C. A comparison of information management issues in the United States of America and the Republic of China. *Information and Management, 18*, 4 (April 1990), 177–88.
21. Hirschheim, R.; Earl, M; Feeny, D.; and Lockett, M. An exploration into the management of the information systems function: key issues and an evolutionary model. *Information Technology Management for Productivity and Competitive Advantage. An IFIP TC–8 Open Conference*, 1988, 4.15–4.38.
22. Hofstede, G. *Cultures and Organization Software of the Mind.* New York: McGrawHill, 1991.
23. Hyman, H. H. *Secondary Analysis of Sampling Surveys: Principles, Procedures, and Potentialities.,* New York: Wiley, 1972.
24. McLean, E. R.; Kappelman, L. A.; and Thompson, J. P. Converging end-user and corporate computing. *Communications of the ACM, 36*, 12 (December 1993), 78–89.
25. Niederman, F.; Brancheau, J. C.; and Wetherbe, J. C. Information systems management issues for the 1990s. *MIS Quarterly, 7*, 4 (December 1991), 475–500.
26. Palvia, P. C. and Palvia, S. MIS issues in India and a comparison with the United States. *International Information Systems* (1992), 101–110.
27. Palvia, P. C.; Palvia, S.; and Zigli, R. M. Global information technology environment: key MIS issues in advanced and less-developed nations. In S. Palvia, P. Palvia and R. Zigli (eds.), *The Global Issues of Information Technology Management.* Harrisburg, PA: Idea Group Publishing, 1992, pp. 2–34.
28. Pervan, G. P. Results from a study of key issues in Australian IS management. *Proceedings of the 4th Australian Conference on Information Systems.* University of Queensland, St. Lucia, Brisbane, Queensland, September 28, 1993, pp. 113–20.
29. Raman, K. S. and Watson, R. T. National culture, IS, and organizational implications. In P. C. Deans and K. R. Karwan (eds), *Global Information Systems and Technology: Focus on the Organization and its Functional Areas.* Harrisburg, PA: Idea Group, 1994, pp. 493–513.
30. Rockart, J. F.; Ball, L.; and Bullen, C. V. The future role of the information systems executive. *MIS Quarterly, 6*, special issue (December 1982), 1–14.
31. Sambamurthy, V. and Zmud, R. W. *Managing IT for Success: The Empowering Business Relationship.* Morristown, NJ: Financial Executives Research Foundation, 1992.
32. Sambamurthy, V. and Zmud, R. W. *Competency Assessment: A Self-Assessment Tool for Creating Business Value Through IT.* Morristown, NJ: Financial Executives Research Foundation, 1994
33. Stewart, D. W. and Kamins, M. A. *Secondary Research: Information Sources and Methods*, 2d ed. Newbury Park, CA: Sage, 1993.
34. Straub, D. W., Jr. and Wetherbe, J. C. Information technologies for the 1990s: an organizational impact perspective. *Communications of the ACM, 32*, 11 (November 1989), 1328–39.
35. Watson, R. T. Key issues in information systems management: an Australian perspective—1988. *Australian Computer Journal, 21*, 3 (1989), 118–29.
36. Watson, R. T. Influences on information systems managers' perceptions of key issues: information scanning and relationship with the CEO. *MIS Quarterly, 14*, 2 (June 1990), 217–31.
37. Watson, R. T. and Brancheau, J. C. Key issues in information systems management: an international perspective. *Information and Management, 20*, 3 (1991), 213–23.

ABOUT THE AUTHORS

Richard T. Watson is an associate professor in the Department of Management at the University of Georgia. His received his Ph.D. in management information systems from the University of Minnesota. He has published in leading journals in MIS, auditing, marketing, business ethics, and communication, and is the author of books on data management and electronic commerce. His current research, which has a strong international flavor, focuses primarily on electronic commerce and management of the MIS function.

Gigi G. Kelly is an assistant professor of Management Information Systems at the College of William and Mary. She holds a B.B.A. from James Madison University, an M.B.A. from Old Dominion University, and a Ph.D. from the University of Georgia. Dr. Kelly teaches in the areas of systems analysis and design, decision support systems, database, and information resource management. Her research interests include group support systems, facilitation, and team development. She has extensive consulting experience in MIS and has published articles in *Small Group Research*, *Computerworld* and various conference proceedings.

Robert D. Galliers is chairman and professor of Information Management at Warwick Business School, University of Warwick, U.K. Prior to his appointment as chairman, he headed Warwick's Doctoral Program in Information Systems and its Information Systems Research Unit. He was previously Foundation Professor and Head of the School of Information Systems at Curtin University, Perth, Western Australia, where he developed Australia's first master's program to emphasize the management issues associated with the introduction and utilization of IT in organizations. He has published widely on aspects of information systems management and is editor-in-chief of the *Journal of Strategic Information Systems*. His recent books include *Information Systems Research: Issues, Methods and Practical Guidelines* and *Strategic Information Management: Challenges and Strategies for Management Information Systems* (with B.S.H. Baker).

James C. Brancheau is associate professor of Information Systems at the University of Colorado, Boulder. He holds a Ph.D. in MIS from the University of Minnesota and has twenty years' experience in the information systems field. Dr. Brancheau conducts research in the general area of managing the IS function and the specific area of implementing new information technologies. His work has been published in leading journals such as *Information Systems Research, MIS Quarterly*, and *ACM Computing Surveys*. He is currently working with industry partners on several Web projects aimed at developing and researching applications in distance education and electronic commerce.

APPENDIX A

DIMENSIONS OF NATIONAL CULTURE

Individualism

Societies differ in their emphasis on individual rights and obligation to society. Individualism describes societies in which the ties between individuals are loose and people are expected to look after themselves and their immediate families. Collectivism describes societies in which people from birth are integrated into strong, cohesive groups that continue to protect the individual throughout life.

Power Distance

Power distance describes a culture's social relationship between superiors and subordinates. It is the extent to which the less powerful members of institutions and organizations in a culture expect and accept that power is distributed unequally.

Uncertainty Avoidance

Uncertainty avoidance is the extent to which the members of a society feel threatened by uncertain or unknown situations and some societies take considerable pain to avoid uncertainty.

Masculinity

The masculinity dimension describes the extent to which social gender roles are differentiated in a society. This is reflected in the way jobs are distributed in a society. Very masculine societies tend to have few women in some occupations. High status positions are usually reserved for men. Feminine societies have a more equal distribution of social gender roles.

Time Orientation

Time orientation can be described by a continuum with long-term orientation at one pole and short-term orientation at the other. Eastern and western societies have different perceptions of time and time orientation. Eastern nations and organizations are more likely to invest in projects whose payoffs are quite distant. In contrast, American society and its organizations have a short-term time orientation and are driven by short-term goals.

The Global Digital Divide: A Sociological Assessment of Trends and Causes

Gili S. Drori and Yong Suk Jang

The authors thank John Meyer, Francisco Ramirez, and the members of Stanford University's Comparative Workshop for their helpful comments on earlier versions of this article. An earlier draft of this article was presented at the 2001 annual meeting of the American Sociological Association, Anaheim, California.

Abstract

This article concerns the comparative dimension of the digital divide, the global digital divide, assessing processes of various sorts that shape this differentiating feature and mapping its trends of change. The authors (a) "map" the global digital divide with multiple indicators to find patterns of varying degrees of differentiation between countries worldwide and between blocs of countries and (b) assess the various national characteristics that contribute to the level of IT connectedness. The authors compare the effects of these various social national conditions on the pervasiveness of IT. Their preliminary results indicate that it is neither political nor economic national characteristics that are the determinants of IT connectedness but that cultural features are the prime causes. In this sense, the global digital divide is more a product of networking into global society than it is a mere reflection of local economic capabilities.

Keywords: digital divide; telecommunications networks; international trade; economic development; democratization; education; social conditions; equality

"The world [at the turn of the 21st century] is divided not by ideology but by technology," argued Jeffrey Sachs (2000, p. 1). His statement highlights a new global divide, the global digital divide, adding a layer of differences to the already immense cross-national inequality. This new divide also enhances our sense of urgency with regard to the issue of global disparities: Yet again global differences in access to resources and in networking with world society further disenfranchise the already marginalized groups and nations. According to Sachs, "Many of the technologically-excluded regions...are caught in a poverty trap" (p. 1). Their condition is further aggravated by the ideals of the current global policy, expecting that any economic growth will rely increasingly on the tenets of the "new economy" and on the promise of information technology (IT). It seems, then, that the IT revolution brings both hope and despair to countries worldwide: It brings the world together, facilitates the transfer of information and goods, and links the countries worldwide in a web of interconnections, yet simultaneously, it adds another layer to the process of global differentiation and exacerbates global inequalities.

The general goal of this article is to further analyze the global digital divide and to comment on its promise. Specifically, the goals of this article are to (a) describe the trends of global IT disparities with multiple cross-national indicators of IT, (b) analyze the country-level features that facilitate IT connectivity, and (c) comment on the potential consequences of the global digital divide. The three parts of this article correspond to the three goals. The first part reviews current literature describing the global digital divide and relying on cross-national data assesses the trends of IT global diffusion since mid-1980s. We find that countries worldwide differ greatly by their level of connectivity with the global IT network but that such differentiation depends very much on the type of IT. In addition, we find that trends of change in this divide also depend on the measure of IT: Whereas overall the global digital divide is expanding, with regard to some dimensions of IT, there is evidence of diminished cross-national differences.

In the second part, we set a cross-national panel regression model to investigate the factors shaping the trends of change in IT connectivity between 1980 and 1995. We thus investigate the effects of several economic, political, and cultural factors on the change in IT connectivity between 1980 and 1995. By contrasting between economic and political factors, on one hand, and cultural factors, on the other hand, we reach two major conclusions: (a) Noneconomic considerations have the prime effect on longitudinal changes in IT connectivity, and (b) education in general and science in particular have a central role in encouraging countries toward greater IT connectivity. Overall, then, despite the economic and political nature of the justifications for IT connections, it is primarily connectedness with global society and local cultural changes—expressed by education and science penetration—that serve as the critical antecedents for any change in the global digital divide.

We conclude this article with a general commentary on the potential consequences that the global digital divide may have on issues of global development, incorporation, and inequality.

Describing The Global Digital Divide And Its Trends of Change

By definition, the global digital divide points to great cross-national differences in IT resources and capabilities. Such IT resources are concentrated in the hands of a few Western, mostly European, affluent countries, whereas the rest of the world lags behind them. Although this global divide is well documented (Compaine, 2001; Main, 2001; United Nations Development Program, 2001), the parameters of the gap are still impressive, and moreover, they differ by the indicator used to measure IT. Figure 1 confirms these impressions, relying on three different cross-national measures of IT: the number of PCs (per 1,000 population in 1998; Figure 1A), the number of Internet hosts (per 10,000 population in 1999; Figure 1B), and the number of National Science Foundation Networks (NSFNets) (by May 1995; Figure 1C).[1] Across these three measures of IT, we clearly see that the distribution of these cross-national measures of IT resources by regional location is highly skewed: IT is condensed within Western Europe and North American countries. In addition, some IT is more unevenly distributed than others: The number of NSFNets is the most concentrated in the West (Figure 1C), whereas the number of PCs is the least concentrated in the West (Figure 1A).

Such uneven cross-national distribution of IT appears even more dramatic when nation-states are differentiated by level of development. As presented in Figure 2, the differences in IT resources between Western countries[2] and non-Western countries are dramatically greater: NSFNets are by far more concentrated in developed countries than in developing countries, and the rate of their proliferation in the duration since initial connection is also dramatically quicker in developed countries than in developing countries.

Although IT is still a feature of mainly the developed world, IT is permeating countries worldwide. Figure 3 indicates that the sophisticated use of Internet[3] changed dramatically between 1991 and 1995. Whereas in September 1991, 119 countries (of 210 valid cases), mostly less developed countries, had a usage score of zero, by June 1995, only 56 countries (of 220 valid cases) received a usage score of zero. During this period, most countries not only joined Internet but also

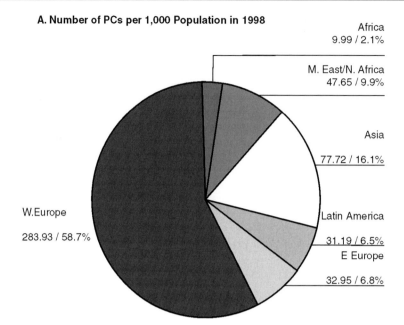

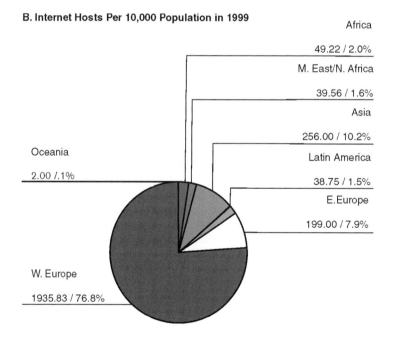

Figure 1 Cross-national IT distribution per region.

Note
IT = information technology; NSFNet = National Science Foundation Network.

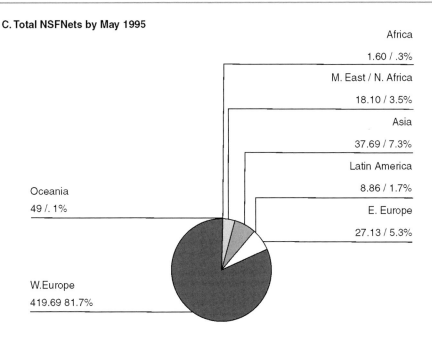

Figure 1 Cross-national IT distribution per region.

Note

IT = information technology; NSFNet = National Science Foundation Network.

increased the intensity of their use of Internet. Although the highest category of usage (value = 8) included only 17 countries in both time periods,[4] there is a dramatic increase in the size of the other categories of high Internet usage. In other words, whereas in September 1991 only 5 countries received a usage score of 6 and 4 countries received a usage score of 7, by June 1995, these categories included 21 and 17 countries, respectively. In summary, there is a dramatic globalization of Internet connections and a dramatic increase in the intensity of Internet usage.

The worldwide diffusion of IT does not, however, occur evenly across countries and world regions. As shown in Figure 4, it is the Eastern European countries that have experienced the most dramatic rates of change in the net-use sophistication. One would expect such dramatic changes in the semiperipheral countries rather than in the core or periphery: The core countries are reaching a "ceiling" of use sophistication,[5] whereas the peripheral countries are lagging behind in this process of IT globalization. Also, in the years 1991–1995, Eastern Europe held the biggest promise for social change, coming out of the wings of sectarian communist regimes.

The changes in the intensity of IT use worldwide are occurring with respect to all IT measures. In this sense, there is a close correspondence between different dimensions of the IT revolution. Table 1 shows evidence of such correspondence as correlation scores among various cross-national measures of IT. The correlation scores are very high and significant, verifying the close interrelations among IT.

In summary, the global digital divide is a well-substantiated phenomenon: countries differ greatly in their access to, and resources of, IT. Whereas on some dimensions of IT there seems to be a closing of this digital gap, most cross-national studies of IT show the gap to be expanding. Moreover, although all countries are engaged in Internet activity of some sort, the distinguishing characteristic is no longer the mere connection or

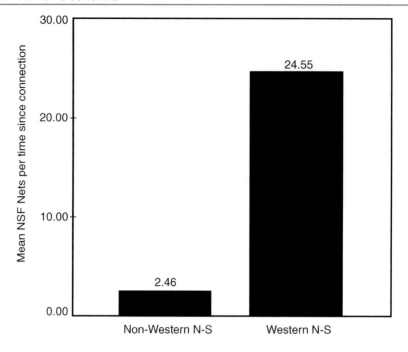

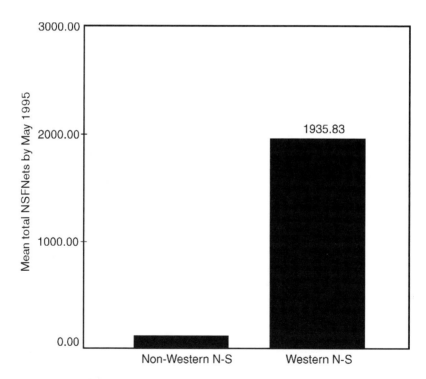

Figure 2 Cross-national IT distribution per development state.

Note
IT = information technology.

Correlations

	Number of PCs per 1,000 population in 1998	Internet Hosts per 10,000 Population 1999	Initial Connection to NSFNet Month	Initial Connection to NSFNet Year	Total NSFNets by May 1991	Net Sophistication Score at September 1995	Net Sophistication Score at June 1995	Score Change Between 1991 and 1995	NSFNets per Time since connection
Number of PCs per 1,000 population in 1998	1.000	.808**	.012	−.756**	.381**	.730**	.6.6**	−.333**	.382**
	—	.000	.928	.000	.002	.000	.000	.001	.002
N	107	107	61	61	61	94	94	94	61
Internet hosts per 10,000 population 1999	.808**	1.000	.095	−.055	.673	.577**	.467**	−.235*	.672**
	.000	—	.452	.663	.000	.000	.000	.010	.000
N	107	144	65	66	65	118	118	118	65
Initial connection to NSFNet month	.012	.095	1.000	−.216	.001	.164	.046	−.210	.005
	.928	.452	—	.051	.994	.141	.682	.059	.968
N	61	65	82	82	82	82	82	82	82
Initial connection to NSFNet year	−.756**	−.055	−.216	1.000	−.285**	−.872**	−.737**	.539**	−.286**
	.000	.663	.051	—	.010	.000	.000	.000	.009
N	61	66	82	83	82	82	82	82	82
Total NSFNets by May 1995	.381**	.673**	.001	−.285**	1.000	.225*	.203	−.123	.999**
	.002	.000	.994	.010	—	.042	.067	.270	.000
N	61	65	82	82	82	82	82	82	82

Table 1 Correlations among IT indicators.

Correlations	Number of PCs per 1,000 population in 1998	Internet Hosts per 10,000 Population 1999	Initial Connection to NSFNet Month	Initial Connection to NSFNet Year	Total NSFNets by May 1991	Net Sophistication Score at September 1995	Net Sophistication Score at June 1995	Score Change Between 1991 and 1995	NSFNets per Time since connection
Net sophistication score at September 1991	.730**	.577**	.164	-.872**	.225*	1.000	.837**	-.198**	.232*
	.000	.000	.141	.000	.042	—	.000	.005	.036
N	94	118	82	82	82	199	198	198	82
Net sophistication score at June 1995	.606**	.467**	.046	-.737**	.203	.837**	1.000	.371**	.216
	.000	.000	.682	.000	.067	.000	—	.000	.051
N	94	118	82	82	82	198	200	198	82
Score change between 1991 and 1995	-.333**	-.235*	-.210	-.539**	-.123	-.198**	.371**	1.000	-.120
	.001	.010	.059	.000	.270	.005	.000	—	.285
N	94	118	82	82	82	198	198	198	82
NSFNets per time since connection	.382**	.672**	.005	-.286**	.999**	.232*	.216	-.120	1.000
	.002	.000	.968	.009	.000	.036	.051	.285	—
N	61	65	82	82	82	82	82	82	82

Note

IT = information technology, NSFNet = National Science Foundation Network.
*Correlation is significant at the .05 level, two-tailed. **Correlation is significant at the .01 level, two-tailed.

initial link with the Internet but rather the intensity and sophistication of such a connection. In other words, today, the crucial dimension of IT is the bridging of the global divide through permeation of IT into whole societies (rather than their elites) and into as many social sectors (rather than merely the export-directed manufacturing sector or the academic sector, which are the first national segments to be connected, if any).

Yet, if national immersion in the global IT culture is to be achieved, what are the national characteristics that are currently holding some countries back? Stated positively, what are the national features that are at the basis of global e-connectivity? To comment on this question, we continue this article with a cross-national, longitudinal study of the national-level factors that shape the change in e-connectivity during the early 1990s.

Factors Shaping The Global Digital Divide

The core question here is, What social factors shape the ability of countries worldwide to absorb, relate to, or connect with IT? Undoubtedly, IT is diffusing worldwide, compelling countries to change their infrastructural relations with IT over time. Yet, what accounts for the differential ways with which countries, once exposed to the IT revolution, change their IT base?

Building on classical sociological theories of change, social scientists commonly cite increasing social complexity as the cause for modernization of infrastructure and practices. The complexity of systems—of trade, of political scope and responsibilities, of production—requires that the tools of managing and controlling these systems be adequate. Therefore, the more complex social systems are, the more their tools for administering such systems are current. IT is just such a tool: It enables complex national production systems to engage with the rapidly changing and highly IT global commerce environment.

We contend, however, that it is the national ideological commitment to social development, rather than social complexity per se, that is at the root of worldwide trends toward greater engagement with IT. Currently, IT is perceived as the "wave of the future," implying that it is the key factor to national development and citing such exemplary cases as the Asian Tigers that have emerged "From Third World to First."[6] Because of this common perception, IT is adopted by countries worldwide, even if its immediate application is doubtful.[7] In this sense, the more committed a society is to the ideology of national development, the more its institutions will incorporate IT. Moreover, such ideological commitment is related to the level of immersion within the world polity: The more embedded a country is in the modern world polity, the more engaged it is in the global IT network.

We contrast these two explanations through a series of empirical propositions. To identify the loci of system complexity, we offer three such sites: national economic development, national political regime, and economic connections with international markets. First, national economic development identifies the complexity of local economic systems. It is indicated by GDP per capita in 1980 (logged). Second, the level of institutional democracy in 1980 identified the complexity of the national political system.[8] Last, the intensity of economic ties with other countries, as indicated by the index of economic openness in 1985, reflects the complexity of international economic connections. These various sites of complexity identify potential sources of pressures on nation-states to more intensely engage in IT networks.

National commitment to the ideology of development is identified by the intensity of education programs and particularly of their emphasis on science. The penetration of global models through education and scientization serves as a carrier of additional global models, such as the notion of development. To indicate such educational and scientization penetration, we rely on a factor of education and science indicators circa 1980.[9] The factor is composed of the enrollment ratio in secondary education, the enrollment ratio in tertiary education, scientific paper publications per capita, and citation counts per capita.[10] Last, the intensity of contacts with the world society (i.e., embeddedness in the world polity) is indicated by a factor of world linkages. The indicators composing this factor are membership in international nongovernmental organizations in 1980, numbers of treaty signing 1981–1985, and membership in international nongovernmental science organizations.[11] Overall, then, we contend that expanded national linkages to the world

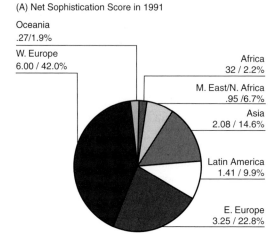
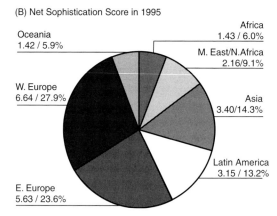
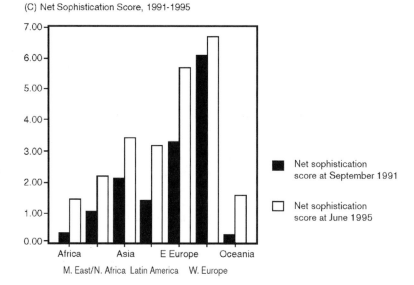

Figure 3 Net sophistication scores, 1991 and 1995.

society with increased emphases on science and education produce higher levels of engagement in global IT.

The indicators are drawn from a variety of sources: Data files are compiled by Banks, Gurr, the United Nations Development Program, and UNESCO. To keep a stable set of cases, we select only those country cases that have data on all dimensions and indicators, thus evaluating the process of IT change for 82 countries. Appendix A lists the names of countries included in the analyses (where they serve as the unit of analysis, as is appropriate for cross-national evaluations).

To overcome potential problems of single-indicator bias, we rely on multiple indicators for each of the model's dimensions,[12] and we alternate such indicators in the models. Despite such change of indicators, our models produce substantively identical results.

Last, we rely on net sophistication score as the dependent variable indicating IT connectivity. We chose this particular variable, out of a growing list of cross-national IT measures, for its two distinctive qualities: Data are available for two time points (September 1991 and June 1995[13]), thus allowing for (a) evaluation of over-time change (and thus a panel regression model) and (b) evaluation of such change at the early stages of the global IT "explosion." Although these scales have "aged," they have done so gracefully: Today, parallel cross-national longitudinal indicators are available for other dimensions of IT (e.g., PC ownership and Internet hosts), but it seems that they share basic features with earlier IT measures (such as skewness) but lack the time perspective.

The Causal Model

Our hypotheses set IT connectivity, indicated by net sophistication score, as the dependent variable in a causal and panel model. The model assesses the cross-national effects of a series of independent variables (measured in early 1980s) on IT connectivity (measured in 1995), while controlling for IT connectivity in 1991. Model specification is as follows:

$$Y_{i(1995)} = \alpha_0 + \beta X_{i(1991)} + \gamma Z_{i(1980)} + \varepsilon_i,$$

where the subscript i is for nation-state and Y_i is the dependent variable, IT connectivity in 1995; X is a lagged dependent variable, IT connectivity in 1985; and Z is a row vector covariate that includes national development, educational and scientific expansion, and world linkage in 1980; α_0 is an intercept; β is a coefficient associated with the lagged dependent variable; γ is a vector of coefficients associated with other covariates; and ε_i is independent random disturbances.

Findings

The results of our panel model, summarized in Table 2, indicate that the level of economic complexity has no significant effect on the 1991–1995 change in a nation's level of IT sophistication. Whereas in Model 1, the effect of the factor score for economic development in 1980 has a positive and significant effect as the sole control variable, such effect becomes nonsignificant and negative in all subsequent models (2–5) when additional factors are introduced into the regression. The negative sign of that regression coefficient is, most probably, a reflection of the fact that the net sophistication scores have increased faster in non-Western countries than in Western countries (see Figure 4). Alternative indicators of system complexity—economic openness (Model 4) and democracy (Model 5)—also have no significant effect on change in IT sophistication. Nevertheless, it is clear from these various models that there is a sharp divide between Western and non-Western countries in their tendency to upgrade their IT usage.

Table 2 indicates, however, that (a) world linkages and (b) permeation of science and education are strong determinants of change in net sophistication scores. Indeed, both factors have consistent strong positive and significant effects on net sophistication score, 1991–1995 (Models 2–5); also, despite various additions to the general model, their significant effects are stable. Scatter plots of net sophistication scores with both these variables reveal clear linear relations (see Appendix B). These robust results indicate that the more links a country has to the global society and that the more a country is permeated by the institutions (and practices) of science and education, the more it upgraded its IT usage during 1991–1995. In this sense, it is factors related to the ideology of development, rather than

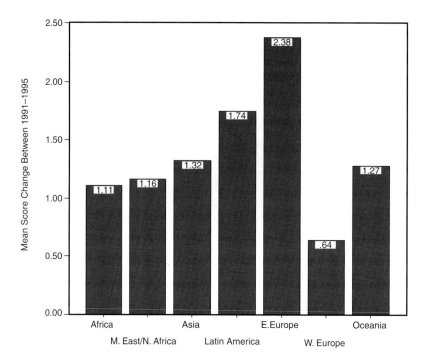

Figure 4 Net sophistication score change, 1991–1995.

development per se, that are crucial for a country's IT connectivity.

Our models for investigating the national factors that shape IT connectivity are, however, preliminary. We intend to verify our results, stable and valid as they are currently, by adding several additional models. First, we intend to transform our dependent variable from a single indicator into a factor score. Currently, few cross-national measures of IT are longitudinal (i.e., available for a duration of time), yet such a multi-time-point indicator is required in order for us to be able to execute a panel model. For this reason, we intend to go on searching for additional cross-national indicators of IT. Second, we intend to probe further into the lack of effects by system complexity factors. Scatter plots for IT and each of the system complexity indicators reveal a series of linear relations; these relations, however, become discrete when added to our causal regression model. This could be a result of the overlap among various indicators of system complexity; simply put, it may be that economic development overlaps with democracy, for example, thus redeeming their additional contributions in the model nonrelevant. Third, case studies of national policies for adopting IT and harnessing it toward national development goals suggest that state involvement and the style of such involvement are key factors in the success of both the adoption and goal achievement (see Evans, 1992a, 1992b). Indications of these factors should also be included into our general model; specifically, cross-national measures of state power and the type of state culture should be included into the empirical models for assessing the social factors shaping the extent of national e-connectivity.

Concluding Comments

Cross-national differences in IT resources and capabilities are sharpening global inequalities and, moreover, setting a new geography of global centrality/marginality (see Sassen, 1997). In this sense, IT joins other national resources (like human capital, natural resources, and manufacturing base) that are crucial for determining a country's ability to engage in global activities, from commerce and production to politics and

	Model 1	Model 2	Model 3	Model 4	Model 5
Net sophistication score, 1991	0.65 (0.08)***	0.53 (0.08)***	0.46 (0.07)***	0.47 (0.08)***	0.46 (0.35)***
Economic development in 1980	0.55 (0.24)**	−0.53 (0.33)	−0.51 (0.31)*	−0.50 (0.32)	−0.51 (0.33)
Educational and scientific expansion in 1980		1.54 (0.35)***	0.91 (0.37)**	0.90 (0.38)**	0.87 (0.39)**
World linkage in 1980			1.09 (0.30)***	1.08 (0.31)***	1.07 (0.31)***
Economic openness in 1985				0.0007(0.004)	0.0007(0.004)
Political regime (scale of democracy) in 1980					0.02 (0.05)
Constant	2.24	2.45	2.52	2.56	2.51
R^2	.73	.77	.80	.80	.80
Number of cases	82	82	82	82	82

Table 2 Results of OLS regressions.

Note
Dependent variable = net sophistication score, 1995. Data in table are unstandardized coefficients and standard errors (in parentheses). OLS = ordinary least squares.
* Significant at the .10 level. ** Significant at the .05 level. ***Significant at the .01 level.

influence. IT is, however, different from such "traditional" resources: To keep up with the rest of the world, a country is required not only to be "Internet ready" but mainly to be "Internet sophisticated." In other words, the world of IT is no longer about making the initial connection; rather, it is currently all about intensity of e-use and sophistication of e-use.

This issue of the timeliness of indicators raises the question: Is the IT revolution similar to earlier information and communication revolutions, on a global scale? Is IT merely another version of the "old" global information and connectivity divide? Some claims are made that Internet has created an information revolution (and all the social consequences of such a social change) similar to that of the telegraph or the print (e.g., Standage, 1998). One way of empirically investigating this claim is to compare degrees of cross-national rates of differences between IT (PCs, Internet hosts) and earlier information technologies (telephone lines and mobiles, fax machines, newspapers, radios); this is doable with current cross-national, commonly available data. Preliminary studies, comparing these various IT indicators of various "technological generations," show clear differences between them: Clearly, the rate of diffusion of Internet technology is by far more rapid than that of telephones, cell phones, and PCs (and more rapid than other household items, such as electricity, VCRs, or microwave ovens; see Compaine, 2001, p. 322; Norris, 2001, p. 33). As eloquently stated by Main (2001), "It took the television 13 years and the telephone 75 years to acquire 50 million users, [whereas] it took the internet [only] five years" (p. 85).

The issue of global IT connectivity is, however, more profound than such empirical propositions about rate of diffusion suggest: It relates to our notions of human progress and to the prospects of social change and social equality (see DiMaggio, Hargittai, Neuman, and Robinson, 2001). The digital revolution carries the great promise of the "new economy," implying a world of networked and integrated economies based on a sophisticated technology global divide. It also holds the promise that geographic location and natural resources are obsolete as the necessary foundations for national economic growth and for social development. This gives hope to less developed countries, yet—at least currently—this hope is not materializing into a reality of global integration and worldwide prosperity. Could the failure be due to the required time lag for such effects to take their shape? Undoubtedly, even developed nations are still struggling with the promise of IT (see "Survey," 2000), and common sense calls for time to pass before any social and economic consequences are visible.[14]

Yet, even in the short duration since the Internet revolution, it is clear that global technology gaps are not disappearing; rather, although cross-national gaps in the distribution of "old" technologies are slowly closing, as the normalization hypothesis describes,[15] the invention of new technologies perpetuates these global technology divides. Moreover, these persistent global technology divides trace for us a core dimension of social and global marginality/centrality: Developing countries, which by definition lag behind the developed countries in many other dimensions of social development, are now also lagging in IT connectivity, which stands as a marker for cutting-edge technology and with it a marker for the prospects of engagement in world affairs.

NOTES

1. We chose these distinctive measures of IT conditions worldwide to account for different stages of the development of the technology (assuming that National Science Foundation Networks [NSFNets] proceeded and were mainly replaced by the Internet) and to account for different dimensions of its "tools" (PC ownership vis-à-vis local active involvement in Internet buildup). By concentrating on different dimensions and different stages of the technology, we address the "normalization hypothesis" (see our concluding comments).
2. The group includes Western Europe, North America, Japan, Australia, and New Zealand.
3. Indicated by a 0–8 scale of intensity of use of four Internet-like systems (Because It's Time Network [BITNET], Unix-to-Unit Copy Program [UUCP], Fidonet, and Internet). The scale was adapted from reports of usage, compiled by the Internet Society and Larry Landweber from the University of Wisconsin–Madison's Computer Science Department. This indicator serves as the dependent variable in our causal models (see Factors Shaping the Global Digital Divide section).
4. Surprisingly, though, the 17 countries named in both time points as having the highest score for Internet use sophistication are not the same. In September 1991, the countries that received a usage score of 8 were Argentina, Denmark, Finland, France, Germany, Greece, Ireland, Israel, Italy, Japan, the Netherlands, Norway, Puerto Rico, Spain, Sweden, Switzerland, and the United States. In June 1995, the list included Argentina, Australia*, Brazil*, Canada*, Chile*, Finland, Germany, Hungary*, Israel, Italy, Japan, South Korea*, Spain, Sweden, Switzerland, and the United States (* denotes countries that are new additions in the highest category of Internet sophistication).
5. Even if they are far from bridging the digital divide among the population within their sovereign boundaries.
6. As is the title for a recent book by Lee Kuan Yew (2000), the Singaporean premier, describing Singapore's "victory over problems of national development."
7. For example, because of lack of other infrastructural features, such as steady electricity supply.
8. We also added to the models indicators for the complexity of other social systems, such as the political systems (indicated by democratization and scales of civil and political liberties). See further discussion in Findings section.
9. By relying on factor scores, we hope to avoid problems of single-indicator bias.
10. All variables are standardized by size of relevant age-group to account for overall size of the nation and thus its potential at producing these national scores.
11. All scores are logged.
12. For example, substituting each of the indicators compiled within the dimension of science and education expansion for the compiled factor score.
13. Calculated from versions 2 and 14 of files at the University of Wisconsin–Madison.
14. Paul David, an Oxford economist, calculates the time lag for the technology-based revolution of electrical power at 40 years (1880–1920).

15 Proponents of the normalization hypothesis focus on the inevitable closing of gaps in differential access to technology over time and often accuse the media in the mythologizing of a soon-to-disappear social divide (see Norris, 2001, pp. 70–71; Powell, 2001; Resnick, 1998; Schement, 2001, p. 305).

REFERENCES

Compaine, B. M. (2001) Declare the war won. In B. M. Compaine (ed.), *The digital divide: Facing a crisis or creating a myth?* (pp. 315–35), Cambridge, MA: MIT Press.

DiMaggio, P., Hargittai, E., Neuman, W. R. and Robinson, J. P. (2001) Social implications of the Internet. *Annual Review of Sociology, 27,* 307–36.

Evans, P. (1992a) Greenhouses and strategic nationalism: A comparative analysis of Brazil's informatics policy. In P. Evans, C. R. Frischtak and P. B. Tigre (eds), *High technology and Third World industrialization: Brazilian computer policy in comparative perspective* (Research Series, No. 85, pp. 1–38). Berkeley: University of California, International and Area Studies Publications.

Evans, P. (1992b) Indian informatics in the eighties: The changing character of state involvement. *World Development, 20*(1), 1–18.

Lee, K. Y. (2000) *From Third World to First.* New York: HarperCollins.

Main, L. (2001) The global information infrastructure: Empowerment or imperialism? *Third World Development, 22*(1), 83–97.

Norris, P. (2001) *Digital divide: Civic engagement, information poverty, and the Internet worldwide.* Cambridge, UK: Cambridge University Press.

Powell, A. C., III. (2001) Falling for the gap: Whatever happened to the digital divide? In B. M. Compaine (ed.), *The digital divide: Facing a crisis or creating a myth?* (pp. 309–14). Cambridge, MA: MIT Press.

Resnick, D. (1998) Politics on the Internet: The normalization of cyberspace. In C. Toulouse and T. W. Luke (eds), *The politics of cyberspace* (pp. 48–68). New York: Routledge.

Sachs, J. (2000, June 24) A new map of the world. *The Economist.* Retrieved February 24, 2003, from <http://www.economist.com/displayStory.cfm?Story_ID=80730#footnote1>

Sassen, S. (1997) Electronic space and power. In *Globalization and its discontents: Essays on the new mobility of people and money* (pp. 177–94). New York: New Press.

Schement, J. R. (2001) Of gaps by which democracy we measure. In B. M. Compaine (Ed.). *The digital divide: Facing a crisis or creating a myth?* (pp. 303–307). Cambridge, MA: MIT Press.

Standage, T. (1998) *The Victorian Internet.* New York: Walker.

Survey: The new economy. (2000, September 23) *The Economist,* pp. 5–40.

United Nations Development Program. (2001) *Human development report 2001: Making new technologies work for human development.* New York: Author.

APPENDIX A

List of Countries in the Analyses ($N = 82$)

Algeria
Argentina
Australia
Austria
Belgium
Benin
Bolivia
Brazil
Burundi
Cameroon
Chile
Colombia
Costa Rica
Denmark
Ecuador
Egypt
El Salvador
Ethiopia
Finland
France
Germany
Ghana
Guatemala
Guinea
Haiti
Honduras
Hungary
India
Indonesia
Iraq
Ireland
Israel
Italy
Ivory Coast

Jamaica
Japan
Kenya
Korea (South)
Kuwait
Liberia
Malawi
Mali
Mauritania
Mexico
Morocco
Nepal
Netherlands
New Zeland
Nicaragua
Niger
Nigeria
Norway
Pakistan
Panama
Paraguay
Peru
Philippines
Portugal
Rwanda
Saudi Arabia
Senegal
Sierra Leone
Singapore
Somalia
Spain
Sri Lanka
Sudan
Sweden
Syria
Tanzania
Thailand

Togo

Trinidad and Tobago

Tunisia

Turkey

Uganda

United Kingdom

Uruguay

Venezuela

Yemen Arab Republic

Yugoslavia

Zaire

Zambia

Appendix B

Scatter Plots of Net Sophistication Scores

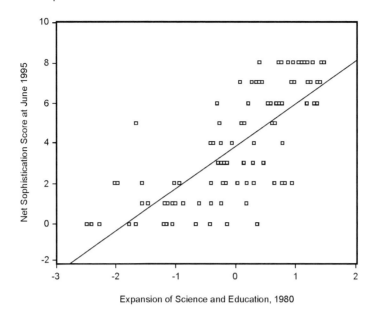

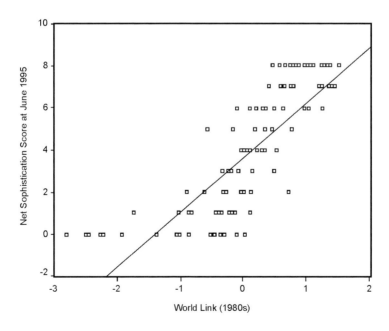

Information Technology and Transitions in the Public Service: A Comparison of Scandinavia and the United States*

K. V. Andersen and K. L. Kraemer

Abstract

New information technologies have the potential to transform the ways governments are organized, the activities they perform, the manner in which such activities are performed and even the nature of the work itself. Governments in the US and Scandinavia have followed fundamentally different approaches to the introduction of computing and to dealing with its effects. In the US, automation has been individualistic – each individual unit of government has introduced the technology according to its own needs. For the most part, the implemented systems were small scale, have followed functional lines, have merely automated existing operations, were implemented incrementally and have evolved slowly over time. In contrast, automation in Scandinavia has been communal – systems have been designed, developed and implemented by communal data processing agencies serving an entire level of government, national or local. The systems introduced were relatively large scale, have crossed functional lines, have involved the reorganization of work, have integrated both data and work processes, and were implemented more or less simultaneously for all units or agencies of government. These differences in approach to automation have influenced each country's view of the role of government in anticipating and dealing with the effects of changes in computer technology on the public workforce.

Introduction

There is nothing more difficult to take in hand, more perilous to conduct, or more uncertain in its success, than to take the lead in the introduction of a new order of things.

Niccolo Machiavelli, *The Prince*

New technologies are frequently described as 'transforming', or 'having the potential to transform' state and local governments and society more generally (Nora and Minc, 1981; Osborne, 1988). In fact, technology has rarely been shown to have such effects. Rather than transforming state and local governments, technology has been adapted by government leaders to fit their perceptions of the opportunities and threats of its application. For the most part, such adaptive application of technology has been incremental and evolutionary precisely because, as Machiavelli's words suggest, dramatic revolutionary change is difficult, perilous and uncertain. Taken together however, incremental, evolutionary change can, and often does, affect the way in which state and local governments operate.

Nowhere is this more the case than with information technology. New information technologies such as computers, airport metal detectors, traffic signal video monitors, radar detection

Citizen information. Expert systems for determining health and social welfare subsidies and software for providing improved, computer-supported information directly to the clients in the municipalities have affected government–citizen interaction and the structure of functions within government in Scandinavian countries. The systems have challenged the municipalities, which traditionally have been split into specialized departments. The new technology has enabled a number of routines to be transferred from various functions to a 'front services' office that interfaces with citizens (Hoff and Stormgaard, 1990).

Automated services delivery. Automatic teller machines (ATMs) and smart cards which permit cash withdrawal and credit charges have in the US reduced the time that welfare recipients must wait in line for payments, and thus the stigma attached to the cashing of welfare cheques, the use of food stamps in making purchases for daily living, etc. But this innovative and client-sensitive method of service delivery currently costs about ten times that of the paper system and is not widely accepted by merchants or banks (Fiordalisi, 1988).

Management information systems. Computerised case management systems have allowed detectives, inspectors, social workers, probation officers and other case workers to be more personally productive, handle a larger caseload and do better casework (for example, better follow-up, to complete more cases, etc.), thereby increasing the diversity of their work, the bottom line performance and the sense of accomplishment in their work. At the same time, the speed-up of work, the errors in case records entered by co-workers and the insatiable appetite of supervisors for more information about the cases from computerized files have sometimes increased worker stress (Danziger and Kraemer, 1986).

Detention monitors. Electronic monitors for detention of both adults and juvenile offenders in homes or in workplaces have reduced the overcrowding of jails, facilitated policing by probation and social workers, and allowed offenders to perform useful work and community service. However, stress has resulted for probation and social workers who are insufficiently trained to use this new technology and for offenders whose lives or work are needlessly disrupted by mistakes and insensitive use of the monitors (Maxfield and Baumer, 1990; Baumer et al., 1991).

Figure 1 Examples of new information technologies in public administration.

jammers, automated teller machines and genetic fingerprints are changing the day-to-day operations of many state and local governments. Figure 1 shows a few examples of these new technologies which illustrate that although these bring new opportunities, they also introduce constraints and dilemmas (Bozeman and Rahm, 1989). Constraints arise because both financial and human resources are seldom sufficient to take advantage of all the opportunities that exist. Dilemmas arise because existing government employees are seldom prepared for the new technologies and the changes they bring to the work pattern.

Thus, we are interested in two broad questions:

(1) What is the nature of information technology use in the public sector?
(2) What will be the effects of IT use on the public service during the 1990s?

This paper addresses these questions with respect to information technology (IT) and computers and information systems in particular. [IT includes computers, office automation, telecommunications and management science techniques. We focus on information technology for several reasons. Firstly, it is pervasive in governments and will become more so during the next decade and beyond. Secondly, it is illustrative of other pervasive technologies such as biotechnology and materials technology which are expected to have major effects in the distant future. Thirdly, information technology is embodied in many other discrete technologies, such as those specific to a particular area like transportation, criminal justice, health care, or infrastructure. Fourthly, more is known about the diffusion, use and effects of information technology than about discrete technologies which tend to have more limited scope of application.]

Comparison	Scandinavia	US
1 Extent of use	All state and local governments	All state and local governments
		Greater sophistication
2 Organization for use	Centralised, shared pattern	Decentralized, diverse
3 Nature of use	Conventional business functions Experiments	Conventional business functions
4 Governmental role in computing affairs	User	User
	Promoter	Promoter
	Regulator for social implications	
5 Worker involvement in automation	Participation in various issues	Participation in design for new applications
	Influences on policy and research	Unions supportive
	Unions supportive	
6 Work organization and IT	Concern and use of work organization and IT relations	Little concern

Table 1 Nature of IT use in Scandinavia and the USA.

For each of these questions we are interested in examining the differences between Scandinavia and the United States.

Nature of IT use in the public sector

Computing in the public sectors in Scandinavia [there are important historical, cultural, economic and social differences between the Scandinavian countries (Denmark, Norway and Sweden), but in this paper we will mainly use examples from Denmark and Sweden, though we believe that similar examples can be found in Norway] and the US historically differs in several important regards (see Table 1). Similarities and differences can be seen with regard to: (1) extent of use; (2) organization for use; (3) nature of use; (4) government role in computing affairs; (5) worker involvement in automation; and (6) work organization and computing. However, these differences might diminish over time as the technology itself changes and as public officials gain greater control over its deployment and use.

Extent of Use

In the US, almost all of the state and local governments have adopted computing, and computing now accounts for about 3% of state and local government operating budgets (Kraemer *et al.*, 1986, 1989; Caudle and Marchand, 1989). Most governments have one central computer installation but the larger state and local governments have multiple installations, or so-called 'departmental computing'. The adoption of PCs has been fairly rapid since the early–mid 1980s, and the PC inventory often equals the dollar value of computing equipment in central and departmental installations. Thus, computing within governments has gradually been extended so that 75% of all functions within government have some kind of IT basis. Moreover, the technology has increasingly been extended to individual users through terminals, PCs and workstations. The ratio of such end-user devices to state and local government employees is currently about 1:200 and expected to reach 1:1 by early in the twenty-first century.

Computing in central and local governments in Scandinavia is at a higher level than most other OECD-countries. While the US has traditionally had more extensive applications (lead users) compared with Scandinavia, a larger proportion of the cities in Sweden and Denmark was using IT in the mid-1970s: 90% of Danish municipalities and 72% of Swedish municipalities, compared with 51% in the US (King and Kraemer, 1985, p. 38). Today all cities use computing, and the technology has also been extended to individual users. The use of computing in the public sector in Scandinavia might have more influence on working conditions because the public sector employs a larger part of the total workforce and spends a larger part of the gross domestic product than do governments in the US.

What accounts for the current relatively low use of computing in Scandinavian governments? Given that the public sector is such a great proportion of the total economy in Scandinavia, one might have expected that automation would be used more widely as a means of achieving efficiency. However, the 'communal' approach to organizing for use and the considerable influences of workers and unions have slowed the pace and extent of automation in Scandinavian governments. In addition, these governments have been concerned about possible social impacts and therefore have been slower to invest, adopt, promote and use technology than might otherwise be the case. These points will become apparent in the following sections.

Organization for Use

Governments at all levels in the US followed a decentralized pattern of organization for computer use in the 1970s. Whether and how computers would be used has been left to each level of government, each unit of government and each agency within a government to decide. While the computing function was initially centralized within government organization and provided as a service to all departments, the advent of minicomputers in the mid 1970s brought about departmental computing, just as the advent of microcomputers in the mid 1980s brought about end-user computing. While most computing services are still provided by one or more computer installations in MIS departments, a plethora of organizational arrangements exists; this includes individual offices for computing, office automation, end-user computing and telecommunications, as well as integrated offices combining these functions in various ways. In addition, as the size of government databases continues to grow yearly, attention has been focused on managing the information in computerized systems, and a new function – information resource management – has been born. This is commonly known as executive information systems (EIS). In short, diversity is the key characteristic of organization for computer use in US state and local governments.

The Scandinavian countries have historically followed a communal or shared pattern of organization for use, through large data centres established to serve each major level of government (national, county and local), although the largest units (such as cities like Copenhagen and Stockholm) have established their own computing centres. The central governments and the associations of local governments in Denmark, Norway and Sweden established the Kommunedata centres that provide most of the computing services to local governments. These centres have had profound effects on the overall diffusion rate of IT-use in those countries. The establishment of the Kommunedata in the Scandinavian countries (in 1965, by the Swedish Union of Local Authorities) was intended to provide central IT-use and a technical organization of sufficient strength to handle advanced applications, to coordinate the technical personnel resources and to develop new application systems for the municipalities. [The public sector employs a large part of the total labour force, and spends more than half of the GNP. Local governments play an important role in managing the welfare state. In Sweden the public sector spends 67% of the GNP, while the public sector in the US spends 38% (1982). Swedish local governments spend 35% of the GNP (1981), the US local governments 8% (1982). The main expenditure in Swedish local governments is health, education and welfare (Bogason, 1987).] The growth problems (unbalanced economic growth, shifting production structure, etc.) in the cities and regions of Sweden directed the development of computing toward specific, restricted fields of application rather than general, abstract ideas such as urban databanks or MIS. It also led to the design of applications intended to serve the country as a whole, rather than the independent development of a large number of scattered, uncoordinated experiments with varying types of applications (OECD, 1974, pp. 109–119). An indicator of the relative strength of the Kommunedata centres is the number of staff trained as computer operators, system planners, etc. In 1986, approximately 1,500 professional staff (systems analysts and programmers) were employed at the Danish Kommunedata centre, and only 108 in the 275 municipalities (Hoff and Stormgaard, 1990, p. 123).

Beginning in the early 1980s, the pattern of centralised computer organization in the government administration underwent change as part of a general transition in the national government. This transition was towards decentralization to local bodies (countries and municipalities). Indeed, computer technology was a major factor facilitating the decentralization of government administration because it permitted data in support of local administration to be collected and restored locally while also sharing and accessing data in the central government (Ingelstam and Palmlund, 1991).

Character of Use

Most applications of computing in US state and local governments are currently conventional and oriented toward business functions and administrative support rather than direct service delivery to citizens (Kling and Kraemer, 1985). A primary reason for this is government emphasis on productivity and administrative control. While control benefits have been achieved, the productivity improvements from these applications have been marginal for the most part, or confounded with other improvements and, therefore, difficult to identify and measure. However, productivity gains are expected to be greater in the future as more emphasis is placed on applications that restructure service delivery, both as a means of reducing costs and meeting special needs. For example, restructuring service delivery towards 'one-stop shopping' and 'little city halls', along with the introduction of computing, are expected to reduce the need for more staff, for multiple service centres and for longer service hours as governments try to respond better to the diverse needs of individuals, households and business. These developments will only increase the pervasiveness of information technology in government.

In Scandinavia, computing has already begun to be oriented towards direct service delivery to citizens, although the main part of the computing is oriented, as in the US, towards business functions and administrative support. Experiments with 'front service' were introduced in the municipality of Ringsted (Denmark), to help the street-level bureaucrats to attend to the problems of citizens better. Citizens need only apply to one place, irrespective of the reason for their application, and a considerable number of matters can be handled there (Hoff and Stormgaard, 1991). The front service has largely been developed to attend to the citizens' rather than bureaucratic interests.

Governmental Role in Computing Affairs

Governments in the US have perceived their role primarily as a user of computers, and to some extent a promoter of greater use through the demonstration of advanced or leading edge applications and the design, development and transfer of mainstream applications. In some instances, federal or state agencies have developed model applications which they sought to have implemented by lower levels through a combination of carrot-and-stick incentives. Buoyed by belief in the benefits of technology, government agencies have not been highly concerned with the social and health aspects of the technology's use. Moreover, because computing has been used primarily to automate existing operations, the social effects have not been great. Most studies of computer use in government indicate that the computer's effect has been neutral, but where there has been an effect it has been beneficial to the quality of worklife of government employees (Danziger and Kraemer, 1986).

In Sweden, the government paid close attention to the social implications of the penetration and development of the use of computing. Studies of social issues related to computer use were carried out with governmental support on learning mechanisms for computer-based systems, the work environment (high ergonomic and health standards) and managerial organizations for work involving computer-based systems (OECD, 1991). Sweden also has a well-organized system of planning and project implementation in IT policy. In addition, government and private sector have had frequent mutual interactions in the planning of programmes for IT development. An incident that triggered the establishment of large-scale IT programmes in Sweden occurred at the beginning of the 1980s. The NATO-member countries refused to provide integrated circuit (IC) chips to Sweden, claiming that Sweden exported some systems listed by the Coordination Committee for Mutual Export controls (COCOM), for construction projects associated with the information systems used in the Moscow Olympics. This incident increased public awareness of the necessity of ensuring a secure supply of key components of manufactured products on which Sweden's international competitiveness relied. Automation and process control equipment were examples of these. This has resulted in the establishment of a range of governmental institutions to deal with the technology. Coordination of the acquisition and utilization of computers in government administration is provided by the Swedish Agency for Administrative Development (SAFAD). Improving the interface between humans and computer-based systems is the responsibility of the Swedish Environment Fund (AMFO), a government agency that belongs to the Ministry of Labour (OECD, 1991). The latter agency, in particular, illustrates the awareness of the social implications of computerization in Sweden.

Worker Involvement in Automation

In the US, worker involvement in computing affairs has focused on participation in the design of new applications. The purpose of their involvement has been to communicate to computer specialists the nature of their operations, the information and processing requirements, and the data definitions in order to facilitate the design of new computerized systems. Worker involvement has extended to the number and nature of screen designs and reports produced by the systems, training of government staff for use, and even evaluation of the system once it became operational. It has not usually been extended to decisions about whether to introduce new systems, as these were generally made by high-level managers or professional staff. However, the decentralization of computing to departments through minicomputers and the introduction of PCs has brought about greater user involvement in the spectrum of decision making about computing matters.

In Scandinavia, the nature of industrial relations implies considerable influence for employees regarding issues of working conditions. Any consequences that may arise from computerization in the workplace must therefore be seen against this background of active and often constructive consultation in the process of implementation. The labour movement in the Scandinavian countries has been highly influential in decisions taken by government, the research conducted and the formulation of strategies for influencing technology development – the Scandinavian approach for studying computerisation is thus highly linked to improving conditions for the workers (Bjerkes *et al.*, 1987; Bermann, 1989; Floyd *et al.*, 1989). The labour movement has, by and large, over the years supported, rather than resisted new technology.

The employee is, by legislation and cooperative arrangement, given the right to receive information about new technology, and its attendant changes in working methods and processes (Mathiassen *et al.*, 1983). However, no general regulations secure employees any influence upon technology change – the employers have the final word. The Swedish unions' attitude towards new technology is more positive than in many other countries. The main reason for this is that unemployment caused directly by technical change has been limited (Ullmark, 1988).

Work Organization and Computing

Because most US computer applications were the automation of existing activities in government agencies, there has been very little concern about the relationship between work organization and computing. Employee unions have generally been supportive of government automation. Experiments were conducted during the 1970s with information and referral (I&R) systems for health, social services and ageing, in an attempt to bring about greater coordination and cooperation among public agencies, but these effects generally failed because insufficient attention was paid to agency incentives for participation in the systems. A brief characterization of work organization and IT-use in the US is thus the general absence of consideration of this important relationship.

In Scandinavia, experiments with new technology have enabled a number of routines to be transferred from various functions to 'front services' offices that interface with citizens (Hoff & Stormgaard, 1990). Expert systems for determining health and social welfare subsidies and software for providing improved, computer-supported information directly to the clients in the municipalities have affected government–citizen interaction and the structure of functions within the government (Karlström, 1986; Khakee, 1985). Such systems have been a challenge to the municipalities which traditionally have been split into specialized departments. A brief characterization of work organization and computing in Scandinavia is thus the awareness of this relationship.

Effects of IT use on The Public Service

Our review of research and practice indicates that the effects of computing can be broadly classified into four general areas: (1) creation of new institutions; (2) organization and distribution of activities performed by government; (3) alteration of work processes; and (4) nature of the work. These effects are generally well documented by individual studies, empirical surveys (both cross-sectional and longitudinal) and/or literature reviews. However, not all effects are equally well understood, as will become apparent below.

Creation of New Institutions

The largest effects of computing and other information technologies have been concentrated in the computing and information systems function in government rather than on the other functions and activities of government (see Table 2). In the US, information technology has resulted in the creation of new government functions and institutions – the information systems function and the MIS department, the telecommunications function and the Office of Telecommunications, the information resources management (IRM) function and the IRM Office (Andersen and Dawes, 1991). Most often these institutions have existed separately, but some governments have integrated them into a single institution. The computing function has undergone the greatest change.

Initially, computing was set up as a centralised function in most governments, usually under the finance department. In time, however, as computer use expanded, the function was often converted into an independent government department outside the finance function. The continued spread of computer use, along with the advent of microcomputers, led to the distribution

New institutions	Primarily in the IT arena
Organization and distribution of activities	Automation follows transitions in administration rather than leads
	Centralization versus decentralization facilitates either
	New organizational forms primarily in IT arena
Alteration of work processes	Coordination and optimization facilitated
	Automation of service delivery in special cases
	Electronic communication with citizens

Table 2 Effects on public service: activities performed by government.

of the computing function among large departments with the former central unit serving finance and administration and the myriad small government functions and agencies. The advent of microcomputers reinforced and hastened this trend towards distribution of computing equipment and expertise to even the smallest functions and activities. In an attempt to manage and facilitate these effects, central IS units created new 'information centres', 'computer stores' and 'end-user computing offices'. At the same time, department users have created their own informal users' groups for sharing information and expertise. These have been independent of the former IS units and sometimes in opposition to their attempts to manage, facilitate, or control computing on an organization-wide basis. These developments have occurred because top managers have not known how to deal with these computer-based developments and/or did not choose to become involved. The disruption and trauma for IS units have been considerable in some instances, and relations between the IS units and the user departments have seriously deteriorated with an overall loss of effectiveness to governments.

In Sweden also, computing resulted in the establishment of a range of governmental institutions. The Swedish Agency for Administrative

Development (SAFAD) was established to coordinate the acquisition and utilization of computers in the central government administration. The Swedish Environment Fund (AMFO), a government agency in the Ministry of Labour, was set up to improve the interface between people and computer-based systems (OECD, 1991). The software houses, referred to as Kommunedata, were established to provide computing and development services to local governments. Kommunedata was established in Sweden in 1965 and in Denmark in 1972. The Kommunedata organizations established a highly centralised activity which provided computing to local governments via terminals to a central mainframe and with development services via a central staff which developed applications intended for use by all local governments. The Kommunedata organizations grew in computing power and number of staff throughout the 1970s and 1980s as computer use expanded throughout local governments. However, computing also grew in the larger municipalities and counties, which were allowed to obtain their own computing equipment and staff because of their size and purported unique requirements.

Organization and Distribution of Activities

Information technology facilitates much wider forms of organization and distribution of governmental activities. Information technology permits either centralised or decentralised organization, and central or local distribution of the activities of government while also permitting greater central monitoring and control. For example, some federal agencies use large centralized information systems to facilitate and monitor state and local government implementation of federal programmes. Some states also extend their operations to regional offices and local administrations through mandated state-wide information systems for health, social services and employment. Some local governments are similar, with city-wide systems for distributed activities like libraries, parks and recreation, and little city halls.

Beginning in the 1950s, with the first introduction of the computer, governments followed a centralized approach in automating government activities, such as budgeting and accounting. This trend continued in the 1960s with time-sharing and inter-governmental systems such as the National Crime Information Center/Computerized Criminal History (NCIC/CCH), and in the 1970s with large-scale computer networks and services integration (Quinn, 1976). Throughout the 1980s, there was a trend towards decentralization of federal government activities to the states and, in turn, from the states to local governments. This decentralization trend has continued and perhaps will even accelerate during the 1990s. It will be facilitated by the increasing availability of computer networks, databases, electronic mail systems and microcomputers at each level of government and throughout the federal system.

In all of these technology deployments, governments choose the approach they will take. Some governments prefer centralized approaches, whereas others opt for decentralized ones (Kraemer *et al.*, 1989). For example, the state of Virginia has a centralized social services information system, whereas California has a decentralized one. Some governments provide one-way information services to citizens, whereas others provide for two-way information and communication (Gurwitt, 1988a). This is illustrated by the different approaches of Kansas City and Santa Monica (California). In Kansas City, a 24-hour city hall based on voice mail, provides information to citizens about government operations and activities, and takes requests or complaints from citizens which are handled and answered within 24 hours. The Santa Monica system, which is based on computer conferencing and electronic mail, allows interactive communication among citizens and between citizens and government officials.

The experience of Sweden also shows that governments select the approach they wish to take to automation, and that automation follows transitions in the administration of government activities rather than leads them. For example, the administration of social welfare in Sweden was traditionally centralized and supported by a nation-wide computerized system serving all social welfare agencies at the national, county and municipality levels. Computerization allowed centralized control of the various social welfare payments for child care, sickness, parental care, unemployment, housing for the aged and handicap care. The first initiatives to computerize the administration of these welfare services were taken in the 1950s, and concentrated from the beginning on developing a centralized, national computer-based information processing system. The system has enabled the government to implement a series of social reforms and provide citizens with a reliable service of payments and different social benefits (Ingelstam and Palmlund, 1991). Beginning in the early 1980s, however, the organization of social welfare was changed dramatically from its centralized form to a more decentralized one, with the distribution of social welfare functions to local administrations along with the required computing equipment, staff and databases (Ingelstam and Palmlund, 1991).

A related change is the distribution of workers themselves. The bulk of government workers will continue to be located in central places like the statehouse (central governmental buildings), the county hall of administration and the city hall. Some will be decentralized to distributed workplaces such as regional offices, metropolitan subcentres, and little city halls. Others will work at home with a link to the office via the computer and telecommunications, that is, they will 'telecommute' (Kraemer, 1982). In the US, the number of such workers is estimated to be around 10 million nationally by the year 2000. Whether it reaches such heights or not, government workers are likely to be among such telecommuters because of the 'services' nature of their jobs. The type of workers who will telecommute are first and foremost those who already work at home, such as computer professionals, writers and editors, handicapped workers and 'piece workers'. For the most part, work at home will not replace work at the office, but will supplement it; that is workers will work certain days (or parts of days) at the office and the remainder at home (Venkatesh and Vitalari, 1992).

In Scandinavia, the introduction of 'front services' into city and county governments has changed both the distribution of activities and workers. In the past, when citizens needed governmental services they had to interact with each functional bureaucracy independently. There was no mechanism to take care of all their needs at one place. The introduction of front services has created a single office which deals with all the needs of the citizens. This office then interfaces with each of the functional bureaucracies and communicates with citizens about questions or issues that arise during the processing of requests. Another illustration is provided by the use of portable computers by the Danish VAT auditors. The VAT auditors bring computer-stored information to the clients (companies) and, in turn, store the information obtained from the clients in their portable computers for transmittal back to the central office (Vittrup, 1989). This has increased the control of the VAT auditors, and also their direct contact with the clients. The auditors, and not the clients, travel more as a result of the computerized system.

Thus, experience in the US and Scandinavia indicates that the technology facilitates either centralised or decentralized organization and distribution of government activities and workers. Historically, mainframe computers have been viewed as facilitating greater centralization. The advent of microcomputers is viewed as facilitating greater decentralization. In fact, computing has always facilitated either approach, or a mix, and still does today (Robey, 1981; Attewell and Rule, 1984). However, people's perceptions of what computers can do have been influenced by these technology developments, and public officials, knowingly or unknowingly, have chosen to implement the technology one way or the other. Often too, those who have chosen a particular approach have rationalized their decisions made on personal or bureaucratic grounds on the technical requirements, technical advantages, or costs of the approach (Danziger et al., 1982).

Alteration of Work Processes within Institutions

The foregoing changes in organization and distribution of activities will be reflected in changes in the processes by which work is carried out within and between institutions. Although the possible changes are many, three are especially important: sophisticated coordination and optimization, automation of direct services to citizens and electronic communication with citizens.

Coordination and optimization. Coordination and optimization refer to the ability of government agencies in far-flung locations to coordinate their activities and to optimize them in terms of some overall interest. While such systems do not currently exist in US state and local governments, the prototypes exist in the federal government and can be extended to state and local governments. The US Army's computerized REQUEST system for assisting recruiters in meeting military occupational special needs while providing incentives for recruits to join the new 'all volunteer' army is an example (Kelman, 1990). Basically, REQUEST starts with a listing of requirements for different military occupational specialities, pay bonuses for signing-up for the specialities, training slots for new recruits and available first assignments for the recruits once trained. This information is available at all army recruiting offices in the US and abroad. The recruiter at each local office inputs information about the recruit's preferences for a speciality, training and first assignment, and shows the availabilities on the computer screen, along with any bonuses, to the recruit. When the recruit makes his choices, they are recorded in the central computer and printed out immediately for the recruit and the local office. REQUEST not only helps the army to meet its personnel needs, but optimises the needs of the army and the desires of the recruits and helps to ensure that recruiters seldom lose a 'sale'.

A modest example at the state level is Colorado's job-bank, a system that exists in other states as well (Ullman and Huber, 1973). Most of Colorado's major cities and counties are linked through a central computer to a system that keeps track of participants in job training programmes and job openings, and allows social service personnel to match job openings to clients' backgrounds and qualifications. A logical extension of this system is to provide terminals for both employers and employees so that they can enter job openings and résumés and do searches for a match on their own (Gurwitt, 1988b, p. 41). Similar systems exist in Scandinavia. Job-banks in Sweden have widened the possibilities of locating a suitable position for applicants (SAFAD, 1980, p. 12). Also the job-banks make it possible to match newly registered positions against applications in a faster way than in the manual system. However, it has not been possible to show that there is any reduction in the period people are unemployed or the number of people who are unemployed.

Automation of service delivery. Automation of service delivery refers to the completely computerised handling of requests for information or service. Here there are many examples already in operation around the US. Several cities have automated citizen access to public services such as building inspections and bibliographic retrieval (from public libraries) and to public records such as land records, tax records, vital records, business licences and other 'public' information (Kahl, 1990). For example, Dallas (Texas) has a system for the scheduling of building permit inspections. Instead of calling a city office that is only open from 8 a.m. to 5 p.m. to schedule an inspection, builders can now call the building inspection office at any time of the day or night. The phone is answered by a microcomputer with a voice response system, which asks for the information about the building to be keyed in on a push-button phone. It then gives the caller a time for the inspection. At the inspection office, that information is then fed automatically to a mainframe computer, which goes on to arrange inspectors' daily schedules and routes (Gurwitt, 1988b, p. 40). Another example is provided by experiments in Ramsey County (St. Paul), Minnesota, the state of Washington, and Berks County, Pennsylvania, that involve rethinking the way in which public assistance payments are made to individuals. Instead of issuing cheques, which the welfare recipients then have to take to the bank to cash, the county is issuing bank cards for welfare recipients, who can then use them at ATMs around the county to draw out cash against their public-assistance accounts. Like any other bank cards, these cards have an expiration date and so the individual's eligibility and assistance are re-examined before a new card is issued. In addition, the cards and/or the ATMs can be programmed with information that limits the amount of any one cash withdrawal, the number of cash withdrawals within any time period, or other user options to encourage cash management (Gurwitt, 1988b; Fiordalisi, 1988).

In Sweden, automation of service delivery is illustrated by a computerized system for administrating various social insurances at the local level. Instead of the citizens having to take the initiative to change their social insurance status, the system issues preprinted forms which are then mailed to citizens when action from their side is needed in order to change the social insurance status. The computer system then takes the information received from the citizens and combines it with the relevant eligibility and benefits rules to produce the new social insurance benefits. While this system provides citizens with better service, the personal contact between citizens and the administration has been reduced (Ingelstam and Palmlund, 1991).

Electronic communication with citizens. Electronic communication with citizens can occur in a variety of ways, but most frequently through the automated handling of citizen requests for information and complaints and through two-way, interactive electronic mail and dialogues. An example is provided by Santa Monica's (California) Public Electronic Network (PEN). Anyone with access to a personal computer, once registered with the city, can use PEN to obtain information about city council hearings, city commission activities, or communicate with city staff, city council members and other city officials, or engage in a 'public dialogue' on community issues such as rent control, the environment, the economy, women's or senior citizens' issues. Computer terminals are located in city hall, public libraries, senior citizen centres, other public buildings and shopping malls, to facilitate access by people without computers (Gurwitt, 1988a; City of Santa Monica, 1989).

Videotex, a combination of computing and television, makes it possible for citizens in Danish municipalities to communicate with databases about government service. The citizens have access to large quantities of information through public videotex terminals and do not need to have a private computer at their disposal (Hoff and Stormgaard, 1991, pp. 228–232). In one experiment, the citizens have videotex terminals located in their homes and can access a number of private services (advertising and home-shopping). In another experiment, terminals have been installed in post offices, libraries and day-centres for old-age pensioners. The citizens have access to information on activities in the municipality, job vacancies, public housing and a 'bulletin board' permitting participation in public debate within the municipality.

The Nature of Work

As might be expected from the foregoing changes, the new information technologies are changing the nature of work in state and local governments. Empirical research has been conducted in both Scandinavia and the US over the last twenty years. The findings are essentially similar and show that increasing automation of work processes is producing several changes, including: (1) speed-up of work; (2) tighter coupling of work; (3) greater independence for professional and staff workers, and greater interdependence for operations workers; (4) greater control over people for managers and professionals, and greater control over jobs for clerical and administrative workers; and (5) greater flexibility in work organization (Attewell and Rule, 1984; Kraemer and King, 1986).

Speed-up of work. Computerization has produced a speed-up of work at all levels within government, ranging from street-level workers to office workers and from professional workers to policy-makers and managers. The speed-up has occurred because the technology allows individuals to work faster, shortens the cycles for processes such as billing, paying and collecting, and records information in real-time, as events and actions occur, and thereby creates an expectation for fast response. An important effect of this speed-up is a general increase in time pressure felt by all types and levels of workers (Danziger and Kraemer, 1986; Irving *et al.*, 1986; Jackson, 1987; Kraemer and Danziger, 1990).

Tighter coupling of work. Information technology is also creating a tighter coupling of work, especially where individuals from several different governmental departments and functions are tied together in a single system such as a financial, personnel, geographic information, or emergency dispatch system. A tighter coupling of work means that what a person does in one part of the organization triggers a decision or action by others, or that what people do in their own parts of the organization creates a picture of something happening that all must respond to in a coordinated fashion. The former is illustrated by the case of a building inspection which discovers serious health, safety and environmental hazards, and triggers the need for response by the fire (hazardous materials), health and police departments. The latter is illustrated when the independent actions of these departments result in determinations that, taken together, suggest that a building must be vacated, sealed-off, or demolished because of the total set of hazards present and the improbability of their amelioration.

Independence/dependence of work groups. As might be expected from the tighter coupling of work, there is a growing interdependence among some work groups as a result of automation, but there is also a growing independence for others. Information technology appears to increase the independence of highly professional and specialized work groups such as engineers, planners, economists, statisticians, management analysts and staff analysts. These groups have always been able to function relatively independently, and computing has only increased their independence at the margins. It has done so by providing them with direct hands-on access to the technology, data and the power to manipulate data in order to produce information relevant to their jobs. This increased capability has tended to heighten their stature and their independence of action (Danziger and Kraemer, 1986).

In contrast, the extension of computing into government has increased the interdependence of office work groups at the operational level, especially when they rely upon one another for input of data (and its accuracy, timeliness and format), for processing cases/clients in a sequence of steps, or for manipulations of data which form the basis for action by others (for example, forecasts or work schedules). The groups most often affected are the clerical, administrative and managerial in both operational and staff functions such as finance and personnel, planning and building, fire and police, and across these functions (for example, geographic information systems, financial systems and personnel systems).

Control/autonomy of individuals and jobs. IT-use has been shown to have several effects related to control of individuals and jobs (Kraemer and Danziger, 1990). Firstly, computing provides a higher level of organizational control and greater capacity for judging performance via computerised monitoring systems built into the operating systems of government. Furthermore, this capacity for work monitoring via the computer is now a reality for professionals, as it has been in the past for clerical/administrative workers (Bjørn-Andersen *et al.*, 1986; Irving *et al.*, 1986).

Secondly, managers and professionals generally enjoy greater increases in control attributed to computing than do clerical/administrative workers (Danziger and Kraemer, 1986; Majchrzak, 1987; Millman and Hartwick, 1987). However, computerized systems also can make the task of control more difficult, especially for those in superordinate roles who themselves become dependent on the technology. For example, a study of supervisors and customer service representatives in a large public utility (Kraut *et al.*, 1989) found that as a result of installing a new customer inquiry system, the supervisors' work was both made more difficult and more technology-dependent. In the past, supervisors had known the job of their subordinates because they themselves had previously been customer service representatives. However, with the introduction of the new computerized system, this knowledge was suddenly obsolete – and the supervisors did not possess nor were they provided with training to develop the skills they needed to operate in the new computerized environment.

Thirdly, computerization had increased workers' sense of control over certain aspects of the job, including mastery over relevant information and improved communications. This has especially been the case for clerical and administrative jobs, and has been accompanied by an increase in time pressures (Kraemer and Danziger, 1990).

Flexibility of work organization. The most significant impact of computing on work organization is that the technology enables managers and policy-makers to choose whatever structural arrangements they desire, including combinations of structural arrangements. IT-use does not determine work organization; computing facilitates it. While information technology may enhance employee skill and autonomy, thereby facilitating decentralization and distribution of work, it also facilitates hierarchical control and task fragmentation (Bjørn-Andersen *et al.*, 1986; Thompson *et al.*, 1989). For example, hierarchical control and task fragmentation are facilitated by information technology when efficiency is the primary goal, the organizational scope is limited, capital cost is low, equipment reliability high, workforce interest low and computerized monitoring effective (for example, in the mail room or central records department of a state or local government organization). This fact highlights the importance of recognising that the organization of work is at least as much a matter of political/managerial choice as it is of function/task necessity. It is a matter of choice about the structure of governance in organizations (Kraemer, 1991).

IT-use can influence the work organization through, for example, automating parts of the production or job-routines and, in some cases, establishing separate organizations that are fully, or highly, automated, while the remaining parts of the 'old' organization are manually oriented. This might lead to a high degree of computer networking in the automated part of the organization, but it also might reduce the degree of integration between automated and manual tasks. Also, establishing separate organizations reduces the possibility of formal rotation between types of tasks in the manual and computerized organizations (Child and Loveridge, 1990).

As all of the foregoing suggests, the effects of change in computing are being felt at all levels of state and local government. IT-use has generated opportunities to reconfigure relationships, including those between levels of government, among subunits of the same jurisdiction, and between levels within state and local governments. The effects of computing on the activities of state and local governments and the organization of work have had ramifications for the nature of work itself. State and local government employees are experiencing greater time pressures, tighter coupling of their work activities, and changes in dependence and autonomy.

However, there are many areas where we do not yet know the effects or have a completely clear idea of the effects. These include the following:

(1) A quantitative indication of the job displacement, new jobs and net employment effects of the adoption of information technology. It is probably impossible to determine these effects across all state and local governments. However, a few carefully constructed empirical case studies over time in highly impacted governments and/or agencies would provide a good indication of the extent of such effects and the key relationships that would help other governments to make their own assessments.
(2) The effects on citizens and public servants of the very new technologies for services automation, computer-assisted service delivery, communication with the public and automatic monitoring. Current knowledge is mainly anecdotal, derived from newspaper and promotional accounts. What is needed is serious study of these technologies in order to draw out more fully their implications for the public service.
(3) A quantitative indication of the numbers and distribution of new technology-related functions and new job classifications in state and local governments. How many technology policy analysts, technology transfer agents, information resource managers, information analysts, GIS specialists, end-user specialists, multi-media specialists and similar new jobs exist in state and local governments? Where are they, and at what rate are they growing?

Conclusion

Computing and other information technologies are part of the general transitions affecting the public service and also bringing about their own transitions. The transitions are evolutionary – not revolutionary. The use of computing is still in the early to middle stages in most governments.

Governments in the US and Scandinavia have followed fundamentally different approaches to the introduction of computing and to dealing with its effects. These differences stem from differences in views about the beneficence of technology, the need for reorganization of work along with the introduction of new technology (the popular word today is 're-engineering') and the role of government in anticipating and mitigating the effects of technology. In the US, the introduction of IT has been individualistic – each individual unit of government has introduced the technology for its own needs. For the most part, the implemented systems were small scale, have followed functional lines, have merely automated existing operations, were implemented incrementally and have evolved slowly over time. While the US has occasionally implemented vertically integrated systems such as NCIC/CCH, these tend to be the exception rather than the rule. Attempts to implement such systems outside the criminal justice area, where a command and control system of authority exists, have generally been unsuccessful.

In contrast, in Scandinavia the introduction of IT has been communal – systems have been designed, developed and implemented by communal data processing agencies serving an entire level of government, national, county, or municipality. The systems introduced have been relatively large scale, have crossed functional lines, have involved the reorganization of work, have integrated both data and work processes, and were implemented more or less simultaneously for all units or agencies of government.

These differences in approach to the introduction of IT have influenced each country's view of the role of government in anticipating and dealing with the effects of changes in computer technology on the public service workforce. In the US, the effects of government IT-use have almost never been disruptive, because the introduction of IT has been incremental and governments have followed a policy of reducing staff through attrition rather than layoffs. Thus, US governments have been relatively unconcerned about effects of IT-use, have responded to each situation in an *ad hoc* fashion and have been reactive rather than proactive in dealing with effects. Scandinavian countries have reorganized work along with the introduction of IT, so that the potential effects on government employees have been more serious. Consequently, they have been concerned about effects of IT-use from the start, have developed plans to deal with these effects, and have been proactive rather than reactive in dealing with the effects of automation.

ACKNOWLEDGEMENTS

During the writing of this paper, we were indebted to Professor James L. Perry of Indiana University and Birgitte Gregersen of Aalborg University for help and advice. This paper is a revision of an earlier version presented at the conference on *Transitions in Public Administration: Comparative Perspectives on Sweden and the United States*, May 24–27, 1992, Örebro, Sweden (published in the *International Journal of Public Administration*). Part of the manuscript has also been presented at the DIAC conference held in San Francisco in May 1992. In addition, the paper contains part of a book recently published in Danish (Andersen *et al.*, 1993).

REFERENCES

Andersen D. F. and Dawes S. S. (1991) *Government Information Management*. Prentice-Hall, New York.

Andersen K. V., Gregersen B. and Kraemer K. L. (1993) *Informationsteknologi i den offentlige sektor* [Information Technology in the Public Sector]. Samfundslitteratur, Copenhagen.

Attewell P. and Rule J. (1984) Computing and organizations: what we know and what we don't know. *Communications of the ACM* **27** (December), 1184–92.

Baumer T. L., Maxfield M G and Mendelsohn R I (1991) *A Comparative Analysis of Three Electronically Monitored Home Detention Programs*. Mimeo, Indianapolis, Indiana.

Bermann T. (ed.) (1989) Scandinavian approaches [special issue]. *Office, Technology, and People* 4(2).

Bjerkes G., Ehn P. and Kyng M. (eds) (1987) *Computers and Democracy: A Scandinavian Challenge*. Avebury, Aldershot.

Bjørn-Andersen N., Earl M., Holst O. and Mumford E. (eds) (1982) *Information Society for Richer, for Poorer*. North-Holland, Amsterdam.

Bjørn-Andersen N, Eason K and Robey D (1986) *Managing Computer Impact: An International Study of Management and Organization*, Ablex, Norwood, New Jersey.

Bogason P. (1987) Capacity for welfare: local governments in Scandinavia and the United States. *Scandinavian Studies* 59(2), 184–202.

Bozeman B. and Rahm D. (1989) The explosion of technology. In *Handbook of Public Administration* (Perry J L, ed.), pp. 54–67. Jossey-Bass, San Francisco, California.

Caudle S. L. and Marchand D. A. (1989) *Managing Information Resources: New Directions in State Government*. Syracuse University, School of Information Studies, Center for Science and Technology, Syracuse, NY.

Child J. and Loveridge R. (1990) *Information Technology in European Services. Towards a Microelectronic Future*. Blackwell, Oxford.

City of Santa Monica (1989) *PEN: Public Electronic Network*. Information Systems Department, Santa Monica, California.

Danziger J. N. and Kraemer K. L. (1986) *People and Computers*. Columbia University Press, New York.

Danziger J. N., Dutton W. H., Kling R. and Kraemer K. L. (1982) *Computers and Politics: High Technology in American Local Governments*. Columbia University Press, New York.

Fiordalisi G. (1988) States endorsing automated services. *City and State* 5(15), 2–25.

Floyd C., Mehl W. M., Reisin F. M., Schmidt G. and Wolf G. (1989) Out of Scandinavia: alternative approaches to software design and system development. *Human–Computer Interaction* 4, 253–350.

Gurwitt R. (1988a) Computers: new ways to govern. *Governing* 1 (May), 34–42.

Gurwitt R. (1988b) The computer revolution: microchipping away at the limits of government. *Governing* 1(8), 35–42.

Hoff J. and Stormgaard K. (1990) A reinforcement strategy for informatization in public administration in Denmark? In *Informatization Strategies in Public Administration* (Frissen P. H. A. and Snellen I. Th M, eds), pp 107–32. Elsevier, Amsterdam.

Hoff J. and Stormgaard K. (1991) Information technology between citizen and administration. *Informatization and the Public Sector* 1(3), 213–36.

Ingelstam L. and Palmlund I. (1991) Computers and people in the welfare state: information technology and social security in Sweden. *Informatization and the Public Sector* 1(1), 5–20.

Irving R. H., Higgins C. A. and Safayeni F. R. (1986) Computerized performance monitoring systems: use and abuse. *Communications of the ACM* 29(8), 794–801.

Jackson L., Jr (1987) Computers and the social psychology of work. *Computers in Human Behavior* 3, 251–62.

Kahl J. (1990) With the government at their fingertips. *Government Technology* October, 19–30.

Karlstrom G. (1986) Information systems in local governments in Sweden. *Computing, Environment, Urban Systems* 11(3), 107–13.

Kelman S. (1990) *The Army and REQUEST*, Kennedy School of Government, Harvard University, Cambridge, Massachusetts.

Khakee A. (1985) Futures-oriented municipal planning. *Technological Forecasting and Social Change* 28, 63–83.

King J. L. and Kraemer K. L. (1985) *The Dynamics of Computing*. Colombia University Press, New York.

Kling R. and Kraemer K. L. (1985) The political character of computerization in service organizations: citizen interest of bureaucratic control. *Computers and the Social Sciences* 1(2), 77–89.

Kraemer K. L. (1982) Telecommunications – transportation substitution and energy productivity. Part I. *Telecommunications Policy* 6(1), 13–29.

Kraemer K. L. (1991) Strategic computing and administrative reform. In *Computerization and Controversy: Value Conflicts and Social Choices* (Dunlop C. and Kling R., eds), pp 167–80. Academic Press, Boston, Massachusetts.

Kraemer K L and Danziger J N (1990) The impacts of computer technology on the worklife of information workers. *Social Science Computer Review* 8(4), 592–613.

Kraemer K. L. and King J. L. (1982) Telecommunications – transportation substitution and energy productivity. Part II. *Telecommunications Policy* 6(2), 87–99.

Kraemer K. L. and King J. L. (1986) Computing and public organizations. *Public Administration Review* 46 (November), 488–496.

Kraemer K. L., King J. L., Dunkle D. E. and Lane J. P. (1986) *The Future of Information Systems in Local Governments*. Public Policy Research Organization, Irvine, California.

Kraemer K. L., King J. L., Dunkle D. E. and Lane J. P. (1989) *Managing Information Systems: Change and Control in Organizational Computing*. Jossey-Bass, San Francisco, California.

Kraut R., Dumais S. and Koch S. (1989) Computerization, productivity, and quality of work-life. *Communications of the ACM* 32(2), 220–38.

Machiavelli N. (1981) *The Prince* (translated by G. Bull) (originally published in 1514), Chapter 6. London.

Majchrzak A. (1987) *Human Aspects of Computer-Aided Design*. Taylor and Francis, Philadelphia, Pennsylvania.

Mathiassen L., Rolskov B. and Vedel E. (1983) Regulating the use of EDP by law and agreement. In *Systems Design for, with, and by the Users* (Briefs U, Ciborra C and Schneider L, Eds), pp 251–64. North-Holland, Amsterdam.

Maxfield M. G. and Baumer T. L. (1990) Home detention with electronic monitoring: comparing pretrial and postconviction programs. *Crime and Delinquency* 36 (October), 521–36.

Millman Z. and Hartwick J. (1987) The impact of automated office systems on middle managers and their work. *MIS Quarterly* 11(4), 479–91.

Nora S. and Minc A. (1981) *The Computerization of Society*. MIT Press, Cambridge, Massachusetts.

OECD (1974) *Information Technology in Local Government. A Survey of the Development of Urban and Regional Information Systems in five European Countries* (Informatics Studies). Organization for Economic, Cooperation, and Development, Paris.

OECD (1991) *A Comparison of Changing Public Policies for Information Technology in Canada, the Netherlands, and Sweden* (Project Report 37599). Organization for Economic, Cooperation, and Development, Directorate for Science, Technology, and Industry, Paris.

Osborne D. (1988) *Laboratories of Democracy*. Harvard Business School Press, Boston, Massachusetts.

Quinn R. E. (1976) The impacts of a computerized information system on the integration and coordination of human services. *Public Administration Review* 36(2), 166–74.

Robey D. (1981) Computer information systems and organization structure. *Communications of the ACM* 24 (October), 679–86.

SAFAD (1980) *Decision-making, Assessment and Evaluation regarding Computerization in the Swedish Governmental Administration*. Swedish Agency for Administrative Development, Stockholm.

Thompson L., Sarbaugh-McCall M. and Norris D. (1989) The social impacts of computing: control in organizations. *Social Science Computer Review* 7(4), 407–17.

Ullman J. C. and Huber G. P. (1973) *The Local Job Bank Program: Performance, Structure, and Direction*. Lexington Books, Lexington, Massachusetts.

Ullmark P. (1988) The Swedish way of introducing new technology. In *New Technology and Industrial Relations in Scandinavia* (Graversen G and Lansbury R. D., eds), pp 80–88. Avebury, Brookfield, Vermont.

Venkatesh A. and Vitalari N. (1992) An emerging distributed work arrangement: an investigation of computer-based supplemental work at home. *Management Science* 38(12), 1687–707.

Vittrup K. (1989) ESCORT: a toolbox for VAT auditors. In *Expert Systems in Public Administration: Evolving Practices and Norms* (Snellen I. Th M., Baquiast J. P. and van de Donk W. B. H. J., Eds), pp 225–42. Elsevier, Amsterdam.

Index

Abco 376–80
Access 302
action research 36, 37–8, 39, 41, 42
Agar, M. H. 58, 60, 63
Alpar and Kim 110–11
American Airlines 193
American Hospital Supply 193
AMFO (Swedish Environment Fund) 435, 437
anonymity 9, 10, 17–23, 24, 25, 37, 39, 40, 42, 48, 51, 56, 66, 74
Appadurai, A. 373
Arrow, K. J. 327
Association of American Railroads 185
ATMs 431, 439
AUCNET 288, 289–96
auctions 285; auto 288–96; "Dutch" 344–5, 360; sample-based 351–2, 353; tele 353–5, 359; video 349–51; *see also* Dutch flower auctions
Australia 392, 393, 394, 396, 397, 398, 399, 401, 402, 403, 405
auto-auction market 288–96
automated service delivery (ATMs) 431, 439
automation, worker involvement 435

Baily, M. N. 188
Bakos, J. Y. 154–5, 288
Balestri, A. 319
Banker, R. D. 153
Barbieri, G. P. *et al.* 328
Barrett, M. 376, 379
Barrett, M. *et al.* 376, 387
Barua, A. *et al.* 110, 112, 124, 155–6, 164, 188
Baskerville, R. L. 36
Beck, U. 372
Beninati, Marie 239
Berndt, E. R. 111, 112, 124
Black, James 93, 94, 98

Blacker, F. *et al* 83, 99
Blackman, I. 240
Boland, R. 83, 99
Boland, R. J., Jr. 57
Boudreau, M. 103
Braa, J. 389
Brancheau, James C. 392, 393, 400, 410
Brannen, M. Y. 388
breakdown resolution 62–3, 71, 71–2
Brenner, R. 327
Bromily, P. 327
Brown, J. S. 83, 84, 94, 95, 97, 98
Brun Passot 196, 199–212
Brynjolfsson, E. 111, 124, 153, 154
Bureau of Economic Analysis 115, 118, 186
Bureautel 199, 201, 204, 205, 207, 208, 209
business process engineering (BPR) 402, 406
business value 215–16; of electronic data interchange 217–30; impact of electronic data interchange 227–30; information exchange 219–20; research 216–18
buyer behaviour 350

Caby, L. 263
California 437
Cantor Financial Futures Exchange (CX) 285, 298, 302–9
Cantor Fitzgerald 302
capacity building 33
car prices 288–96
career reward structures 89–90
Casson, M. 319, 325
Castells, M. 372
centralization 130, 437, 437–8
CFTC, the 298
Challenger disaster 10

Chan, A. P. 262
Chan, Yolande E. 172
changeable time 39, 40
Checkland, P. 41
Chicago Board of Trade 299, 300, 300–1, 303, 309
Chicago Mercantile Exchange 285, 302
Chief Information Officers (CIO) 33–5, 36–8, 39–42
Chisholm, R. F. 36
choice shift 12, 15, 16, 20–1, 24, 26
Chrysler 196, 216, 218–30
Ciborra, C. U. 332
citizen information 431
Clemens, E. K. 342
Cline, Melinda K. 190
Coase, R.H. 259
COGEMA 205
collaboration 94, 95, 156
Colorado 439
COMET 300–1
communication: apprehension 17; cues 12–17, 24; forums 83; and group polarization 10, 12–17; richness 330; and social presence 12–13, 24
Communications of the ACM 153, 159–65, 173–4, 285
communities of practice 82–99, 84, 97–9
competing agendas 91–2
competitive advantage theory 315
Compound UK 83, 85–99
compromise 11
computer-mediated communication (CMC) 3; and anonymity 10, 17, 18–23, 24, 51; and communication cues 12, 13, 14–17; effectiveness 56; and gender equality 51–2; and group decision-processes 4; and group polarization 4, 9, 10–11, 24–6; and group size

25; information levels 58;
interpretive analysis 47–75;
and organizational decisions
26; and risk behaviour 25, 26;
and social presence 13
computers: and productivity 110,
125; rates of return 119–21;
service life 118–19; spending
115, 116–17, 117–23, 122, 124;
see also PCs
consciousness 57
consciousness change 62–3
contact recording 87, 88, 88–9,
91, 92
content anonymity 17
convergence contexts 131, 132,
137, 143, 144
Cooper, Robin 253
coordination 260–3, 268, 273–4,
438–9; costs 258; and cultural
differences 377; and electronic
networks 263–4, 267, 271–3;
success 264
Coordination Committee for
Mutual Export Controls
(COCOM) 435
Corbetta, G. 328
critical social theory 330
Cron, W. L. 243–4
Cross, Robert 88, 89
cross-cultural communication
367–8, 373
cross-cultural conflict 375, 379–
80, 384, 385
cross-cultural interaction 378,
388–9
cross-cultural software
development 367–8, 372–3,
376–8; contradiction and
conflict 379–80, 384, 385;
reflexivity 380, 384;
structurational analysis 378–80,
381–4, 385, 387, 387–8, 389;
and technology transfer 380–4
cross-cultural working 385–8, 389
Crystal Ball 4.0 303
Culture and cultural context 325,
329–31, 373–4, 379, 383, 385–
8, 389, 398–400, 406, 411
Cummings, L. L. 327

databases 87–8, 88–9, 90, 91, 291
Datini, Francesco 320, 330
Davison, Robert 35
Dean, C. P. 397
decentralization 437, 438
decision bias 11
decision making 3, 4, 53, 54–6, 130

Demattè, C. 328
democratization 415, 419, 422
Denmark 432, 433, 434, 437, 438
Dennis, A. R. 25
depersonalization 13
detention monitors 431
Deutsche Terminbörse (DTB) 300
dialectic of control, the 384
Dickson, G. W. *et al* 406
digital divide, the 368–9, 412–19;
analysis 421–4; the causal
model 421; factors shaping
419–21; net sophistication
scores 420, 421, 422, 429
Digital Equipment Corporation
(DEC) 203, 206, 208, 211
direct service delivery 434
Dore, R. 329
Douglas Aircraft 226
Dubrovsky, V. J. 24, 26
Duguid, P. 83, 84, 94, 95, 97, 98
Dutch flower auctions 342–3,
343–5; the Aalsmeer sample-
based auction 351–2, 353;
application analysis 355–9;
challenges 347, 349; evaluation
347, 348–9; image representations and negotiated trading
352, 354; IT initiatives 349–60;
process-stakeholder analysis
349–55; Tele Flower Auction
system 350, 353–5, 356, 357,
358–9; transaction costs 347;
the Vidifleur initiative 349,
349–51
dynamic networks 315–16

e-mail 87, 197, 274
East African Flower (EAF) 353–5,
358–9
Economat 204, 205
economic development 419, 421,
422, 424
economic structure 398–9
Eden, C. 36
education 419, 421
Effective Channel Response 238
Effective Customer Response 238
Ein Dor, P. 397
Elden, M. 36
electronic brokerage effects 261
electronic data interchange 193,
194–5, 197, 199–212; benefits
204–7, 208; and business value
215–30; cost/benefit analysis
229, 230; diffusion 207–9;
financial applications 210;
impact of 227–30, 274; and

information handling costs
226–7; and inventory
management 219, 220, 221–2,
223–5, 230; investment 204,
209; limitations 262; and quick
response technology 238; and
the Single European Market
196, 199, 210–11, 212;
strategic business initiatives
209–10; and transportation
costs 222, 225–6
electronic futures trading 285,
298–309
electronic hierarchies 283
electronic marketplaces 283–7,
341–2; advantages 292, 293;
buyer numbers 295–6;
characteristics 292–3; Dutch
flower auction IT initiatives
349–60; elimination of
middleman margins 293;
exchange processes 343, 345,
346; market power of sellers
294; and prices 288–96;
process-stakeholder analysis
342–3; products 293; reasons
for higher prices 293–6;
regulation 293
electronic networks 197, 258,
260–1, 262, 263–75
empowerment 39
enterprise performance, IT
investment and 183–9
equality 424
Estonia 392, 393, 394, 396, 398,
399, 401, 403
ethnography 48, 57, 60, 71, 166
Eurex 308
Europe 200, 392, 393, 394, 396,
398, 399, 400, 401, 403
European Community 210
exchange organizations; and
costs 343; and IT 341–2
exchange processes 343, 344
executive information systems
(EIS) 433

Fiducial 210–11
Figon, Olivier 214
flaming 13
Flower Auction Aalsmeer (VBA)
344, 349, 350, 351–2
Flower Auction Holland (BVH)
344, 349, 349–51
flower auctions *see* Dutch flower
auctions
France, office supplies market
200, 200–1

Franchi, M. 328
Fukuyama, F. 326, 327
Futures Commission Merchants 300–1
futures markets 285, 298–309

Galliers, Robert D. 410
gang delinquency 10
Garratt, Martin 89
Gaspard 200
Geertz, C. 57
gender equality 49–52, 53, 55–6, 56, 58–75
geographical information systems software 380–4, 385, 388, 389
Germany, office supplies market 200
Giddens, A. 373–6, 384, 389
Glaser, B. 58
globalization 365, 367, 372–3, 376, 388, 389, 415, 419, 421
Globex 302
Goffmann, E. 85, 98
Goodall, K. 388
Goras 376–80
government role 434–5, 442
Granovetter, M. 262
Great Britain 200, 392, 393, 394, 396, 397, 398, 399, 401, 402–3, 403, 406
grounded theory 48
group decision making 4, 53, 54–6
group membership 12
group polarization; and anonymity 10, 17–23, 24; background 11–12; and communication 10; and communication cues 12–17; and computer-mediated communication 4, 9, 10–11, 24–6; definition 10; examples 10, 11; impact of anonymity 18–23; impact of communication cues 14–17; measurement of 12, 15; and organizational decisions 26; and organizational setting 25; and risk behaviour 25, 26; situations that may benefit from 26; and social presence 12–13, 17–18
group support systems (GSS) 3, 4, 36; advantages 48; and alternative generation 55–6, 68–9; and anonymity 37, 39, 40, 42, 48, 51, 56, 66, 74; application of 42; and consensus 36–7, 53, 56, 74; effectiveness 42; and gender equality 51–2, 53, 56, 58–75;

and idea generation/categorization 36–7, 38, 53, 54–5; and information exchange 67–8, 74–5; interpretive analysis 6, 47–75; as 'memory' 38; positivist analysis 52, 53–6, 66, 67, 72–5; and problem-solving 6, 48, 53; and process improvement 33–43; research and development 33, 48
GroupSystems for Windows v.1.1d. 33–43
groupware 3, 15, 22; alternative means of maintaining relationships 92–4; and collaboration 94; and knowledge work 7, 84–5, 99; mitigating activities 95–6, 97; and participation 82–99, 95, 96–9; politicising 88, 89–90, 96; safe-guarding activities 91–2; spread of 82; and surveillance and control 96
Gtec 376–80
Guilbert 200, 208
GUILTEX 400 208
Gulf Cooperative Council 392, 393, 394, 396, 398, 399–400, 401, 403, 404–5

Hamfelt, C. 325
Harris, Sidney E. 244
Hayes, N. 83, 84, 85
Henderson, J. C. 240
hermeneutics 48, 57, 60, 60–1, 71, 74, 75
Hill, C. W. L. 327
Hirschheim, R. *et al.* 394
Hislop, D. 83
Hofstede, G. 325, 331, 385, 386, 398
Holland, C. 240
Holland Supply Bank (HSB) 349, 350, 352, 354, 355, 357, 358
Hong Kong 392, 393, 394, 396, 398, 399, 403, 405
Hong Kong Management Association 35
Hoos, I. R. 130
Huxham, C. 36

Icarus paradox 106, 144–6
immediacy 12–13
IMP Group 325
India: correspondence analysis 400, 401; culture 383, 386, 398; economic status 399;

geographical information systems software 380–4, 385, 388, 389; IS management key issues 396, 403, 404; key issues studies 392, 393, 394; political/legal environment 399; software development 376–80; work patterns 387
Industrial marketing and Purchasing group 324
industrial networks 324–5, 325–6
inequality *see* digital divide, the
information handling costs 226–7
information resource management 433
Information Richness Theory 330
information systems: analysis 332; computer capital 115, 116–17, 117–23, 124; and cross-cultural working 385–9; and cultural differences 379, 386–7; effects of investment 124–5; evaluation 128–46; hidden costs 122–3; key issues 392–408; marginal products of computer capital 120, 122; and productivity 109–25; rates of return 119–21; societal aspects 365–9; and spending 109–25; staff spending 115, 116–17, 118, 119, 121, 122; and structuration theory 373, 374, 375; structurational analysis 386–7, 389; *see also* information technology
information systems management: correspondence analysis 400–1; cultural context 398–400; family-developed 405; family-developing 403–4; fashion swings 406, 407; international issues 395–6; key issues studies 392–400, 405–8; local studies 407; pyramid of people-developing 404–5; regional analysis 401–5; regional rankings 397–400; village market-developed 401–3
Information Systems Research 153, 159–65, 175–6
information technology: benefits 321; and business value 215–30; and centralization 130; cross-cultural transfer 373; cross-national distribution 413–19; cross-national measures 422; and decision making 130; and

Dutch flower auctions 349–60; and electronic markets 341–2; and enterprise performance 183–9; evaluation approaches 167; and globalization 372–3; government adoption of 430–2; Icarus paradox 106, 144–6; ideological commitment to 419, 421–2; impact measurement 156–7; investment 129, 183–9, 215–30, 236–7; investment categories 184–5, 186, 187, 187–8, 188–9; investment context 157–8; investment measurement 186–8; investment temporal lag effects 189; investment value 152–67; and managerial work 128–46; networks 258, 260–1, 262, 263–75; and organizational change 131–2; organizational control 441; and organizational culture and context 103, 106; and performance 103, 106, 107, 235–53; and productivity 103, 103–4, 109–25, 129, 152–67, 153–8, 183–9, 217; productivity paradox: *see* productivity paradox; productivity research 110–11, 123–4; reasons for investment in 112; reduction in information availability 357; regional development 397–400, 401–5; service life 118–19; social factors affecting access 419–21; societal aspects 365–9; and spending 103, 107, 109–25; strategic alignment 239–42, 244, 249–51, 252, 253; strategic use 235–53; strategy 333; use in the public sector 432–43; usage analysis 134–43; usage measurement 133–4; value article trends 164–5; value research 155–6, 159, 162–5, 166; value studies 107; and work process 129; *see also* information systems
information technology indicators 417–18
inhibition 13
inter-firm transactions 262
inter-organizational systems 193
interactivity 3
intermediate production processes 185

international computing: correspondence analysis 400–1, 402; international issues 395–6, 397; key issues studies 392–400, 405–8; *see also* digital divide, the
International Data Corporation 115
International Data Group 114–15
international diplomacy 33, 41, 42
International Federation for Information Processing (IFIP) 365, 367
Internet, the 83, 292–3, 413, 414, 415, 417–18, 419, 423–4
interorganizational systems 219–20, 239, 260–1, 274, 283, 285; failure of 313, 322, 322–4, 329; interpretive analysis 324–33; Prato 318–20; theoretical perspectives 311, 312, 314–16, 322–4, 331–3; and trust 326–9
interpretive analysis 57; and anonymity 66; assessment 72–5; authenticity 70–1; behavioural information 62–3; breakdown resolution 62–3, 71, 71–2; and computer-mediated communication 47–75; and conflict 73–4; documentation 71; emotional information 61–2; evaluation criteria 69–72; group support systems 6, 47–75; interorganizational systems 324–33; interpretation 64–9, 73; interpreting information types 58–64; language 58; member checking 70; methods 57–8, 75; replication 71–2; triangulation 69–70, 74
interpretive methods 48–9
Interstate Commerce Commission 185
inventory holding costs 221, 224–5
inventory management 219, 220, 221–2, 223–5, 230
Investment, IT 129, 183–9, 215–30, 236–7; categories 184–5, 186, 187, 187–8, 188–9; computer capital 115, 116–17, 117–23, 124; context 157–8; information systems 109–25; IS staff 115, 116–17, 118, 119, 121, 122; measurement 186–8; strategic and tactical 185, 186, 187, 188, 189; tactical IT 185, 186, 187, 188, 189; temporal

lag effects 189; threshold 185, 186, 187, 188, 189; value 152–67
Inzerilli, G. 318, 319, 324, 327
IT enabled competitive advantage 199–212

Jagis Group 376, 378
Jamaica, software development 376–80
Japan: auto-auction market 288–96; work patterns 387
Jelassi, Tawfik 214
Jessup, L. M. *et al.* 53
Jessup, Leonard M. 80
Johanson, J. 325
Johnston, R. 316, 324, 327
Journal of Management Information Systems 153, 159–65, 177–9
just-in-time purchasing system 209, 216, 218, 227–8, 229, 238

Kalathur, Suresh 234
Kambil, A. 261, 343
Kansas City 437
Katz, Joseph L. 244
Kauffman, R. J. 153, 166
Kekre, Sunder 234
Kelly, Gigi G. 410
Khare, A. 387
King, Martin Luther 10
Kling, R. 311–12, 315, 323, 332–3
knowledge management/work 7, 82–3, 84–5, 99
Koenig, C. 326
Kommunedata 437
Konsynski, B. *et al* 342
Korpela, M. 386
Krajca, E. 218–19
Kraut, R. *et al.* 285
Krishna, S. 388

labour relations 435
Lam, A. 387
language 57, 66
Lave, J. 83, 84, 84–5, 94, 97, 98, 98–9, 99
Lawrence, P. R. 316, 324, 327
Lazerson, M. H. 327, 328
learning 84, 97–9, 98–9
Leavitt, H. J. 130
Lee, A. S. 75, 313, 330, 331
Lee, H. G. 342
Lester, S. 103
Levi-Strauss, Claude 41
Lindberg, A. K. 325
liquidity 342
local government 433–4, 437

Lockett, G. 240
logistics 211, 218–19
London Stock Exchange 298–9, 342
loss of face 13
Lotus Notes 84, 86, 87–8, 88–9
Loveman, G. W. 110, 111, 111–12, 124
Luftman, Jerry 252
Lyytinen, K. *et al.* 42

McCall, M. W. Jr. 133
McGuire, T. W. 24
McKinlay, A. 83
Macquarie Graduate School of Management 35
Mahmood, M. A. 153
Malone, T. W. 316
Malone, T. W. *et al.* 260–1, 263, 268, 283, 355
management information systems 431
Management of the Productivity of Information Technology database 111
managerial work: and the Icarus paradox 144–6; IT and 128–46; IT usage analysis 134–43; and IT value 166–7; nature of 132; overuse of IT 145; roles 133, 137, 140, 142, 143, 145, 149–50; and specialisation spirals 145
Marche à Terme International deFrance (Matif) 300
market intermediaries 293
Markus, L. 323
Markus, M. L. 285
master/apprentice relationships 97–8
materials planning 220
Matra Espace 205–6
Mattsson, G. 325, 326, 327
middleman margins 293
Miles, R. 313, 315–16, 321, 324, 326
Miller, D. 145
Mintel 201, 208, 209
Mintzberg, H. 132, 133, 149–50
MIS literature 154–6; analysis 159–65, 173–82
MIS Quarterly 153, 159–65, 180–2, 193–6
Morrison, C. J. 111, 124
motivation 38, 39, 41, 94, 96–9, 156, 183
Mukhopadhyay, Tridas 219–20, 234
Myers, M. D. 385

NASDAQ 292, 299
National Crime Information Center/Computerized Criminal History (NCIC/CCH) 437, 442
National Health Service 86
National Science Foundation Networks 413, 417–18
negotiated culture 388
New York Board of Trade 302
New York Mercantile Exchange 302
Newell, S. 83
Nigeria 386
Norway 432, 433
novel arguments 17, 22–3, 24
novelty 11, 16–17
NSC 308

obsolete inventory costs 221–2, 225
one-upmanship 11, 13, 16–17, 22–3, 24
open coding 57–8
open-outcry practices 300–1, 303
operations managers 133
order frequency 266
order matching systems 299, 305, 305–7, 308
Orgad, M. 397
organization development literature 156–8, 163–4
organizational change 131–2
organizational culture and context 25, 38, 42, 42–3, 57, 67, 73, 86–7, 103, 106, 129, 130–1, 137, 140, 142–4, 153–8, 166, 189, 207, 229, 239–40, 267, 399
organizational decisions 26
organizational forms 259–61
organizational goals 156
organizational performance 103, 183–9, 236; measurement of 152–67
Orlikowski, W. J. 96, 313
output statistics 111–12, 117, 117–18, 125, 129
outsourcing 406

packaging costs 352
Palmer, J.W. 285
Palvia, P. C. 369
Palvia, S. C. 369
Pannicini, I. 319, 325
Papp, Raymond 252
Parsons, T. 328, 329
participation 37, 38, 39, 40, 41, 42, 54, 55, 56, 66, 68, 73; career enhancing 96; discussion databases 90; full 98; and groupware 82–99, 95, 96–9; legitimate 85, 98; legitimate peripheral 84–5; motivation 96–9; non-career oriented 91; peripheral 85, 97–8; politicising 89–90, 91, 96; and surveillance and control 96
Passot, Jean Philippe 201, 207
path analysis 134–5
PCs: cross-national distribution 413, 414, 417–18; government adoption of 432
Péchiney 206
Pedhazur, E. J. 135
PEN (Public Electronic Network) 440
performance: and IT 103, 106, 107, 235–53; measurement of 243–4; and quick response technology 246–9, 251–3; and strategic alignment 249–51, 251–3
personal relationships 258, 261–3, 263, 264, 267, 268, 273, 274–5
persuasive argumentation 17, 18
persuasive arguments theory 11–12, 16–17, 22–3, 24, 25
Pinsonneault, Alain 148
pluralistic balance 11, 13, 16–17, 17–18, 22, 24
political enclaves 7, 85, 88–90, 91, 95–6
political/legal environment 399
politicising 88, 89–90, 91, 96
politics 312
Porter, M. E. 313, 318
positive methods 48, 49
positivist analysis 52, 53–6, 57, 66, 67, 72–5
Prato (Italy) 311, 313–14, 316–18, 329; cultural context 325, 329–31; industrial networks 325; interpretive analysis 324–33; introduction of IT 320–2; organizational model 318–20; social bonds 325–6, 327, 328–9; technical-economic analysis 322–4, 331–2; trust relationships 326–7, 327–9
preference change 12, 15, 16, 20–1, 24, 26
prices, effect of electronic marketplaces on 288–96
problem-solving, and group support systems 6, 48, 53

process anonymity 17
process improvement, group support systems and 33–43
process intervention 40, 41
process reviews 35
process-stakeholder analysis 342, 342–3, 349–55, 360
production function 112
productivity: and computer capital 117–23; and information systems spending 109–25; and IS staff spending 118, 119; and IT 103, 103–4, 109–25, 129, 152–67, 153–8, 183–9, 217; and IT investment 129, 183–9; output statistics 111–12; research 110–11, 123–4; theoretical issues 112–14
productivity paradox 103, 103–4, 106–7, 109–25, 129–44, 152–8, 158, 164, 196
profitability, tactical IT investment and 189
Project A 302
project reports 35
Public Electronic Network (PEN) 440
public sector, the 369, 430–2; automated service delivery (ATMs) 431, 439; character of IT use 434; coordination and optimization 438–9; creation of new institutions 436–7; effects of IT use 436–43, 442–3; electronic communication 440; extent of IT use 432–3; government role 434–5; IT deployment 437–8; nature of IT use 432–6; organization of computer use 433–4; spread of computer use 436; staff 433–4, 438, 442–3; work processes 438–42
public service, effects of IT use 436–43
Puri, S. K. 384

Quick Response 196–7
quick response technology 238; impact of 246–9; levels 236, 252; in retailing 235–53; in speciality retailing 237–42
Quinn, J. B. 188

Rai, A. *et al.* 163
rail industry 183–9
Raj 376–80, 377, 378, 379
RCA 226
reflection 36

relationships, reinforcing 92–4
relationships and trust perspective 312, 313–14
reorientation contexts 131, 132, 137, 142–3, 144
REQUEST 438–9
requirements engineering 33
Reseau 320–1, 328
resource dependency theory 315
retailing: impact of quick response technology 236–53; IT use 235–7; performance construct 243–4; retail pyramid 237; speciality 237–42, 245–6, 249–51, 252–3; strategic alignment 239–42, 244, 249–51
RETEX 242–3, 244, 246
reward structures 94, 95
Ricks, D. A. 397
Rieser, V. 328
risk behaviour 25, 26
Ritaine, E. 317, 318, 319
Rivard, Suzanne 148
Robertson, R. 373
Robey, D. 103, 323
Rockart, J. F. 316

Sachs, Jeffrey 412
SACI 200, 210
safe enclaves 7, 85, 90–4, 95–6, 97
Sahay, S. 57, 70, 380, 383, 384, 388, 389
Salancik, G. R. 13, 17
sales volume 189
Salk, J. E. 388
Salomon, Kurt, and Associates 242
Sambamurthy, V. 407
Santa Monica 437, 440
SATELITE 210
Saunders, Tom 86, 89
Scandinavia 432–43
Scarborough, H. 83
Schein, E. H. 156–7, 166
screen based trading 298–309
SEAQ 292, 299
SEC, the 298
Segev, E. 397
Segmented Institutionalism perspective 312, 315, 323, 332–3
Segrist, C. A. 133
service costs 288, 293
Sethi *et al.* 244
shared meanings 57
Shils, E. A. 328, 329
SICLAD 201–2, 203, 204, 205, 205–6, 207, 208, 209

Siegel, J. 17, 24
single European Market 196, 199, 210–11, 212
situated learning 98–9
Slide Auctions 294
Slovenia 392, 393, 394, 396, 398, 399, 400, 401, 403, 404, 405
SMS 4
Snow, C. 313, 315–16, 321, 324, 326
Sobol, M. G. 244
social bonds 325–6, 327, 328–9
social capital 327, 333
social change 415
social comparison 13, 17, 17–18
social comparison theory 11–12, 16–17, 22–3, 24, 25
social complexity 419
social constructions 57
social cues 17
social development 419
social impact theory 25
social presence 10; and anonymity 17–18, 21, 24, 25, 26; and communication cues 12–13, 24; definition 12; and interactivity 24
social support systems 10
socio-economic theories, empirical tests 312–14
socio-political conflict perspective 323
Soft Systems Methodology 332
software development 184; cross-cultural 367–8, 372–89; virtual 367–8
software inspections 33
specialist knowledge and resources 82
Spending: computer capital 115, 116–17, 117–23, 124; information systems 109–25; IS staff 115, 116–17, 118, 119, 121, 122; telepurchasing 204
Spie-Batignolles 207
SPRINT 313, 320–2, 323
SPRINTEL 286, 311, 321, 323, 327, 328, 329, 331
staff professionals 133
State University 49–52
Stechschulte, Jim 229
Steinfield, C. 263
strategic alignment 239–42, 244, 249–51, 251–3
strategic investment 184
strategic IT investment 185, 186, 187, 188, 189
strategic management 33

strategic selling 87, 89, 90, 91, 92
Strauss, A. 58
structural properties 375
structuration theory 373–6, 389
suppliers, manufactures relationships with 197, 204–7, 208, 209–10, 218, 220, 230
Surface transportation Board 185
surveillance and control 96
Swan, J. 83
Sweden 432, 433, 433–4, 434–5, 435, 436–7, 437–8, 439
Swedish Agency for Administrative Development (SAFAD) 435, 436–7
Swedish Environment Fund (AMFO) 435, 437
symbolic analytical workers 82
System Rationalism perspective 311–12, 332–3
system rationalization 315

tactical investment 184
tactical IT investment 185, 186, 187, 188, 189
Taiwan 392, 393, 394, 396, 398, 399, 400, 401, 403, 403–4
Tan, F. B. 385
Tavistock Institute 365
TC9 365, 367
team meeting, group support systems and 4
team work 3
technical-economic rationality 311, 312, 314–16, 322–4
technical-economic theories 331–2
technological status 399–400
technology strategy 120–1
technology transfer, cross cultural 380–1; contradiction and conflict 384, 385; reflexivity 384; structurational analysis 381–4, 385, 387, 389
Tele Flower Auction system (TFA) 349, 350, 353–5, 357, 358–9
telecommuters 438
telepurchasing 199–212; benefits 204–7, 208; diffusion 207–9; investment 204, 209; and the Single European Market 196, 199, 210–11, 212; strategic business initiatives 209–10; users' perspectives 205–7
Tenkasi, R. V. 83, 99

terminology, misuse of 34
textile industry, Italy 316–18
textual cues 12, 13, 16
Thorelli, H. B. 327
threat settings 17
threshold IT investment 185, 186, 187, 188, 189
time pressure 266, 268
Toennies, F. 329
TOPS 300–1
trading automation 298–309
trading costs, futures market 305–7
trading strategy, futures market 305–7
Transaction Cost Economics 313
transaction cost predictors 270
transaction cost theory 315, 332
transaction costs 342, 343, 347
transactional IT investment 185, 186, 187, 188, 189
TRANSPAC 209
transportation costs 222, 225–6, 233–4
Transportation in America (1991) 225
Trauth, Eileen M. 80, 387
Trompenaars, F. 325, 328, 329
trust 312, 325, 326–9, 332, 333, 357
Tsui, A. 133

UNESCO 421
United Kingdom 200, 392, 393, 394, 396, 397, 398, 399, 401, 402–3, 403, 406
United Nations Development Program 421
United States of America 392, 393, 397, 398, 399, 401, 403, 432–43
US Army 438–9
US multinational study (US MNC) 393, 394, 396, 400, 402, 403, 406
US Society of Information Management (US SIM) 392, 393, 394, 396, 401, 405, 406
U.S. Treasury Securities market 299–302, 303
USS Kyushu 295
Uzzi, B. 262

valid arguments 17, 22–3

validity 11, 16–17
value-added networks 203
value-added partnerships 316, 318–19
van Wijk, G. 326
Venkatraman, N. 240
verbal cues 12, 13, 16, 24
Videotex 440
videotext 321, 322, 323
Vietnam War 10
Virginia 437
virtual groups 75
virtual organizations 197, 257–60; coordination mechanisms 260–3, 267, 268, 271–3, 273–4; and coordination success 264; and electronic networks 263–75; and IT networks 258, 259; and order quality 271–2; transaction cost predictors 270; and trust 326
virtual storefronts 293
virtual workplace 75
visual cues 12, 16, 17, 24, 26

Wade, R. H. 369
Walsham, Geoff 57, 70, 83, 84, 85, 373, 375, 376, 379, 380, 384, 389, 391
Watson, Richard T. 392, 393, 400, 409–10
Weber, B. 342
Weill, P. 166, 244
Weisband, S. P. 24, 26
Wenger, E. 83, 84, 84–5, 94, 97, 98, 98–9, 99
Wetherbe, J. C. 394
WG9.1-4 365, 367
Whisler, T. L. 130
Willcocks, L. 103
Williamson, O. 259, 260
Williamson, O. E. 313, 327
women, anonymity and 40
Wood-Harper, A. T. 36
work flexibility 441–2
work groups 440–1
work patterns, cross-cultural working and 387–8
World-Wide Web 292–3

Zack, M. H. 330
Zeta 33–43
Zmud, R. W. 407
Zuber, J. A. 24

eBooks – at www.eBookstore.tandf.co.uk

A library at your fingertips!

eBooks are electronic versions of printed books. You can store them on your PC/laptop or browse them online.

They have advantages for anyone needing rapid access to a wide variety of published, copyright information.

eBooks can help your research by enabling you to bookmark chapters, annotate text and use instant searches to find specific words or phrases. Several eBook files would fit on even a small laptop or PDA.

NEW: Save money by eSubscribing: cheap, online access to any eBook for as long as you need it.

Annual subscription packages

We now offer special low-cost bulk subscriptions to packages of eBooks in certain subject areas. These are available to libraries or to individuals.

For more information please contact webmaster.ebooks@tandf.co.uk

We're continually developing the eBook concept, so keep up to date by visiting the website.

www.eBookstore.tandf.co.uk